Lecture Notes in Physics

Edited by H. Araki, Kyoto, J. Ehlers, München, K. Hepp, Zürich
R. Kippenhahn, München, D. Ruelle, Bures-sur-Yvette
H.A. Weidenmüller, Heidelberg, J. Wess, Karlsruhe and J. Zittartz, Köln

Managing Editor: W. Beiglböck

351

A. Dervieux B. Larrouturou (Eds.)

Numerical Combustion

Proceedings of the Third International Conference
on Numerical Combustion
Held in Juan les Pins, Antibes,
May 23–26, 1989

Springer-Verlag Berlin Heidelberg GmbH

Editors

Alain Dervieux
Bernard Larrouturou
INRIA, Sophia-Antipolis
F-06560 Valbonne, France

ISBN 978-3-662-13749-9   ISBN 978-3-540-46866-0 (eBook)
DOI 10.1007/978-3-540-46866-0

© Springer-Verlag Berlin Heidelberg 1989
Originally published by Springer-Verlag Berlin Heidelberg New York in 1989
Softcover reprint of the hardcover 1st edition 1989

2153/3140-543210 – Printed on acid-free paper

FOREWORD

This book contains the texts of the lectures that were presented at the Third International Conference on Numerical Combustion, held in Antibes–Juan les Pins, May 23–26, 1989.

This conference followed the meetings on numerical combustion held at Sophia–Antipolis (France) in 1985 and San Francisco (USA) in 1987 organized by INRIA (Institut National de Recherche en Informatique et Automatique) and SIAM (Society for Industrial and Applied Mathematics); the next meeting on this theme will be held in the United States in 1991. This series of conferences is being organized in response to the growing interest in the subject shown by scientists from various fields: applied mathematicians, physicists, chemists, and engine manufacturers and other industrialists.

Around 120 scientists and engineers from more than 10 countries attended this meeting, where nine invited lectures and twenty-seven selected contributions were presented.

The examination of the contributions in this volume leads to several remarks. Firstly, it seems that the numerical simulation of transonic and supersonic combustion phenomena has now become a theme of major interest, although it was not so developed in former conferences in this series. In this domain, detonation occupies an important place, in particular with investigations of the transition to detonation. Whereas the main simplified models used in the past few years were essentially reaction–diffusion models or low–Mach–number models, we now observe that hyperbolic models (and all related numerical tools, such as Riemann solvers) are more and more widely used in the combustion community: for detonations, transonic combustion (reactive jets, shock-induced ignition), but also for other phenomena (granular flow, two-phase reactive flows, etc).

Another recently emerging theme concerns the study of (essentially supersonic) reacting mixing layers, while the modelling of turbulent combustion remains a major and challenging issue. Besides these, numerical simulations of hydrocarbon flames with a complete set of chemical reactions are now carried out in two-dimensional geometries, and complex–reactive–flow simulations, in particular for simulations of the internal combustion engine, still retain the attention of many researchers, from both the industrial and scientific communities.

* * *

We thank all the following institutions, which financially supported or participated in the organization of the conference:

CNES (Centre National d'Etudes Spatiales),

CNRS (Centre National de la Recherche Scientifique),

DRET (Direction des Recherches, Etudes et Techniques),

ERCOFTAC (European Research Community On Flow Turbulence And Combustion),

IFP (Institut Français du Pétrôle),

INRIA (Institut National de Recherche en Informatique et Automatique),

SIAM (Society for Industrial and Applied Mathematics).

We are indebted to H. Guillard for his participation in the Scientific Secretariat of the Conference, and to P. Clavin, R. Glowinski, R. J. Kee, N. Peters, W. A. Sirignano, D. B. Spalding, R. Temam, and F. A. Williams as members of the Scientific Committee organizing this Conference. We also wish to address our thanks to the chairmen of the different sessions, and to all speakers.

Finally, sincere gratitude is expressed to T. Bricheteau, S. Gosset, and E. Many of the Public Relations Service at INRIA, whose help greatly contributed to the success of the Conference, and to F. Trucas who helped us in the preparation of this volume.

Valbonne, July 1989

A. Dervieux
B. Larrouturou

CONTENTS

PART 1: INVITED LECTURES

PART 2: CONTRIBUTED LECTURES

PART 1: INVITED LECTURES

VORTICITY GENERATION IN A NONPREMIXED FLAME SHEET

Wm. T. Ashurst
Combustion Research Facility
Sandia National Laboratories
Livermore, CA 94551-0969

ABSTRACT
Marble's problem of a vortex winding up a diffusion flame sheet is simulated (Marble, 1985), with the effects of variable density and vorticity generation included. The initial flame is straight and passes through the vortex core, but not through the center of the vortex as in Marble's work. During the first eddy turn-over time, the swirling flow forms a flame tip region, and the amount of tip rotation around the vortex depends on the diffusion coefficient. During the next eddy turn-over time, the flame shape is quasi-steady with minor differences due to heat release and vorticity production over the parameter range examined. The gradient of mixture fraction at the flame sheet determines the amount of reaction in this Burke-Schumann flame model. This swirling flow maintains a uniform gradient along most of the flame surface, with the exception of a low-gradient region at the flame tip. If finite-rate chemistry is considered, then the tip region would likely be at higher temperature than the rest of the flame surface.

INTRODUCTION

Turbulent mixing of a gaseous fuel with air is a much applied, but little understood combustion application. Supercomputers have allowed the simulation of detailed kinetic flame structure in simple flow fields, and laser diagnostics can provide experimental information without disturbing the flame structure. However, the turbulence problem itself is not yet solved in detail and turbulent reacting flow is a further complication which is usually modeled with constant density assumptions. In this paper we focus on a few aspects of chemical reaction in an unsteady flow: the interaction of a nonpremixed flame with a two-dimensional vortex. Analysis of non-reacting turbulence simulations show that the strain rate which is parallel to the vorticity axis is usually smaller than strain rates which are normal to the local vorticity direction (Ashurst et al., 1987). Thus, a two-dimensional flow pattern can represent the major strain effect on a reaction front. In this paper, all the chemical details have been replaced by a zero-thickness flame-sheet, the Burke-Schumann model, and so the volume expansion due to heat release couples the chemistry to the flow. In addition to the velocity field created by this volume expansion, there is also new vorticity generated in those locations where the density gradient is not aligned with the pressure gradient, the baroclinic effect. Previous simulations of premixed flame propagation in the field of a vortex showed significant production of vorticity when the vortex turn-over time is shorter than the flame passage time (Ashurst

& McMurtry, 1989). However, there is no propagation mechanism in nonpremixed flames, and this difference gives a completely different character to the flame-vortex interaction. Instead of the flame speed, it is the fuel and oxidizer boundary conditions that determine the duration of the nonpremixed interaction. In the constant-density flame limit, Marble (1985) has shown analytically that a vortex winds up a diffusion flame in such a manner that the reacted core area grows linearly in time. Laverdant & Candel (1988) have confirmed these results in nonlinear numerical simulations. Our objective is the inclusion of volume expansion and vorticity generation in this flame-vortex problem for nonpremixed combustion.

Our numerical technique in previous vortex-flame work is a Lagrangian description of vorticity combined with an Eulerian reacting scalar (Ashurst & Barr, 1983). This same recipe is less desirable for diffusion reactions which may last for long times while maintaining a small vortical region between irrotational boundary conditions. The diffusive aspect of the discrete vortex method is not its strong point, and requiring a close match between vorticity diffusion in Lagrangian coordinates and scalar diffusion on an Eulerian grid highlights this blemish in the vortex method. (The propagation effect in premixed combustion means that vorticity generated in the flame zone is convected away from the region of interest in a short time, and hence may be removed from the calculation.) Therefore, in the current nonpremixed work, both the vorticity and a conserved scalar are described on an Eulerian grid.

Laverdant & Candel (1988) have used implicit time integration with the Flux-Corrected-Transport scheme to remove the diffusive time-step restriction and to suppress new extrema in the solution adjacent to regions with steep gradients. We have used explicit time integration and only the high-order convection scheme from our previous FCT work. Our motivation was not a general, robust code, but a simple and adequate tool for this particular diffusive problem. In previous examinations of the FCT scheme for convective-diffusive problems (Barr & Ashurst, 1982), we noticed that the FCT procedure introduces numerical wiggles, which if used in this nonpremixed combustion simulation, would affect the estimate of volume expansion (as described below, the volume expansion is proportional to the second derivative of the mixture fraction variable). If one can afford the grid resolution, the FCT procedure is not needed and is not desirable. Because in this work the combustion is described by a conserved scalar that has a smooth variation in the vicinity of the flame, there is no need to resolve minor species in thin reaction zones, and we use a fixed grid spacing. This would not be true for detailed kinetic simulations, where some sort of adaptative grid would be necessary.

NUMERICAL METHOD

We use the Burke-Schumann flame sheet model in order to consider nonpremixed combustion in the swirling flow of a vortex, the problem selected by Marble (1985). Marble used the Gaussian vortex velocity field to convect the flame-sheet elements in his constant density solution, and thus, has an isolated vortex flow. For numerical convenience we use a periodic boundary in one coordinate direction, and so, have an array of vortices interacting with the flame. The early formation of the flame shape about a single vortex is not greatly affected by the periodic image vortices, but at later times the periodic condition dominates the flow pattern. In turbulent reacting flow applications, the vortices will be neither isolated nor periodic, nor all aligned in one direction. In the following simulations we focus on the early behavior of the flame-vortex interaction.

The combustion is described by the convection and diffusion of the mixture fraction Z. The boundary values of Z are unity on the fuel side and zero on the oxidizer side, cf. page 75 in Williams (1985). With the flame-sheet assumption the temperature is a function of Z, as is the density ρ, since we assume constant thermodynamic pressure - the low Mach number limit. The mixture fraction at the flame sheet is Z_{fl}, the adiabatic flame temperature is T_{ad} and the ambient temperature is T_o. The linear relations for temperature and mass fractions in terms of Z are: for $Z < Z_{fl}$, no fuel present,

$$Y_O = Y_{O,0}(1 - Z/Z_{fl}); \quad T/T_o = 1 + \tau Z/Z_{fl}$$

and for $Z > Z_{fl}$, no oxidizer present,

$$Y_F = Y_{F,0}(Z - Z_{fl})/(1 - Z_{fl}); \quad T/T_o = 1 + \tau(1 - Z)/(1 - Z_{fl})$$

where $\tau = T_{ad}/T_o - 1$. The value of τ determines the amount of heat release. We set $Z_{fl} = 0.2$ and $Y_{O,0} = Y_{F,0} = 0.2$, corresponding to a diluted fuel reacting with air.

We solve transport equations for the mixture fraction Z and the vorticity ω. The velocity field is defined by the scalar and vector potentials $\vec{u} = \nabla\phi + \nabla \times \psi$. These potentials are given by two Poisson equations $\nabla^2\phi = \nabla \cdot \vec{u}$ and $\nabla^2\psi = -\omega$. The boundary conditions for these Poisson equations are based on the desire for combustion at constant pressure and not constant volume. With periodicity in one direction we use the other direction to impose the constant-pressure combustion condition by allowing inflow/outflow velocities. The outflow velocity is set to the sum of the velocity divergence $\int\int \nabla \cdot \vec{u} \, dA = \int \vec{u} \cdot d\vec{n}$, where the latter integral is composed of the two edges which are not in the periodic direction. Either inflow or outflow due to the vortex is set by the circulation within the domain $\int\int \nabla \times \vec{u} \, dA = \int_C \vec{u} \cdot d\vec{l}$, where in the contour integral the two periodic edges cancel identically. This condition imposes an average velocity difference across the system.

With the velocity defined by scalar and vector potentials, the above conditions supply Neumann conditions along the non-periodic edges: $< \partial\phi/\partial n >$ is the average velocity leaving the domain and $< \partial\psi/\partial n >$ is the average velocity difference across the domain. Thus, while we have not imposed impermeable slip walls (and hence constant-volume combustion as in McMurtry et al., 1986), we have imposed an average condition. At greater numerical cost, we could use the grid values of $\nabla \cdot \vec{u}$ and ω as source strengths in the Biot-Savart integral for the distribution of ϕ and ψ along the non-periodic edges; for a square with N^2 grid points, the cost is $2N \times N^2$ for each function to obtain Dirichlet conditions. In this case it might be easier not to use a periodic boundary because numerous images of the grid sources will be needed to match the infinite numer of images of the periodic boundary. In the average sense we have a free boundary condition within a fixed numerical domain without using a grid transformation.

In the flame sheet model, the chemical reaction rate is infinite and the amount of reaction depends on the diffusive flux of fuel and oxidizer at the flame location. Carrier *et al.* (1975) evaluated the effect of strain rate on the flame sheet and the changes due to finite, but large, reaction rates. Their Damköhler number D_1 is the reaction rate divided by strain rate. They show that corrections to the flame-sheet results for temperature and mass fractions scale with an inverse power of D_1 and are confined to a small zone, whose width has the same scaling. To first order, there are no corrections to the convective terms.

We assume an infinite reaction rate, and so the energy equation, with the product of ρT constant, relates the divergence of the velocity field to the divergence of the heat flux: $\nabla \cdot \vec{u} = \nabla \cdot (D\nabla T/T_o)$, Lewis number is unity $\lambda = \rho_o c_p D$. There is no reaction term because the reaction occurs at the flame temperature in a zone of zero thickness and does not affect the velocity. It is the heat flux away from the flame which affects the fluid dynamics because all of the thermal expansion occurs as a material element diffuses towards the flame. In the numerical scheme the heat flux divergence is evaluated on either side of the flame, but not across the flame. The amount of reaction determines the jump in temperature gradient across the flame but the reaction is not explicitly calculated in this formulation.

The finite difference variables are cell-centered for Z and ω, but the velocities are defined at the cell face. For the scalar potential the source term is cell-centered, the divergence of the heat flux is proportional to $\nabla^2 Z$. For the vector potential the vorticity source term is defined at the cell corners by using the average of four neighboring cell-centered values of ω. Appropriate differencing of ϕ and ψ gives the cell-face velocities.

In two dimensions, the vorticity is a scalar and the transport equation is

$$\frac{\partial \omega}{\partial t} + \nabla \cdot \omega \vec{u} = \frac{1}{\rho^2} \nabla \rho \times \nabla P + \nu \nabla^2 \omega$$

where the first term on the right side is the baroclinic production of vorticity caused by misalignment between density and pressure gradients. This pressure P is the dynamical pressure associated with the velocity $\vec{u}$ and not the constant thermodynamic pressure P_o. In a constant density flow, or in a flow where the density is only a function of P, this baroclinic term is zero. In those cases the pressure is not needed to advance the solution. In this work we only need the pressure gradient where the density gradient is nonzero, and not the pressure throughout the domain to estimate the vorticity production term. From the velocity field at two discrete times we can form the spatial and temporal derivatives in the equation of motion, and determine the pressure gradient as

$$\frac{-1}{\rho} \nabla P = \frac{D\vec{u}}{Dt} - \nu \nabla^2 \vec{u}$$

For time advancement we use first-order differencing in the viscous and baroclinic terms, and a predictor-corrector algorithm in the convective terms (predict ω^* and Z^* at time $t + dt/2$ with all terms in the equations, and use ω^* and Z^* in the convective terms to obtain ω and Z at time $t + dt$). The pressure gradient is based on information at times t and $t - dt$, and may not be consistent with the vorticity solution at $t + dt$. Thus, this procedure does not guarantee conservation of momentum, and we do find small changes in the total circulation as described next.

To judge the magnitude of possible errors in this scheme, a check of the momentum flux was done and this indicated imbalances of order $\pm dt$. Another measure of this error is the conservation of circulation, since $d\Gamma/dt$ is related to a force summation and the total circulation is constant in an isolated system. One way to express the time rate of change of circulation is by a contour integral of dp/ρ around the reacting region. If this contour is in a constant-density fluid, the circulation is constant even though vorticity generation occurs within the contour. In the simulations, over time periods of interest, there are small changes in Γ of a few percent and proportional to the heat release magnitude. If we estimate the pressure gradient without the viscous term $(\nu\nabla^2\vec{u})$, we obtain an opposite and much larger change in Γ over the same time period. However in these two simulations just described, one in which we increase the pressure gradient error, we do not observe any difference in the flame shape that evolves. Thus, the conservation of circulation, or its lack of, does not appear to have important effects on the general development of the flame-vortex interaction. Similar lack of conserved quantities is encountered in the Lagrangian vortex dynamics methods which use core smoothing (conserve Γ but not energy), but the features of the evolving flow seem to have very good correspondence with experiment (see Ashurst & Meiburg, 1988 and Meiburg & Lasheras, 1988). We have not tried to improve our pressure gradient estimate, but note that McMurtry *et al.* (1986) stress that a second-order accurate density is used in their scheme of finite-rate combustion in a constant volume.

The advantage of our procedure is that the boundary pressure is not needed and the boundary velocities satisfy constant-pressure combustion. We use this scheme to explore possible flame-vortex interactions and present the simulations next.

RESULTS

As in Marble's work we are interested in the vortex effect on the distortion and convection of a flame sheet. Marble placed the vortex center on the flame at the start of the computation and assumed that each flame element is rotated about the core while the varying strain rate affected the diffusive flux of fuel and oxidizer at the element. This configuration yields two spiral arms of flame, centered about a growing core of burnt products. We select a different initial condition: the straight flame is near a vortex, but does not pass through the vortex center. We chose this configuration based on experimental observations that the flame location is not within the vorticity distribution at the start of jet and mixing layer flows, *cf.* Roberts (1985) and Dimotakis (1989). This off-center configuration creates a fold in the flame sheet, the sharpness of the fold depending on the relative magnitude of vortex rotation to diffusion, the Péclet number Γ/D. In the simulations presented below, we first examine the flow with weak heat release and vary the diffusion coefficient. A quasi-steady flame configuration occurs during the first few eddy turn-over times. Then we show results with different amounts of heat release and vorticity generation, and finish with the variation of reaction due to strain-rate effects.

Vortex-Flame Initial Conditions

The computational domain is $\pm L$ in x, y, with 128 grid cells in each direction. The periodic direction is in y and so there are image vortices spaced at $2L$ in that direction. Initially the flame is straight and located at $x = 0.125L$, $-L < y < L$. The distribution of mixture fraction is the error function solution, for $0.99 < Z < 0.01$, and is spread over seven grid cells.

We start all flows with a Gaussian distribution of vorticity centered at the origin with a core size of $0.25L$. The vorticity distribution for an isolated vortex is $\omega(r, t) = \Gamma/(\pi 4\nu t)\exp(-\eta^2)$ with $\eta^2 = r^2/(4\nu t)$. The core size is defined so that η^2 has the value of three when r is $0.25L$ at the start of the computation, which gives 95% of the circulation within this radius. In all cases we use $\Gamma = 2\Gamma_o$ and set both Γ_o and L to unity, time is expressed in units of L^2/Γ_o. The Schmidt number is unity, $\nu = D$.

A vortex has a swirling flow with zero rotation at the core center, rising to a maximum and then decaying to zero with further distance from the vortex. With the Gaussian distribution, the maximum swirl speed is located where the vorticity is 0.285 of the peak vorticity, η is 1.121. For this problem we can express time in terms of a turn-over period: the time t_u for one rotation about the core at a particular radius. This is not a precise definition, since the rotation time increases with radial distance from the vortex center and changes with viscous spreading of the vortex. Ignoring these effects, we pick a representative turn-over time in the initial flow and regard that value as constant for the next one or two turn-over periods.

As a comment, we note that a weak out-of-plane strain rate would maintain a constant core size and a constant turn-over time with small modifications to the in-plane strain rates which are included in the current simulations. This modified flow pattern would correspond to the stretching vortex flow that is inferred to be typical of high energy dissipation regions in Navier-Stokes turbulence (Ashurst *et al.*, 1987).

Formation of a Folded Flame

For weak heat release, $T_{ad} = 1.1T_o$ and circulation-to-diffusion ratio of $\Gamma/D = 800$, we show the development of the flame shape in Figure 1. We see a fold appearing by half of a turn-over time, where t_u is $0.6L^2/\Gamma_o$. At one turn-over time, we show in dashed lines the cusp that occurs in a non-diffusive material line which starts at the same initial location as the flame sheet. Notice that the main curvature around the vortex is the same, and only in the tip region does diffusion make a difference in flame shape at this time. The amount of tip rotation around the core does not increase during the next turn-over period, as can be seen by comparing the last two configurations. Of course, the material line will continue to rotate about the core, whereas the diffusive tip approaches a quasi-steady angular location. During this last turn-over period, the flame shape has expanded in a radial direction due to viscous spreading of the vortex core. The peak vorticity decreases to less than two-thirds of its initial value during the time period shown in Figure 1. This decay follows the Gaussian behavior. Beyond the time shown, the tip location develops a new quasi-steady shape in this dwindling flow.

Now we investigate the flame shape dependence on diffusion. Keeping the weak heat release of $T_{ad} = 1.1T_o$, we change the ratio of Γ/D, and examine the amount of flame tip rotation at a fixed time after starting with a straight flame. In Figure 2, the values of Γ/D are 1600, 800 and 200. In each case the elapsed time is approximately

L^2/Γ_o and corresponds to one or two turn-over times, depending on which radius is used. As far as convection of the flame sheet is concerned, these velocity fields are essentially the same even though the value of ν also changes by a factor of eight ($\nu = D$). The effect of ν is reduced because in each calculation the starting distribution of vorticity is the same; they only differ in the amount of viscous spreading from the same initial profile. The flows with the largest and smallest values of ν convect a material line in the same way over this time period with the exception of a region near the vortex center. Within this region the amount of rotation of the cusp in the material line is less with the larger ν value. Since the flame tip of the diffusive interface does not follow the material-line cusp, we can regard these flows to be the same for convection of the flame sheet.

With the assumption that the convective flow is similar in these three flows, we see in Figure 2 that the amount of flame tip rotation about the vortex depends on the magnitude of diffusion. Using a triangular control volume, we show the convection of oxidizer into the volume and the diffusion of oxidizer out of the volume, that is diffusion to the flame location where oxidizer combines with fuel to form product. As the scaled flux arrows indicate, a smaller diffusion coefficient results in more tip rotation in order to consume the oxidizer. So, in the first few turn-over times, the amount of tip rotation increases with the ratio of Γ/D.

Variation of Heat Release

All of the above simulations were essentially at constant density in that the adiabatic flame temperature is only $1.1T_o$. We now vary the heat release with $\Gamma/D = 800$. At similar times, actually at the same number of computational steps, we compare and find that the flame shape for T_{ad} of 1.1, 3 and $5T_o$ are not very different, see Figure 3. With increasing $\nabla \cdot \vec{u}$, the computational time step is reduced, but we achieve about the same displacement in the same number of steps. The increasing heat release broadens the flame tip region but does not change the amount of tip rotation around the vortex core. Thus, in the folded flame region, the volume expansion pushes apart the flames that are back-to-back. The penetrating finger of oxidizer is broader for larger heat release and so the tip is more rounded and has a larger radius of curvature.

Vorticity Generation

The baroclinic generation of vorticity during the transient from the initial conditions of a straight flame to the folded flame is given next. The generation of vorticity along the flame is due to misalignment of density and pressure gradients. The dominate pressure gradient is that of the swirling flow, and so it is in the outward radial direction. The temperature is maximum at the flame, which means that the density gradient points away from the density minimum located at the flame surface. The component of pressure gradient tangent to the flame surface will form a dipole layer of vorticity about the flame, assuming a uniform pressure gradient over the local width of the density gradient. This creation of two signs of vorticity differs from that in a premixed flame, which has a uni-directional density gradient and locally creates only one sign of vorticity.

We use one of the simulations from Figure 3, T_{ad} is $3T_o$ and Γ/D is 800. The initial vorticity is negative, clockwise rotation, centered at the origin and 95% of the vorticity is contained within the radius of 0.25. The initial turn-over time is $0.6L^2/\Gamma_o$. The initial flame is straight and located at $x = 0.125$.

Figures 4a-d present the vorticity contours with the negative vorticity shown in dotted lines, the contour intervals are one-tenth of the maximum absolute vorticity value at the time shown, in all cases this maximum occurs in the negative vorticity. The arrangement of positive and negative vorticity seen in Figure 4a is a result of the initially straight flame near the vortex. Notice the lack of generated vorticity in the region near the y axis, where the vector cross product of density gradient and pressure gradient goes to zero in the initial configuration.

In Figure 4b, the lower dipole layer, which started along the flame initially at $y < 0$, has been rotated to the left side of the vortex core, while the vorticity generated along the upper flame is moving to the bottom of the core region. The initial vorticity distribution has been disrupted so that there are now two negative extrema in the core region. The maximum of the positive vorticity is less than half the negative magnitude. The time development of the positive vorticity indicates that the largest value has already occured before this time of $t_u/2$ and while the magnitude of the negative vorticity decays approximately with the Gaussian behavior the maximum positive vorticity continues to become a smaller fraction of the negative amount. At $2t_u$, the positive maximum is only 1/7 of the negative magnitude.

While the vorticity is generated antisymmetricaly with respect to the flame location, there is no restriction that the flame must remain between the generated vorticity. The vorticity and the mixture fraction may have different gradients, and so even with unit Schmidt number the diffusive fluxes may be unequal. It is evident that the flame sheet is not located at the zero vorticity contour near the fold in the flame.

Since this transient vorticity generation is dependent on the selected initial condition of a straight flame near a Gaussian vortex, we are not sure how general this behavior may be. Roll-up of a shear layer will generate smaller pressure gradients in comparison to those of the concentrated vortex. In the simulations done to date, the vorticity generation does not make overwhelming perturbations to the dynamical evolution of the constant-density simulations.

Strain-Rate Effect on Reaction

Using the nearly isothermal case, we investigate the effect of strain rate on the width of the diffusive zone. Carrier *et al.* (1975) examined the flame sheet model subject to both constant strain and time-dependent strain. The vital result is that compressive strains have the ability to stop the diffusive spreading of the mixture fraction which describes the reaction zone. For constant strain rate, the relevant width is several multiples of the Batchelor length scale, $l_B = \sqrt{D/\epsilon}$. In our simulations, the maximum strain rate in the isolated vortex flow can be used to estimate the diffusive zone thickness. The shear strain in this swirling flow has a broad maximum just outside the radius of maximum swirl velocity. This shear strain can be expressed as two normal strain-rates, one extensive and one compressive, with directions that are 45 degrees from the local radius vector. Using this maximum strain-rate, the diffusive width is estimated to be less than $0.1L$ when Γ/D is greater than 800, but twice this width for Γ/D of 200. During the first few turn-over times these width estimates appear reasonable. Differences are to be expected, since instead of a compressive strain always directed normal to the scalar gradient, as analyzed by Carrier *et al.*, there is rotation and straining of a connected line and some sections of this line do not fit the assumed one-dimensional problem. An example of the strain field is given in Figure 5. From the simulation with $\Gamma/D = 200$ at time $1.33L^2/\Gamma_o$ ($\approx 2t_u$), the extensive strain-rates are shown by drawing a line at every fourth grid point, the line length is scaled to the strain-rate magnitude and the line direction is that of the extensive strain. The compressive direction is perpendicular to the extensive direction and equal in magnitude in this nearly isothermal flow, T_{ad} is $1.1T_o$. At this time the vorticity region has an elliptic shape with an axis ratio of 1.2 caused by the straining from the periodic images, but in general the extensive direction is 45 degrees from the radius vector.

In addition to the width of the diffusive zone, Carrier *et al.* calculated changes in the flame structure when the reaction rate is not infinite in comparison to the strain rate. Using a one-dimensional analysis for corrections to the flame sheet model; and assuming a single-step, finite-rate reaction; they find that the temperature is decreased below the adiabatic value by an amount proportional to $(\epsilon/B)^p$, where B is the finite reaction-rate, ϵ is the strain rate and p is related to the stoichiometric coefficients of the reaction.

We can extrapolate this strain effect on temperature to our flame sheet simulations by checking the variation of the scalar gradient along the flame surface. To show both the flame, and the scalar gradient at the flame, in Figure 5 we draw five contour lines of mixture fraction: Z_{fl}, $Z_{fl} \pm 0.01$ and $Z_{fl} \pm 0.02$. Over most of the flame, these five contours appear as one dark line with constant width, but in the region where the flame is folded we see the spreading of these five contour lines, indicating a much smaller scalar gradient at the flame tip. The reduction in gradient is about a factor of four, and this configuration is not changing rapidly with time. This reduction in gradient is not caused by a local reduction in strain, the dash lines have about the same length in that region, but the reduction is caused by the topology of the flame surface. The flame tip is the only section of the flame surface where we see the extensive direction normal to the flame. We might expect this topology in a flow where the flame extends beyond the separation distance of two vortices which have the same rotation sense, as in a jet or mixing layer. Between those sections of flame surface which are pulled in by one vortex and expelled by the neighboring vortex the flame will become folded and have a reduced scalar gradient. We infer such flame shapes in the experimental results of Dibble *et al.* (1986), see their Figure 4b. Using the one-dimensional strain analysis by Carrier *et al.* (1975), this gradient reduction effect would result in a higher temperature at the flame tip in comparison to the rest of the flame. This estimate ignores gradients and convection along the flame surface (see page 77 in Williams, 1985).

CONCLUSIONS

Two-dimensional simulations of nonpremixed reaction in the field of a vortex have been done at constant pressure, and include the effects of volume expansion and baroclinic vorticity generation. We start with the vortex on the fuel side of the zero-thickness flame sheet and observe that a fold develops in the flame shape, but the flame is not drawn into the vortex core. The duration of the vortex-flame interaction is dependent on the fuel and oxidizer boundary conditions, because the reaction front is trapped at the stoichiometric location as it lacks the propagation mechanism of premixed combustion.

Nearly isothermal flames, ten percent temperature rise, have shapes similar to those calculated with much larger heat release, with maximum temperature of three or five times the ambient temperature. We do not find any dramatic effects caused by the vorticity generation. While the overall shape is similar, the curvature at the fold in the flame does depend on the magnitude of heat release. The volume expansion associated with heat release increases the back-to-back distance between the flame surfaces adjacent to the fold. Hence, a constant-density simulation underestimates the distance between these two flame surfaces, and thus underestimates the time for reaction to consume these features.

During the first eddy turn-over time, the amount of flame folding increases with the ratio of vortex circulation to the diffusion coefficient. In a reacting turbulent flow the diffusion coefficient may vary due to its temperature dependence, but the larger variation in Γ/D results from the various vortex circulations that describe the particular turbulent flow. The gradient of mixture fraction is reduced by the folding of the flame as this is a region where the extensive strain-rate is normal, instead of parallel to the flame. The square of this gradient is the scalar dissipation, which is a parameter in flamelet models of turbulent combustion. These flamelet models usually ignore gradients along the flame surface, but in our simulations these gradients are not small. So, in the folded region there is the possibility that the flame structure will be different from a straight, strained flame that occurs in stagnation flow regions, a configuration that has been extensively analyzed and measured to date.

ACKNOWLEDGMENTS

This work was supported by the United States Department of Energy, through the Office of Basic Energy Sciences, Division of Chemical Sciences. Discussions of this work with Sébastien Candel, Paul Dimotakis, Frank Marble, Steve Margolis, Eckart Meiburg, Reggie Mitchell and Forman Williams have been beneficial.

REFERENCES

Ashurst, Wm. T. and Barr, P.K. (1983). Stochastic Calculation of Laminar Wrinkled Flame Propagation via Vortex Dynamics. *Combust. Sci. Tech.* **34**, 227.

Ashurst, Wm. T. and McMurtry, P. A. (1989). Flame Generation of Vorticity: Vortex Dipoles from Monopoles. *Combust. Sci. Tech.* in press.

Ashurst, Wm. T. and Meiburg, Eckart (1988). Three-dimensional shear layers via vortex dynamics. *J. Fluid Mech.* **189**, 87.

Ashurst, Wm. T., Kerstein, A. R., Kerr, R. M. and Gibson, C. H. (1987). Alignment of vorticity and scalar gradient with stain rate in simulated Navier-Stokes turbulence. *Phys. Fluids* **30**, 2343 and 3293.

Barr, P. K. and Ashurst, W. T. (1982). Evaluation of Zalesak's flux-corrected transport algorithm for convection and diffusion. Sandia Report, SAND-81-8233.

Carrier, C. F., Fendell, F. E. and Marble, F. E. (1975). The Effect of Strain Rate on Diffusion Flames. *SIAM J. Appl. Math.* **28**, 463.

Dibble, R. W., Long, M. B. and Masri, A. (1986). Two-Dimensional Imaging of C_2 in Turbulent Nonpremixed Jet Flames. *Dynamics of Reactive Systems Part II: Modeling and Heterogeneous Combustion*, edited by J. R. Bowen *et al.*, Vol. 105 of Progress in Astronautics and Aeronautics, AIAA. pp. 99-109.

Dimotakis, P. E. (1989). Turbulent Free Shear Layer Mixing. AIAA-89-0262.

Laverdant, A. M. and Candel, S. M. (1988). A Numerical Analysis of a Diffusion Flame-Vortex Interaction. *Combust. Sci. Tech.* **60** 79. see also Interaction of Diffusion and Premixed Flames with a Vortex. *Rech. Aérosp.* – n° 1988-3.

Marble, F.E. (1985). Growth of a Diffusion Flame in the Field of a Vortex. *Recent Advances In The Aerospace Sciences*, Plenum Publishing, pp. 395-413.

McMurtry, P. A., Jou, W.-H., Riley, J. J. and Metcalfe, R. W. (1986). Direct Numerical Simulations of a Reacting Mixing Layer with Chemical Heat Release. *AIAA J.* **24**, 962.

Meiburg, E. and Lasheras, J. C. (1988). Experimental and numerical investigation of the three-dimensional transition in plane wakes. *J. Fluid Mech.* **190**, 1.

Roberts, F. A. (1985). Effects of a Periodic Disturbance on Structure and Mixing in Turbulent Shear Layers and Wakes. Ph. D. thesis, California Institute of Technology.

Williams, F.A. (1985). *Combustion Theory*, 2nd Ed., The Benjamin/Cummings Publishing Company, Menlo Park, California.

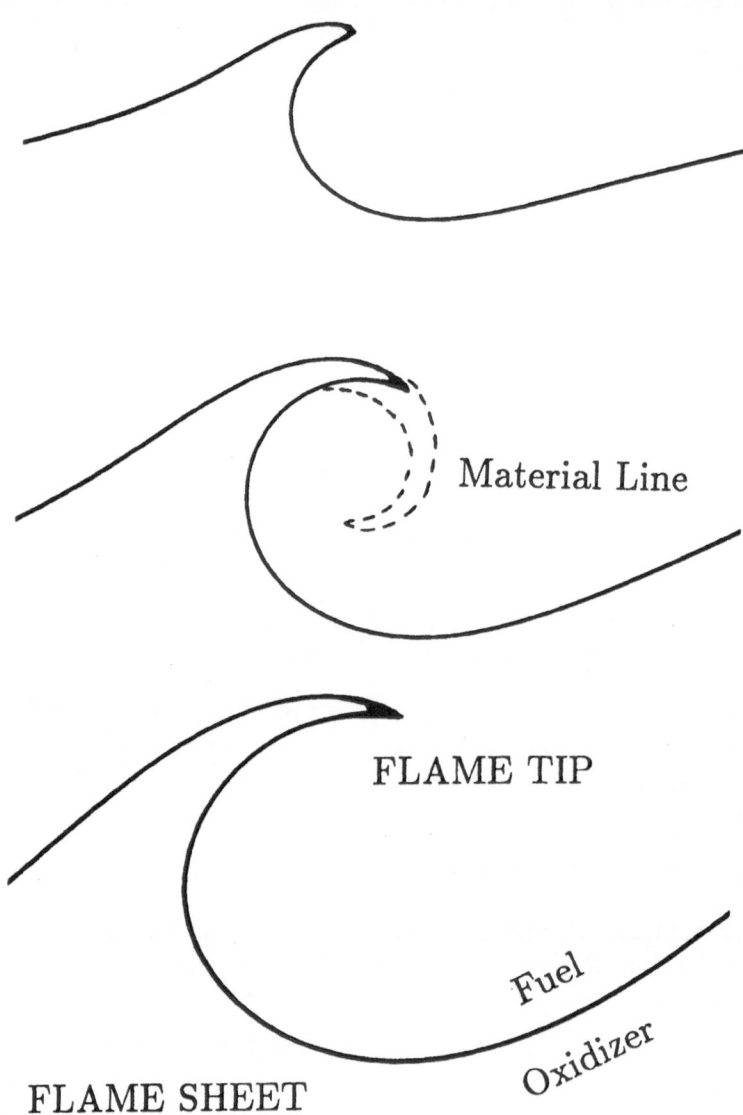

Figure 1. Folding of a flame sheet by the swirl flow of a vortex during the first two eddy turn-over times, from top to bottom, the times are $1/2$, 1 and $2t_u$. The initially straight flame is swept around the vortex core (not shown) and acquires a flame tip. Shown at one turn-over time are dash lines representing the location of a nonreacting material line which started congruent with the flame sheet. Thus, the effect of diffusion is to change the cusp of the material line to the rounded tip of the flame sheet. The flame tip does not continue to rotate around the core as the material line does.

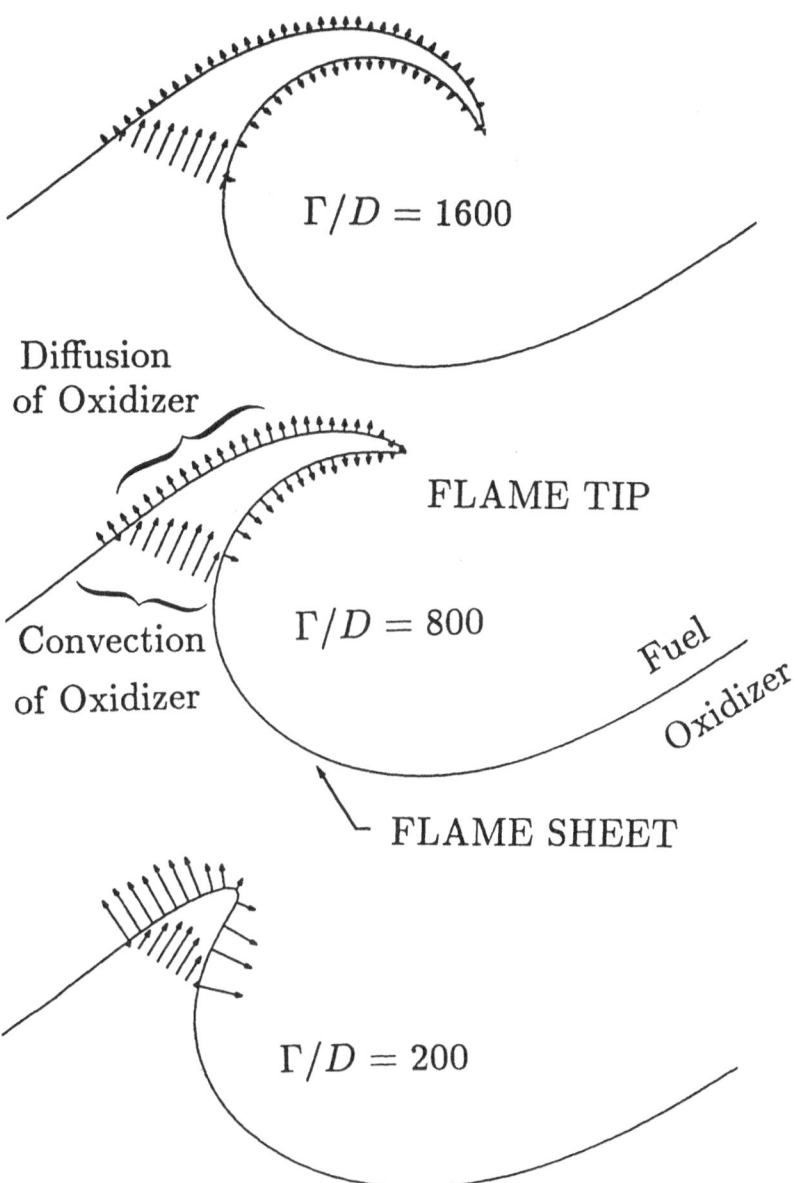

$\Gamma/D = 1600$

Diffusion
of Oxidizer

FLAME TIP

$\Gamma/D = 800$

Convection
of Oxidizer

Fuel

Oxidizer

FLAME SHEET

$\Gamma/D = 200$

Figure 2. The amount of flame tip rotation around the vortex depends on diffusion because oxidizer that is convected to the tip region is consumed at the flame by a rate proportional to the diffusive flux. In these three flows, shown at the same time, the convective velocities are similar, but the diffusion differs by factors of two and four. The arrows denote the fluxes and for visibility the scaling of the diffusive flux is twice that of the convective flux of oxidizer.

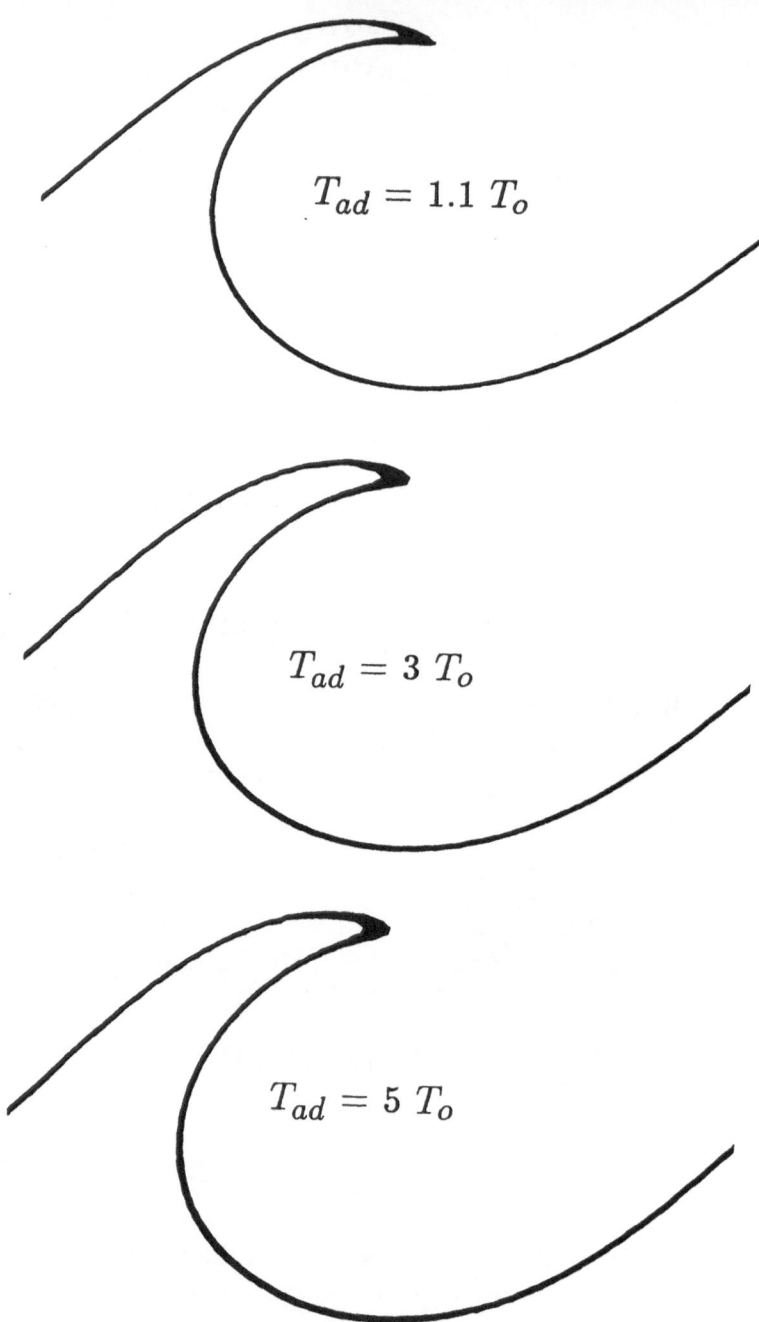

Figure 3. The flame tip region has a larger radius with larger heat release, resulting in a larger distance between the two flame surfaces bordering the peninsula of oxidizer. In these three cases the adiabatic flame temperature is 1.1, 3 and 5 times the ambient temperature (from top to bottom).

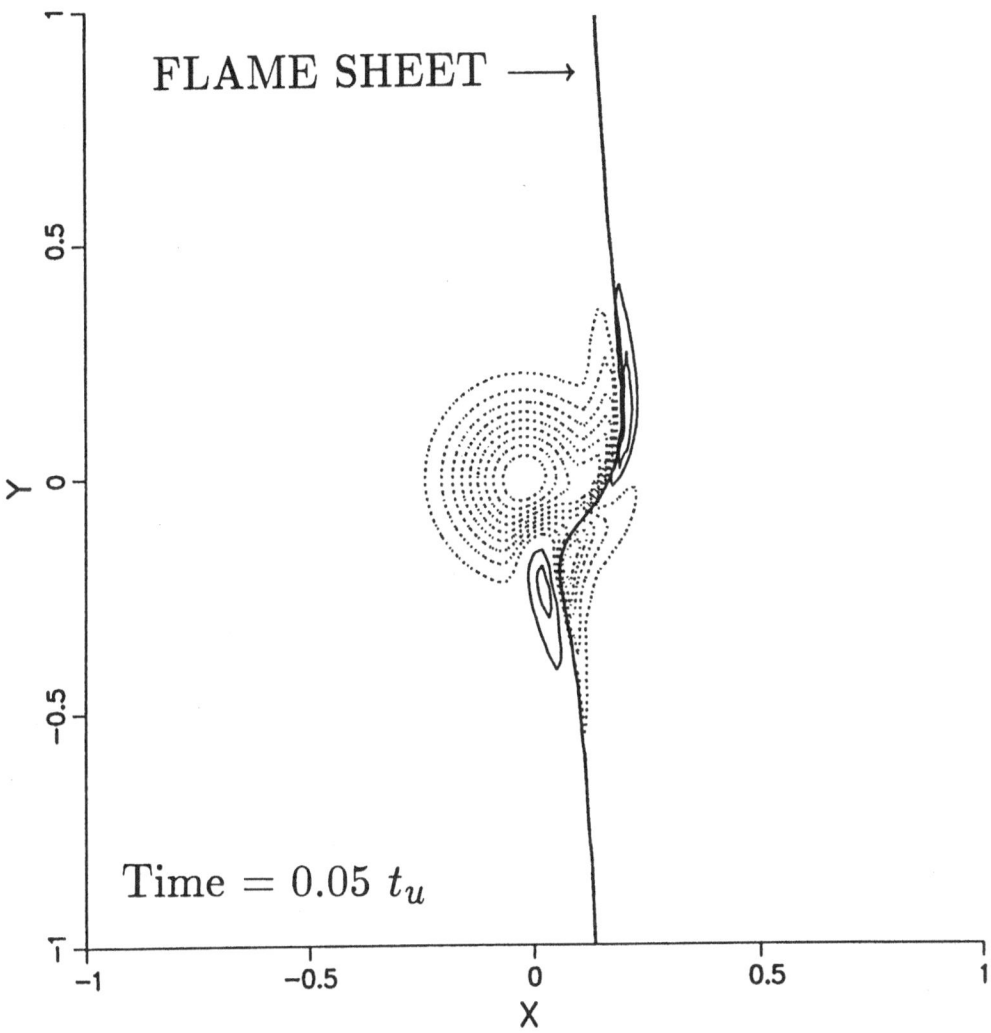

Figure 4a. Baroclinic generation of vorticity in the diffusion flame is characterized by dipole layers of vorticity, the sign of generated vorticity is antisymmetic with respect to the flame surface. The initial vorticity is a Gaussian distribution, with 95% of the vorticity within the radius of 0.25. The straight flame starts at x of 0.125. Contours of negative vorticity (clockwise rotation) are in dotted lines, intervals between contours are ten percent of the maximum vorticity magnitude.

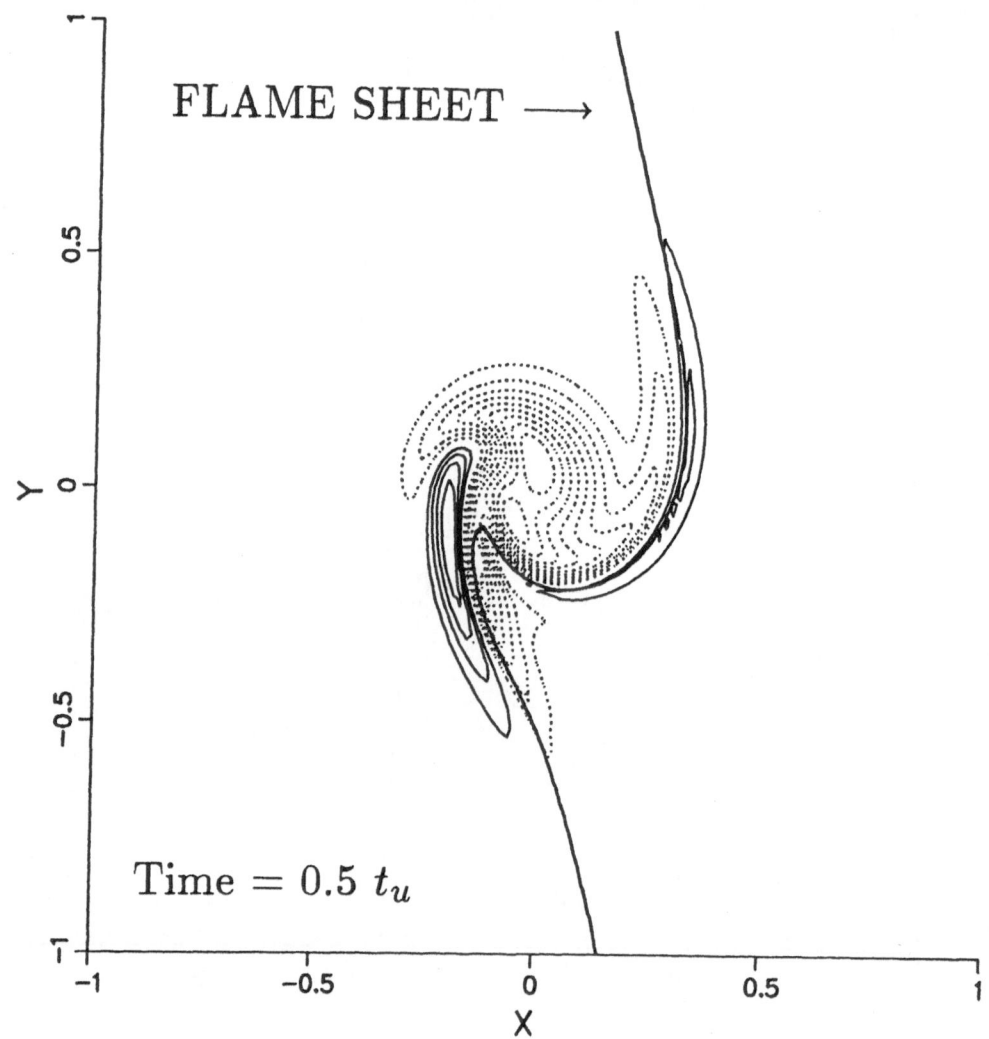

Figure 4b. Vorticity contours and flame shape at one-half of a turnover time. At this time the generated positive-vorticity maximum is almost half of the negative-vorticity magnitude. There are two negative extrema in the core region.

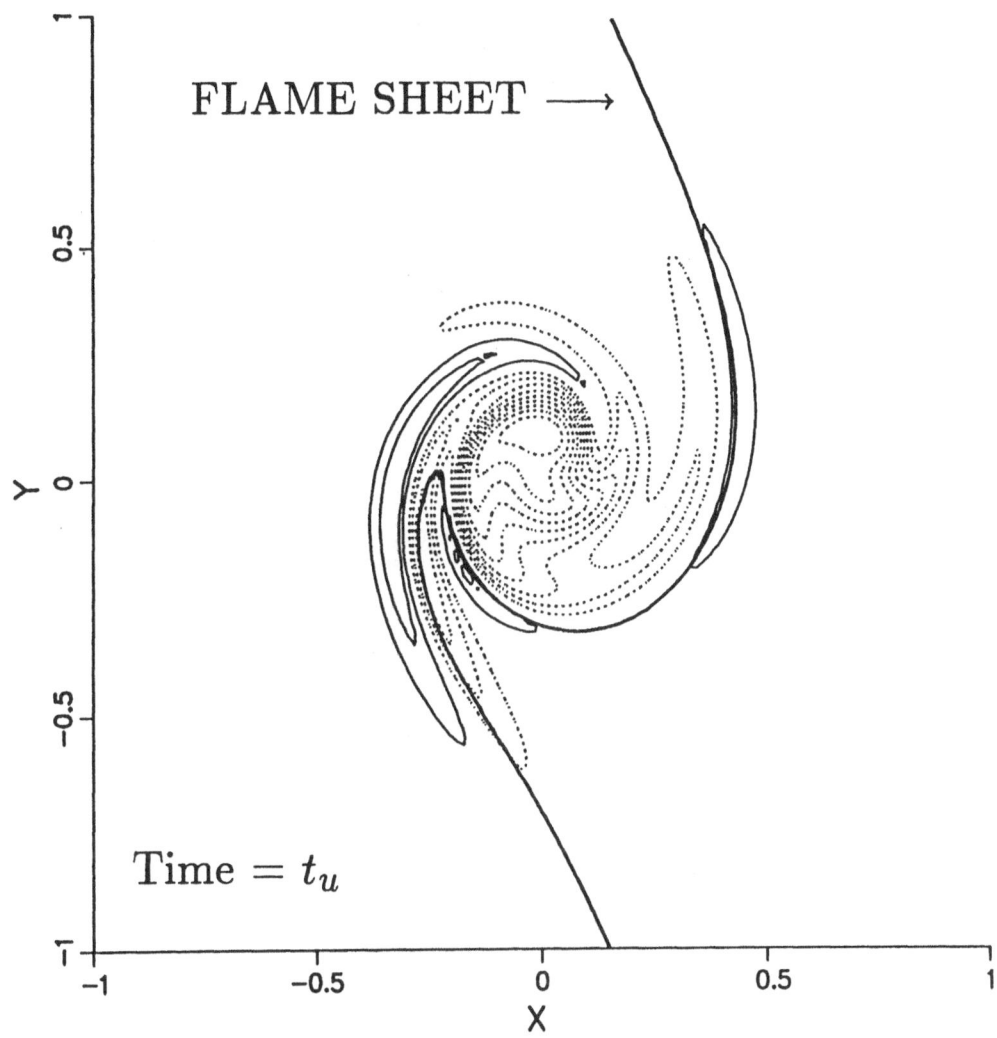

Figure 4c. Vorticity contours and flame shape at one turnover time. The two negative extrema in the core at the previous time have combined into one minimum.

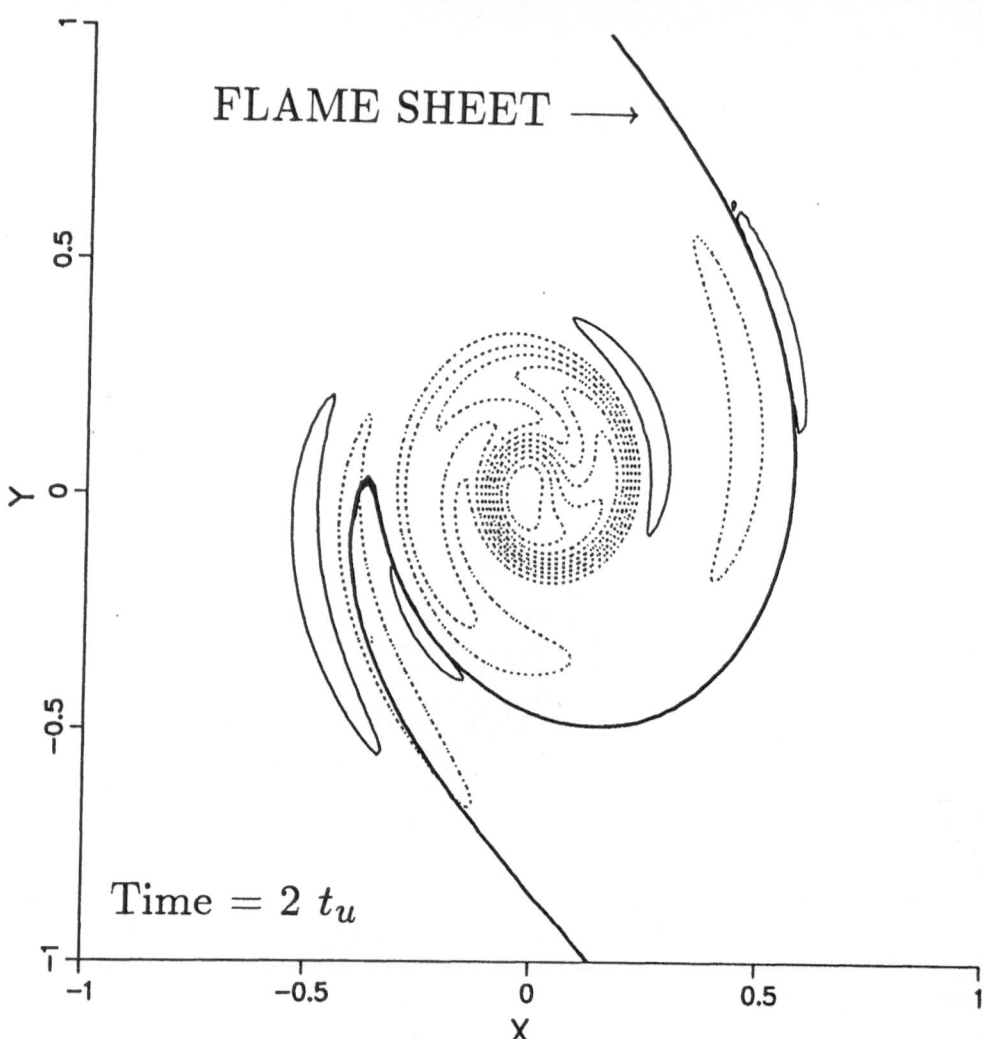

Figure 4d. Vorticity contours and flame shape at two turnover times. At this time the generated positive-vorticity maximum is 1/7 of the negative-vorticity magnitude. The adiabatic flame temperature is three times the ambient temperature in this simulation and the initial vortex circulation is 800 times the scalar diffusivity.

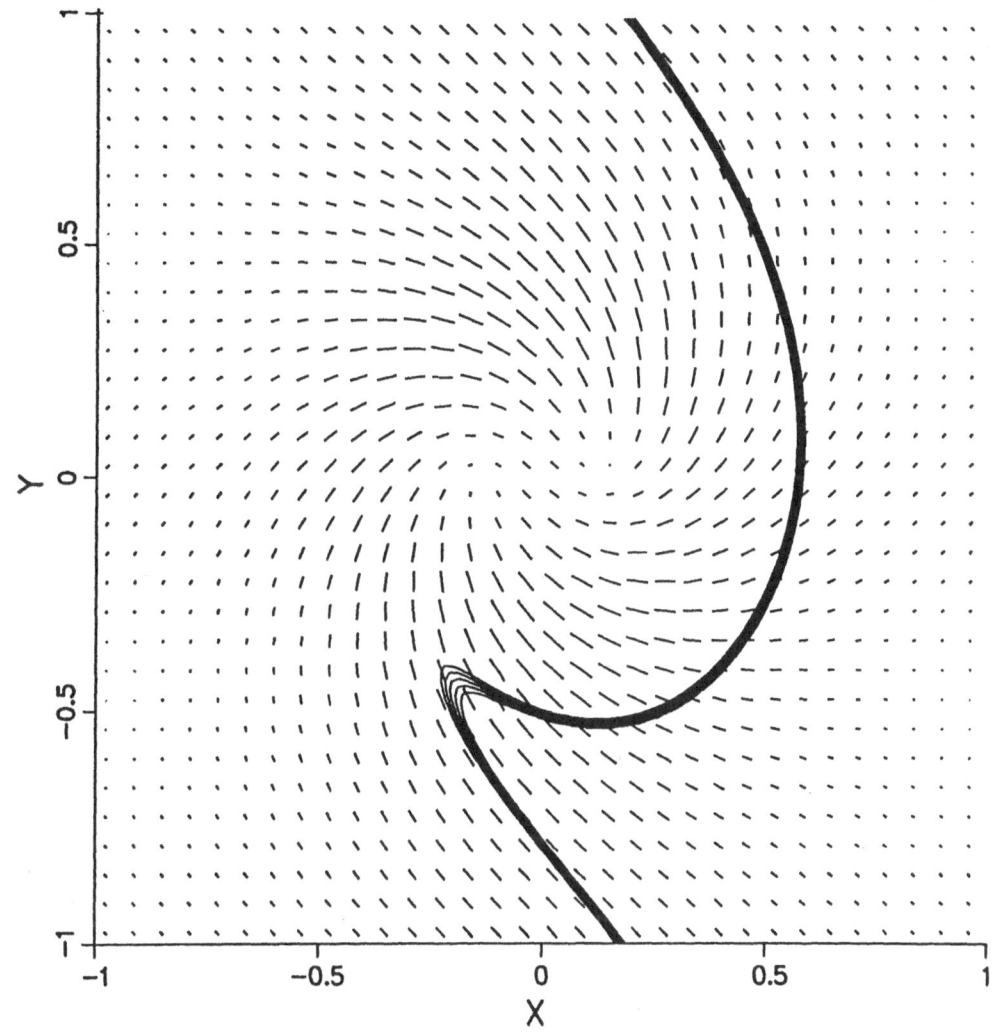

Figure 5. Flame shape and extensive strain-rate after two turn-over times, $\Gamma/D = 200$, $T_{ad} = 1.1T_o$. The line segments are scaled to the strain-rate magnitude and are in the extensive direction. The flow is periodic in the y direction, and so the flame is pulled in by the swirling clockwise rotation, but expelled by the image vortex at $y = -2$. This flow pattern forms a flame tip, and in the tip region we see that the extensive strain-rate is normal to the flame instead of more parallel. The mixture fraction gradient is reduced where the flame traverses the region of extensive strain normal to the flame – as shown by plotting five contour lines of the mixture fraction, notice that only in the flame tip region is the spacing between these contours discernable (the middle line is the flame sheet location).

A NUMERICAL SIMULATION OF SHOCK GENERATED
IGNITION USING THE RANDOM CHOICE METHOD

J.F. Clarke & G. Singh

Aerodynamics, Cranfield Institute of Technology,
Bedford, MK43 OAL, U.K.

Abstract

Exothermically reactive gas flow in the region between a piston and a strong
shock wave, (Mach 3.0), is modelled numerically using the Random Choice Method to solve
the Euler equations. The numerical algorithm consists of the Random Choice scheme,
formulated in a Lagrangian configuration, coupled with time-operator splitting to
treat combustion chemistry. The combustion reaction is assumed to be of simple
irreversible Arrhenius type. Numerical results show that ignition first occurs close
to the piston face, and is followed by the formation of a compression pulse that
finally contains a shock, an unsteady induction domain and a fast flame prior to
transition of this system into a ZND-structured detonation.

1. Introduction

Recent work on the evolution of detonation waves in gases has been designed to
look closely at the processes that take place in the simplest geometrical situation,
namely unsteady motion in one space dimension. The ways in which chemical and gas-
dynamical events interact with one another are thereby revealed in the absence of the
complicating effects of geometrical attenuation or focussing. The intention is to
make sure that the fundamental aspects of such interactions are thoroughly understood
before embarking on the more complex issues posed by multi-dimensional behaviour. The
present paper is primarily to report on the results of one particular phase of this
enquiry into the fundamentals of wave/chemistry interaction.

Of course, the general topic is one with a long history which is both well
known and not in any need of exposition here. It is appropriate to set the present
scene by referring to work by Clarke, Kassoy and Riley (1986, hereinafter referred to
as CKR) on detonation-initiation as a consequence of substantial heat-power addition
through a wall bounding a right-hand half-infinite space of initially cold combustible
gas. Heat-conduction is obviously an essential element in the resultant processes in
such a system, and the theoretical model necessitated solution of the full Navier-
Stokes equations for the reacting mixture. Calculations were conveniently made in a
Lagrangian frame of reference and were, of course, numerical. The method used was
iterative, based initially on extrapolations of dependent variables from one time

level to the next, with gas velocity u, absolute temperature T and reactant mass
fraction c calculated by using a Crank-Nicolson method to solve diffusion equations
satisfied by these variables at successive stages of any individual iteration;
density ρ and pressure p are updated at appropriate points in the process by means of
the mass conservation equation and thermal equation of state respectively. Convergence
is tested by evaluating the sum of the absolute values of differences between
successive iterates at all spatial nodes, and is deemed to have been achieved when
this sum is less than 10^{-4}. Dependent variables are intially non-dimensionalised
and normalised.

The method is robust, sufficiently accurate for our purposes, and quite slow;
dependability has been the requirement, not economy of computer time.

Results show that after some complex early-time behaviour in the thermal boundary
layer adjacent to the wall a particular triplet pattern of waves emerges and propagates
out into the cold unreacted fluid. The leading wave is an adiabatic gas-dynamical
shock; it is followed by an induction domain in which small-amplitude disturbances
to both chemical and gas-dynamical fields interact in a manner that is essentially
non-uniform in both space and time (cf work by Jackson and Kapila, 1985; Clarke, 1981;
Clarke and Cant, 1984); finally, the third element of the triplet is identifiable as
a quasi-steady fast flame. A paper (Clarke, et al, 1989) that sets out specifically
to examine this last proposition in some detail is in preparation and is, in effect,
a postscript to CKR since it uses precisely the same numerical scheme. A wide-ranging
discussion, review and analysis of the whole field of study of fast-flames, unsteady
induction-domains and shocks has recently been put together by Clarke (1988), and
provides more evidence of an analytical kind in favour of the existence of the kind
of triplet of waves that we have just described.

Despite all this, the evidence is not yet complete. Gaps, albeit diminishing
ones, still exist in the (necessarily) approximate-analytical work. However, these
studies point clearly towards triplet structures that are, outside the interiors of
adiabatic shocks, essentially free from the influences of diffusion in its widest
sense. The model of gas behaviour used by CKR includes real, physical, diffusion by
its very nature, and more will be added by the finite- difference scheme used for its
numerical resolution.

Although we believe that the numerical calculations of CKR and Clarke, et al,
(1989) do also point to predominantly diffusion-free triplet-wave structures there is
one fairly conclusive way to check on such beliefs. That is, to make numerical
calculations of solutions to the Euler, not Navier-Stokes, equations by a method that
does not itself contain numerical diffusion and which captures shock-waves without
smoothing and without spurious oscillations, which brings us directly to the point
of our present investigation.

The Random Choice Method (RCM) is established as a reliable numerical technique for the solution of gasdynamical problems in plane unsteady flows [Saito and Glass (1984); Takano (1986)]. In this paper the RCM is applied to an unsteady one-dimensional piston problem for a reacting gas. The advantage of the RCM over conventional finite difference methods is its ability to capture discontinuities (shocks and contacts) with infinite resolution; it also handles wave interactions efficiently.

The problem to be solved assumes an unperturbed atmosphere consisting of a uniform combustible gas, occupying a semi infinite space, bounded on the left by a piston. At time t=0, the piston is instantaneously accelerated to a constant velocity, thereby creating a strong shock wave. The shock wave is of sufficient strength to make induction times immediately downstream of the initial shock very much shorter than the 'long' time that exists in the ambient atmosphere. Thus the precursor shock effectively switches on chemical activity in the region between itself and the piston surface.

A description of events as they evolve in the induction domain for early times is given by Clarke & Cant (1984) and Jackson & Kapila (1985). However these analyses are only valid up to certain times t_{is}, indicative of the onset of ignition at the piston face. The present task is a natural numerical extension of this work for later times, thus capturing ignition and post-ignition phenomena.

2. Numerical Model

It is mathematically and computationally advantageous to work with a Lagrangian formulation of the Eulerian conservation equations for the chemically reacting mixture. The equations are made dimensionless with the following set of variables: temperature T_{is}, pressure P_{is}, density ρ_{is}, sound speed a_{is}, induction time t_{is}, and fuel mass fraction c_{is}, where subscript 'is' indicates a value taken just downstream of the initial shock. The dimensionless Lagrangian co-ordinate can then be defined by

$$\rho_{is}\, a_{is}\, t_{is}\, \psi \;=\; \int_{x'_p(t')}^{x'(t')} \rho'(x',t')dx'$$

where $x'_p(t')$ is the piston path in dimensional x',t' space. The quantity t_{is} is the induction time for the shocked state behind the initial shock at constant pressure. The chemical reaction is assumed to be of the simple form fuel F → products P, so that the conservation equations for the reactive flow become:

$$U_t \;+\; G\, U_\psi \;=\; S, \tag{1}$$

where

$$U \;=\; (\rho,\, u,\, p,\, c)^T,$$

$$G = \begin{pmatrix} 0 & \rho^2 & 0 & 0 \\ 0 & 0 & 1/\gamma & 0 \\ 0 & \gamma p \rho & 0 & 0 \\ 0 & 0 & 0 & 0 \end{pmatrix}, \quad S = \begin{pmatrix} 0 \\ 0 \\ \gamma\rho\, c\Omega Q \exp(-E/T) \\ -c\Omega \exp(-E/T) \end{pmatrix}.$$

S is the matrix of source terms and the thermal equation of state

$$p = \rho T ,$$

provides the fifth relation necessary to describe the behaviour of the five independent variables. E is the dimensional molar activation energy divided by RT_{is}, where R is the molar gas constant. Q is the chemical energy released by full decomposition of fuel to products, and is measured in units of $C_p T_{is}$, where C_p is the specific heat capacity at constant pressure (assumed constant). The product WBt_{is} is written as Ω, where W is the molecular weight of the fuel species and B is the pre-exponential factor in the Arrenhius reaction; γ is the (constant) ratio of specific heats.

2.1 Numerical Integration and the Splitting Technique

Numerical integration of (1) is performed by application of the RCM and a time operator splitting (TOS) technique. The system of equations (1) can be split into two sets. The first set, called the hydrodynamic equations, describes the dynamics of a chemically inert gas and the second set, the rate equations, describes the chemical processes. The solution U^{n+1} at the time level t_{n+1} of the partial differential equations (1) with initial data U^n at the previous time level t_n is constructed as follows. Initial data at each time level t_n is in discretised form so that the computational domain consists of a sequence of Riemann problems. The hydrodynamic equations, given by

$$U_t + G U_\psi = 0 \tag{2}$$

with initial data U^n at time level t_n, are integrated by the RCM over an appropriate time step Δt_n. The result of this integration is a single value U_H^{n+1}, say, sampled from the Riemann problem data. Subscript and superscript H and n+1, respectively, denote the solution of the homogeneous system (2) at time t_{n+1} To incorporate chemical reactions, and to complete the integration of (1) we use the following TOS technique. The solution U_H^{n+1} just obtained is now used as initial data for integration of the rate equations

$$U_t = S \tag{3}$$

over the same step size Δt_n. This yields the 'final' solution U^{n+1} at time t_{n+1}. Justification of TOS for discontinuous flows is difficult (for smooth flows see Yanenko (1971)); it is generally justified in an ad-hoc fashion (e.g. Clarke and Toro (1985), Takano (1986), Toro (1989)).

2.2 The Random Choice Method: A Lagrangian Formulation.

The spatial domain is discretised into M computational cells of size $\Delta\psi$. In general $\Delta\psi$ need not be constant but we assume that it is so here. Integration of the system of equations (2) is therefore performed along particle paths. The complete problem at specific time level in the discretised domain consists of a sequence of Riemann problems whose solutions are aggregated to form the solution for the problem at the next time level.

Solution of a typical Riemann problem in ψ,t space with initial data ρ_L, u_L, p_L for $\psi<\psi^*$ and ρ_R, u_R, p_R for $\psi>\psi^*$ consists of four uniform regions with constant properties divided by a right wave (RW) and a left wave (LW), either of which can be a shock wave (S) or a rarefaction wave (R), together with a contact surface C that lies between LW and RW. The characteristic wave directions in ψ-t space are eigen-values of the matrix G and are given by the lines $d\psi/dt = \pm\rho a$ and $\psi = $ constant. Analogies between Lagrangian and Eulerian wave systems in a typical Riemann problem are now apparent. The lines whose slopes have magnitude ρa (acoustic impedance) in the ψ-t plane correspond to acoustic-wave paths $u\pm a$ in the x,t plane. A line $\psi = $ constant (i.e. a particle path) may be a slip line $dx/dt = u$ in the x-t plane.

From this point onwards the technique is similar to the one described by Toro (1987) and we also employ the Van Der Corput sampling routine, due to Colella (1982). The size of the time step to be taken at each time level is limited by the criterion

$$\Delta t = CFL \ \Delta\psi \ / \ MAX(\rho a)$$

where CFL is a Courant number in the interval [0, 1/2].

Calculations of the effects of chemical reactions over time interval Δt are included in the modelling by solving (3) using the TOS technique outlined above. From (3) it can be seen that chemical reactions proceed under conditions of constant density and velocity but both pressure and fuel mass fraction vary due to the release of chemical energy. The chemical kinetic equations (3) are numerically integrated with initial conditions U_H^{n+1} at time t_{n+1} supplied from the RCM operating on the homogeneous equations (2). A schematic representation of the time marching procedure during one complete time step is shown in Figure 1; ϑ_{n+1} is the random number generated at t_{n+1}. The main difficulties in integrating (3) stem from the extreme sensitivity of the Arrenhius factor to local changes in the temperature. The reaction terms become stiff as ignition time is approached. A numerical integration algorithm D02EJF, selected from the NAG library, was employed to cope with these problems. The chemical-kinetics code may easily be extended to combustible gases undergoing more complex chemical-kinetic behaviour.

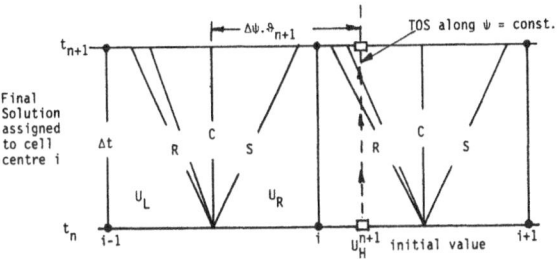

Figure 1. Time Marching Solution from t_n to t_{n+1}

3. Boundary Conditions and Parameters

A list of the typical parameters used in the numerical analysis is given in Table 1 below

W	30	T_a	300K
B	$10^{10} s^{-1}$	p_a	10^5 Nm^{-2}
E	14.99	ρ_a	1.2kgm^{-3}
C_p	10^4 Jkgmol^{-1}K^{-1}	u_a	0 ms^{-1}
Q	10^8 Jkgmol^{-1}	c_a	0.1
T_{is}	803K	a	342ms^{-1}

Table 1. Ambient atmospheric conditions are denoted by the subscript a.

The appropriate conditions at the piston face are given by,

$$p(0) = p(1), \quad u(0) = -u(1) + 2u_p, \quad \rho(0) = \rho(1),$$

where u_p is the piston velocity and numbers in brackets denote cell position in the computational domain. The zeroth cell is a fictitious state that lies to the left of the piston. The Courant number used throughout the computations is 0.25. The initial shocked values for temperature, pressure, density and the initial piston velocity required to create a Mach 3.0 shock wave, given the ambient atmospheric conditions, are readily obtained using the normal shock relations (Liepmann Roshko [1957]). Typical induction times for the gas ahead of and behind a Mach 3.0 shock travelling into gas at 300 K are (e.g. Strehlow [1969]) 1.9 hrs, 5.87 x 10^{-7} secs respectively. The computational mesh contained 2400 fine cells, in the neighbourhood of the piston-face (in anticipation of rapidly-changing events in this domain), and 49 coarse cells.

4. Results and Discussion

The inert-gas-flow part of the computer code was tested with Sod's shock tube problem [1978]. The results, not presented in this paper due to limitation of space, can be translated back into an Eulerian co-ordinate system and compared with the exact solution; excellent agreement between the two sets of results is achieved.

The results presented here are concerned with ignition and post-ignition events. Fig.2 is a ψ,t diagram that illustrates the overall scale of events; it indicates the path taken by the maximum reaction-rate and shows that a reaction-generated shock has appeared by time $t \simeq 1.136$ units, just ahead of this path. The initially broad, but shrinking, region within which reaction is significant is also depicted. By the time this region has shrunk to the point where it then travels in unison with the reaction-generated shock, this combination of processes has become a ZND detonation, as will be demonstrated.

Figs. 3, 4, 5, 6, 7 and 8 show, respectively, pressure, temperature, fuel mass-fraction, density, absolute gas velocity (zero ahead of the precursor shock) and chemical reaction-rate, plotted as profiles versus Lagrangian distance ψ for times t_n (n=1,...12) with $t_1 = 1.10(0.01)1.13 = t_4$; $t_5 = 1.14(0.005)1.15 = t_7$; $t_8 = 1.16$, $t_9 = 1.165$; $t_{10} = 1.175$, $t_{11} = 1.18$; $t_{12} = 1.19$, $t_{13} = 1.195$; $t_{14} = 1.20$, $t_{15} = 1.21$; $t_{16} = 1.215$, $t_{17} = 1.22$; $t_{18} = 1.225$, $t_{19} = 1.23$. Perusal of these sets of results allows one to build up a comprehensive picture of events within the system. Remarks in this paper must be confined to observations about one or two of the system's salient features.

It is first of all important to comment that the sequence of events is initiated by the purely mechanical act of pushing a piston into cold unreacted gas. Addition of heat from external sources has no part to play.

Fig. 3 shows, particularly at t_3 and t_4 , how the "wave-like" character of the rapid-reaction domain (c.f. Figs. 5 and 8 at these times) gives rise to a pulse of high pressure, whose peak value grows with time. Fig. 6 shows that density exhibits a similar pulse-like character at these times, whilst Fig. 7 indicates that the absolute gas velocity is broadly consistent with the local compression-pulse character; only the temperature distribution (Fig.4) shows that the pulse is not isentropic, since it contains a reaction wave.

Between t_4 and t_5 a shock-wave is formed in the leading parts of the pulse. This is seen most clearly in Fig.9, which plots p vs. specific volume v throughout the interesting parts of the field adjacent to the piston. The RCM provides unequivocal evidence for the existence of a shock (at t_5 and for all subsequent times) that originates from the regions of intense chemical action. The shock-jump occurs from one computational cell to the next and is not smeared across several cells.

To reiterate it is clear that the shock-wave, whose presence is manifest in Fig.9 at t_5, has formed at some earlier time within the domain of significant chemical reaction. There is a superficial resemblance between the p,v-locus at, for example, time t_5 here and a similar locus described by Clarke (1988) and by Clarke, et al, (1989). In these latter situations, which are initiated by an input of thermal power, the only shock-wave within the system at early times is the precursor shock. The main body of the reaction-wave, the region where reaction is most intense, lies between S and E (Fig.9) in such a situation and is distinguished on a p,v-picture as a straight (Rayleigh) line right up to the point at which all reaction ceases. The implication of this fact is that the reaction wave is a quasi-steady fast flame, linked to the precursor shock by an essentially unsteady induction domain near S, as explained in our Introduction.

It is evident from Fig.9 that no such Rayleigh line exists, taking time t_5 as exemplary, near the "all-burnt" or "end-of-reaction" regions in present circumstances. If an attempt is made to fit a straight line to the segment of the p,v-locus between S' and E' (Fig.9) for example, the logic of the triplet system of waves found in CKR, and expounded by Clarke (1988) and Clarke, et al, (1989), breaks down; in particular, the all-burnt Hugoniot curve that is implied by the instantaneous existence of such a Rayleigh line simply does not lie in any physically acceptable relationship to that Rayleigh line.

In short, the present configuration does not contain a triplet-wave system, with one element behaving as a quasi-steady fast flame. What we observe at early times such as t_5 is essentially unsteady in all its character, save for the shock itself. (The latter is quasi-steady in character by virtue of being treated as a discontinuity in consequence of both the Euler formulation and the RCM used to acquire our solutions).

It is most important to remark that the criterion used to detect a newly-formed shock in our implementation of the RCM is rather rough. We can be confident that a shock has formed by time t_5 but not quite so confident that a weak shock does not already exist at t_4 or even earlier. Quite a useful indicator of the presence of a shock wave is the sudden change of slope in fuel mass-fraction versus ψ (Fig.5) as temperature-jump across the shock switches on increased rates of reaction. It appears to be unlikely that a shock exists at t_4, but the matter could only be unequivocally resolved by more refined numerical work, probably in conjunction with local analysis.

To reiterate, events prior to formation of a detonation in the present mechanic-ally driven system appear to be uniformly unsteady. In contrast to the quasi-steady features found at early times in thermally-driven events we find a situation here that is much more reminiscent of the one, described in elementary model terms by Clarke (1979), that exists within a pulse of compression travelling through a reactive atmosphere. The extreme rapidity of amplification of the shock-wave is one very noticeable common feature of both model and present situations, and we see it here playing a crucial role lifting reaction rates and pressure levels towards detonation levels.

The thermally-initiated system does eventually go through a phase of activity that is, in all its essentials, exactly the same as the unsteady compression-pulse/ reaction combination that we see here at times around t_5, but it appears as a culmination or conclusion of triplet-wave activity. Just as in the present case, detonation emerges directly from the localised pulse-like unsteady events when thermal input is the initiator. Initiation by mechanical input appears to by-pass the (quite lengthy) triplet-wave phase that is such a feature of the CKR and Clarke, et al, (1989) results.

That a ZND detonation wave does indeed ultimately form in present circumstances is confirmed as follows. First, Fig.10 shows that as time progresses past t_5 the amplitudes and speeds of propagation of the reaction-induced shocks grow. Simultaneously the part of the p,v-locus that represents conditions behind the reaction-induced shock begins to approach more and more closely the Rayleigh line that represents shock transition. Furthermore this post-shock region begins more and more to resemble a quasi-steady fast-flame as more and more of it lies on a straight p vs.v line (cf Fig.10). By time t_{12} shock and fast-flame Rayleigh lines are coincident all the way to the all-burnt condition; shock and fast-flame have the same Lagrange speed and so make up a ZND detonation; this condition appears to be sustained for $t > t_{12}$.

The RCM enables us to calculate very accurate shock-wave Lagrange speeds. As a result, equally accurate values of gas speed, and hence local Mach number, relative to the shock are available. Fig.11 illustrates some typical results and makes it clear that all-burnt conditions coincide with sonic flow out of the detonation-wave system.

We have therefore demonstrated the creation of a quasi-steady CJ, ZND, detonation by purely mechanical action on the set of reactive Euler equations.

We turn finally to Fig. 12, which displays a magnified view of that part of Fig.2 near time t = 1.11 and ψ = 0. The locus of maximum reaction rate is shown to depart from the piston face shortly after time t = 1.105 and to do so with small, perhaps even zero, value of dt/dψ. The implications of this observation are quite profound and relate directly to some recent and as yet unpublished asymptotic analysis (E → ∞ in our notation) by Dold and Kapila (Kapila and Dold, (1989); Dold and Kapila (1989)). This work is intrinsically important for the light that it sheds on the whole theory of detonation formation and shows, inter alia, how careful one must be with numerical studies. Without Dr. Dold's advice we would not have subjected our numerical data to the kind of close scrutiny that is implicit in Fig.12, and we are grateful to him for pointing out that this important feature of compressible reactive flow should be found in our study.

5. Conclusions

A combination of the Random Choice Method (RCM) and time operator splitting (TOS) has been used to acquire numerical results for a mechanically driven ignition event. Both the Euler model of the process, and the method used to evaluate them, rigorously exclude diffusion of all kinds. As a result the effects that are observed, namely the spontaneous appearance of a reaction wave and an associated shock wave (embryo ZND-detonation), arise as a result of combustion and compressibility inter-actions alone and add weight to some other recent numerical and analytical work.

References

Clarke, J.F., (1979) "On the evolution of compression pulses in an exploding atmosphere; initial behaviour". J. Fluid Mech., 94, 195-208.

Clarke, J.F., (1981). "Propagation of gasdynamic disturbances in an explosive atmosphere". Prog. in Astro. and Aeronautics, 76, 383-402.

Clarke, J.F., (1988). "Flames, waves and detonation". Prog. in Energy and Combust. Sci., (submitted).

Clarke, J.F. & Cant, R.S., (1984). "Non-steady gasdynamic effects in the induction domain behind a strong shock wave". Progress in Aeronautics and Astronautics 95, 142-163.

Clarke, J.F., Kassoy, D.R., Riley, N., (1986). "Direct initiation of a plane detonation wave". Proc. R. Soc., Lond., A408, 129-148.

Clarke, J.F., Toro, E.F., (1985). "Gas flows generated by solid-propellant burning". Lecture notes in Physics, 241, 192-205.

Clarke, J.F., Kassoy, D.R., Riley, N., Meharzi, N., & Vasantha, S. "On the evolution of plane detonation", (1989), (in preparation).

Colella, P. (1982). "Glimm's method for gas dynamics". SIAM J. Sci Stat. Comput., 3, 76-92.

Dold, J.W. & Kapila, A. (1989). "Evolution to detonation in a preconditioned homogeneous explosive", (in preparation).

Jackson, T.L., Kapila, A. (1985). "Shock-generated ignition". SIAM J. Appl. Math., 45, 130-137.

Kapila, A. & Dold, J.W. (1989) "A theoretical picture of shock-to-detonation transition in a homogeneous explosive". Detonation Symposium (1989; Proceedings, to appear).

Liepmann, H.W. & Roshko, A. (1957). "Elements of gasdynamics", John Wiley, New York.

Saito, T. & Glass, I.I., (1984). "Applications of the Random Choice Method to problems in gasdynamics". Prog. Aerospace Sci, 21, 201-247.

Sod, G.A. (1978). "A survey of several finite difference methods for systems of non-linear hyperbolic conservation law". J. Comp. Phys. 27, 1-31.

Strehlow, R.A.(1969). "Fundamentals of combustion". International Text Book Co., Swanton, Pa.

Takano, Y. (1986). "An application of the Random Choice Method to a reacting gas with many chemical species". J. Comp. Phys., 67, 173-187.

Toro, E.F. (1987). "The Random Choice Method on a non-staggered grid utilising an efficient Riemann solver". College of Aeronautics Report 8708, Cranfield Institute of Technology, England.

Toro, E.F., (1989) "Riemann-problem based Techniques for computing reactive two-phase flows", (this Symposium).

Yanenko, N.N., (1971). "The method of fractional steps". Springer Verlag.

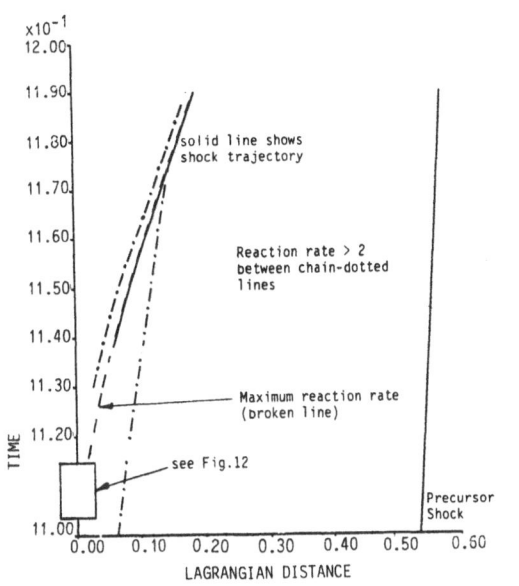

Fig. 2

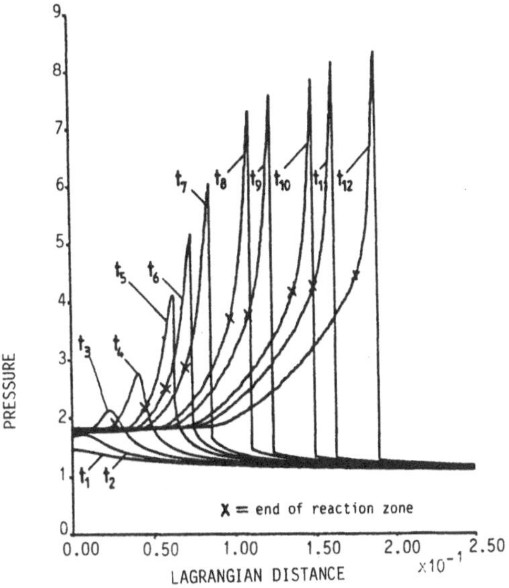

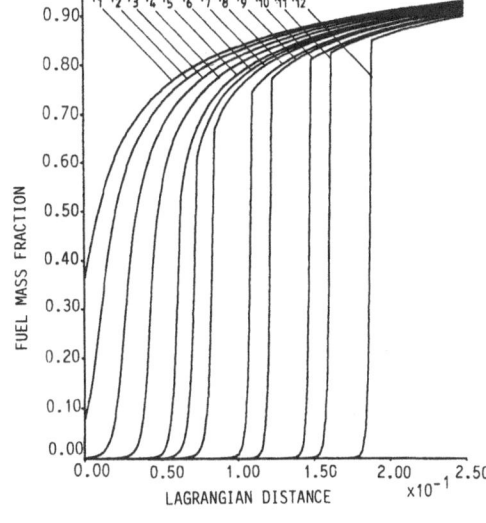

Fig. 3

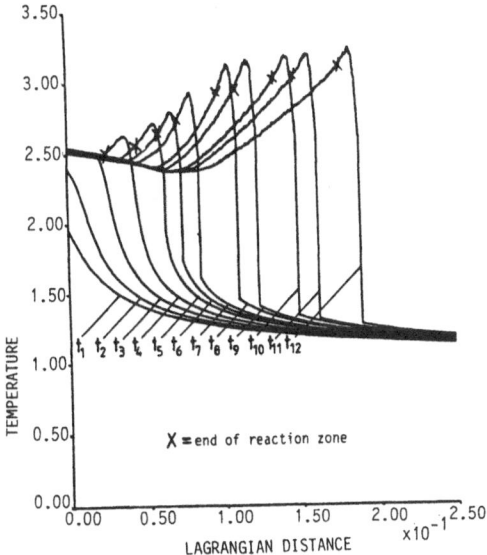

Fig. 4

Fig. 5

34

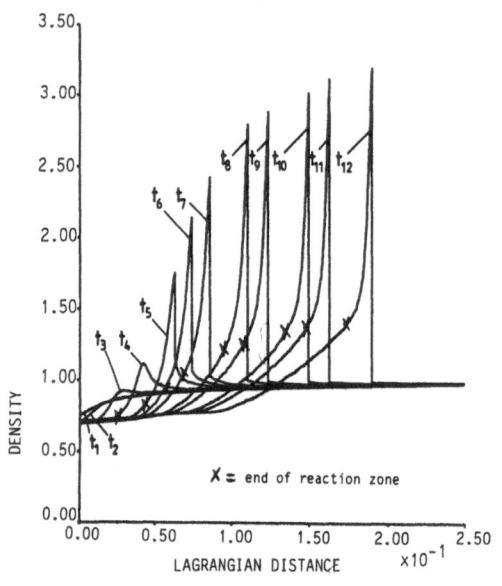

Fig. 6

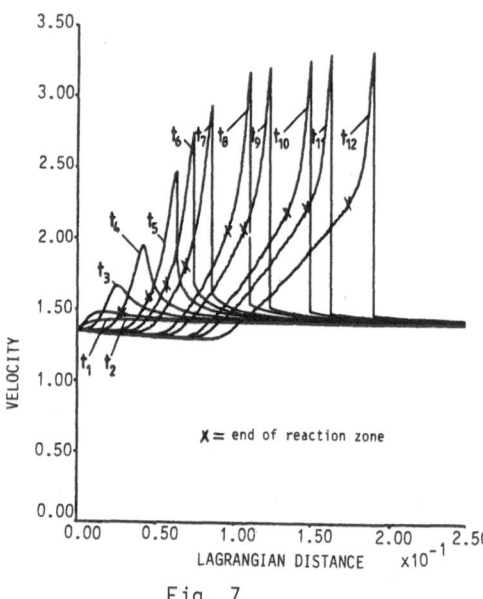

Fig. 7

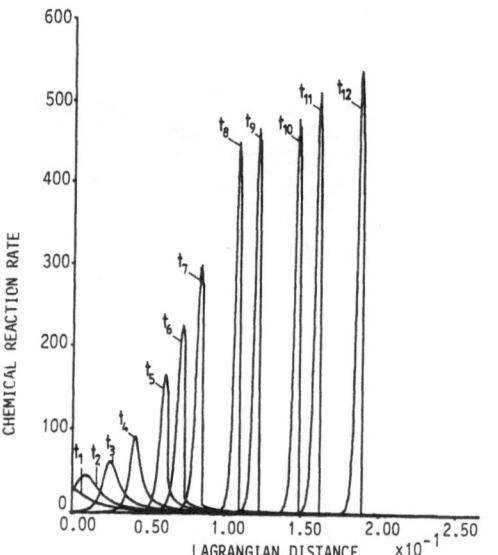

Fig. 8

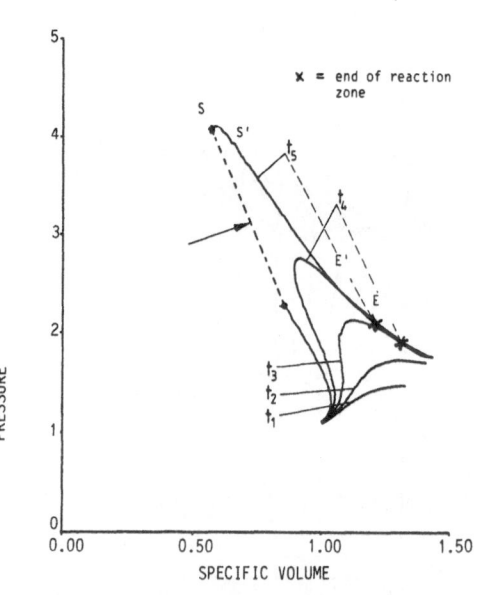

Fig. 9

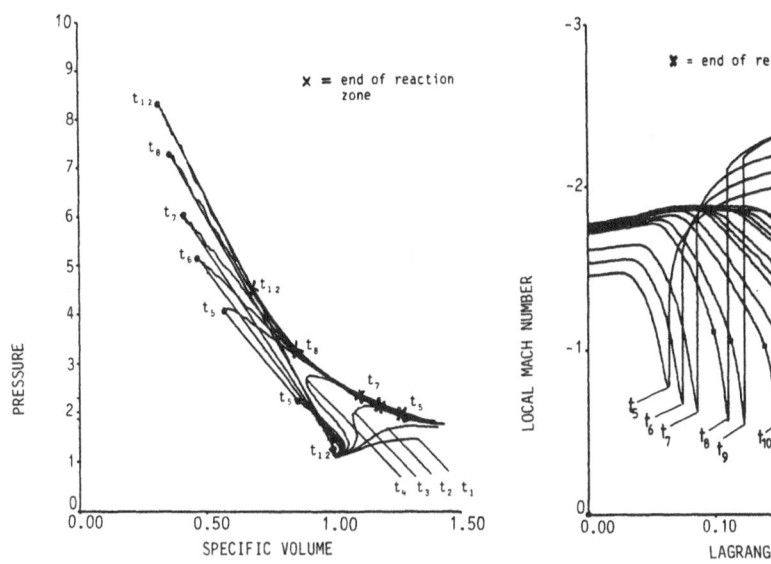

Fig.10

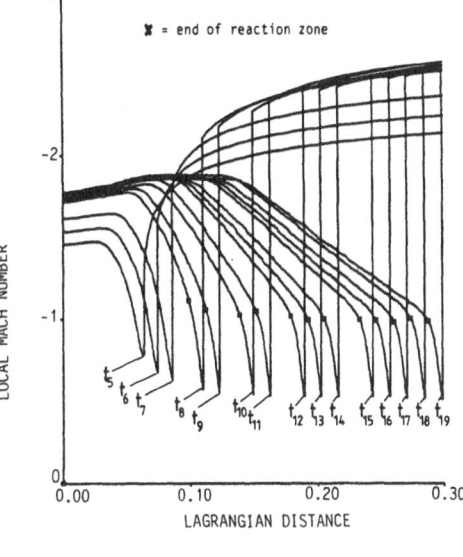

Fig.11

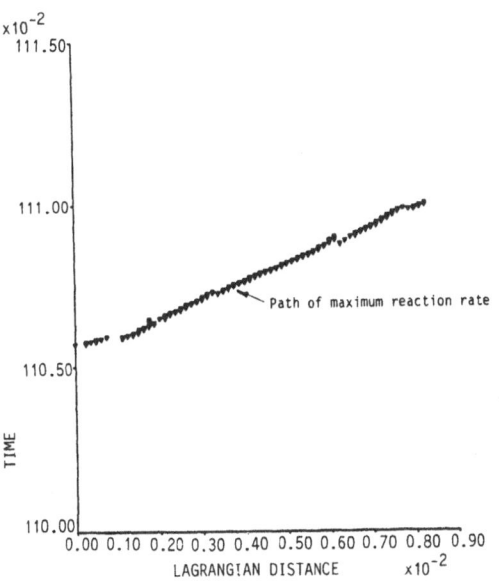

Fig.12

A Numerical Study of Mixing Enhancement in Supersonic Reacting Flow Fields

J. Philip Drummond and H. S. Mukunda
NASA Langley Research Center
Hampton, Virginia

Abstract

Work has been underway for a number of years at the NASA Langley Research Center to develop a supersonic combustion ramjet or scramjet that is capable of propelling a vehicle at hypersonic speeds in the atmosphere or beyond. A recent part of that research has been directed toward the optimization of the scramjet combustor, and in particular the efficiency of fuel-air mixing and reaction in the engine. A supersonic, spatially developing and reacting mixing layer serves as an excellent physical model for the mixing and reaction processes that take place in a scramjet combustor. This paper describes a study of fuel-air mixing and reaction in a supersonic mixing layer and discusses several techniques that were applied for enhancing the mixing processes and the overall combustion efficiency in the layer. Based on the results of this study, an alternate fuel injector configuration was computationally designed, and that configuration significantly increased the amount of fuel-air mixing and combustion over a given combustor length that was achieved.

Nomenclature

A constant in Arrhenius law

a speed of sound, m/s

c_p specific heat at constant pressure, J/kg-K

D_T thermal diffusion coefficient, m^2/s

D_{ij} binary diffusion coefficient, m^2/s

E activation energy, cal/gm-mole; total
internal energy, J/kg

$\vec{F}$	flux vector
f_i	mass fraction of species i
G_R	Gibbs energy of reaction, J/kg-mole
$\vec{H}$	source vector
h_i	enthalpy of species i, J/kg
h_o	total enthalpy, J/kg
h_i^o	reference enthalpy at standard conditions, J/kg
I	identity matrix
K	equilibrium constant
k_b	reverse reaction rate
k_f	forward reaction rate
M	molecular weight, kg/kg-mole
N	number of nodes
N_S	number of species
n_i	moles of species i
p	static pressure, n/m^2
$\dot{q}$	heat flux, J/m^2-s
R	steady-state residual; gas constant, J/kg-K
R^o	universal gas constant, J/kg-mole-K
$R^{o,}$	universal gas constant, cm^3-atm/gm-mole-K
T	static temperature, K
t	time, s
Δt	time step, s
$\vec{U}$	dependent variable vector
u	streamwise velocity, m/s
v	transverse velocity, m/s
$\tilde{v}_i$	diffusion velocity of species i, m/s

$\tilde{u}_i$ streamwise diffusion velocity of species i, m/s

$\tilde{v}_i$ transverse diffusion velocity of species i, m/s

$\dot{w}_i$ species production rate of species i, kg/m^3-s

x   streamwise spatial variable, m

y   transverse spatial variable, m

Δx streamwise spatial step size, m

Δy transverse spatial step size, m

ξ computational streamwise coordinate

η computational transverse coordinate

μ laminar viscosity, kg/m-s

λ second coefficient of viscosity, kg/m-s

ρ density, kg/m^3

σ normal stress, N/m^2

τ shear stress, N/m^2

Subscripts

e   edge value

i,j   species indices

R   reactions, reference value

s   species

Superscript

-     mass weighted value, mean value

'   fluctuating quantity

Introduction

Work has been underway for a number of years, both in the United States and abroad, to develop advanced aerospace propulsion systems for use late in this century and beyond. One such program is now underway at the NASA Langley Research Center to develop a hydrogen-fueled supersonic combustion ramjet, also known as a scramjet, that is capable of propelling a vehicle at hypersonic speeds in the atmosphere or beyond the atmosphere into orbit. A part of that research has recently been directed toward the optimization of the scramjet combustor, and in particular the efficiency of fuel-air mixing and reaction taking place in the engine. In the very high speed vehicle configurations currently being considered, achieving a high combustor efficiency becomes particularly difficult. This is a consequence of the fact that with increasing vehicle Mach number, the average Mach number in the combustor also increases. As the combustor Mach number increases, the degree of fuel-air mixing that can be achieved through natural convective and diffusive processes is reduced, leading to an overall decrease in combustion efficiency and thrust.

From the above discussion, it is clear that a detailed understanding of the scramjet com bustor flow field is critical to the achievement of a successful design. Even though the combustor flow field is quite complex, it can be realistically viewed as a collection of spatially developing and reacting supersonic mixing layers that are initially discrete, but that ultimately merge into larger more complex zones. These mixing layers begin downstream of a set of fuel injectors that introduce gaseous hydrogen in both a parallel and transverse direction into a supersonic air stream entering from the engine inlet. The behavior of the initial portion of the combustor flow, in the mixing layers near the fuel injectors, appears to be most critical, since this is where the mechanism for efficient high speed mixing must be established to achieve the required degree of combustion downstream. Because of the structure of the flow field in this initial portion of the com bustor, a single supersonic, spatially developing and reacting mixing layer serves as an excellent physical model for the overall flow field. Even though this reacting mixing layer flow is geometrically simple, it can still be made to retain all of the fluid mechanical and chemical complexities present in the actual combustor flow field.

Prior studies on supersonic reacting mixing layers have been quite limited. A fair amount of work was carried out, however, on nonreacting mixing layers, both at subsonic and supersonic speeds. Even without combustion, the results of these studies provided a significant amount of useful information for understanding reacting layers. A review of nonreacting mixing layer studies was given in references 1 and 2 and the reader is referred to those papers for further discussion. Many of the important features described for nonreacting mixing layers also occurred in reacting layers. A majority of the studies performed on reacting mixing layers were carried out, however, only at subsonic speeds. These papers were also reviewed in references 1 and 2. As noted earlier, the supersonic reacting mixing layer problem has received limited attention to date. Menon, Anderson, and Pai[3] studied the stability of a laminar premixed, spatially developing, supersonic mixing layer undergoing chemical reaction. They introduced an infinitesimal disturbance into the layer and examined its spatial stability for both reacting and nonreacting flows. Chemical reaction was shown to produce an increased disturbance amplification rate resulting in a decrease in flow stability. Drummond[4] simulated a supersonic spatially developing and reacting mixing layer by using a finite difference procedure to solve the Navier-Stokes equations coupled to a system of species continuity equations. Reaction was described using a finite rate chemistry scheme. He showed that the vortical structure present in subsonic reacting mixing layer flows was predominant in supersonic flows as well. This vortical structure had a pronounced effect upon mixing and chemical reaction. Significant burning took place on the edge of the layer in the vortices, significantly broadening the flame zone. The vortical flow also produced a rolling up of unburned reactants producing a state of unmixedness that reduced the overall mixing and combustion efficiency in the layer. Drummond and Hussaini[2] also simulated the supersonic reacting mixing layer case considered in reference 4 using a Chebyshev pseudospectral method that offered higher spatial accuracy than that provided by the finite difference solution.

Because of the difficulties described earlier for achieving a high supersonic combustion efficiency, attention has now turned to the development of techniques for enhancing the rate of fuel-air mixing in the combustor. The supersonic mixing layer again provided an excellent physical model for carrying out these studies. In an earlier study of this problem, Brown and Roshko[5] showed that the spreading rate of a supersonic mixing layer decreased with increasing Mach number, exhibiting a factor of three decrease in spread rate as compared with an incompressible mixing layer with the same density ratio. They concluded that the reduced spread rate was primarily a function of compressibility. Papamoschou and Roshko[6] also observed that the spreading rate of compressible mixing layers was significantly reduced over that of incompressible layers. They defined a convective Mach number $M_c = (U - U_c)/a$ where U is the free-stream velocity, U_c is the convection velocity of the large-scale structures in the layer, and a is the local speed of sound. The reduction in mixing layer spreading rate (by approximately a factor of three or four) was shown in their experiment to correlate well with increasing convective Mach number. The authors therefore concluded that the reduced spreading rate was attributable to a stabilizing effect of the convective Mach number. Ragab and Wu[7] analyzed the spatially developing supersonic mixing layer using linear spatial stability theory. They also found that the decreased spreading rate of the mixing layer corre-lated well with convective Mach number, and their predictions agreed well over most of the Mach number range with the results of Papamoschou and Roshko. The work presented in references 5 through 7 thus shows, through both experiment and theory, the difficulty in achieving a high degree of mixing in unenhanced supersonic mixing layers.

Faced with this challenge, several authors have examined potential techniques for enhancing the mixing rates in supersonic mixing layers. Guirguis, et al.[8] showed that the spreading rate of a confined mixing layer could be improved if the pressure of the two streams was different. Encouraged by this result, Guirguis[9] inserted a bluff body at the base of the splitter plate separating the two streams. The body produced an in-stability further upstream in the layer and resulted in a more rapid rate of spread. Kumar, Bushnell, and Hussaini[10] discussed a number of mixing problems that may exist in scramjet combustors. Several techniques for enhancing turbulence and mixing in combustor flow fields were suggested and one enhancement technique that employed an oscillating shock was studied numerically. In this case, a premixed stoichiometric hydrogen-air flow was processed through a spatially and temporally oscillating shock wave and the resulting flow was studied with and without chemical reaction. The oscillating shock was shown to increase the level of turbulence in the flow field, and the degree of turbulence enhancement was seen to increase with a decreasing frequency of shock oscillation. Chemical reaction as defined by an equilibrium model was shown to have little effect relative to the nonreacting results.

The present paper describes a numerical study of fuel-air mixing in a two-dimensional supersonic, spatially developing and reacting mixing layer, and discusses several techniques that were applied for enhancing the mixing processes and the overall combustion efficiency in the layer. Based on the results of this study, an al-ternate fuel injector configuration was designed computationally, and that design sig-nificantly increased the amount of fuel-air mixing and combustion that was achieved. The next section describes the theory and solution techniques upon which this study is based.

Theory

Governing Equations

The two-dimensional, Navier-Stokes, energy, and species continuity equations governing multiple species undergoing chemical reaction are given by

$$\frac{\partial \vec{U}}{\partial t} + \frac{\partial \vec{F}(\vec{U})}{\partial x} + \frac{\partial \vec{G}(\vec{U})}{\partial y} = \vec{H} \qquad (1)$$

where

$$\vec{U} = \left\{ \begin{array}{c} \rho \\ \rho u \\ \rho v \\ \rho E \\ \rho f_i \end{array} \right\}, \quad \vec{F} = \left\{ \begin{array}{c} \rho u \\ \rho u u - \sigma_x \\ \rho u v - \tau_{yx} \\ (\rho E - \sigma_x) u - \tau_{xy} v + q_x \\ \rho u f_i + \rho \tilde{u}_i f_i \end{array} \right\}$$

$$\vec{G} = \left\{ \begin{array}{c} \rho v \\ \rho u v - \tau_{xy} \\ \rho v v - \sigma_y \\ (\rho E - \sigma_y) v - \tau_{yx} u + q_y \\ \rho v f_i + \rho \tilde{v}_i f_i \end{array} \right\} \quad \vec{H} = \left\{ \begin{array}{c} 0 \\ \rho \sum_i f_i b_{ix} \\ \rho \sum_i f_i b_{iy} \\ \rho \sum_i f_i \vec{b}_i (\vec{V} + \vec{V}_i) \\ \dot{w}_i \end{array} \right\}$$

and

$$\sigma_x = -p + \lambda\left(\frac{\partial u}{\partial x} + \frac{\partial v}{\partial y}\right) + 2\mu \frac{\partial u}{\partial x}$$

$$\sigma_y = -p + \lambda\left(\frac{\partial u}{\partial x} + \frac{\partial v}{\partial y}\right) + 2\mu \frac{\partial v}{\partial y}$$

$$\tau_{xy} = \tau_{yx} = \mu\left(\frac{\partial u}{\partial y} + \frac{\partial v}{\partial x}\right)$$

$$q_x = -k\frac{\partial T}{\partial x} + \rho \sum_{i=1}^{N_s} h_i f_i \tilde{u}_i + R^o T \sum_{i=1}^{N_s} \sum_{j=1}^{N_s} \left(\frac{X_j D_{Ti}}{M_i D_{ij}}\right) (\tilde{u}_i - \tilde{u}_j)$$

$$q_y = -k\frac{\partial T}{\partial y} + \rho \sum_{i=1}^{N_s} h_i f_i \tilde{v}_i + R^o T \sum_{i=1}^{N_s} \sum_{j=1}^{N_s} \left(\frac{X_j D_{Ti}}{M_i D_{ij}}\right) (\tilde{v}_i - \tilde{v}_j)$$

$$E = \sum_{i=1}^{N_s} h_i f_i - \frac{p}{\rho} + \frac{u^2 + v^2}{2} \tag{2}$$

$$h_i = h_i^o + \int_{T_R}^{T} C_{p_i} \, dT \tag{3}$$

$$p = \rho R^o T \sum_{i=1}^{N_s} f_i / M_i \tag{4}$$

The diffusion velocities are found by solving

$$\vec{\nabla} X_i = \sum_{j=1}^{N_s} \left(\frac{X_i X_j}{D_{ij}}\right) (\vec{V}_j - \vec{V}_i) + (f_i - x_i)\left(\frac{\vec{\nabla} p}{p}\right)$$

$$+ \left(\frac{\rho}{p}\right) \sum_{j=1}^{N_s} f_i f_j (\vec{b}_i - \vec{b}_j) + \sum_{j=1}^{N_s} \left(\frac{X_i X_j}{\rho D_{ij}}\right)\left(\frac{D_{Tj}}{f_j} - \frac{D_{Ti}}{f_i}\right)\left(\frac{\vec{\nabla} T}{T}\right) \tag{5}$$

Note that if there are N_s chemical species, then
$i = 1,2,...,(N_s - 1)$ and $(N_s - 1)$ equations must be solved for the species f_i. The final species mass fraction f_{N_s} can then be found by conservation of mass since

$$\sum_{i=1}^{N_s} f_i = 1$$

Thermodynamics Model

To calculate the required thermodynamic quantities, the specific heat for each species is first defined by a fourth-order polynomial in temperature

$$\frac{C_{P_i}}{R} = A_i + B_i T + C_i T^2 + D_i T^3 + E_i T^4 \qquad (6)$$

The coefficients are found by a curve fit of the data tabulated in reference 11. Knowing the specific heat of each species, the enthalpy of each species can then be found from equation (3), and the total internal energy is computed from equation (2). To determine the equilibrium constant (required in the next section) for each chemical reaction being considered, the Gibbs energy of each species must first be found. For a constant pressure process, (c_p/T) from equation (6) is first integrated over temperature to define the entropy of the species, and then the resulting expression is integrated again over temperature to obtain a fifth-order polynomial in temperature for the Gibbs energy of each species. The Gibbs energy of reaction, ΔG_R, can then be calculated as the difference between the Gibbs energy of product and reactant species. The equilibrium constant for each reaction can then be found from

$$K = \left(\frac{1}{R^o \cdot T}\right)^{\Delta n} \exp\left(-\frac{\Delta G_R}{R^o T}\right)$$

where Δn is the change in the number of moles when going from reactants to products.

Chemistry Models

In this study, the finite-rate chemical reaction of gaseous hydrogen fuel and air must be modeled. The reaction is modeled here by either a four species, single reaction model or a nine species, eighteen reaction model described in Table 1. In either case, the forward rate of each reaction j is given by the Arrhenius law

$$k_{f_j} = A_j T^{N_j} \exp(-E_j / R^o T)$$

Values for A, N, and E are also given in Table 1. Knowing the forward rate, the reverse rate is then given by

$$k_{b_j} = k_{f_j} / K_j$$

Once the forward and reverse reaction rates have been determined, the production rates of the species can be found from the law of mass action. Details of this calculation are given in references 1, 2 and 4.

Diffusion Models

The coefficients governing the diffusion of momentum, energy, and mass are determined from models based on kinetic theory. These models are now described briefly; further details are again given in references 1, 2, and 4. Sutherland's law is employed to compute the individual species viscosities. Once the viscosity of each species has been computed, the mixture viscosity can than be computed from Wilke's law. An alternate form of Sutherland's law is also used to compute the individual species thermal conductivities. The mixture thermal conductivity is then found from these individual values and Wassilewa's formula. For dilute gases, the Chapman and Cowling law can be used to determine the binary diffusion coefficient Dij that controls the diffusion of each species i into the remaining species j. Once the binary diffusion coefficients for all species combinations are known, the diffusion velocities of each species can be computed from equation (5). The diffusion velocity is the velocity induced upon each species by all diffusion processes that are present in the flow. When computing the diffusion velocities, it is assumed that the thermal diffusion coefficient, D_T, is negligible compared with the binary diffusion coefficient. This assumption is true in general, but it may be somewhat suspect for low molecular weight molecules such as hydrogen. The solution of equation (5) requires solving a simultaneous equation system, with the number of equations equivalent to the number of species present for each component of the diffusion velocity. It should be noted that for i species, however, the system of i equations defined by (5) is not linearly independent. One of the equations must be replaced by the constraint $\sum_{i=1}^{N_s} \rho f_i \vec{V_i} = 0$ to make the system linearly independent. The resulting system of equations is solved for the diffusion velocities using the Householder method.[12]

Solution of the Governing Equations

The governing equations (1) are written in the physical domain (x,y) and they must be transformed to an appropriate uniform computational domain (ξ,η) for solution. The equations are solved on a grid which is highly compressed in both x and y in the physical domain near regions where high gradients exist. In the computational domain, equations (1) become

$$\frac{\partial \hat{U}}{\partial t} + \frac{\partial \hat{F}}{\partial \xi} + \frac{\partial \hat{G}}{\partial \eta} = \hat{H} \qquad (10)$$

where

$$\hat{U} = JU, \quad \hat{H} = JH$$

$$\hat{F} = y_\eta F - x_\eta G$$

$$\hat{G} = x_\xi G - y_\xi F$$

$$J = x_\xi y_\eta - y_\xi x_\eta$$

Here $(x_\xi, x_\eta, y_\xi, y_\eta)$ are the transformation metrics and J is the Jacobian of the transformation. The metrics are computed numerically once the physical coordinate grid has been prescribed.

The governing equations can be stiff due to the kinetic source terms contained in the vector H and due to diffusive terms in the vectors F and G. Only the kinetic terms introduce stiffness in this work. Therefore, the kinetic source terms were computed implicitly.[13,14] In a temporally discrete form, equation (10) then becomes

$$\hat{U}^{n+1} = \hat{U}^n - \Delta t \left[\left(\frac{\partial \hat{F}}{\partial \xi} \right)^n + \left(\frac{\partial \hat{G}}{\partial \eta} \right)^n - \hat{H}^{n+1} \right] \qquad (11)$$

After employing a Newton linearization for H and rewriting in delta form, equation (11) becomes

$$[I - \Delta t K^n] \Delta \hat{U}^{n+1} = -\Delta t \hat{R}^n \qquad (12)$$

where

$$\hat{R}^n = \left(\frac{\partial \hat{F}}{\partial \xi}\right)^n + \left(\frac{\partial \hat{G}}{\partial \eta}\right)^n - \hat{H}^n$$

is the steady-state residual, I is the identity matrix, K^n is the Jacobian of $\hat{H}$ with respect to $\hat{U}$, $\left(\partial \hat{H}/\partial \hat{U}\right)$, and $\hat{U}^{n+1} = (U^{n+1} - U^n)$. The simultaneous system of equations (12) is then solved using a hybrid MacCormack-Householder scheme. Details of the method are given in references 1 and 4.

Boundary and Initial Conditions

The governing equations are spatially elliptic and require boundary conditions along all four boundaries. In the present work, the inflow boundary is always supersonic, so the velocities, static temperature, static pressure, and species are specified and fixed there. When the upper and lower boundaries lie in the free stream, the normal gradient of the above variables is required to vanish along those boundaries. (The gradient conditions not only satisfy free stream conditions, but also provide nonreflective condi-tions that pass disturbances through the boundary rather than reflecting them back into the domain.) In one problem to be considered in the results, oblique shocks enter the upper and lower boundary from the far field. Since the shock angles and flow condi-tions are known in this case, post shock conditions are set along the upper and lower boundaries downstream of the shocks. The outflow boundary is also supersonic, and values of the velocities, static temperature, static pressure, and species are determined by extrapolation from upstream values. Finally, when solid boundaries lie in the com-putational domain, no slip (u=0, v=0) boundary conditions are used to specify velocity components, adiabatic conditions ($\partial T/\partial y = 0$) are assumed, the boundary layer assump-tion on pressure ($\partial p/\partial y = 0$) is enforced, and the walls are assumed to be noncatalytic ($\partial f/\partial y = 0$).

The governing equations also require a set of initial conditions. The equations are initialized at the flow boundary by setting values of the velocities, static temperature, static pressure, and species identical to the values chosen for the inflow boundary conditions. The remainder of the domain downstream of the inflow boundary is chosen to be made up of air at rest with ambient conditions for the temperature and pressure.

When the initial conditions are specified in this manner, the solution when properly advanced in time[1,4] can be made to evolve in a real time sense with the inflow propagating into an undisturbed region downstream. These initial conditions are therefore equivalent to opening a valve at t=0 and then allowing the flow exiting the valve to proceed into an undisturbed medium downstream. Having now specified all required initial and boundary data, the equations are marched in time from the initial time to some final specified integration time.

Results

Once the theory and solution procedure described above had been developed and coded, several spatially developing mixing layer flows were simulated using the computer program. The code was first evaluated by comparing with a known solution for spatially developing mixing layer. Lock[15] solved the incompressible boundary-layer equations for a laminar mixing layer with upper stream velocity U1 and lower stream velocity U2. The configuration is shown in figure 1 along with his solution for air with a velocity ratio U2/U1 = 0.5. The solution is similar and is plotted in terms of a similarity variable versus the nondimensional streamwise velocity u/U1. The definition was modified using the Howarth-Dorodnitzyn transformation $\tilde{y} = \int_0^y \frac{\rho}{\rho_e} \, dy$ (13) to allow comparison with the solution for the same problem based on the compressible Navier-Stokes equations. Equation (13) transforms the compressible results from the program to the equivalent incompressible results for this problem. The definition for η reduces in incompressible flow to that used by Lock. Results from the computation for U2/U1 = 0.5 are also shown in figure 1. The calculations were made on a computational grid with 201 nodes in the streamwise direction and 51 nodes in the transverse direction across the layer. The grid was highly compressed in the transverse direction with a minimum grid spacing of 0.1 mm. The results become similar at a value of x/L equal to or greater than 0.1. The comparison of the computation with the exact solution of Lock is excellent for all values of η. The transformation coordinate η is highly stretched relative to the physical coordinate y. It is therefore clear from the comparison that the large velocity gradient across the mixing layer is well resolved.

Once the computer code developed in this effort was evaulated for nonreacting flow, it was then applied to several reacting flow cases. The purpose of this study was to assess several candidate configurations for enhancing fuel-air mixing and reaction in a mixing layer that would then lead to a better understanding for achieving improved mixing and combustion in the supersonic combustor of a scramjet engine. The first three cases involved a supersonic, spatially developing and chemically reacting mixing layer. The first of these cases, shown in figure 2, served as a benchmark calculation in that it contained no enhancement mechanism. Case 1 involved a mixing layer developing between a fuel stream and an air stream that were initially separated by an infinitely thin splitter plate. The fuel stream was made up of a mixture of ten percent hydrogen and ninety percent nitrogen introduced above the plate at a velocity of 2672 m/s, a static temperature of 2000 K, and a static pressure of .1 01 MPa (1 atm). Nitrogen gas was included to reduce the speed of sound of the fuel mixture. Air was introduced below the plate at a velocity of 1729 m/s, a static temperature of 2000 K, and a static pressure of .1 01 MPa. These conditions resulted in a Mach number of 2 for both streams. The physical domain considered in this case was 0.1 m long and 0.1 m high. The computational grid was identical to that used in the validation case that was described above. The simulation of this case was begun at t=0 with static conditions (u=0, v=0) in the flow domain. At this time fuel and air flows were initiated off of the trailing edge of the splitter plate (in a method analogous to opening two valves) at the conditions given above, and the gases then proceeded downstream. The mixing layer then evolved between the two gases in both space and time. The calculation was then advanced in a real time sense until an integration time of approximately 0.1 ms was reached. This time represented 14 computational sweeps of the flow field and allowed a periodic solution to develop. Chemical reaction of the hydrogen and air occurred after a sufficient degree of mixing had occurred. The chemical reaction in this case (and the next two cases) was modeled using the one step hydrogen-air model described in the previous section and in Table 1.

Results for Case 1 are shown in figure 3. The first result given is a contour plot of the velocity field. The mixing layer is seen to develop slowly in a smooth laminar fashion. A Kelvin-Helmholtz instability begins downstream at an x/L of approximately 0.8, but there is an insufficient length in the region of interest for the instability to evolve significantly. A similar result showing temperature contours in the layer is also given in figure 3. The temperature is also seen to rise smoothly in the layer due to both viscous heating and chemical reaction. The instability is also apparent downstream, and an increased amount of water production and an associated temperature rise to 2353 K result due to increased reaction in the vortical structures that evolve in this region. This can be seen more clearly in the final plot in the figure that gives a contour plot of water mass fraction in the layer. Water begins appearing at an x/L of about 0.1, then evolves at a slowly increasing rate until an x/L of 0.8 is reached, and the rate increases some- what in the instability region reaching a peak value of 2.4 percent by mass. Even with the increased water production that occurs in the region of the instability, however, the overall degree of reaction is still quite limited in this case. This difficulty is compounded even further by the limited transverse spread of the layer in the y coordinate direction with increasing values of the streamwise coordinate x. The mixing layer must exhibit significantly more transverse spread within the limit of the streamwise coordinate if an acceptable level of mixing and combustion efficiency is to be achieved.

In an attempt to improve the level of mixing and reaction in the benchmark mixing layer case considered above, two mixing enhancement mechanisms were employed. The first approach (denoted as Case 2) is described in figure 4. Here, the mixing layer considered in Case 1 is processed through two shocks entering the flow domain from the upper and lower boundaries. Each shock is set at an angle of ten degrees by choosing appropriate boundary conditions along the upper and lower boundaries beyond the shock that are consistent with a ten degree shock in a Mach 2 flow of hydrogen and air, respectively. The two shocks then propagate from the upper and lower boundaries into the flow and across the mixing layer. The resulting flow field, taken again at about 0.1 ms, is shown in figure 5. Inspection of the velocity field shows no marked enhancement in mixing layer spread rate due to the stationary shocks. The instability appears slightly further upstream at x/L = 0.7, just behind the location of shock interaction with the layer. The amplitude of the instability is not increased, however, and the viscous region of the layer as defined by the velocity gradient is not thickened relative to the previous case. The temperature contours given in figure 5 show identical trends. Temperatures reach a peak value of 3178 K in the center of the layer near x/L = 1.0, but the increase is due primarily to the temperature rise through the shocks and some small increase in water production resulting from the higher tempera- ture of the reactants. Even with this further increase in water production, giving peak values of about 6 percent by mass, the overall degree of reaction and the amount of product that is produced is still quite low. This can be seen by viewing the water contour plot in figure 5 which shows a layer thickness defined by water that is not any greater than that observed in the previous unenhanced case.

The second enhancement study, designated as Case 3, is described in figure 6. Conditions are again the same as in the previous two studies. In this case, a small square cylinder is placed along the fuel-air interface at x/L = 0.2. The body is 0.0012 m high and 0.002 m long. When the body is placed in the flow, it results in the formation of a bow shock just ahead of the body. The shock has strong curvature in the immediate neighborhood of the body. When the high velocity gradient region of the mixing layer is processed by this curved shock, vorticity is produced. The vorticity is then convected downstream where it produces enhanced macromixing of fuel and air. This effect can be seen in the results given in figure 7. The velocity field can be seen to become unstable near the trailing edge of the interference body, and the thickness of the layer as defined by the velocity gradient grows rapidly with increasing streamwise coordinate. Rapid growth of the layer can also be seen in the temperature contour plot given in figure 7. A significant amount of vortical structure is also apparent in the temperature field, and the individual vortices grow and amalgomate with one another as the layer develops with increasing streamwise distance. A peak temperature of 2500 K was observed at the center of several of the vortical structures. The water mass fraction also peaked at the centers with values as high as 3.4 percent by mass. The layer thickness defined by water mass fraction that is shown in the figure is also significantly greater at all values of x relative to the previous two cases in this study.

While spread rate gives a good visual indication of the development of a mixing layer, a more useful and practical indication of mixing is given by the mixing efficiency of the layer. The mixing efficiency is defined as a number between 0 and 1 that specifies the amount of fuel that could react at any x station if chemical reaction was taking place. Therefore, if all fuel could be consumed at a given x station, the mixing efficiency at that station would be unity. The mixing efficiency plotted as a function of the streamwise coordinate for Cases 1 through 3 is given in figure 8. The efficiency values for the three cases should be viewed in a relative sense because there are significant regions of fuel and air that can never mix since they are located at large transverse distances away from the mixing layer. What is important in this comparison, then, is the relative degree to which efficient mixing takes place between the cases. All three cases show a similar streamwise development of mixing efficiency to the 0.02 m station. Beyond that location, Cases 1 and 2 exhibit a similar slow development in efficiency, while Case 3 experiences a significant growth in mixing efficiency. The more rapid spread of the mixing layer due to the higher level of induced vorticity in case 3, that was observed in the earlier results, translates directly into a significantly higher level of mixing efficiency. Near the outflow station, the peak mixing efficiency of Case 3 is approximately four times that of Case 1 and three times that of Case 2. The oscillatory nature of the mixing efficiency plots for the three cases is related to the instantaneous structure present in each of the mixing layers at the time that the results are plotted. The structure of the plots can be directly tied to the representation of layers given in figures 3, 5, and 7. At that instant in time (0.1 ms), the highest rate of chemical reaction and the largest amount of product at a given x station is present in the neighborhood of the largest vortical structures, whereas less product is present at stations where the layer pinches between the vortices.

Once fuel and air have mixed, the degree to which they then react is defined at any streamwise station by the combustion efficiency. The combustion efficiency is again a number between 0 and 1 that defines the degree of chemical reaction that has taken place. The combustion efficiency that results for Cases 1 through 3 is given in figure 9. Comparison between the three cases should again be made in a relative sense due to the geometry of the mixing layer. In addition, the one-step chemistry used to model reaction for Cases 1, 2, and 3 underestimates water production relative to the more complete model used later in this study. The model is still useful for this relative comparison, however. The combustion efficiency results given in figure 9 follow trends quite similar to those observed for mixing efficiencies given in figure 8. All three cases exhibit similar increases in combustion efficiency up to the 0.02 m station, but then Case 3 shows a significant increase over the other two cases. This trend continues until the 0.08 m station, where the combustion efficiency for Case 2 increases rapidly. Reexamination of the mixing efficiency for Case 2 shows, however, that this increase is not related to improved mixing, but rather to an increased reaction rate following a significant temperature rise through the shocks present in Case 2. Overall, combustion efficiencies for Case 3 are nearly three times higher than Case 1 and 1.5 times higher than Case 2. It is also important to note that the higher levels of mixing efficiency achieved in Case 3 occur well upstream in the layer relative to efficiency increases in the other two cases. Therefore, enhancement techniques of the type used in Case 3 may be an effective means of shortening the overall combustor length while still retaining a high degree of chemical reaction and the associated thrust.

To better understand the success of enhancement on fuel-air mixing, statistics of the resulting flow field in Case 3 were next examined. The results of the Case 1 benchmark analysis were also included in this study to allow comparison with Case 3. To extract these statistics from the simulations, values of the flow variables were collected at selected spatial locations over 50 time steps spanning approximately two computational sweeps of the flow field. The resulting statistical features of the flow are summarized in Table 2. Included in the table are peak values of root mean square total velocity, streamwise velocity, temperature, density, and water mass fraction normalized by their respective mean values. Also included are peak values of the of the mean total velocity, the peak Reynolds stress normalized by the free stream dynamic pressure, and the layer vorticity thickness normalized by streamwise distance. In addition, a spectral analysis of the fluctuating variables was also conducted. Results of the analysis were correlated in terms of a frequency nondimensionalized by the ratio of local velocity and local vorticity thickness. Significant energy was found to exist in the upstream fluctuating velocity field at frequencies of 0.01, 0.06, 0.09, and 0.12. Well downstream in the layer, the energy spectrum was quite broad, and there were no significant local peaks. This distribution indicated that the flow was transitioning, thus justifying the collection of statistical data. The tabulated values of vorticity thickness in Table 2 indicate that the mean growth rate is enhanced by about 40 percent in Case 3 as compared to Case 1. The fluctuating quantities also exhibit significantly higher values for Case 3. The Reynolds stress, and the fluctuating values of temperature, density, and water mass fraction are approximately twice as large in the enhanced case as compared to the unenhanced case, and the fluctuating total velocity is nearly four times as high. The statistical results, therefore, also indicate a significant improvement in mixing and combustion in Case 3.

The success achieved with the mixing enhancement technique employed in Case 3 motivated the authors to apply this approach to a more realistic configuration. The fuel injection strut of a conventional scramjet engine was chosen for study. The fuel injection struts provide locations for instream injection of gaseous hydrogen fuel into the air coming from the inlet of the engine. Fuel is injected in both parallel and transverse directions to the incoming air stream. Transverse injection predominates over parallel injection when the engine is operating in the high Mach number regime to hasten fuel-air mixing and combustion. At lower Mach numbers, more parallel injection is used to slow the mixing process and achieve a heat release schedule similar to that achieved at higher Mach number. A schematic of a typical injector configuration on the trailing edge of a fuel injection strut is given in figure 10a. Transverse injection takes place following a rearward facing step that provides improved flameholding, and parallel injection occurs at the strut base. The results achieved in Case 3 suggested that a reorientation of the injectors might improve the rate of fuel-air mixing and chemical reaction that could be achieved downstream of the strut. That change is reflected in figure 10b. In the new design, the parallel injector has been moved from the base of the strut to the vertical wall of the rearward facing step. The transverse injector now produces a curved bow shock that interacts with the high velocity gradient of the jet from the parallel injector resulting in vorticity production. The transverse jet thus serves a similar function to that provided by the interference body employed in Case 3. The present design is more practical, however. The solid interference body would produce significant losses in an engine, and aerodynamic heating would also pose a problem. The transverse injector is present in both strut designs, however, and so it introduces no significant losses in the new design as compared to the old one.

To assess the new strut design, a computational study was again performed. The calculation was begun at the rearward facing step. A parallel slot fuel injector was located on the face of the step that injected a mixture of ten percent hydrogen and ninety percent nitrogen gas at a velocity of 2672 m/s, a temperature of 2000 K, and a pressure of .1 01 MPa (1 atm). The injector was 0.064 cm high and located 0.032 cm above the strut wall. A transverse slot fuel injector was located 0.26 cm downstream of the step. An identical hydrogen-nitrogen mixture was injected there at 2672 m/s, a temperature of 2000 K, and a pressure of .5 05 MPa. The slot was sized to be one-fifth the width of the parallel jet so that the same amount of fuel was introduced from each injector. The strength of the transverse jet insured that it would produce a shock of sufficient strength to result in significant vorticity production.

With the new strut now configured, two cases were considered. The first case (identified as Case 4) considered operation of only the parallel injector. Case 5 involved operation of both the parallel and transverse injectors. Both cases were computed on a computational grid of 218 streamwise points and 51 transverse points. The grid was highly compressed in the transverse direction about the parallel injector and highly compressed in the streamwise direction about the transverse injector. Chemistry was modeled in both cases using the nine species, eighteen reaction, hydrogen-air scheme described in Table 1. This model provided a more realistic description of reaction than the one-step model, but it was computationally more expensive. However, it seemed more appropriate to apply the detailed model to these practical cases to more accurately represent product production and heat release in the simulation. Results for Case 4 at a time of 0.1 ms are shown in figure 11. The unenhanced parallel jet behaves much like the unenhanced mixing layer in Case 1. The velocity contour plot exhibits a mild instability further upstream in this case, but there is no significant growth of the jet. Lack of growth can also be observed in the temperature contours in figure 11. The thermal layer is somewhat thicker than the velocity thickness of the previous plot due to burning, heat release, and a resulting temperature rise on the edges of the layer. A peak temperature of 2636 K is reached near the end of the jet. Water contours, that result from the complex reaction process, are shown in the final plot. As expected, the water contours closely resemble the temperature contours, and only a moderate degree of spread in the water field is observed. Peak values of around 22 percent by mass of water are achieved near the end of the layer, however, indicating that a significant degree of chemical reaction does occur where fuel and air are able to sufficiently mix.

To improve the degree of fuel-air mixing, the transverse fuel injector was activated so that it might interact with the parallel injector. The resulting flow field, again at a time of 0.1 ms, is shown in figure 12. The degree of mixing enhancement induced when the parallel and transverse jets interact is significant. The interaction of the parallel jet with the curved bow shock ahead of the transverse jet again produces vorticity, resulting in increased mixing. The bow shock can be seen in both the velocity and temperature contour plots. The more rapid development of the fuel jets can also be seen in those plots along with a significant increase in the spread of the jet. A peak temperature of 2705 K is reached downstream in the reaction zone. The water contour plot in figure 12 also shows markedly more jet development due to the interaction. The jet spreads much more rapidly than in the previous case and water mass fractions of as high as 24 percent occur across appreciable portions of the jet.

A more quantitative comparison of the last two cases is again made by examining their mixing and combustion efficiencies. A comparison of mixing efficiencies for Cases 4 and 5 is given in figure 13. The mixing efficiency of Case 5 increases much more rapidly than Case 4 due to enhancement. Efficiencies of around 90 percent are achieved in only 40 percent of the solution domain length, whereas the unenhanced Case 4 requires 75 percent of the solution length to achieve the same level of mixing. This is especially noteworthy since twice as much fuel is being injected in Case 5 as compared to Case 4. Similar results are also observed when combustion efficiencies for the two cases are compared in figure 14. Combustion efficiencies of about 80 percent are obtained for the enhanced case within the first 40 percent of the solution length whereas the unenhanced case requires 75 percent of the solution length to achieve the same level of reaction. The higher combustion efficiencies that occur at earlier streamwise stations are again achieved with twice the amount of injected fuel. It is clear, then, that the simple enhancement technique employed in this study is quite effective in achieving a higher level of fuel-air mixing and combustion in shorter stream-wise distances as compared to more conventional unenhanced approaches. The same or similar approaches should therefore offer an efficient means for achieving improved levels of combustion efficiency with a shorter overall combustor length.

Concluding Remarks

Techniques for enhancing fuel-air mixing and combustion in a scramjet engine have been explored in this paper. A supersonic spatially developing and chemically reacting mixing layer well represents the early stages of mixing and reaction that take place in a scramjet combustor. The mixing layer was therefore considered as a model problem for the initial phases of this study. Once this problem was chosen, a simulation of a reacting mixing layer without enhancement was first performed to serve as a benchmark for the enhancement studies that followed. Two calculations employing enhancement were then carried out, the first employing planar shocks and the second employing a shock with curvature. The second enhancement approach proved more attractive. The curved shock produced vorticity when it interacted with the high velocity gradient in mixing layer and a higher degree of mixing and reaction then resulted. Based on the results of this study, an alternate fuel injector configuration employing interacting jets was computationally designed. That configuration significantly increased the amount of fuel-air mixing and reaction that was achieved over a given combustor length.

References

1. Drummond, J. P.; Rogers, R. C.; and Hussaini, M. Y.: A Numerical Model for Super-sonic Reacting Mixing Layers. Computer Methods in Applied Mechanics and Engineering, v. 64, 1987, pp. 39-60.

2. Drummond, J. P.; and Hussaini, M. Y.: Numerical Simulation of a Supersonic Reacting Mixing Layer. AIAA Paper No. 87-1325, June 1987.

3. Menon, S.; Anderson, J. D.; and Pai, S. I.: Stability of a Laminar Premixed Super-

sonic Free Shear Layer with Chemical Reactions. Int. Journal Eng. Sci., v. 22, no. 4, 1984, pp. 361-374.

4. Drummond, J. P.: Two-Dimensional Numerical Simulation of a Supersonic Chemically Reacting Mixing Layer. NASA-TM, 1988.

5. Brown, G. L.; and Roshko, A.: On Density Effects and Large Structure in Turbulent Mixing Layers. J. Fluid Mechanics, V. 64, no. 4, 1974, pp. 775-816.

6. Papamoschou, D.; and Roshko, A.: Observations of Supersonic Free Shear Layers. AIAA Paper No. 86-0162, Jan. 1986.

7. Ragab, S. A.; and Wu, J. L.: Instabilities in the Free Shear Layer Formed by Two Supersonic Streams. AIAA Paper No. 88-0038, Jan. 1988.

8. Guirguis, R. H.; Grinstein, F. F.; Young, T. R.; Oran, E. S.; Kailasanath, K.; and Boris, J. P.: Mixing Enhancement in Supersonic Shear Layers. AIAA Paper No. 87-0373, Jan. 1987.

9. Guirguis, R. H.: Mixing Enhancement in Supersonic Shear Layers: III. Effect of Convective Mach Number. AIAA Paper No. 88-0701, Jan. 1988.

10. Kumar, A.; Bushnell, D. M.; and Hussaini, M. Y.: A Mixing Augmentation Technique for Hypervelocity Scramjets. AIAA Paper No. 87-1182, June 1987.

11. McBride, B. J.; Heimel, S.; Ehlens, J. G.; and Gordon, S.: Thermodynamic Properties to 6000 K for 210 Substances Involving the First 18 Elements. NASA SP-3001, 1963.

12. Householder, A. S.: The Theory of Matrices in Numerical Analysis. Dover Publications, New York, 1964, pp. 122-140.

13. Bussing, T. R. A.; and Murman, E. M.: A Finite Volume Method for Calculation of Compressed Chemically Reacting Flows. AIAA Paper No. 85-0331, 1985.

14. Widhopf, G. F.; and Victoria, K. J.: On the Solutions of the Unsteady Navier-Stokes Equations Including Multicomponent Finite Rate Chemistry. Computers and Fluids, v. 1, 1973, pp. 159-184.

15. White, M. W.: Viscous Fluid Flow. McGraw-Hill, New York, 1974, pp. 288-290.

TABLE 1. - Finite-Rate Chemistry Model and Arrhenius Rate Coefficients for Each Reaction

Reaction number	Reaction	A	N	E, kJ/g-mole
1.	$H_2+O_2=OH+OH$	0.1700E+14	0.00	201.5
2.	$H+O_2=OH+O$	0.1420E+15	0.00	68.6
3.	$OH+H_2=H_2O+H$	0.3160E+08	1.80	12.78
4.	$O+H_2=OH+H$	0.2070E+15	0.00	57.5
5.	$OH+OH=H_2O+O$	0.5500E+14	0.00	29.3
6.	$H+OH=H_2O+M$	0.2210E+23	-2.00	0.0
7.	$H+H=H_2+M$	0.6530E+18	-1.00	0.0
8.	$H+O_2=HO_2+M$	0.3200E+19	-1.00	0.0
9.	$HO_2+OH=H_2O+O_2$	0.5000E+14	0.00	4.2
10.	$HO_2+H=H_2+O_2$	0.2530E+14	0.00	2.9
11.	$HO_2+H=OH+OH$	0.1990E+15	0.00	7.5
12.	$HO_2+O=OH+O_2$	0.5000E+14	0.00	4.2
13.	$HO_2+HO_2=H_2O_2+O_2$	0.1990E+13	0.00	0.0
14.	$HO_2+H_2= H_2O_2+H$	0.3010E+12	0.00	78.2
15.	$H_2O_2+OH=HO_2+H_2O$	0.1020E+14	0.00	7.9
16.	$H_2O_2+H=OH+H_2O$	0.5000E+15	0.00	41.9
17.	$H_2O_2+O=OH+HO_2$	0.1990E+14	0.00	24.7
18.	$M+H_2O_2=OH+OH$	0.1210E+18	0.00	190.4
For the single step reaction		0.5510E+15	0.00	30.2

TABLE 2. - Statistical Features of Cases 1 and 3

CASE	(1)	(3)
CONFIGURATION	M=2, FUEL, 2000K	M=2, FUEL, 2000K
	M=2, AIR, 2000K	M=2, AIR, 2000K
δ_ω/x	0.04 TO 0.044	0.056 TO 0.058
$\sqrt{u'^2}/U$	3.5%	10%
$\sqrt{T'^2}/T$	3.6%	6.7%
∇/U_e	~ 0.4%	~ 1%
$\sqrt{v'^2}/\nabla$	3.5%	12%
$\sqrt{\rho'^2}/\bar{\rho}$	17%	34%
$\sqrt{f_{H_2O}'^2}/\overline{f_{H_2O}}$	2.5%	4.7%
$\overline{\rho u'v'}/U_e^2 \rho_e^2$	9.1×10^{-4}	1.84×10^{-3}

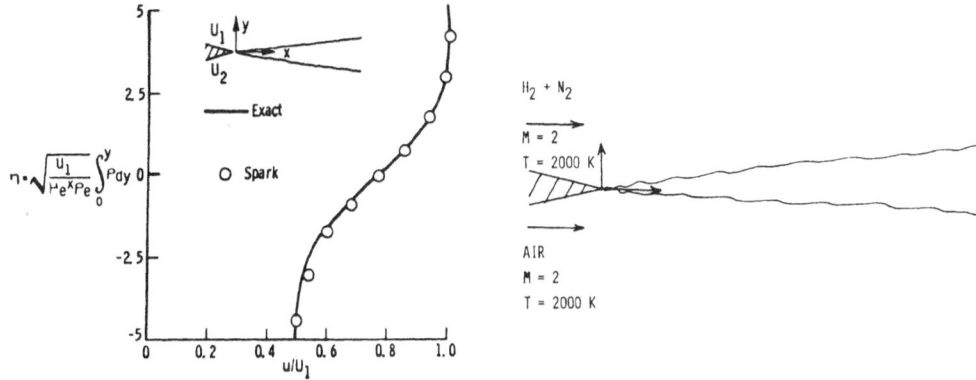

Fig.1 - Comparison of the computation with the exact solution of Lock.

Fig.2 - Schematic of the supersonic reacting mixing layer in Case 1.

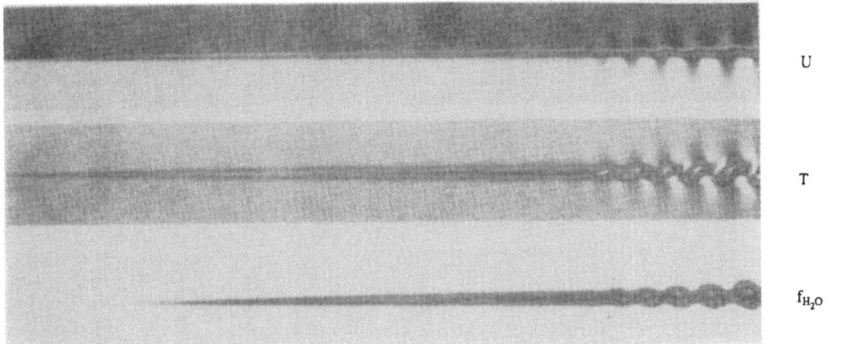

Fig.3 - Velocity, temperature, and water mass fraction contours in Case 1.

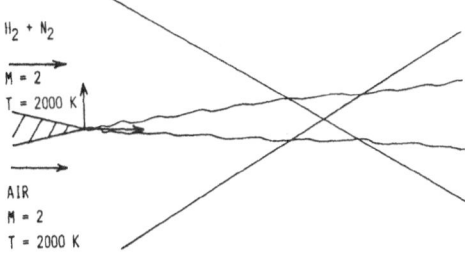

Fig.4 - Schematic of the supersonic reacting mixing layer interacting with two shocks in Case 2.

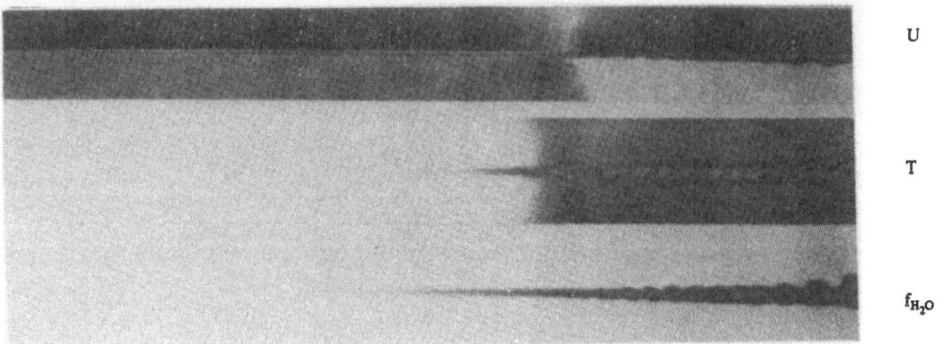

Fig.5 - Velocity, temperature, and water mass fraction contours in Case 2.

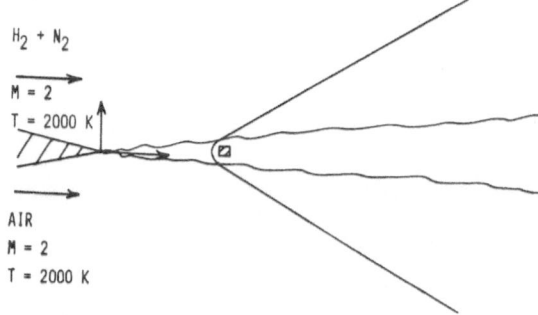

Fig.6 - Schematic of the supersonic reacting
mixing layer interacting with a curved
shock in Case 3.

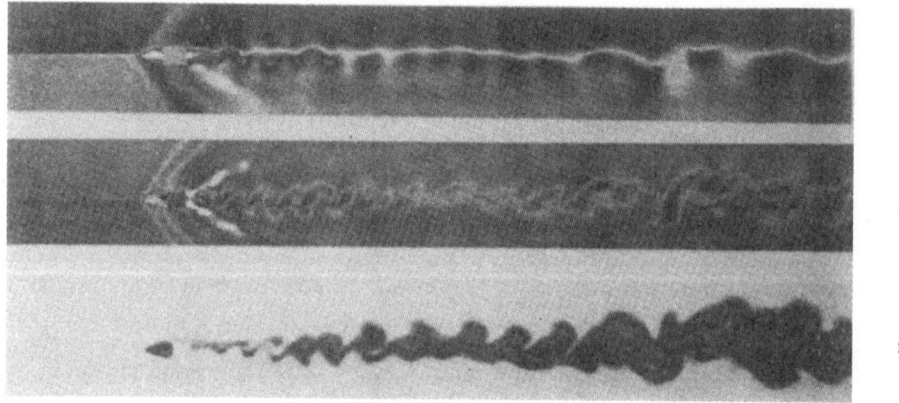

Fig.7 - Velocity, temperature, and water mass fraction contours in Case 3.

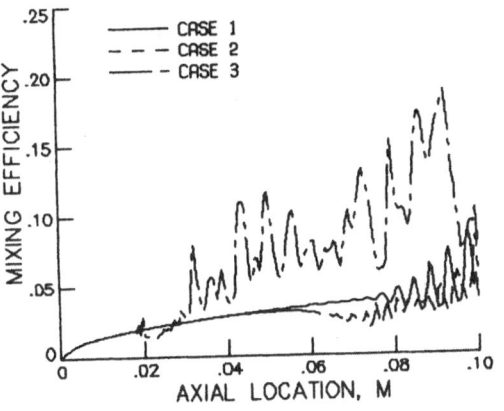

Fig.8 - Mixing efficiency versus streamwise
station for Cases 1 through 3.

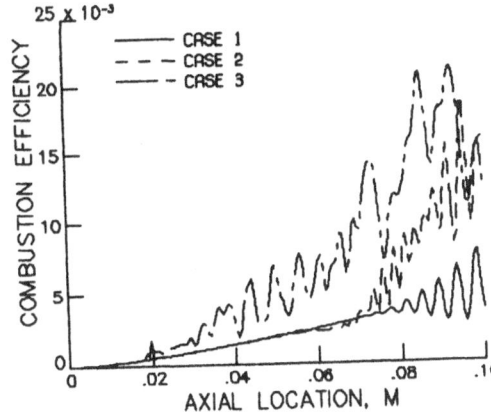

Fig.9 - Combustion efficiency versus streamwise
station for Cases 1, 2 and 3.

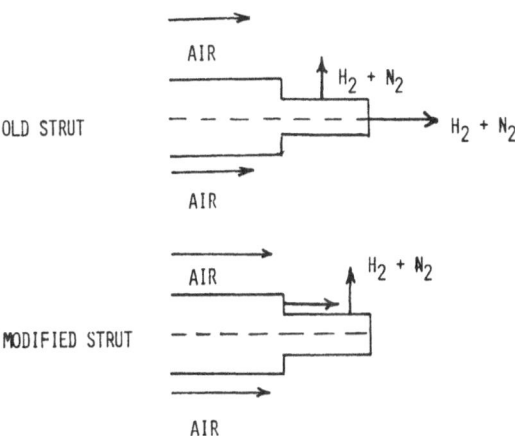

Fig.10 - Schematic of conventional and modified
fuel injector strut configurations.

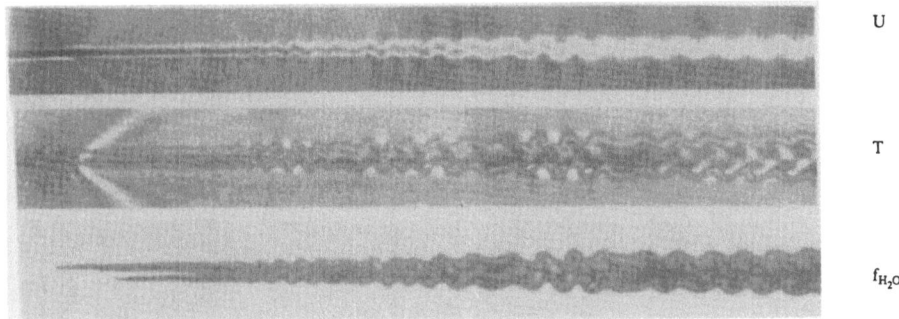

Fig.11 - Velocity, temperature, and water mass fraction contours in Case 4 with only the parallel injector.

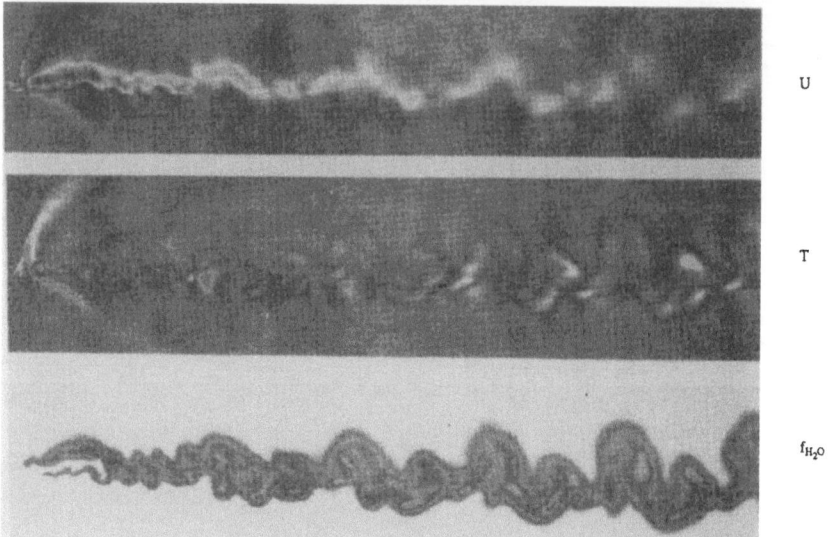

Fig.12 - Velocity, temperature, and water mass fraction contours in Case 5 with interaction
of the parallel and transverse injectors.

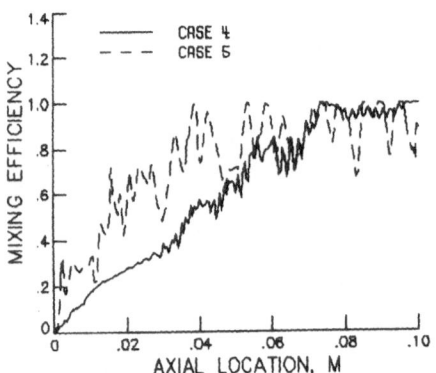

Fig.13 - Mixing efficiency versus streamwise station
for Cases 4 and 5.

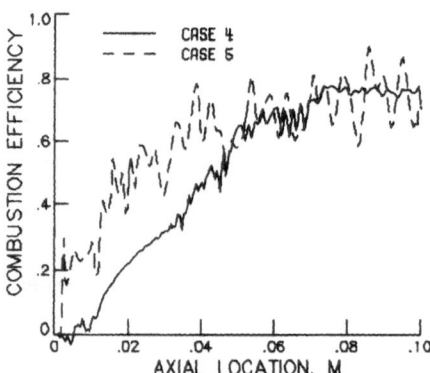

Fig.14 - Combustion efficiency versus streamwise
station for Cases 4 and 5.

Calculation of Low Mach Number

Reacting Flows*

Harry A. Dwyer - Professor
University of California
Davis, California

Abstract

A numerical method has been developed to calculate low Mach number and reacting flows in general and the burning and vaporization of a hydrocarbon droplet in particular. The method is time accurate and it makes use of the continuity equation to form a pressure correction. The basic transport equations have been given in finite volume form, and a predictor/corrector algorithm has been employed. The continuty equation and the pressure correction step have been solved to high numerical accuracy with the use of a direct solver technique, which also has the advantage of improved efficiency. The physical problem of the burning and vaporization of a high molecular weight hydrocarbon droplet is a very difficult one due to the very large temperature and density gradients in the flow, and also due to the complicated flow structures which develop during a droplet's lifetime. The methods used in this research have successfully solved these problems, and a detailed history of the droplet has been calculated and studied.

Introduction

The purpose of this article is to describe progress that has been made in the numerical simulation of the unsteady processes that occur during the vaporization and combustion of individual hydrocarbon fuel droplets in a hot oxidizing environment. It is the premise of this type of research that a good understanding of the most important physical processes in droplet dynamics can be obtained with a digital computer simulation, and that the studies will be valuable in other areas of combustion science such as spray combustion. It is also the purpose of the article to describe the methods that have been used in detail so that they can be utilized and improved as the power of the digital computer grows.

Although there has been a large number of developments in the use of the digital computer to simulate many fluid flow problems in such fields as aerodynamics and simple turbulent flow, there has not been as much activity in multi-phase and reacting

---

*This research was partially supported by Sandia National Laboratories, Livermore.

flows. The primary reason is due to the fact that a vaporizing and burning droplet involves the interplay of a large number of different physical time scales. These time scales are characterized by the dimensionless parameters which appear in the equations of motion and are well known by the names of the famous men who have contributed to their use, such as Reynolds, Domkohler, Mach, Prandlt and Lewis. The resulting difficulties from the multiple time scales in reacting flow problems can be comprehended when one is reminded of the fact, that many gifted scientist spend their entire careers in studying only a small portion of the time scale variation associated with just one dimensionless parameter, such as Reynolds number.

An early and current leader in low Mach number flows has been the group at Imperial College in London under the original leadership of Brian Spalding, and which has developed a large number of outstanding researchers. A good survey of the present state of the methods, which have been used to solve a wide variety of reacting and nonreacting flows, can be found in the textbook of Patankar, Reference[1]. A more general view of the development of the field can be found in the pioneering textbook of Roache, Reference [2], and this book describes in detail many of the early research efforts at Los Alamos National Laboratories and the efforts by Alexander Chorin, Reference [3].   The problems that developed in the solution of low Mach number flows were due to a sensitivity to errors in the calculation of the pressure field from the continuity equation, and a possible decoupling of the pressure field from the velocity field if certain types of difference schemes are utilized. It is now known that there are many ways to avoid these problems, and a modern new approach will be presented in this article.

One of the first numerical efforts to study the vaporization of a hydrocarbon droplet was the steady state work of Renksizbulut, Reference [4]. In this investigation a steady state correlation for the total drag and Nusselt number was investigated and compared with experimental results and found to have merit. Unsteady investigations of vaporizing hydrocarbon droplets were carried out by Dwyer and Sanders, Reference [5] and [6], with an averaged density approximation. Further and more complete studies by Patnaik, Reference [7] and Haywood, Reference [8], confirmed these results, and the investigation of Haywood showed that a modified form of the steady state correlation of Renksizbulut may have validity in complex hydrocarbon studies.

Problem Definition

A description of the physical problem to be addressed is exhibited in figure (1) where the flow over a vaporizing droplet is shown in a schematic way. As can be seen from this figure the flowfield is characterized by the freestream conditions far from the surface and upstream, the droplet itself and its interface, and the conditions downstream and far from the surface. In its most complete form the problem involves an interaction between the gas and liquid phases, and the complete system must be modeled and calculated.   The basic tasks are made much more difficult for the burning of hydrocarbon fuel droplets because of the very large variation of physical and transport properties which occur due to differences in species molecular weight and the large temperature differences caused by chemical reactions and freestream conditions. An example of typical property variations caused by individual effects for a decane droplet is presented in the lower portion of figure (1). If the cold surface characteristics of pure decane are used for the density and the viscosity rather than the hot gas freestream properties, the Reynolds number varies by more than a factor of fifty times. This variation in Reynolds number is very large and presents serious problems if an approximate calculation of drag and heat transfer is attempted with a semi-empirical correlation.

System Equations

A very significant finding of this research effort is the importance of the use of control volume techniques to generate the basic system of finite difference equations for mass, momentum and energy transport. It has been the experience of the present investigator that a large number of mistakes are continually being made in going from the differential equations to a finite size volume when one is trying to describe a complex flow system. A much more physically satisfying approach and one which leads to a reduced number of errors is to start with the basic laws of physics written in finite form or the so called control volume form. The transport equations in control volume form can be written as:

Control Volume Equations

Continuity Eq.
$$\partial/\partial t \left[\iiint_v \rho \, dV \right] + \iint_s \rho \, \mathbf{V} \cdot \mathbf{dA} = 0$$

Momentum Eq.
$$\partial/\partial t \left[\iiint_v \rho \mathbf{V} \, dV \right] + \iint_s \rho \mathbf{V} \, \mathbf{V} \cdot \mathbf{dA} =$$
$$- \iint_s p \mathbf{dA} - \iint_s \tau'' \cdot \mathbf{dA}$$

Energy Eq.
$$\partial/\partial t \left[\iiint_v \rho e \, dV \right] + \iint_s \rho h \, \mathbf{V} \cdot \mathbf{dA} =$$
$$- \iint_s \mathbf{q_e} \cdot \mathbf{dA} + \iiint_s \rho S_e'' \, dV$$

Species Eq.
$$\partial/\partial t \left[\iiint_v \rho Y_k \, dV \right] + \iint_s \rho Y_k \, \mathbf{V} \cdot \mathbf{dA} =$$
$$- \iint_s \mathbf{q_k} \cdot \mathbf{dA} \quad + \quad \iiint_v \rho S_k'' \, dV$$

where the surface, $\mathbf{dA}$, and volume integrals, dV, are defined by the midpoints between the cells, t - time, ρ - density, $\mathbf{V}$ - the velocity vector in cartesian form, τ'' - the stress tensor, q_e - heat flux vector, S_e - the heat release source term, Y_k - the mass fraction of specie k, q_k - diffusion flux of specie k, and S_k - the chemical rate term for species k.

The Low Mach Number Limit

For the problems which will be treated in this article it will be useful to take the low Mach number limit of the above equations. A clear description of this process can be found in the article by Merkle, Reference [11], and the primary purpose of taking this limit is to eliminate acoustic wave propagation from the system of equations. The result of this process is that the pressure field can be separated into a thermodynamic part, P_r , and a dynamic part, p, which appears in the momentum equations. The thermodynamic part of the pressure is constant in space, and is utilized in the equation of state to calculate the local density for a gas (For an incompressible fluid the thermodynamic pressure does not appear directly in the basic equations). The dynamic part of the pressure field is needed for the solution of the momentum equations, and the lack of an explicit equation to calculate this pressure field has been one of the principal difficulties with this limit of the Navier-Stokes equations.

Most methods for the calculation of the dynamic pressure field have been based on the use of the continuity equation to generate an elliptic Poisson equation. The pioneers in this field have been the researchers at the Los Alamos Research Laboratories and A.J. Chorin, References [2] and [3], and the researchers at Imperial College have introduce many variations to the basic methods. Although there has been a significant number of problems solved, there does not seem to be a large amount of effort spent on time dependent or variable density problems. In the present paper a time-dependent method will be given, and many of the difficulties associated with this model in the past will be explained.

Another good reason to utilize the low Mach number model is due of the large temperature differences in a problem with chemical reactions and heat transfer. Local errors in the temperature field of just a couple of degrees centigrade can be converted into rather large pressure changes when compared to the stagnation pressure for a flow with a maximum velocity of less than ten meters per second (The reader can check this with the use of the isentropic equations of gas dynamics). When the low Mach number approximation is applied this possibility is eliminated since the thermodynamic pressure is constant in space.

Boundary Conditions

The application of boundary conditions is made much simpler with the use of the control volume formulation of the basic equations. The boundary conditions can usually be obtained by taking the limit as the volume of a cell approaches zero, and by applying results from physical observation. For example, the conditions of a continuous distribution of velocity, temperature and other thermodynamic variables follow from physical considerations and observations, while the conditions for the pressure field, the stresses, the heat fluxes, and the mass fluxes at the interface locations can be found from the surface limit of the control volume equations.

Numerical Methods

A basic purpose of the article is to present in detail the numerical methods that have been developed and used to solve the present class of problems. The challenging droplet dynamics problem represents a severe test for the numerical methods. The methods used can be described in three parts, and these are given below.

A. Basic Numerical Method

The fundamental philosophy that was considered in the choice of the numerical method was to employ the most implicit method possible within the capability of the memory and speed of the digital computer. The direct solution of the entire system of resulting algebraic equations with a banded solver, Reference [12], is a very implicit and accurate method, but it is beyond the capability of even the future generation of digital computers for multi-dimensional problems with chemical reactions and time-dependent conditions. The first choice considered was the approximate factorization methods developed at NASA Ames, Reference [13], and which are used widely in the Aerospace industry under high Reynolds number conditions. These methods have good stability characteristics for two-dimensional problems and can employ a variety of time accurate forms, however they do have some problems with complex grids for lower Reynolds numbers. A better alternative for problems with strong diffusion or elliptic influences is to use predictor/corrector methods with the updating of all the terms in the equations, Reference [14].

In order to achieve a factored form for the space coordinates a predictor/corrector two step algorithm is utilized to solve the convective space terms and the Laplacian like operators. On the predictor step one coordinate direction is implicit and on the corrector step the other coordinate direction is made the implicit one. During both steps the new values of T at time $n+1$ are used immediately in the algorithm after being calculated. The finite difference forms for the diffusion operator on a uniform grid become

Predictor - x direction implicit
$$\nabla^2 T = [T^{n'+1}_{i+1,j} - 2\,T^{n'+1}_{i,j} + T^{n'+1}_{i-1,j}]/\Delta x^2 + [T^{n}_{i,j+1} - 2\,T^{n'+1}_{i,j} + T^{n'+1}_{i,j-1}]/\Delta y^2$$

Corrector - y direction implicit
$$\nabla^2 T = [T^{n'+1}_{i+1,j} - 2\,T^{n+1}_{i,j} + T^{n+1}_{i-1,j}]/\Delta x^2 + [T^{n+1}_{i,j+1} - 2\,T^{n+1}_{i,j} + T^{n+1}_{i,j-1}]/\Delta y^2$$

where the index $(n'+1)$ denotes the predicted values from the first part of the algorithm.

The main advantage of a predictor/corrector algorithm is for problems with large differences in grid cell size or fast chemistry. In this situation the added implicitness of the corrector and the updating of all terms in the equation can make a substantial improvement in the efficiency and stability of the resulting solution (There is some degradation of time accuracy for the predictor, but this can be improved with an ADI type of predictor).

The application of the above algorithm to the finite volume form of the Navier-Stokes equations is straight forward but tedious, and can be illustrated with the heat flux term in the energy equation

$$- \iint \mathbf{q_e} \cdot \mathbf{dA} = k \, [\partial T/\partial x] \, dA_x + [\partial T/\partial y] \, dA_y + [\partial T/\partial z] \, dA_z$$

If the control volumes are defined by a nonorthogonal distribution of cell centers organized with the generalized coordinates ξ, η, and ζ, the above expression becomes

$$- \iint \mathbf{q_e} \cdot \mathbf{dA} = k \, \{ \, [\partial T/\partial \xi]\xi_x + [\partial T/\partial \eta]\eta_x + [\partial T/\partial \zeta]\zeta_x \, \} \, dA_x$$
$$+ \, \{ \, [\partial T/\partial \xi]\xi_y + [\partial T/\partial \eta]\eta_y + [\partial T/\partial \zeta]\zeta_y \, \} \, dA_y$$
$$+ \, \{ \, [\partial T/\partial \xi]\xi_z + [\partial T/\partial \eta]\eta_z + [\partial T/\partial \zeta]\zeta_z \, \} \, dA_z$$

where a local coordinate transformation is used between the x y z and the ξ η ζ coordinate directions (This transformation is always found by numerical methods). The predictor/corrector algorithm is then utilized along the directions ξ, η and ζ, if the grid points are set up in an ordered fashion. The other terms in the control volume equations which involve derivatives, such as the stress tensor and the diffusion fluxes, are evaluated in a similar way.

A brief discussion will now be presented for the numerical treatment of the various terms which appear in the basic system of equations. The general rule is that all volume terms are evaluated at the center of the cell, and all surface terms at the surface of the cell (the transport properties included). The surface terms can only be found be averaging between cell centers, and for a second order method (in the finite difference sense) this implies a linear variation between cell centers. For the convective flux terms and the pressure force in the momentum equations there is no coordinate transformation necessary to calculate these terms and only vector dot products are required to form the system of algebraic equations. The major problem that one faces is the calculation of the pressure field from the continuity equation and the sensitivity of the momentum equations to errors in the pressure field. This type of sensitivity is not encountered in high Mach number flows where there seems to be no need to take special precautions. However, for low Mach number flows with or without density variation special precautions must be taken to avoid decoupling of the pressure field and the build up of errors. The methods presented above have been used for a wide variety of heat and mass transfer problems, Reference [18] , as well as three dimensional flows, Reference [19].

B. Continuity Equation and the Pressure Field

One of the major difficulties with the low Mach number and incompressible flow models has been the calculation of the pressure field. This difficulty is due to a little understood sensitivity of the pressure field and the convective velocity term to errors in the convergence of the continuity equation. This sensitivity seems to be much greater than in the full compressible equations, and the problem seems to be amplified further when low Mach number variable density solutions are required. It is the present authors opinion that the problem is not one in principle, but only a practical one of sensitivity to error and poor convergence. In this article a new method is described (The methods borrows heavily on the previous work of Chorin) and has successfully treated reacting, and low Mach number flows in an efficient and time accurate manner. The method involves two parts and these are respectively the solution of the continuity equation and the pressure correction at the unknown time level $n+1$.

The method can be thought of as a two variable method rather than a two grid method of the staggered grid type, which has been used for incompressible flows in the past, Reference [1]. The staggered grid method loses much of its attractiveness in variable density flow due to the fact that the density must be extrapolated to the cell boundaries, and this involves an inaccuracy in the description of the mass flux into a cell volume. The present solution is to define a velocity at the cell center, V, used in the inertia volume term in the momentum equations, and a velocity at the cell boundaries, V_b, which will be used to satisfy the continuity equation and to calculate fluxes into cells. The first step of the procedure is to solve the momentum and other transport equations for intermediate values of the variables with the old pressure field. This procedure is outlined below in a typical finite difference form, which is more familiar for most readers.

Step I Solution of Momentum Equations with old pressure field

$$\rho (V' - V^n) / \Delta t = -\nabla p^n + RHS^{n+1}$$

where the following notation has been used

V' = Intermediate velocity at cell center

V^n = Previous velocity at the cell center

$p^{n+1} = p^n + \alpha$

RHS^{n+1} - Viscous plus the convective terms at $n+1$

Step II Correction of the momentum equations with the pressure correction, α, to be determined from the continuity equation.

$$\rho (V^{n+1} - V') / \Delta t = - \nabla \alpha$$

The addition of steps I and II yields

$$\rho \, (V^{n+1} - V^n) = - \nabla(p+\alpha) + RHS^{n+1}$$

which is a consistent form of the full Naiver Stokes equations, and the overall method can be made second order accurate in time if iteration is employed.

The pressure correction, α, is to be determined from the use of the continuity equation in the following fashion

Basic Idea: A velocity correction at the cell boundaries will be introduced to satisfy the continuity equation, and which will determine the pressure correction during a time step. This is an alternative to having staggered grids, and it seems to be advantageous for time-dependent problems on complex grids.

The specific procedure is to form the average density and velocities at the cell boundaries and apply the continuity equation, which results in the following equations

Integrate $\rho_b \, V_b$ around the cell

(I) $\iint_b \rho_b \, [\, V_b = V'_b + V_c \,] \cdot dA$

Apply Continuity Equation

$\iint_b [\, \rho_b \, V_b \cdot dA \,] = - \partial \rho \, / \, \partial t \; dV$

therefore $\iint_b [\, \rho_b \, V'_b \cdot dA \,] = - \iint_b [\, \rho_b \, V_c \cdot dA \,] - \partial \rho / \partial t \; dV$

where the following definitions are used

V'_b = Averaged velocity at cell boundaries obtained from Step I of
the momentum equations

V_b = Corrected velocity at the cell boundaries after the velocity correction

V_c = Velocity correction at the cell boundaries

$V_b = V'_b + V_c$

ρ_b = Average density at cell boundaries, obtained from the energy equation and the equation of state.

ρ = density at the center of the cell

dV = Volume of cell

The next step is to define a velocity potential ϕ

$- \nabla \phi = \rho_b \, V_c$

If the velocity correction comes from a potential function, then all the rotational components of velocity come only from the momentum equations where viscosity plays its roll. The velocity potential is substituted into the continuity equation, and the following integral Poisson equation for ϕ is obtained

(II.) $\partial \rho / \partial t \; dV + \iint_b [\, \rho_b \, V'_b \cdot dA \,] = \iint_b [\, \nabla \phi \cdot dA \,]$

If the half cells which appear near the computation boundaries for a non-staggered grid are included in the mass balance there is no need to apply boundary conditions for the velocity potential ϕ (Except for the downstream boundary). In general the half cells near the wall are not treated for the transport equations, but if the half cells are not include in the mass balances, it is necessary to apply boundary conditions for the pressure correction α. The boundary conditions should be obtained from the limit of the momentum equation near a surface, or approximations related to them. The typical high Reynolds number boundary conditions for ϕ are the following, but they should be modified for specific applications

Boundary Conditions for ϕ

Solid Wall

$V_c \cdot dA]_w = 0$ (No wall mass transfer)

or $\nabla\phi \cdot dA = 0$ or $\partial\phi/\partial n = 0$

Inflow - Constant Velocity

$\partial\phi/\partial n = 0$

Outflow

$\phi = 0$ This boundary condition can be confusing, but corresponds to the analytical condition. Also we need a value of ϕ specified somewhere on the boundaries.

The pressure correction, α, is obtained directly from the velocity correction potential, ϕ, and the justification for this can be seen by applying the partial form of the momentum equation at the grid boundary (The same as a staggered grid)

$\rho_b \, V_c / \Delta t \, dV_b \;=\; - \iint_b \alpha_b \, \mathbf{n} \cdot d\mathbf{A}$ Split Momentum equation applied at the boundary of the cell and where the following notation is used.

dV_b = Volume of hypothetical staggered grid

$\alpha = \phi / \Delta t$

Apply Gauss's Theorem

$\rho_b \, V_c / \Delta t \, dV_b \;=\; -\nabla \alpha_b \, dV_b = -\nabla\phi / \Delta t \, dV_b$

Thus is seen that the velocity potential, ϕ, and the pressure correction, α, are directly related by the time step.

The above formulation of the calculation of the pressure field has the advantages that there is not any decoupling of the pressure field, the continuity equation can be solved to machine accuracy at every time step, and the method is time accurate. This combination represents an advance over most older methods, and it accomplishes it without the complexity of a staggered grid. Another significant new feature is the use of a direct solver to calculate the velocity and pressure corrections.

Results

The results presentation of the article will be concerned with a complex droplet burning problem, which involved the vaporization and burning of a 50 micron radius octane droplet at 10 atmospheres. The initial conditions for the problem consisted of the injection of the droplet into high temperature air, Re=100 and T_g=1250 K, and the initial droplet temperature was 470 K. In order to simplify the calculation the internal droplet circulation was neglected, and it was assumed that the droplet temperature was uniform in space and increased in time due to droplet heating (This is a reasonable approximation since a large percentage of the droplet mass is contained in the outer portion of the droplet). Under these conditions there is an initial droplet heatup period until the total surface heat transfer is eventually consumed by mass transfer. The concentration of the fuel at the droplet surface is determined by the Clasius-Clapeyron relationship, Reference [16], and the surface mass transfer determine by diffusion (Note: The transport properties of Reference [16] have been utilized and the single step chemical mechanism of Reference [17]).

It is instructive to begin the study with some isotherm contours around the droplet, figures (2) and (3), which illustrate clearly the influence of Reynolds number on the chemistry and flame structure (The coordinate system has been normalized with respect to the local droplet radius and the Reynolds number is based on the local radius and relative velocity between the gas and droplet center). The highest Reynolds number condition in figure (2) shows an isotherm pattern similar to the steady state results presented in Reference [6] (Note: The isotherm distributions are uniform over the ranges stated in the figures). The gas phase of the flow appears to be essentially quasi-steady under the present flow conditions, and the chemistry is not fast enough to stabilize the flame on the forward portion of the droplet. As the droplet velocity decays with respect to the gas, the flame structure begins to approach the droplet in the vicinity of the well mixed and relatively hot separated shear layer which leaves the droplet surface. As the Reynolds number approaches twenty five, figures (2), the flame begins to move around the droplet to the front stagnation point, and at the Reynolds number of near twenty, figure (3), the flame completely surrounds the droplet. The time given in all of the figures has been made dimensionless with the initial droplet radius and the thermal diffusivity of the freestream air.

The distribution of oxygen mass fraction at an intermediate Reynolds numbers is shown in figure (4), and it is seen that the mass fraction contours reflect the local value of the Reynolds number very closely. It is clear that very little oxygen is present near the symmetry line of the wake. At the lower Reynolds number condition the influence of the low mass diffusion coefficient relative to heat transfer plays a significant role, and this Lewis number effect in the gas phase causes the fuel mass fraction contours to be bunched near the droplet surface, and this is a direct result of the higher Peclet number.

Another potential problem with any multidimensional numerical simulation is caused by the flame approaching the upstream computational boundary of the gas flowfield at the lower Reynolds number condition. In order to avoid this problem the physical boundary conditions of the computational zone must be extended further upstream, and the transition from the inflow to the outflow boundary condition must be made further upstream. Also, from the numerical point of view it should be mentioned again that the dynamic pressure field calculation is very sensitive to fast chemistry through the time-dependent density term in the continuity equation. If the mass fractions begin to oscillate between neighboring cells, the pressure oscillations can become quite large, and the solutions can degrade seriously. Further studies are needed in this area to determine if the low Mach number model amplifies this problem because of the infinite sound speed assumption that is applied.

The global characteristics of this flow as a function of time are shown in figures (5) through (8). A particular interesting characteristic is the total drag coefficient and its components, which are shown in figure (5). In general the value of the drag coefficient is much lower than the standard drag curve for flow over a sphere, but the general level is predicted in an approximate way by the correlation of Renksizbulut and Haywood (entitled the RH correlation in figure (5)), Reference [8]. This steady state correlation assumes that all the heat transfer is utilized for mass transfer, and this is not true early in the droplet lifetime because of internal heating. The calculated drag coefficent is higher than the correlation due to reduced mass transfer, and the calculated results approach the correlation as the droplet temperature increases due to heating. The correlation also does not take into account the dynamic motion of the flame around the droplet, and the increase in drag coefficient during flame engulfment is not reproduced. However, in general the correlation does give a rough time dependent average of the drag coefficient, and it can be useful in design applications. The pressure and thrust drag components shown in figure (5) make it quite clear that the major component of drag is being contributed by the pressure drag. The thrust drag and the friction drag (that contributed by the shear stress) are small relative to the pressure drag, and the most dramatic influence of the surface mass transfer is the lowering of the friction drag.

The Nusselt number curve in figure (6) does not agree with conventional results or the correlation developed by Renksizbulut, Reference [8], due to the influence of the flame structure. The definition of Nusselt number contains the difference in temperature between the gas freestream and the droplet surface temperature, but this underestimates the potential for heat transfer caused by the higher temperatures of the chemical reactions. However, it is not a simple question of using a different temperature in the definition of the Nusselt number, such as the adiabatic flame temperature, since the flame position is changing continuously in time. As the Reynolds number decreases the flame first contributes to heat transfer in the wake of the droplet, and then gradually heats the entire droplet as the flame envelops the droplet at lower Reynolds numbers. It is seen that there is a rather long transition in the heat transfer processes as the flame moves around the droplet, and the peak in Nusselt number is caused when the flame first surrounds the droplet. It is the opinion of the present investigators that this type of process can only be treated well with a complete numerical simulation.

The significant influence that variable transport properties can have on a vaporizing and burning droplet problem can be seen in figures (7). These two definitions are based on the gas and surface properties respectively, and they are defined as

$$Re_g = \rho_g V_g D / \mu_g \qquad Re_s = \rho_s V_g D / \mu_s$$

The surface Reynolds number is much larger than the gas values, and this is caused by the lower temperature, higher molecular weight, and lower viscosity near the octane droplet surface. In order to use a correlation for the drag coefficient the large density change at the surface is not important, since the inviscid flow outside the boundary layer is only influenced by the freestream conditions. However, for the interior of a droplet spray the density change caused by cooling and high fuel concentrations could easily cause Reynolds number differences of the order shown in figure (7). Therefore, it appears that in many droplet flows it will be very difficult to apply semi-empirical correlations because of the lack of knowledge about the flow transport properties. In order to obtain these properties a numerical simulation seems to be required, and the numerical simulation will yield the flowfield characteristics without the need for a correlation. Also, since the capabilities of the digital computer are increasing rapidly, it will be more feasible in the future to carry out difficult calculations of this type. Therefore, it appears that simulations of the type presented in this article will play an increasing role in droplet dynamics and combustion flow systems in general.

Conclusions

The major conclusions of the present research effort have been the following:

1. A general control volume formulation of the basic equations of mass, momentum and energy has been developed and presented. The method reduces to generalized coordinates on a structured grid, but it has the advantageous of clear physics for all equations and boundary conditions, as well as extensions to complex cell shape.

2. A two velocity variable formulation of the pressure correction algorithm for low Mach number flows has been developed and applied to reacting flows. This new method is an alternative to the use of a staggered grid and is applicable to unsteady and variable density flows.

3. The direct solution of the pressure correction Poisson equation has been accomplished with a banded solver of the LUD type. The banded solver is independent of the transport properties and time step, and a matrix inversion only needs to be inverted once for use at all time steps. The use of this technique insures that the

continuity equation will be converged to machine zero at each time step, and nonlinear errors in the pressure correction algorithm will be suppressed.

4. Very fast chemical reaction rates can cause oscillations in the pressure field, and this may be caused by the low Mach number formulation itself. This subject as well as the uncertainty in chemical reaction rates should be studied more in the future.

5. For a fuel droplet starting at a high Reynolds number condition the flow regimes can change quickly due to an interplay with the flame location. As the flame moves from the rear to the front of the droplet there are large changes in droplet heating rates, and smaller changes in the drag coefficient.

6. The general ability of the computed results to simulate droplet burning and flow seems to be good at the present time, and the future of this type of research seems to be promising. However, there is much work left to be done on the chemical rate terms from the point of view of the numerical methods and the physical accuracy.

References

1. Patankar, **Numerical Heat Transfer and Fluid Flow**, Hemisphere Publishing Corporation, New York, 1980.

2. Roache, P., **Computational Fluid Dynamics**, Hermosa Publishers, Albuquerque, New Mexico, 1979.

3. Chorin, A.J., "Numerical Solution of the Naiver-Stokes Eqs.", Math. Computations, Vol.22, pp. 745-762, 1968.

4. Renksizbulut, M., and Yuen, M.C., J. of Heat Transfer, 105, 389-397 (1983).

5. Dwyer, H.A. and Sanders, B.R., "Detailed Computation of Unsteady Droplet Dynamics", Twentieth Symposium on Combustion (Inter), the Combustion Institute, 1984/pp. 1743-1749.

6. Dwyer, H.A., and Sanders, B.R.,"A Detailed Study of Burning Fuel Droplets", Twenty First Symposium on Combustion, The Combustion Institute, 1986.

7. Patnaik, G., Sirignano, W.A., Dwyer, H.A. and Sanders, B.S., "A Numerical Technique for the Solution of Vaporizing Fuel Droplet", Proceedings of the 10th International Coll. on the Dynamics of Explosions and Reactive Systems, Berkeley, Ca., 1985, Also Progress in Aero. and Astro. AIAA, 1986.

8. Haywood, R.J., **Variable-Property, Blowing and Transient Effects in Convective Droplet Evaporation with Internal Circulation**, M.S. Thesis, University of Waterloo, 1986.

9. Steger, J.L., Implicit Finite Difference Simulation of Flow About Arbitrary Two-Dimensional Geometries", **AIAA J.**, Vol. 16, No.7, July 1978, pp. 679-686.

10. Anderson, D.A., Tannehill, J.C., and Pletcher, R.H., **Computational Fluid Mechanics and Heat Transfer**, Hemisphere Publishing Corporation, New York, 1984.

11. Merkle, C.L., and Choi, Y., "Computation of Compressible Flows at Very Low Mach Numbers", AIAA 24th Aerospace Sciences Meeting, Jan. 1986, Reno.

12. Dongarra, J.J., et. al., LINPACK-USER'S GUIDE, SIAM, Philadelphia, 1979.

13. Beam, R.M. and Warming, R.F., "An Implicit Factored Scheme for the Compressible Navier-Stokes Equations", AIAA J., Vol.16, No. 4 (1978) pp.393-402.

14. Ibrani, S., and Dwyer, H.A., "A Study of Flow Interactions During Axisymmetric Spin-Up", To be Published by the AIAA Journal 1987, Also AIAA Paper Aerospace Sciences Meeting, January 1986, Reno.

15. Bird, R.B., Stewart, W.E., and Lightfoot, L.N., *Transport Phenomena*, John Wiley and Sons, 1960.

16. Abramson, B. and Sirignano, W.A., "Approximate Theory of a Single Droplet Vaporization in a Convective Field", 2nd ASME-JSME Joint Thermal Eng'g Conf., Hawaii, March, 1987.

17. Westbrook, C. and Dryer, F.L., "Simplified Reaction Mechanisms for the Oxidation of Hydrocarbon Fuels in Flames", Combustion Science and Technology, 27, 1981, pp.31-43.

18. Soliman, M., **Solution of the Navier-Stokes Equations with Variable Density, Heat and Mass Transfer**, PhD Thesis, Department of Mechanical Engineering, University of California, Davis, June 1988.

19. Dandy, D., and Dwyer, H. A., "The Influence of Freestream Vorticity on Drag Lift and Heat Transfer", To be presented at the AIAA Aerospace Sciences Meeting, Reno, January, 1989. Also submitted to the Journal of Fluid Mechanics.

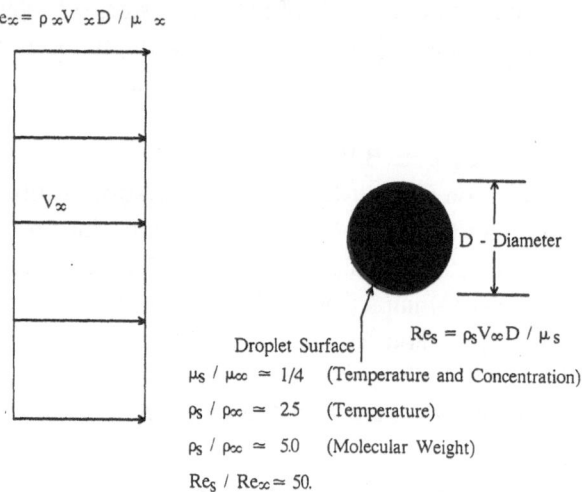

Figure (1) Flow system and Geometry

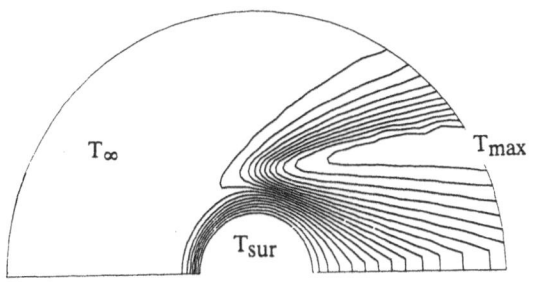

Figure (2) Temperature Contours, R_e = 31.9, T_s = 494 K, τ = 19.5
Contour Range - 494 K ≤ T ≤ 2438 K

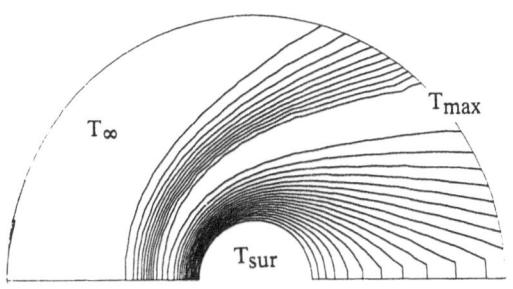

Figure (3) Temperature Contours, R_e = 22.7, T_s = 503 K, τ = 24.5
Contour Range - 503 K ≤ T ≤ 2538 K

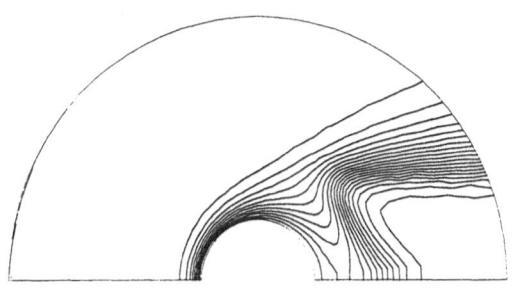

Figure (4) Oxygen Contours, R_e = 49.1, T_s = 479 K, τ = 11.9
Contour Range - 0 ≤ T ≤ .21

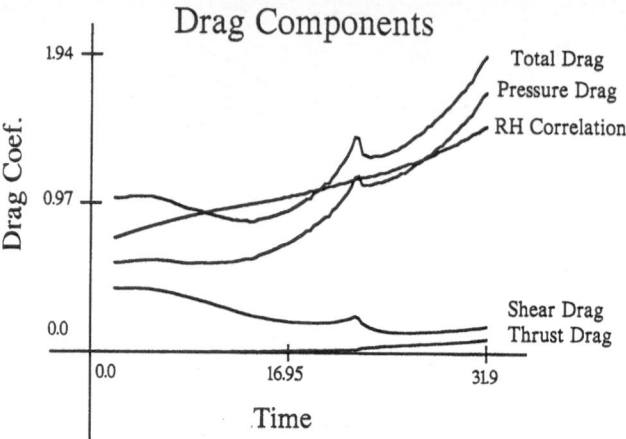

Figure (5) Droplet Drag Components Versus Time

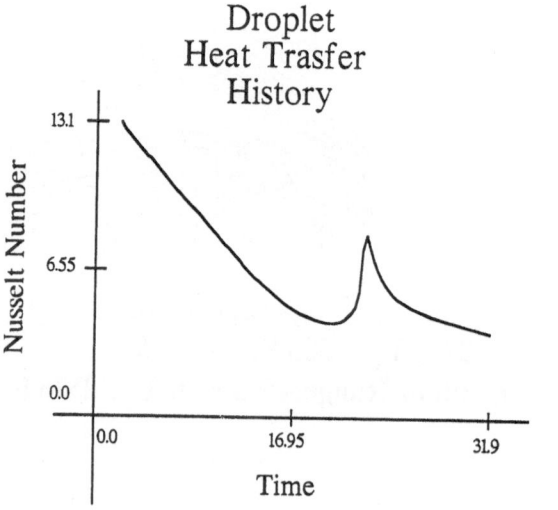

Figure (6) Nusselt Number Versus Time

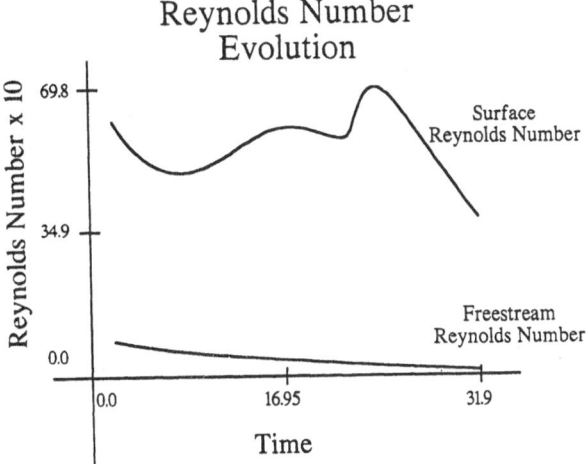

Figure (7) Reynolds Number Versus Time

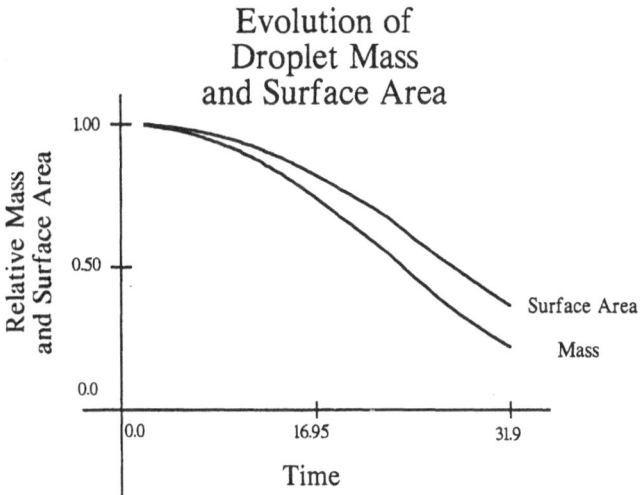

Figure (8) Variation of Droplet Surface Area and Mass

NUMERICAL SIMULATIONS OF FLAMES AND DETONATIONS

K. Kailasanath, E.S. Oran and J.P. Boris

Laboratory for Computational Physics and Fluid Dynamics
Naval Research Laboratory
Washington, D.C. 20375, USA

Abstract

Time-dependent numerical simulations of multidimensional flames and detonations are discussed in this paper. The differences in the processes which must be modelled and the approaches adopted in simulating flames and detonations are highlighted. A two-dimensional flame model is described and then results of calculations are presented that show the effects of gravity on the structure and propagation of laminar, premixed flames. Similarly, a two-dimensional detonation model is described and used to simulate the cellular structure of detonations. Finally, some current developments in computers and algorithms and their impact on the numerical modelling of reactive flows is discussed.

Introduction

Laminar flames and detonations are among the oldest and most fundamental problems in combustion. They both involve the coupling between fluid dynamics and chemical reactions. In many situations both of them can be idealized as reaction fronts or waves. A distinguishing characteristic between flames and detonations is their propagation velocities, with flames typically travelling at cm/s and detonations at km/s in premixed gases. This difference in propagation velocities between flames and detonations makes significant differences in both the physical processes which must be included in models and the approaches adopted in their numerical simulation.

For the high-speed flows in detonations, it is generally sufficient to consider the interactions between fluid convection and chemical energy release. Diffusive transport processes such as thermal conduction, species diffusion, thermal diffusion and viscous effects can be neglected in regions away from walls. However, in the low-speed flows involved in flames, the diffusive transport effects play a major role in the propagation process and must be considered. Because physical diffusion plays an important role in the propagation of flames, it is essential to minimize numerical diffusion and perhaps even advantageous to consider Lagrangian methods. On the other hand, in the numerical simulations of detonations, it is generally necessary to go to lower-order methods near shock fronts in order to prevent spurious undershoots and overshoots. In a reactive flow problem, it is particularly important to minimize spurious oscillations because these oscillations can couple with chemistry and grow rapidly. Furthermore, explicit methods are prohibitively expensive for simulating flames unlike for detonations because the flow is evolving slowly in flames.

This paper presents a brief review of the early efforts in the detailed modelling of flames and then discusses two-dimensional simulations of cellular flames. Then, a similar approach is adopted in discussing detonations. Finally, some of the current trends in computers and computational modelling of fluid dynamics and its impact on the numerical simulation of reactive flows are also briefly discussed.

Laminar Flames in Premixed Gases

Mallard and LeChatelier published results on the propagation velocity of a deflagration wave as early as 1881 [1]. Since then there have been numerous analytical studies of various flame related phenomena and many of these are discussed by Williams [2] and Clavin [3]. With the advent of computers it became possible to study the detailed structure and evolution of flames. One of the first attempts to solve the laminar flame problem numerically was that of Hirschfelder et al. [4]. They formulated the unsteady flame problem as a system of three-dimensional nonlinear partial differential equations and solved the one-dimensional steady flame as a two point boundary value problem. Although the first problem they studied involved only single step kinetics, they later applied the same solution procedure to study flames for which the kinetics involved chain reactions [5,6].

Since the work of Hirschfelder et al.[4], there have been far too many one-dimensional numerical studies of flames to discuss in this short paper. Many of the early efforts in modelling laminar flames has been discussed by Kailasanath et al. [7]. Several of the more recent approaches to modelling these flames are discussed in detail in the proceedings of a workshop on modelling laminar flame propagation [8]. The different approaches adopted include explicit, implicit, method-of-lines, fractional-step, Eulerian and Lagrangian techniques to solve either the steady state or time-dependent problem.

One of the models discussed at the workshop was FLAME1D, which is an implicit, Lagrangian approach to solve unsteady flame problems [7, 8]. A detailed model of a flame must contain accurate representations of the convective, diffusive, and chemical processes. The individual importance of these processes varies from rich to lean flames, and is especially notable near the flammability limits [9] where the exact behavior of these flames depends on a delicate balance among the processes. FLAME1D uses detailed chemical kinetics without any steady state or quasi-equilibrium assumptions. It uses an asymptotic coupling method in conjunction with timestep splitting to couple the various physical and chemical processes. This approach allows the use of entirely different algorithms for the processes represented by the different terms in the conservation equations. Thus it is numerically very efficient and inexpensive. A Lagrangian formulation is used in order to maintain steep gradients for a long period of time. The model has been used for a variety of flame studies such as calculations of minimum ignition energies [7,10], effects of curvature and dilution on flame propagation [11] and flammability limits [9]. More recently, a two-dimensional flame model, FLIC [12] has been developed. This model is described below in some detail and then an application of the model to the study of the effects of gravity on the structure of multidimensional flames is discussed.

FLIC, a Multidimensional Flame Model

In FLAME1D, a Lagrangian method was chosen for convective transport because eliminating the advection term in effect means eliminating numerical diffusion from the calculation. Extending, this Lagrangian approach to multidimensions is extremely difficult and expensive. Therefore an Eulerian approach was chosen for the two dimensional flame model. However, most Eulerian methods are either more numerically diffusive than what is acceptable in a flame model, or they are explicit and hence extremely inefficient at the very low velocities associated with laminar flames. To circumvent these numerical problems, BIC-FCT, the Barely Implicit Correction to Flux-Corrected Transport [13] was developed. BIC-FCT combines an explicit high-order, nonlinear FCT method [14,15] with an implicit correction process. This combination maintains high-order accuracy and yet removes the timestep limit imposed by the speed of sound. By using FCT for the explicit step, BIC-FCT is accurate enough to compute with sharp gradients without overshoots and undershoots. Thus spurious numerical oscillations that would lead to unphysical chemical reactions do not occur. The development of this new algorithm has made it possible to model multidimensional flames in detail.

Thermal conductivity of the individual species is modeled by a polynomial fit in temperature to existing experimental data. Individual conductivities are then averaged using a mixture rule [7,16] to get the thermal conductivity coefficient of the gas mixture. A similar process is used to obtain the mixture viscosity from individual viscosities. Heat and momentum diffusion are then calculated explicitly using these coefficients.

Mass diffusion also plays a major role in determining the properties of laminar flames. Binary mass diffusion coefficients are represented by an exponential fit to experimental data, and the individual species diffusion coefficients are obtained by applying mixture rules [7]. The individual species diffusion velocities are solved for explicitly by applying Fick's law followed by a correction procedure to ensure zero net flux [16]. This procedure is equivalent to using the iterative algorithm DFLUX [17] to second order. This method is substantially faster than one that uses matrix inversions and is well suited for a vector computer. This algorithm is also explicit, but because the effective Lewis number of the mixture is close to one, the timestep suitable for heat conduction is adequate for mass diffusion as well.

Chemistry of the hydrogen–oxygen flame is modeled by a set of 24 reversible reaction rates describing the interaction of eight species, H_2, O_2, H, O, OH, HO_2, H_2O_2, H_2O, and N_2 is considered a non-reacting diluent [18]. This reaction set is solved at each timestep with a vectorized version of CHEMEQ, an integrator for stiff ordinary differential equations [19]. Because of the complexity of the reaction scheme and the large number of cells in a two-dimensional calculation, the solution of the chemical rate equations takes a large fraction of the total computational time. A special version of CHEMEQ called TBA was developed to exploit the special hardware features of the CRAY X-MP vector computer.

All of the chemical and physical processes are solved sequentially and then are coupled asymptotically by timestep splitting [20]. This modular approach greatly simplifies the model and makes it easier to test and change the model. Individual modules were tested against known analytic and other previously verified numerical solutions. One-dimensional predictions of the complete model were compared to those from the Lagrangian model FLAME1D [7].

Effects of Gravity On Multidimensional Flame Structure

It is well known from flammability studies in normal (earth) gravity that a flame propagating upward in a tube propagates in a wider range of mixture stoichiometries and dilutions than a flame propagating downward. This suggests that the flammability limits depend on gravity. Instabilities are also often observed in the propagation of flames near the flammability limits. Upward-propagating flames in a tube are usually curved so that they are convex toward the cold gases, whereas downward-propagating flames can be nearly planar [21]. The characteristic convex shape may be a result of the Rayleigh-Taylor instability, which results when a heavier fluid is on top of a lighter fluid. In an upward-propagating flame, the light, hot burned material is on the bottom, and the dense, cold unburned material is on the top. If only the Rayleigh-Taylor instability were present, a flame would be one-dimensional in a zero-gravity environment. However, flames in many mixtures in reduced-gravity exhibit multidimensional cellular structures.

Experimental studies of cellular flames have provided a qualitative understanding of cellular flames in earth gravity [21-23]. These experimental observations, the theoretical analysis of Zeldovich [24], Barenblatt et al. [25], Sivashinsky [26,27], and recent numerical simulations [28] all suggest that the cellular structure is due to a thermo-diffusive instability, which occurs because of the differences in the rate of heat and mass diffusion. On earth, the physical mechanisms causing the thermo-diffusive instability and the Rayleigh-Taylor instability can be important simultaneously. Numerical simulations using the model described above have been used to study the effects of gravity on flame propagation in a mixture which exhibits cellular structure in a zero-gravity environment [29, 30].

Initial conditions for the two-dimensional calculations were taken from one-dimensional calculations that give the conditions for steady, propagating flames. Fresh unburnt gas flows in from one side and the products of chemical reaction at the flame front flow out at the other side. If the inflow velocity is set equal to the burning velocity of the planar flame, the flame zone is stationary in space and a steady solution is obtained. Thus, the transient effects arising from the ignition process can be eliminated and the one-dimensional solution provides a relaxed initial condition for the two-dimensional calculation. The two-dimensional computational domain was 2.0 cm × 4.5 cm, which was resolved by a 56 × 96 variably spaced grid. Fine zones, 0.36 mm × 0.15 mm, were clustered around the flame front. The planar flame is then perturbed and the subsequent evolution of the flame front is studied. Similar calculations are also performed with the buoyancy forces in the direction of flame propagation and against the direction of flame propagation.

Flames in a $H_2:O_2:N_2$ / 1:1:10 Mixture

In Fig. 1, the multidimensional flame is depicted at a series of times using OH concentration contours. The evolution of upward, zero-gravity and downward propagating flames is compared in the figure. The differences among the three cases are dramatic. The zero-gravity case shows a cellular structure. At 20ms, the shape and size of the three flames are comparable. The zero-gravity flame exhibits a stable cellular structure with little change from 20 to 100 ms. The upward-propagating flame becomes more and more curved and the central portion of the flame moves more rapidly than the sides. In this case, the buoyancy-induced and the thermo-diffusive mechanisms are both destabilizing. At 100 ms, the upward-propagating flame exhibits a bubble-like appearance characteristic of near-limit upward-propagating flames. In the downward-propagating case, buoyancy forces tend to stabilize and return a perturbed flame back to its initial planar configuration. However, at later times, the flame front begins to oscillate around its initial planar configuration. The fact that early in the calculations, at 20 ms, all three flames look alike suggests that the instability leading to cellular structure is growing more rapidly then the buoyancy-induced instability. By 100 ms, the buoyancy-induced instability has had enough time to strongly interact with the cellular structure and in the case of the downward-propagating flame essentially nullify the effects of the thermo-diffusive instability.

A similar comparison as in Fig. 1 has been made for flames propagating in a $H_2:O_2:N_2$ / 1.5:1:10 mixture. In that case, the differences between the three cases were small. From such studies it can be concluded that the effect of gravity on flame propagation depends strongly on the stoichiometry of the mixture. Currently, the effect of heat and radiation losses on the upward and downward propagation of flames is being studied.

Multidimensional Detonations

Some of the earliest scientific studies of detonations were reported by Berthelet and Vielle [31] and Mallard and LeChatelier [32] around 1881. Since then the major structural features of self-sustained gaseous detonations have been studied extensively both experimentally (for example Ref. 33-36) and numerically [37-41]. It is now understood that the detonation front is not one-dimensional even though the time-averaged detonation velocity may be close to the Chapman-Jouguet detonation velocity. Detonation propagation is a complex, multidimensional process involving interactions between incident shocks, Mach stems, transverse waves and boundaries of the regions through which the detonation is moving. Experiments have also shown that a propagating detonation leaves a very regular, cell-like pattern on the sidewalls of the confining chamber. These patterns are etched by the triple-point formed at the front of the detonation by the intersection of the transverse wave with the incident shock and the Mach stem. Thus the cell patterns are histories of the location of the triple point. The size and regularity of this cell structure is characteristic of the particular combination of initial material conditions, such as composition, density and pressure and the geometry of the confining chamber. In most mixtures, the cell structure is irregular and even in mixtures which show a regular structure, many detailed sub-structures are also evident in experimental observations.

Numerical simulations of propagating two-dimensional detonations have shown that detonations cells can be formed by perturbing a planar one-dimensional detonation [37,38,40,42]. For argon diluted mixtures, the calculations have also shown that the final structure is independent of mechanism by which the initial one-dimensional detonation is perturbed, though in some cases the evolution to the final state may take propagation through many transverse-wave cycles [40]. Simulations have also been able to predict irregular structures in mixtures which are expected to show such structures based on experimental observations [42]. Detailed comparisons with experimental observations are not possible when irregular structures are observed because the source of irregularity may be quite different in the two cases. Some of the simulations have also shown some detailed structures such as the formation of unburned pockets behind the detonation front [39,40] as well as some sub-structures [42,43] within the detonation cell. Below a numerical model used to simulate detonations is described and then a simulation using the model is discussed in some detail.

Numerical Model to Simulate Detonations

The model used to simulate multidimensional detonations is different in many aspects from that used to simulate flames because of the inherent differences in the processes being modelled. For the high speed flows involved in detonations, molecular transport processes such as thermal conduction and molecular diffusion can be neglected. Viscous effects are also not important in idealized situations where the effects of confining walls can be neglected. The two processes which must be modelled are the fluid convection and chemistry. As before the fluid dynamics and chemistry terms in the conservation equations can be solved using different algorithms and coupled using timestep splitting [20]. In the model used for the simulations described here, the algorithm used for convective transport is the explicit Flux-Corrected Transport (FCT) algorithm [14,15] mentioned earlier.

The chemistry is modelled using a two-step parametric model [44,45]. The first step is an induction step in which no energy release occurs. The second step models the energy release process and starts only after the elapse of the induction time. An induction reaction progress variable is then defined which is a measure of how long the material has remained at a given temperature and pressure. In the calculations, this quantity is convected with the fluid and is used to indicate when the available chemical energy should be released. For this parametric model, three quantities must be specified: the time before any energy is released (the chemical induction time), the time it takes to release the energy, and the total amount of energy released. Induction times as a function of temperature and pressure can be obtained either from experiments or from detailed chemical kinetic calculations. We have already synthesized this information for hydrogen-air mixtures from our earlier calculations of reactive flows. The energy release times and the amount of energy release can be varied to gain a qualitative understanding of the effects of these parameters on the propagation and structure of detonations. These parameters can also be calibrated using information from detailed chemistry solutions.

Simulations of Two-dimensional Detonations

The calculations are initiated by placing a large pocket of unburned material behind a planar propagating detonation, i.e., one without any initial transverse structure. As the pocket burns, it provides the perturbation necessary to allow the planar detonation to go unstable and form a multidimensional structure. Previous simulations [40] have suggested that this structure is unique because changing the size and orientation of the initial perturbing pocket did not affect the final state. The resulting detonation is also an equilibrium configuration detonation [46] because its structure repeats itself at equally spaced intervals as the detonation propagates down the channel. The simulations have also indicated that the curvature of the transverse waves play an important role in determining the cell size as well as in the formation of unburned gas pockets behind the detonation front.

The early stages in the formation of a pair of transverse waves are shown in Fig. 2. This figure is a composite of eight "snapshots" of the pressure contours near the detonation front at intervals of 10 microseconds. For this calculation, an unburnt gas pocket was placed symmetrically behind a planar detonation in a 5 cm wide channel. Pressure waves are generated when this pocket burns. These waves interact with the incident shock front causing the front to curve outwards. Furthermore the pressure waves also reflect from the side walls of the channel and move transverse to the incident shock front, as seen in frame 2 of Fig. 2. These pressure waves are strengthened due to collisions with each other and further increase the curvature of the incident shock front, as can be seen in frame 7. After a short time a portion of the incident shock reflects from the side walls of the channel. Frame 8 of Fig.2 shows that a Mach reflection takes place since a pair of triple points are seen. The refected shock waves, which are the transverse waves, are weak initially but are later strengthed due to collisions with each other and the walls.

The further evolution of a pair of triple points is shown in Fig. 3. This figure is a composite of eight pressure contours at intervals of 10 microseconds. The path and the direction of movement of the triple points are indicated by the lines with arrows. In the first frame the transverse waves are moving towards each other and away from the wall. By the fourth frame they have collided with each other and are moving away from the center of the channel. In frame 7 they are again moving towards the center after colliding with the walls. Frames 7 and 8 are similar to frames 1 and 2, respectively, showing that an equillibrium configuration detonation has been established.

Figure 3 shows that the triple point structure does not appear to immediately bounce off when it hits the wall, indicating that a complete detonation cell has not been formed. The pattern of the triple points for this calculation is clearer in Fig. 4A, which shows a time and space gap at the walls as the structure reforms. The formation of an incomplete cell is discussed in detail in Ref. [40], where it is shown that the curvature of the transverse wave at the time of reflection is very important in determining the cell structure. Increasing the width of the channel to 7 cm, as shown in Fig. 4B, results in a considerably reduced gap in the path of the triple points near the walls. Finally, the locus of the triple points for a 9 cm wide channel, shown in Fig. 4C, forms a complete detonation cell and what appears to be partial structures above and below it. From the figure, the cell height and length are estimated to be about 8.5 cm and 19.6 cm, respectively.

Currently it is possible to do three-dimensional detonation calculations with simplified chemistry models or two-dimensional simulations with complex chemistry. Simulations have also been done of detonations in liquid nitromethane [42] and the propagation of detonations from a tube into an unconfined environment [47].

New Directions

The simulations discussed above are computer intensive even though they are in simple geometries and ignore many complicating factors such as the effects of walls and radiations. There are many advances being made currently that may alleviate some of the difficulties in doing more practical problems. The advances can be broadly classified into three categories: new numerical algorithms, new computer architectures, and more complex phenomenological models. The discussion below is based on a recent paper by Oran and Boris [48].

New Algorithms

New algorithmic developments focus on sophisticated adaptive gridding to maintain high-resolution at gradients, finite-element methods with unstructured grids, very fast near-neighbors and constraint algorithms for particle dynamics, more sophisticated methods for coupling physical processes with disparate time scales, and software especially optimized for the new computer architectures. Two relatively new numerical developments which could have a great impact in the future of reactive flow simulations are finite-element formulations and molecular dynamics.

Finite-element techniques have been used for years to solve complicated problems in structural mechanics. It is relatively recent that finite-element methods have been used to solve problems in fluid dynamics. The finite-element Flux-Corrected Transport (FEM-FCT) combines the advantages of the unstructured grid approach of a finite-element method with the accuracy of a nonlinear, monotone finite-difference fluid dynamic algorithm [49]. A major disadvantage of finite-element methods is that they are expensive in both time and memory. However, for problems involving reactive flows in complex geometries, finite-element based methods may well prove to be cost effective. Examples of reactive flow calculations using finite-element methods can be found in the report by Habbal et al. [50] and other examples involving complex shock interactions has been reported by Baum et al. [51].

A major part of a particle dynamics or Lagrangian fluid dynamics calculation is the continual reevaluation of which nodes are near neighbors in a large set of nodes that often seem to be moving randomly. The expense arises because each node can potentially interact with any of N-1 other nodes in the system. This is called the N^2 problem because only a few of the N^2 interactions possible between pairs of nodes are potentially important. The Monotonic Lagrangian Grid (MLG) which scales as $N \log N$ is a new method designed to beat the N^2 problem [52, 53]. The MLG is really a data structure for storing the positions and other data needed to describe N moving nodes. These N nodes could, for example, represent fluid elements, droplets or molecules that must be tracked in a multidimensional calculation. Preliminary calculations using MLG in a molecular dynamics framework for simulating condensed phase detonations have been reported [54].

Developments in Computers

Highly parallel and hybrid computer systems are becoming increasingly common as an alternate or supplement to the supercomputers. Examples of highly parallel computers are the TMC Connection Machine, the BBN Butterfly Machine and the various hyercube machines such as the NCUBE system. These machines have operation counts of 50-2500 megaflops and 16-256 processors. Hybrid computer systems are usually a combination of conventional computers (such as Vaxes) and some parallel processors. An example of such a system is the Graphical Array Processing System (GAPS) at the U.S. Naval Research Laboratory.

The GAPS is an asynchronous, multi-tasking, high-performance. parallel processing system consisting of an APTEC 2400 Computer connected to a VAX 11/780 with 12 megabytes of additional fast memory and about 2 gigabytes of online disc storage. It currently includes six Numerix MARS 432 array processors which have a maximum performance of 30 megaflops each. Currently, a Reactive Flow Model (RFM) has been implemented on the GAPS and results from these simulations can be displayed as they are calculated. This provides a means of doing numerical experiments which are analogous to laboratory experiments. However, much faster processors are needed to be really able to simulate flow events on the physical time scale.

More Complex Problems

Even with faster and newer computer architectures and algorithms, the level of complexity of the problems that can be solved are very low. Three-dimensional simulations are increasingly becoming possible but with these larger computations more and more of the processes involved must be phenomenalogically modelled. For reactive flow, chemistry is usually one of the most time consuming processes to be accurately resolved. The calculations described earlier on multi-dimensional flames are one of the few examples where a detailed chemical kinectis mechanism is used in a time-dependent two-dimensional calculation. The simplified two-step model used in the detonation calculations described above is an example where useful results can be obtained using a carefully calibrated model. There are many current efforts (for example,[55]) to derive systematically reduced reaction mechanisms from complex kinetics. These methods must still be validated in complex multidimensional fluid problems. Physical and chemical processes involving droplets, sprays, and soot, to name just a few, are usually modelled phenomenologically. When using such models, it is important to remember that the resulting physical description is limited by the range of validity of the phenomenologies.

Concluding Remarks

The simulations discussed here are just examples of what has been done in the detailed modelling of reactive flows. They also show how numerical simulations can be used to gain a better insight into problems in combustion. For example, the flame simulations show that the effects of gravity are greater in mixtures with low burning velocities which are generally closer to the flammability limits. The simulations also indicate that the instability leading to cellular flames grows more rapidly than the buoyancy-induced Rayleigh-Taylor instability. The detonation simulations show that the interactions between chemistry and complex-shock structures can be resolved numerically. Such calculations have also been used to understand the mechanisms leading to self-sustained detonations and the tracing of cell-like patterns. With the improvements that are being made in computer hardware and software, more and more complex problems with fewer phenomenologies will be simulated in the future.

Acknowledgements

This work was sponsored by NASA in the Microgravity Science and Applications Program and by the Naval Research Laboratory. The authors would also like to acknowledge numerous discussions and research colloborations with Drs Gopal Patnaik, Theodore Young, and Raafat Guirguis on topics discussed in this paper.

References

1. Mallard, E., and LeChatelier, H.L., *Ann. Mines* 4, 379 (1883).
2. Williams, F.A., *Combustion Theory* , Addison Wesley, Reading, MA, 1965.
3. Clavin, P., *Prog. Energy Combust. Sci.*, 11, 1 (1985).
4. Hirschfelder, J.O., Curtiss, C.F., Henkel, M.J., Spaulding, W.P., and Hummel, H., *Third Symposium on Combustion and Flame and Explosion Phenomena* , The Williams and Wilkins Co., Baltimore, MD, 121-139 (1949).
5. Hirschfelder, J.O., Curtiss, C.F., and Campbell, D.E., *Fourth Symposium (International) on Combustion* , The Williams and Wilkins Co., Baltimore, MD, 190-211 (1953).
6. Gidding, J.C., and Hirschfelder, J.O., *Sixth Symposium (International) on Combustion* , Reinhold Publishing Co., NY, 199-213 (1957).
7. Kailasanath, K., Oran, E.S., and Boris,J.P. *A One-Dimensional Time-Dependent Model for Flame Initiation, Propagation and Quenching*, Naval Research Laboratory Memorandum Report 4910, 1982.
8. Peters, N., and Warnatz, J., Eds., *Numerical Methods in Laminar Flame Propagation,* Friedr. Vieweg & Sohn, Wiesbaden, W. Germany, 1982.
9. Kailasanath, K., and Oran, E.S., In *Complex Chemical Reaction Systems*, Eds. J. Warnatz and W. Jager, pp. 243-252, Springer-Verlag, Heidelberg, West Germany, 1987.
10. Kailasanath, K., Oran, E.S., and Boris, J.P., *Combust. Flame,* 47, 173, 1982.
11. Kailasanath, K., and Oran, E.S., *Prog. Aero. and Astro.* 105, Part 1, 167-179, 1986.
12. Patnaik, G., Laskey, K.J., Kailasanath, K., Oran, E.S., and Brun, T.A., *FLIC-A Detailed Two-Dimensional Flame Model,* to be published as Naval Research Laboratory Memorandum Report , 1989.
13. G. Patnaik, R.H. Guirguis, J.P. Boris, and E.S. Oran, *J. Comput. Phys.* **71**, 1 (1987).
14. J.P. Boris and D.L. Book, *Meth. Comput. Phys.* **16**, 85 (1976).
15. J.P. Boris, *Flux-Corrected Transport Modules for Solving Generalized Continuity Equations*, Naval Research Laboratory Memorandum Report 3237, 1976.
16. R.J. Kee, G. Dixon-Lewis, J. Warnatz, M.E. Coltrin, and J.A. Miller, *A Fortran Computer Code Package for the Evaluation of Gas-Phase Multi-Component Transport Properties*, Sandia National Laboratories Report SAND86-8246, 1986.
17. W.W. Jones, and J.P. Boris, *Comp. Chem.* **5**, 139 (1981).
18. T.L. Burks and E.S. Oran, *A Computational Study of the Chemical Kinetics of Hydrogen Combustion* , Naval Research Laboratory Memorandum Report 4446, 1981.

19. T.R. Young and J.P. Boris, *J. Phys. Chem.* **81**, 2424 (1977).

20. E.S. Oran and J.P. Boris, *Numerical Simulation of Reactive Flow*, Elsevier, New York, 1987.

21. Bregon, B., Gordon, A. S. and Williams, F.A., *Combust. Flame.* 33: 33-45 (1978).

22. Markstein, G.H., Non Steady Flame Propagation, Macmillan, New York, 1964.

23. Mitani, T., and Williams F.A., *Combust. Flame.* 39: 169–190 (1980).

24. Zeldovich, Y. B., The Theory of Combustion and Detonation, Academy of Sciences, USSR, 1944. Also, Zeldovich, Y.B., and Drosdov, N.P., *J. Experi. Theoret. Physics.* 17: 134 (1943).

25. Barenblatt, G. I., Zeldovich, Y. B. and Istratov, A. G., *Zh. Prikl. Mekh. Tekh. Fiz.* 4: 21-26 (1962).

26. Sivashinsky, G.I., *Combust. Sci. Tech.* 15: 137–145 (1977).

27. Sivashinsky, G.I., *Ann. Rev. Fluid Mech.* 15: 179–199 (1983).

28. Patnaik, G., Kailasanath, K., Laskey, K.J., and Oran, E.S., Detailed Numerical Simulations of Cellular Flames, *to be published in the Proceedings of the 22th Symposium (International) on Combustion*, Seattle, WA., August, 1988.

29. Kailasanath, K., Patnaik, G.P., and Oran, E.S., Effect of Gravity on Multi-Dimensional Laminar Premixed Flame Structure, IAF Paper No. 88-354, 39th IAF Congress, Bangalore, India, International Astronautical Federation, Paris, France, 1988.

30. Patnaik, G.P., Kailasanath, K., and Oran, E.S., Effect of Gravity on Flame Instabilities in Premixed Gases, AIAA Paper No. 89-0502, AIAA 27th Aerospace Sciences Meeting, Reno, NV, AIAA, Washington, D.C., 1989.

31. Berthelot, M., and Vieille, P., *Compt. Rend. Acad. Sci., Paris* 93, 18 (1881).

32. Mallard, E., and LeChatelier, H.L., *Compt. Rend. Acad. Sci., Paris* 93, 145 (1881).

33. Strehlow, R.A., Fundamentals of Combustion, Krieger Pub. Co., New York, 1979, Chapt. 9.

34. Fickett, W. and Davis, W.C. Detonation, University of California Press, Berkeley 1979, Chapt. 7.

35. Oppenheim, A.K. Astro. Acta 11:391 (1965).

36. Strehlow, R.A. Combust. Flame 12:81 (1968).

37. Taki, S., and Fujiwara, T., AIAA J. 16:73-77 (1978).

38. Taki, S., and Fujiwara, T., Eighteenth Symposium (International) on Combustion, The Combustion Institute, 1981, pp. 1671-1681.

39. Oran, E.S., Young, T.R., Boris, J.P., Picone, J.M., and Edwards, D.H., Nineteenth Symposium (International) on Combustion, The Combustion Institute, 1982, pp. 573-582.

40. Kailasanath, K., Oran, E.S., Boris, J.P. and Young, T.R., Combust. Flame 61:199-209 (1985).

41. Kailasanath, K., Oran, E.S., Boris, J.P., and Young, T.R., A Computational Method for Determining Detonation Cell Size, AIAA Paper No. 85-0236, AIAA 23rd Aerospace Sciences Meeting, Reno, NV, AIAA, Washington, D.C., 1985.

42. Guirguis, R., Oran, E.S., and Kailasanath, K., Combust. Flame 65:339-365 (1986).

43. Guirguis, R., Oran, E.S., and Kailasanath, K., Twenty-first Symposium (International) on Combustion, The Combustion Institute, 1988, pp. 1659-1668.

44. Oran, E.S., Boris, J.P., Young, T., Flanigan, M., Burks, T., and Picone, M., Eighteenth Symposium (International) on Combustion, The Combustion Institute, Pittsburgh, 1981, pp. 1641-1649.

45. Kailasanath, K., Gardner, J.H., Oran, E.S., and Boris, J.P., Numerical Simulations of Unsteady Reactive Flows in a Combustion Chamber, to be published in *Combust. Flame*, 1989.

46. Strehlow, R.A., *Astro. Acta* 15:345 (1970).

47. Hiramatsu, K., Fujiwara, T., and Taki, S., Proceedings of the Fourteenth International Symposium on Space Technology and Science, 549 (1984)

48. Oran, E.S., and Boris, J.P., *Computers & Structures* 30:69-77 (1988).

49. Lohner, R., Morgan, K., Peraire, J., and Vahdati, M., *Comm.Appl.Num Meth.* 4:717-730 (1988).

50. Habbal, A., Dervieux, A., Guillard, H., and Larrouturou, B., Explicit Calculation of Reactive Flows with an Upwind Finite Element Hydrodynamical Code, INRIA Report No. 690, INRIA-Sophia Antipolis, France, June 1987.

51. Baum, J.D., and Lohner, R., Numerical Simulation of Shock-Elevated Box Interaction Using an Adaptive Finite-Element Shock Capturing Scheme, AIAA Paper No. 89-0653, AIAA 27th Aerospace Sciences Meeting, Reno, NV, AIAA, Washington, D.C., 1989.

52. Boris, J.P., *J. Comput. Phys.* 66:1-20 (1986).

53. Lambrakos, S., and Boris, J.P., *J. Comput. Phys.* 73:2546 (1987).

54. Lambrakos, S.G., Peyrard, M., Oran, E.S., and Boris, J.P., *Physical Review B* ,39:993-1005 (1989).

55. Paczko, G., Lefdal, P.M., and Peters, N., Twenty-first Symposium (International) on Combustion, The Combustion Institute, 1988, pp. 739-748.

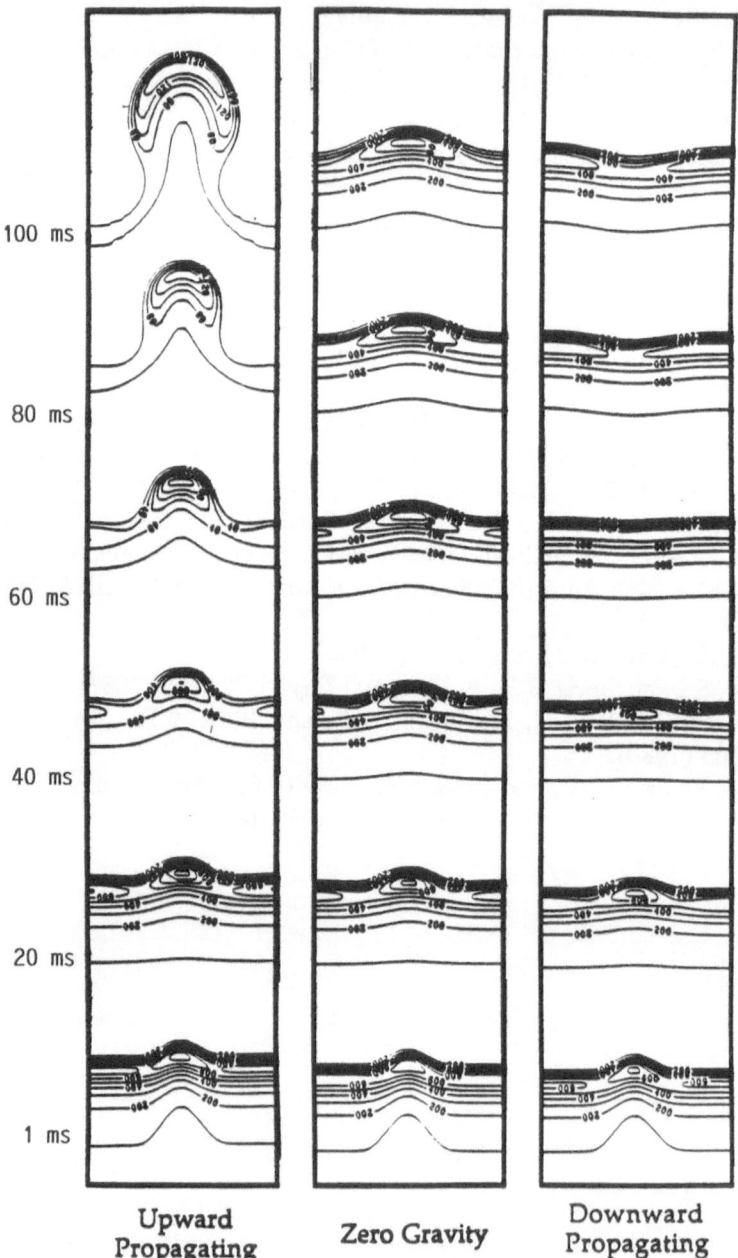

100 ms

80 ms

60 ms

40 ms

20 ms

1 ms

Upward
Propagating

Zero Gravity

Downward
Propagating

Figure 1. A comparison of the evolution of upward, downward and zero-gravity flames in a
$H_2:O_2:N_2/1:1:10$ mixture.

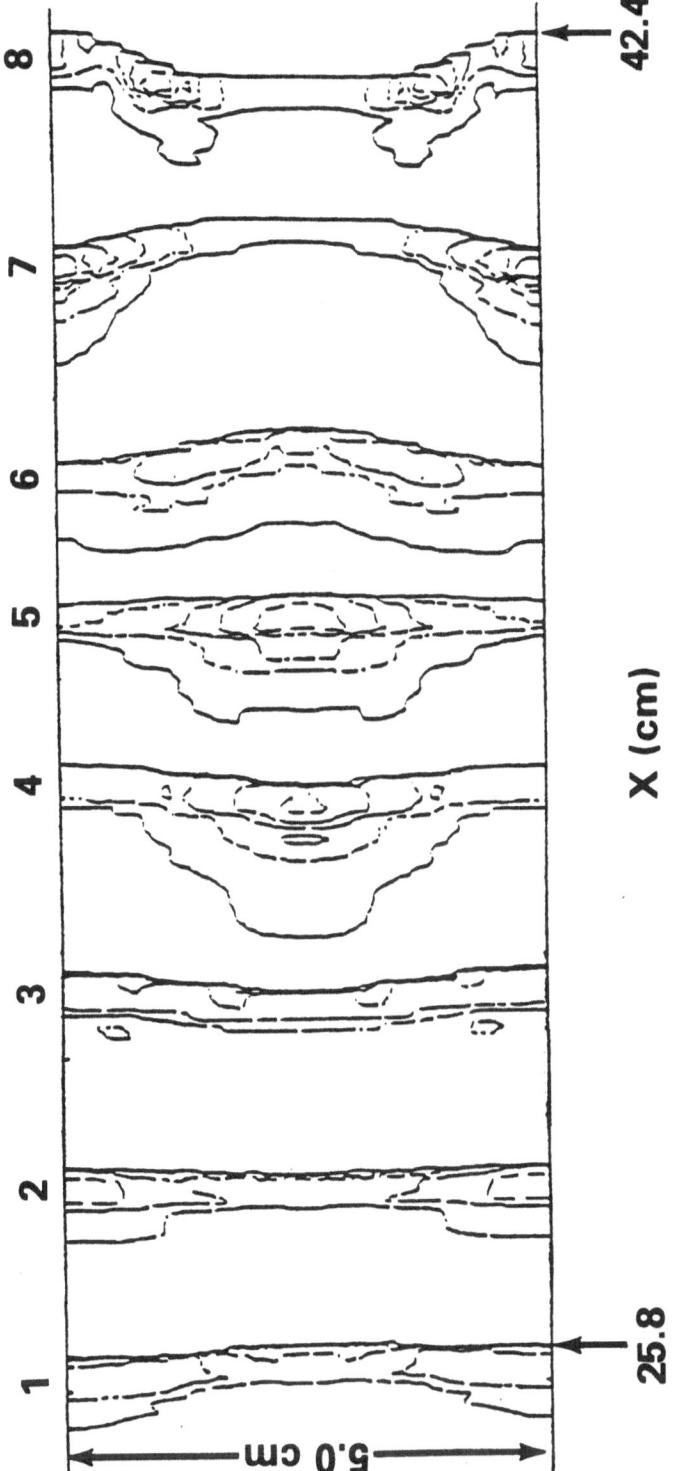

Figure 2. Composite of pressure contours showing the early stages in the formation of a pair of triple points.

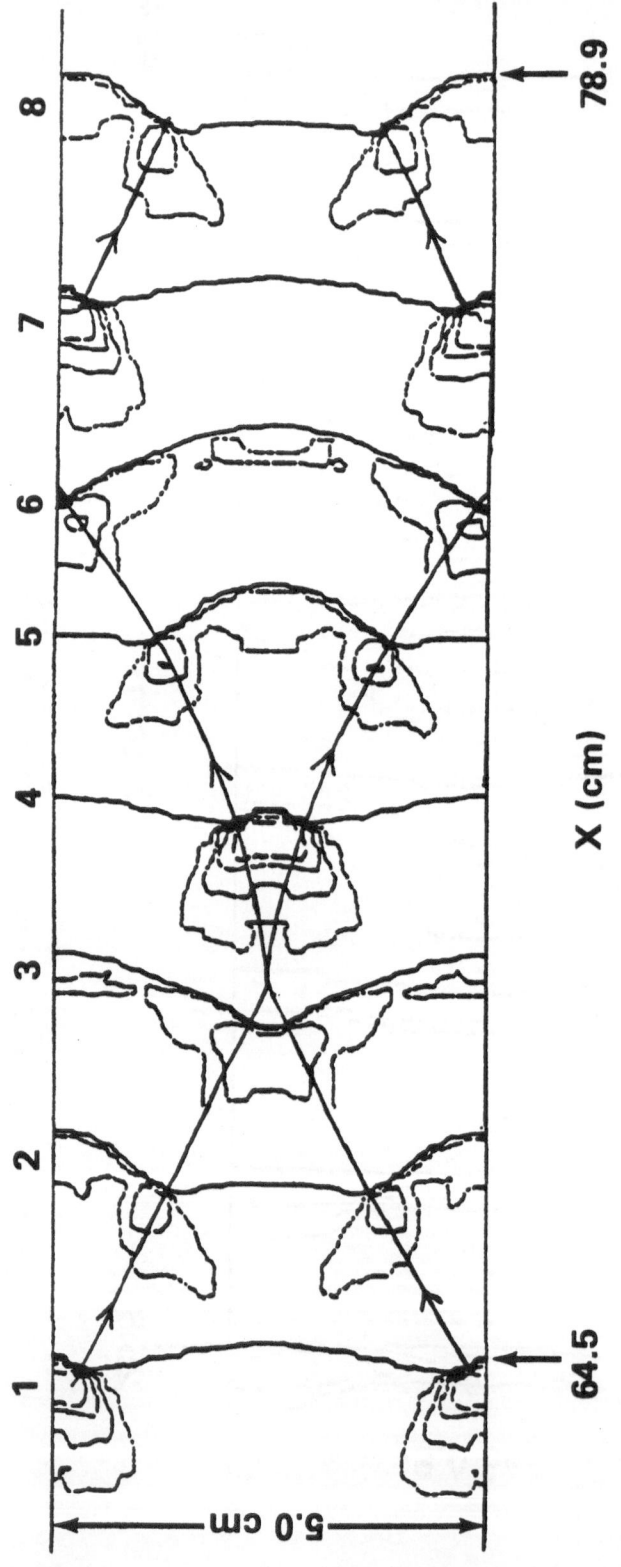

Figure 3. Composite of pressure contours from eight timesteps. The direction of movement of the triple points is indicate by the lines with arrows.

97

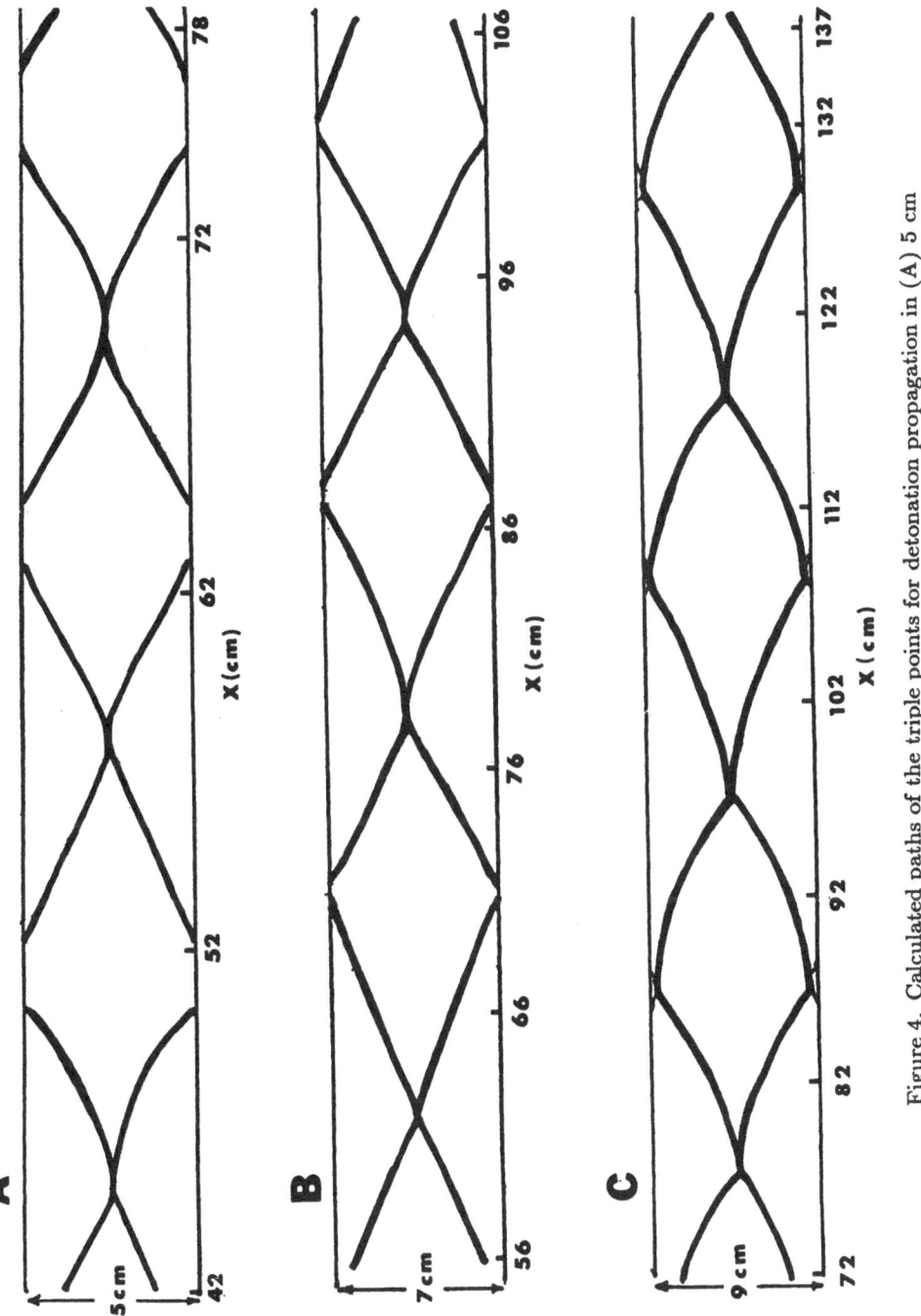

Figure 4. Calculated paths of the triple points for detonation propagation in (A) 5 cm wide channel, (B) 7 cm wide channel and (C) 9 cm wide channel.

RECENT DEVELOPMENTS IN THE COHERENT FLAMELET DESCRIPTION OF TURBULENT COMBUSTION

Eric Maistret*, Nasser Darabiha, Thierry Poinsot, Denis Veynante, François Lacas, Sébastien Candel and Emile Esposito

Laboratoire E.M2.C., CNRS, Ecole Centrale Paris, 92295 Châtenay Malabry Cedex, FRANCE
* and SNECMA - Centre de VILLAROCHE - 77550 Moissy Cramayel, FRANCE

Turbulent combustion may be modeled in many different ways. The classical approach is based on statistical techniques and the determination of the mean reaction rates usually involves a probability density function. Alternative descriptions rely on the flamelet concept. The reactive flow field is viewed as a collection of flame elements. The structure of these flamelets may be identified and analysed separately. An important advantage of this concept is that it essentially decouples the complex chemistry problems from the modeling of the turbulent flow field. There is already a great variety of flamelet models. In certain cases a probability density function is used to couple the local flamelet analysis to the flow description. Other models use the flamelet passage frequency to evaluate the mean reaction terms. Another approach is based on a balance equation for the flame area. This equation describes the transport of the flame surface by the turbulent flow field and the physical mechanisms which produce and destroy the reactive surface. This approach originally proposed by Marble and Broadwell has been the subject of numerous developments and extensions in our laboratory. The coherent flamelet description which has evolved from these studies is briefly described in this paper. Illustrative calculations are provided in the case of a turbulent flame stabilized in a duct.

In fact, the turbulent ducted flame constitutes one of the simplest geometry that may be studied and it provides a suitable test for models. Our recent work, described in this paper, tries to complement the available experimental information and refine the numerical models used to describe such flames. The modeling effort is guided by systematic observations providing spatial distributions of the local heat release source terms. These distributions are markedly influenced by the equivalence ratio and by the flow velocity at the combustor inlet indicating that the interaction between the chemical kinetics and the flow turbulence is important. These new data allow detailed comparisons between numerical predictions and experiments. It is found that classical combustion descriptions such as the Eddy Break-Up model are unable to reproduce the trends observed. The problem is in fact quite challenging and it has been overlooked in the past because little experimental information was available on the mean spatial distribution of heat release in turbulent flames. Predictions obtained with the Coherent Flamelet Model are already in good agreement with measurements, but further improvements are required.

This work is supported by SNECMA and DRET

1. INTRODUCTION

The problem of turbulent combustion has attracted considerable attention in recent years and some progress has been accomplished in this field. A variety of models is now available and successful calculations have been conducted in simple and more complex configurations (see Libby and Williams 1980 and Williams 1985 for basic presentations, and Correa and Shyy 1987 for a recent survey of computational models of continuous turbulent combustion). Standard models have mostly relied on statistical techniques and the determination of the local average reaction rates has usually involved a probability density function. This probability density function is often specified empirically or it may be obtained as a solution of an evolution equation. The latter approach is less practical especially if complex chemistry is to be considered.

Alternative descriptions of turbulent combustion have been founded on the flamelet concept. The reactive flow field is viewed as a collection of flame elements embedded in the turbulent stream (Carrier et al. 1975, Williams 1975, Marble and Broadwell 1977, 1979, Peters 1984, 1986, Spalding 1978, Bray et al. 1984, Clavin and Williams 1982). The structure of these flamelets may be identified and analysed separately. An important advantage of the flamelet concept is that it essentially decouples the complex chemistry problems from the modeling of the turbulent flow field. In practical applications a flamelet library is first constructed and provides specific properties such as the consumption rates per unit flame area, ignition and extinction conditions that are required in the computation of the turbulent flow field.

While the domain of application of flamelet models is not well defined at this time and remains a subject of discussion it is believed that the concept is applicable in the range of large Damköhler numbers and for characteristic turbulent scales much larger than a typical flame thickness. The two conditions are satisfied in many practical situations and flamelet descriptions are relevant in many cases and at least in certain ranges of operation of internal combustion engines and continuous flow combustors. Because this debate cannot be settled at the moment it is left aside and we will simply judge the model from its predictions.

Now, there are diverse approaches to the flamelet description. However they all share the following ingredients :

(1)   A laminar flamelet submodel (or submodels) providing the local structure and properties of the reactive elements.

(2)   A description of the turbulent flow comprising mass average equations describing the mean flow variables and the main species mass fractions and relevant closure equations.

(2)   A rule or a set of rules which couple the flamelet submodels to the turbulent flow description.

(4)   Additional submodels accounting for chemical reactions taking place outside the flamelets.

There is considerable flexibility in specifying and combining these elements and many alternative ways are being explored (see the reviews of Peters 1984, 1986, and Bray 1987).

In certain cases a probability density function is used to couple the local flamelet analysis to the flow description. Other models rely on the flamelet passage frequency to evaluate the mean reaction terms (see Bray 1987 for a survey of this method). Another approach is based on a balance equation for the flame area.

This equation describes the transport of the flame surface by the turbulent flow field and the physical mechanisms which produce and destroy the reactive surface. Our own work is based on this type of description. It extends ideas developed by Marble and his co-workers in their analysis of non-premixed combustion (Carrier, Fendell and Marble 1975, Marble and Broadwell 1977, 1979). It is suggested in these studies that turbulent combustion of unmixed reactants is controlled in the early stages by a competition between straining of the flame elements and mutual annihilation of the flame area due to interactive destruction of neighbouring flame sheets.

The Coherent Flame Model identifies important physical mechanisms of turbulent combustion such as the production of flame area by stretching, its destruction by flame shortening and the central influence of the strain rate acting on the local flamelets. Because these aspects are important in the large scale coherent motions found in shear flows, the model accounts in some sense for the presence of these organized fluctuations.

Our own effort has been to extend the coherent flame description to premixed flow configurations (Candel et al. 1982, Darabiha 1985, Darabiha et al. 1987,) and to explore its potential in non-premixed situations (Veynante et al. 1987, Lacas et al. 1987). A new description which accounts for premixed and non-premixed flamelets was recently presented (Veynante et al. 1989a). The case of propagating turbulent fronts is treated in separate papers (Lacas et al. 1989, Veynante et al. 1989b). Non-uniformly premixed flames are considered by Veynante et al. (1989c). Our progress in developing a model for premixed flames is reported in the present paper. While the basic formulation is still that described in Darabiha et al. (1987), new ideas have been formulated during the last years and are currently being tested. Comparisons between calculations and experiments have now been conducted systematically and will be discussed in the last section of this article. We will specifically consider the turbulent ducted flame configuration. This case constitutes one of the simplest geometry that may be studied and that has both fundamental and practical interest.

Early investigations of this situation were conducted some thirty years ago by Zukoski and Marble (1955), Wright (1959) and provided considerable information on the complex process of flame stabilization. A qualitative understanding of the problem was also derived from a wide set of careful experiments. Simple scaling laws were obtained which effectively described blowoff conditions. These results provided basic tools in the design of jet engine afterburners. Another landmark investigation carried during the same period was due to Marble and Adamson (1954). This well known article treated the problem of flame ignition and stabilization in a laminar mixing region. It was shown how the heat released by the chemical reaction could produce a temperature maximum in the mixing layer leading to ignition and flame spreading. The turbulent ducted flame stabilized on a bluff body still remains a fundamental configuration and its modeling is a subject of continued interest and technological significance. Much of the current work in this field has been directed at the numerical modeling of turbulent flames (see Correa and Shyy 1987 for a recent survey). Modern optical diagnostics and digital signal processing methods have also allowed detailed experiments which could not be carried out in the early fifties.

Our own work described in this paper and in some recent articles (Darabiha et al. 1987) tries to complement the information on turbulent ducted flames and develop numerical descriptions of this basic configuration. In distinction with most of the recent experiments (see Libby et al. 1986 for a review of premixed turbulent flame experiments) which concentrate on measurements of temperature, velocity or concentration profiles, our effort has been to obtain information on the mean source terms and especially on the mean heat release distribution in the flow. There are important reasons which justify this choice. First, it is shown that the spatial distributions of the mean heat release are strongly influenced by the equivalence ratio and the flow velocity thus providing a unique view of the interaction between the chemical kinetics and the flow turbulence. Second the data also allow a direct evaluation of the theoretical expressions used to model the mean consumption rates appearing in the averaged balance equations.

The detailed examination of the mean source term differs from the more common tests performed in the literature on velocity, temperature and mass fraction profiles all of which are only indirectly related to the modeled reaction rates. Because the flow variables are obtained by integrating the dynamic equations they are less sensitive to the modeling assumptions and do not allow a direct assessment of the combustion models. It is a fact that reasonable mean flow profiles may be obtained with the simplest assumptions.

However such comparisons are insufficient and do not distinguish the fine points of the models. If one wishes to describe the effects of finite rate chemistry and turbulence on the structure of the flame it appears clearly that the mean source terms should be examined and precisely represented. With this objective in mind a large set of experiments has been carried out to measure the distribution of light emission from C_2 and CH radicals. The spatial distributions of the radiated light may be interpreted as giving a qualitative mapping of the local mean heat release in the turbulent flame. Because finite-rate chemistry effects are quite pronounced the modeling of the flow is a challenging problem. Indeed standard "fast" chemistry models are unable to reproduce the trends observed. Improved descriptions are needed which account explicitly for the interaction between the chemical kinetics and the flow.

We begin with a rapid survey of the Coherent Flame Model (Section 2). The special case of premixed combustion is considered in Section 3. Experimental results are described in Section 4 and compared with numerical calculations in Section 5. The calculations are performed with a 3-D code, "ECRIN", provided by SNECMA.

2. GENERAL BACKGROUND ON THE COHERENT FLAME MODEL

The Coherent Flame Model is based on the conceptual view that the turbulent reaction field may be described as a collection of laminar flame elements. These elements are convected and distorted by the turbulent motion but retain an identifiable structure. In this sense, the flamelets remain "coherent" (i.e. "organized"). This is the case if the thickness of the reaction zone δ_r is small in comparison with the typical scales l_t of the energy containing turbulent motion.

A rough sketch of this situation is given in Figure 1. A flame element placed in the flowfield is mainly affected by the local strain rate which acts in the plane of the flame, modifies its structure and changes the local reaction rate. As a consequence, the local consumption rate per unit of flame area may be obtained from an analysis of strained laminar diffusion flames. The simplest geometry allowing this analysis is shown in Figure 1-c. The strain rate imposed by the turbulent motion has also the effect of increasing the available flame area. This production process is balanced by various destruction mechanisms such as flame shortening (mutual annihilation of adjacent flamelets by consumption of one of the reactant, Figure 1-b) or flame quenching (due, for example, to an excessively large strain rate).

It follows from these general considerations that the mean consumption rate of a species i per unit volume at a point of the flow, $\overline{W}_i$, may be determined as the product of the mean flame surface density at that point (i.e. the flame surface per unit of volume), $\overline{\Sigma}_f$, by the consumption rate per unit of flame area, V_{Di}, obtained from the analysis of local strained laminar flamelets :

$$\overline{W}_i = V_{Di}\,\overline{\Sigma}_f$$

In its simplest form, the coherent flame description of turbulent reactive flows combines the three following elements :

1) a model for the turbulent flow comprising a standard set of Reynolds or mass averaged dynamic equations and a turbulence closure model.

2) a local model for the laminar strained flame elements providing the consumption rates per unit flame area. Complex molecular transport and detailed chemistry may be incorporated into this model which is essentially decoupled from the turbulent flow calculation.

3) a balance equation for the flame surface density accounting for transport, diffusion, production and destruction of flame area.

The specificity of the Coherent Flame Model lies in the use of the *flame surface density* to relate the local flamelet analysis to the global turbulent flow field calculation. This equation takes into account transport, diffusion, production and destruction of flame area and has the general form :

$$\{\,transport\,\} \;=\; \{\,turbulent\ diffusion\,\} + \{\,production\,\} - \{\,destruction\,\}$$

A balance equation of this type is proposed by Marble and Broadwell to describe the evolution of the flame surface density in the case of a turbulent diffusion flame. For a global reaction between the two main species F and O :

$$F + sO \longrightarrow P$$

in a turbulent fluid field with mean velocity components $\overline{V}_k$, scalar turbulent diffusivity D and local mean strain rate ε_s, the flame surface density satisfies the partial differential equation

$$\frac{\partial \overline{\Sigma}_f}{\partial t} + \overline{V}_k \frac{\partial \overline{\Sigma}_f}{\partial x_k} = \frac{\partial}{\partial x_k}\left[D \frac{\partial \overline{\Sigma}_f}{\partial x_k}\right] + \varepsilon_s\,\overline{\Sigma}_f - \left(\frac{V_{D_F}}{\overline{X}_F} + \frac{V_{D_O}}{\overline{X}_O}\right)\overline{\Sigma}_f^{\,2}$$

where $\overline{X}_i$ is the volume fraction of the species i (i is either O or F).

The production term describes the increase of flame area due to the local strain rate ε_s whereas the destruction term only takes into account the flame shortening mechanism due to mutual interaction of adjacent flame elements.

The local strain rate plays an important role in this flamelet model and must be derived from known variables characterizing the turbulent flow (turbulent kinetic energy, dissipation, typical turbulent length scale, ...). In the case of a two dimensional mixing layer such as that shown in Figure 1-a, Marble and Broadwell propose an estimate of the local strain rate ε_s based on the transverse gradient of the mean axial velocity :

$$\varepsilon_s = \alpha \left| \frac{\partial \overline{U}}{\partial y} \right|$$

This expression may be used in simple flows like shear layers but it is not adequate in more complex situations.

3. APPLICATION TO PREMIXED COMBUSTION

3.1 Balance equation for the flame area per unit mass

The balance equation for the flame area plays a key role in this model. This equation describes the competition between the various processes which create or destroy flame area. It may be derived from first principles but some additional closure assumptions are also required.

In the basic Coherent Flame Model the balance equation relating the local and global description levels was written for the flame area per unit volume designated as Σ_f. Now recent applications to reactive flows with nonsteady changes of the mean pressure indicate that a slightly different quantity is more suitable. In fact the flame area per unit mass S_f is physically more significant. The two quantities are of course related by $\Sigma_f = \rho\, S_f$.

The balance equation for the mean flame area per unit mass may be derived by adopting the same procedure as that used for Σ_f (see Darabiha et al. 1987). One starts from the basic transport equation for a material surface per unit volume Σ :

$$\frac{1}{\Sigma}\frac{d\Sigma}{dt} = -\, n_i\, n_j\, s_{ij}$$

where $s_{ij} = \frac{1}{2}\left(\frac{\partial V_i}{\partial x_j} + \frac{\partial V_j}{\partial x_i} \right)$ designates the rate of strain tensor and n_i are the components of the normal to Σ.

Replacing Σ by ρS one obtains

$$\frac{d}{dt}(\rho S) = \frac{\partial}{\partial t}(\rho S) + V_k \frac{\partial}{\partial x_k}(\rho S) = -\rho\, n_i\, n_j\, s_{ij}\, S$$

and then

$$\frac{\partial}{\partial t}(\rho S) + \frac{\partial}{\partial x_k}(\rho V_k S) = -\rho\left(n_i\, n_j\, s_{ij} - s_{kk} \right) S$$

Introducing the mass averaged decomposition of the different variables one gets

$$\frac{\partial}{\partial t}\left(\bar{\rho}\widetilde{S}\right)+\frac{\partial}{\partial x_k}\left(\bar{\rho}\widetilde{V}_k\widetilde{S}\right)=-\bar{\rho}\left[\overline{\left(n_i\,n_j\,s_{ij}-s_{kk}\right)S}\right]+\frac{\partial}{\partial x_k}\left(-\bar{\rho}\,\widetilde{V_k''S''}\right)$$

At this point it is necessary to make use of closure assumptions. A standard expression for the

turbulent flux term is $-\bar{\rho}\,\widetilde{V_k''S''}=\dfrac{\mu_t}{\sigma_s}\dfrac{\partial\widetilde{S}}{\partial x_k}$ where μ_t is the turbulent viscosity and σ_s is the Prandtl/Schmidt

number for the flame surface.

Now the first term may be estimated as $-\bar{\rho}\left[\overline{\left(n_i\,n_j\,s_{ij}-s_{kk}\right)S}\right]=\alpha\,\bar{\rho}\,\varepsilon_s\,\widetilde{S}$ where ε_s is a mean

strain rate. This term describes the *augmentation of material surface by the rate of strain*. This expression of
the production term differs slightly from that previously used because we now consider the flame surface
per unit mass, which enables us to take into account large pressure variations.

These assumptions are sufficient to close the balance of mean material surface per unit mass S.
However *flame elements are not material surfaces* and it is necessary to include a *chemical shortening term*
similar to that appearing in the balance equation for the flame surface density Σ_f. This destruction term
describing the process of flame shortening (the destruction of flame area by consumption of the reactants
separating adjacent flame sheets) may take different forms. For premixed flames, we consider that the rate
of annihilation of flame surface is proportional to the mass of fuel burnt per second and inversely
proportional to the mass of fuel in the small volume V considered :

$$-\left(\frac{1}{\bar{\rho}\widetilde{S}_f}\frac{d\bar{\rho}\widetilde{S}_f}{dt}\right)_{annihilation} \qquad proportional\ to \qquad \frac{\bar{\rho}\dfrac{\dot{Q}}{\left(-\Delta h_f^0\right)}\widetilde{S}_f V}{\bar{\rho}\widetilde{Y}_F V}$$

where the heat release rate per unit flame area $\dot{Q}$ is obtained from the local flamelet calculations; $\left(-\Delta h_F^0\right)$
designates the heat release per unit mass of fuel and Y_F is the fuel mass fraction.

The annihilation term is then written as $\qquad -\beta\bar{\rho}\dfrac{\dot{Q}}{\left(-\Delta h_F^0\right)\widetilde{Y}_F}\widetilde{S}_f^2$

and the balance equation of flame surface per unit mass becomes

$$\frac{\partial}{\partial t}\left(\bar{\rho}\widetilde{S}_f\right)+\frac{\partial}{\partial x_k}\left(\bar{\rho}\widetilde{V}_k\widetilde{S}_f\right)=\frac{\partial}{\partial x_k}\left[\frac{\mu_t}{\sigma_{s_f}}\frac{\partial\widetilde{S}_f}{\partial x_k}\right]+\alpha\bar{\rho}\,\varepsilon_s\,\widetilde{S}_f-\beta\bar{\rho}\frac{\dot{Q}}{\left(-\Delta h_F^0\right)\widetilde{Y}_F}\widetilde{S}_f^2$$

However it is now believed that the production term should be modified to take into account limiting
mechanisms like *extinction*. Furthermore the determination of the effective strain rate should take into
account the scale of the turbulent eddies which are acting on the flame surface. Below a certain scale the
strain rate associated with the small eddies becomes large and leads to a local extinction of the flame. In this
case we have to modify the production term if the strain rate becomes too large. This can be taken into
account in different formulations. For instance, one may consider a certain critical strain rate ε_{se} (which

depends on the local conditions and is determined from the local analysis) and write the destruction term as $-2\alpha\bar{\rho}\left(\varepsilon_s - \varepsilon_{se}\right) h\left(\varepsilon_s - \varepsilon_{se}\right) \widetilde{S}_f$ where h is the Heavyside function ($h(x) = 1$ if $x>0$ and $h(x) = 0$ if $x \leq 0$).

Another possibility is to multiply the production term by a function representing the efficiency of the strain rate (close to 1 for low strain rates and close to 0 for high strain rates). The function $\dfrac{\dot{Q}}{\dot{Q}_o}$ (where $\dot{Q}$ is the heat release rate per unit flame area -obtained from the local flamelet analysis- and $\dot{Q}_0$ is the value of $\dot{Q}$ for $\varepsilon_s = 0$) is well representative of this influence of the strain rate. The production term becomes $\alpha\bar{\rho}\,\varepsilon_s\widetilde{S}_f\dfrac{\dot{Q}}{\dot{Q}_0}$. This expression is used in the calculations presented in this paper.

The balance equation for $\widetilde{S}_f$ is finally written as

$$\frac{\partial}{\partial t}\left(\bar{\rho}\widetilde{S}_f\right) + \frac{\partial}{\partial x_k}\left(\bar{\rho}\widetilde{V}_k\widetilde{S}_f\right) = \frac{\partial}{\partial x_k}\left[\frac{\mu_t}{\sigma_{S_f}}\frac{\partial\widetilde{S}_f}{\partial x_k}\right] + \alpha\bar{\rho}\ \varepsilon_s\widetilde{S}_f\frac{\dot{Q}}{\dot{Q}_0} - \beta\bar{\rho}\frac{\dot{Q}}{\left(-\Delta h_r^0\right)\widetilde{Y}_F}\widetilde{S}_f^2$$

Taking into account the effect of turbulence, the strain rate appearing in this equation may be evaluated from $\varepsilon_s = C_s\dfrac{\varepsilon}{k}$ and the constants take the following values :

$$\sigma_{S_f} = 1, \ \alpha = 10, \ \beta = 0.4 \ \text{and} \ C_s = 0.17.$$

It is probable that the final form of the balance equation for the flame area will also differ from that presented in this paper. This is a subject of current research and modified expressions have been derived by us and by others (see for example Peters et al. 1989). A stochastic flamelet model comprising an evolution equation for the flame area has been proposed by Pope and Cheng (1988). A coupled set of equations describing the effects of the coherent fluctuations and fine grained turbulence on the material surface density is established by Liu (1988). The evolution of flame fronts in turbulent flows is also investigated by Ashurst, Sivashinsky and Yakhot (1988).

3.2 The local flamelet structure

Consumption rates per unit flame area and the corresponding heat release are obtained from an examination of the local flame structure.

Many configurations may be used as local models (Figure 2). One may for example consider the counterflow of fresh reactants and burnt gases as shown in Figure 2-a. An alternative model is that of twin flames formed by the counterflow of two fresh streams as sketched in Figure 2-b.

Both models are being considered but only the first is used in the present calculations. Computational methods are described in detail by Giovangigli (1988) and results for propane-air strained flames are contained in Darabiha et al. (1988).The flamelet structure and properties depend on pressure, fresh gas temperature T_o, equivalence ratio Φ, burnt gas temperature T_b and strain rate ε_s.

A large number of flamelets has been determined for a constant pressure of 1 bar and a constant fresh gas temperature $T_o = 300$ K (see Darabiha et al. 1988). For a given hot stream temperature T_b the heat release per unit flame area $\dot{Q}$ may be obtained as a function of the equivalence ratio and the strain rate. A typical plot of $\dot{Q}(\Phi, \varepsilon_s)$ is shown in Figure 3. In this particular case the temperature T_b is equal to the adiabatic flame temperature $T_b = T_{ad}(\Phi)$. These data are used in the numerical simulations included in this paper. More generally all the flamelet structures and characteristics may be stored in a large data base and the values of the consumption and heat release rates may be retrieved during the calculation of the turbulent flow.

3.3 Determination of the mean reaction rates

The mean consumption and heat release rates are determined from the flame surface density $\widetilde{S_f}$ and from the rates obtained in the local flamelet analysis. One writes, for example, that the average rates per unit volume are :

$$\begin{cases} \overline{\dot{W}} = \bar{\rho}\,\dot{Q}\,\widetilde{S_f} \\ \overline{\dot{W}}_F = -\,\bar{\rho}\,\dfrac{\dot{Q}}{\left(-\Delta h_F^o\right)}\,\widetilde{S_f} \end{cases}$$

where $\overline{\dot{W}}$ and $\overline{\dot{W}}_F$ are respectively the mean heat release rate per unit volume and the mean fuel consumption rate.

These expressions may also be refined by taking into account the fact that the local flamelets correspond to a distribution of strain rates. It is then necessary to specify or presume a probability density function describing this distribution .

4. EXPERIMENTS ON TURBULENT DUCTED FLAMES

4.1. Experimental configuration

The experimental set up used in this study is shown in Figure 4. A mixture of air and propane is injected through a long duct into a rectangular combustor. The inflow stream has a pressure of 1 bar and a temperature of 300 K. The maximum mass flow rate is 120 g/s. The height, depth and length of the combustion chamber are respectively 50, 80 and 300 mm. The upper and lower combustor walls are made of thick ceramic material while the lateral walls are transparent artificial quartz windows which allow the visualization of the whole chamber.

The inlet plane comprises a V-gutter flame holder placed at the duct center and producing a 50 % blockage. Ignition of the premixed stream is obtained with a spark plug. Combustion is then stabilized by hot gases recirculating behind the V-gutter. This configuration is simulating a simplified SNECMA afterburner.

4.2. C_2 and CH radical light emission

In the present experiment no attempt is made to obtain a detailed description of the chemical kinetics in the reaction zone. Our scope is to get an information on the local heat release source term distribution. It is important to note that the heat release source term is far more interesting than temperature for modeling purposes. Indeed the temperature is less sensitive to the combustion model because it is integrated. On the contrary, the heat release source term is directly dependent on the combustion model and allows an interesting evaluation of it.

While exact measurements of the local heat release are not available, many observations indicate that the light emitted by the combustion zone is related to the reaction intensity and hence to the heat-release process. Certain radicals like C_2 and CH appear almost exclusively in reactive zones and their concentration is always small. Hence the self absorption of the light emitted by these radicals is not important and the radiated light is directly related to the reaction rate or equivalently to the heat release rate. While a linear relation between the heat release source term and the light emission from free radicals has been proved in some special situations it may be safely stated that a monotonic -and more or less linear- relation exists between these two quantities. The measurement of the heat release source term relies on this assumption. The calibration of the radical light emission may be obtained from an additional information (for instance gas analysis in the exit plane of the combustor) providing the global heat release in the combustion chamber.

The emission bands used to this purpose are the (0,0) C_2 and (0,0) CH bands at wavelengths of 516 and 430 nm respectively. Each wavelength is isolated in the emission spectrum with a narrowband interferential filter ($\Delta\lambda = 5$ nm). The light emitted by the flame is collected by a $f = 100$ mm convex lens located at 200 mm from the combustor center plane. The light beam is detected by a photomultiplier through a 2 mm diameter pinhole to provide local measurement : the emitting area actually seen by the detector has a diameter of about 2 mm. The detector output is amplified and transmitted to a Masscomp-5600 computer. This computer also controls the optical displacement in the vertical and horizontal directions (see Figure 5). The photomultiplier scans a grid comprising 600 points (30 points horizontally and 20 vertically). At each point of the grid, 30000 data samples are acquired at a sampling frequency of 30 kHz (after a low pass 6 kHz filtering to prevent aliasing) and averaged. This process yields the spatial distribution of the mean light emission from C_2 and CH radicals.

4.3. Experimental results : effects of the equivalence ratio and flow rate

As already indicated, the C_2 and CH light emission provides information on the local rate of heat release. Spatial distributions of the light emitted by C_2 and CH radicals are displayed in Figures 6 and 7 respectively. These maps correspond to the same flow rate (75 g/s) but to different values of the equivalence ratio. The Figure 8 represents a set of CH light emission maps obtained by varying the mass flow rate for a fixed value of the equivalence ratio ($\Phi = 0.75$).

One first observes that C_2 and CH maps have the same shape. This is not a surprise because both C_2 and CH are tracers of the local heat release in the flame. Next one may examine the evolution of the mean flame structure with the equivalence ratio (Figures 6 and 7). At low values of Φ, concentrated regions of reaction are observed in the vicinity of the flame holder. For $\Phi = 0.60$ these regions form a merged reaction zone. When the equivalence ratio increases, two flames develop from the lips of the flameholder. These flames become more and more separated when Φ increases. The C_2 map for $\Phi = 0.65$ is very interesting because it appears to be a critical equivalence ratio : there are two flames but they are not completely separated. For the highest equivalence ratios, the light emission reaches its maximum in the vicinity of the two sidewalls. This peculiar phenomenon is observed because the flame now touches the boundaries which are in the present case nearly adiabatic. At the points where the flame reaches the walls, the flow in the boundary layer is decelerated and its velocity is less than that existing in the main flow and the flame angle increases (this angle is measured with respect to the axial direction). The wall temperature takes large values and the wall region acts like a secondary stabilization zone for the incoming stream of fresh mixture. In this region combustion is activated and the light emission is enhanced. For $\Phi = 0.90$ the two flames are separated by a region of reduced reaction. In each flame region the radical light emission increases progressively and reaches its maximum in the vicinity of the combustor walls.

Consider now the maps obtained for different values of the flow rate (Figure 8). For low values of the mass flow rate the two flame regions are well separated and a maximum rate of heat release is found near the walls. As the mass flow rate is increased the flame angle decreases slightly and the maximum is shifted upstream.

These data clearly show that the equivalence ratio and the mass flow rate have a notable influence on the flame structure. The modifications associated with these two parameters are not extensively documented in the previous literature (see Libby et al. 1986 for a review of the available data, Zukoski and Marble 1955 for an early analysis of the influence of the mass flow rate on the wake transition behind a flame stabilizer).

5. NUMERICAL RESULTS AND COMPARISON WITH EXPERIMENTS

Numerical results corresponding to the experimental configuration are now described. The calculations are carried out with a single set of constants. Standard values are retained for the k-ε model ($C_\mu = 0.09$, $C_1 = 1.44$, $C_2 = 1.92$, $\sigma_k = 1$, $\sigma_\varepsilon = 1.3$).

Calculations have been carried out with two distinct numerical codes :
- an iterative elliptic 3-D code (ECRIN) developed by SNECMA, based on the "SIMPLE" technique providing the steady state solution
- a time-dependent finite-difference algorithm providing transient as well as steady state solution for two-dimensional flows (MOCROI, see Dupoirieux 1985).

Systematic comparisons between the results obtained with these two methods show only minor differences indicating that the numerical solution is independent of the code. Tests were also carried out to study grid effect and make sure that the solution did not depend on the grid selected. Conditions retained for the numerical simulation are gathered in Table 1. Starting from steady state cold flow conditions a converged solution is obtained with ECRIN after about 700 iterations.

Numerical results are displayed in Figure 9. Figure 9-a displays the mean axial velocity as a contour plot (using a grey scale display). This velocity component is slightly negative in the wake of the flame stabilizer. It takes positive values further downstream. The hot gases produced by the chemical reaction are accelerated and the axial velocity reaches its maximum values on the combustor axis. The stream function and pressure distribution are plotted in Figures 9-b and 9-c respectively. The length of the recirculation is nearly three flameholder widths and corresponds to the visual observations. The mean temperature distribution appears in Figure 9-d. The temperature changes rapidly across the two reaction regions originating from the stabilizer lips. These regions may be visualized by plotting the distributions of flame area (Figure 9-e) and the mean heat release source term (Figure 9-f).

For the operating conditions selected for this simulation the two reaction regions are separated and the pattern obtained is close to that found in the corresponding experiments (Figures 6-c and 7-c). The agreement is not perfect but the main features are retrieved. Similar results are obtained for flow conditions corresponding to separated reaction regions.

As indicated before, most models provide reasonable temperature distributions but the heat release terms are not well predicted. This discrepancy is illustrated in Figure 10 which displays the mean source distribution and the temperature corresponding to a standard "Eddy Break Up" model. While the temperature distributions are not far from those obtained with the Coherent Flamelet Model, the distribution of the heat release rate is extremely different. The regions where the source term reach a maximum are very narrow. The reaction rapidly goes to completion so that the maximum is close to the stabilizer lip. A tuning of the constants of the EBU model could probably lead to an improved pattern but the result would have a restricted domain of validity.

Turning back to the Coherent Flamelet Model, it is possible to look at the effect of a variation of Φ. Figure 11 displays the mean heat release rate obtained for $\Phi = 0.60$, 0.75 and 0.85. The trend observed experimentally is also found numerically. For low values of Φ, the two flames are closer than for higher equivalence ratios. The merged regime corresponding to low values of the equivalence ratio is not fully retrieved but the pattern is similar to that observed in the experiment.

6. CONCLUSION

The Coherent Flamelet Model is briefly reviewed in this paper. The fundamental configuration of a premixed turbulent flame stabilized in a duct is studied. This configuration is simulating a simplified SNECMA afterburner. The mean heat release term determined numerically is compared with experimental measurements of the mean light emission from the reaction zone. This comparison is essentially qualitative but it does provide indications that the model describes some of the controlling mechanisms of turbulent combustion.

ACKNOWLEDGMENTS

We wish to acknowledge the continuous support provided to this work by SNECMA. This research is also funded by DRET as a part of the combustor modeling project A3C. We also wish to address our grateful thanks to Professor Frank Marble for many discussions and suggestions.

REFERENCES

Ashurst W.T., Sivashinsky G.I. and Yakhot V. (1988). Flame front propagation in nonsteady hydrodynamic fields. Sandia Rep. SAND88-8891.

Bray K.N.C. (1987). Methods on including realistic chemical reaction mechanisms in turbulent combustion model. 2nd Workshop on modelling of chemical reaction systems, J. Warnatz and W. Jager ed., Springer Verlag, Heidelberg.

Bray K.N.C., Libby P.A. and Moss J.B. (1984). Flamelet crossing frequencies and mean reaction rates in premixed turbulent combustion. Comb. Sci. and Tech. 41, 143-172.

Candel S., Darabiha N. and Esposito E. (1982) Models for a turbulent premixed turbulent combustor. AIAA Paper 82-1261.

Candel S.M., Maistret E., Darabiha N., Poinsot T., Veynante D.and Lacas F. (1988). Experimental and numerical studies of turbulent ducted flames. Proc. of the Marble Symp., Caltech, Pasadena, August 1988.

Carrier G.F., Fendell F.E. and Marble F.E. (1975). The effect of strain rate on diffusion flames. SIAM J. of Appl. Math. 28, 463-500.

Clavin P. and Williams F.A. (1982). Effects of molecular diffusion and thermal expansion on the structure and dynamic of premixed flames in turbulent flows. J. Fluid Mech. 116, 215.

Correa S.M. and Shyvy W. (1987). Computational models and methods for continuous gaseous turbulent combustion. Progr. Energy and Comb. Sci. 13, 249-292.

Darabiha N. (1985). Un modèle de flamme cohérente pour la combustion prémélangée. Analyse d'un foyer turbulent à élargissement brusque. Thèse de Docteur Ingénieur, Ecole Centrale de Paris.

Darabiha N., Candel S. and Marble F.E. (1985). The effect of strain rate on a premixed laminar flame. Comb. and Flame 64, 203-217.

Darabiha N., Giovangigli V., Candel S. and Smooke M.D. (1988). Extinction of strained premixed propane-air flames with complex chemistry. Comb. Sci. and Tech.

Darabiha N., Giovangigli V., Trouvé A., Candel S. and Esposito E. (1987). Coherent flame description of turbulent premixed ducted flames. Proc. of the France USA Joint Workshop on Turbulent Combustion, Rouen 1987.

Dupoirieux F., Scherrer D. (1985). Méthodes numériques à convergence rapide utilisées pour le calcul des écoulements réactifs. Numerical Simulation of Combustion phenomena, Sophia-Antipolis, May 21-24, 1985.

Giovangigli V. (1988). Structure et extinction de flammes laminaires prémélangées. Thèse de Doctorat ès Sciences Mathématiques. Université Pierre et Marie Curie, Paris 6.

Kuo K.K. (1986). Principles of Combustion. John Wiley, New York.

Lacas F. Zikikout S. and Candel S. (1987). A comparison between calculated experimental mean source terms in non premixed turbulent combustion. AIAA/SAE/ASME/ASEE 23rd Joint Propulsion Conference, AIAA-87-1782, July.

Lacas F., Veynante D. and Candel S.M. (1989). A numerical study of propagating premixed turbulent flames. Numerical Combustion, Antibes, May 23-26, 1989.

Libby P.A. and Williams F.A., ed. (1980). Turbulent reacting flows. Springer Verlag, New York

Libby P.A., Sivasegram S. and Whitelaw J.A. (1986). Premixed combustion. Prog. Energy and Comb. Sci. 12, 393-405.

Liew S.K., Bray K.N.C. and Moss J.B. (1981). A flamelet model of turbulent non premixed combustion. Comb. Sci. and Tech. 27, 69-73.

Liu J.T.C. (1988). Coherent mode interaction in developing free shear layers with application to coherent flames. Proc. of the Marble Symposium, Caltech, Pasadena, Aug.1988.

Marble F.E. and Broadwell J.E. (1977). The coherent flame model for turbulent chemical reactions. Project squid headquarters, Chaffee Hall, Purdue Univ. Tech. Rep. TRW.

Marble F.E. and Adamson T.C. (1954). Ignition and combustion in a laminar mixing zone. Jet Propulsion 24, 85-94.

Marble F.E. and Broadwell J.E. (1979). A theoretical analysis of nitric oxide production in a methane-air turbulent diffusion flame. EPA Tech. Report.

Peters N. (1984) Laminar diffusion flamelets in non-premixed turbulent combustion. Progr. Energy and Comb. Sci. 10, 319-339.

Peters N. (1986). Laminar flamelet concepts in turbulent combustion. 21st Symp. (Int.) on Combustion, The Combustion Institute.

Peters N., Göttgens J., Klein R. and Clavin P. (1989). Flame surface and burning velocity in premixed turbulent flames. Preprint.

Pope S.B. and Cheng W.K. (1988). The stochastic flamelet model of turbulent combustion. Presented at the 22nd Symp.(Int.) on Combustion, Seattle .

Spalding D.B. (1978). The influence of laminar transport and chemical kinetics on the time mean reaction rate in a turbulent flame. Seventeenth International Symposium on Combustion. The Combustion Institute, Pittsburgh, 431.

Veynante D., Candel S.M. and Martin J.P. (1987). Coherent flame modelling of chemical reaction in a turbulent mixing layer. 2nd Workshop on modelling of chemical reaction systems, J. Warnatz ans W. Jager ed., Springer Verlag, Heidelberg.

Veynante D., Lacas F. and Candel S.M. (1989a). A new flamelet combustion model combining premixed and nonpremixed turbulent flames. AIAA Paper 89-487.

Veynante D., Lacas F. and Candel S.M. (1989b). Numerical simulation of the ignition of a turbulent diffusion flame. Submitted to the 12th ICDERS, U. of Michigan, July 1989

Veynante D., Lacas F., Maistret E. and Candel S.M. (1989c). Coherent flame model in non uniformly premixed turbulent flames. Submitted to 7th Turbulent Shear Flows, Stanford, Aug. 21-23.

Williams F.A. (1975) . A review of some theoretical combustions of turbulent flame structure. AGARD Conf. Proc. 164 II 1.1.

Williams F.A. (1985). Combustion Theory. Benjamin Cummings, Palo Alto.

Wright F.H. (1959). Bluff body stabilization : blockage effects. Comb. and Flame 3, 319.

Zukoski E.E. and Marble F.E. (1955). The role of wake transition in the process of flame stabilization on bluff bodies. Comb. Researches and Rev. Butterworths, London.

Fuel A
Oxidizer B

Flame element at time t

Flame element at time t + δt

Fuel A
Oxidizer B

a) Production of flame surface by the strain rate

b) Flame annihilation process

c) Local model of the strained diffusion flame element

Figure 1
Schematic representation of the physical processes
described in the Coherent Flame Model.

Code : ECRIN (iterative elliptic algorithm developped by SNECMA)

Grid : 27 vertical and 73 horizontal nodes (considering the plane of symmetry, the calculations are performed on one half of the combustion chamber). The grid is non uniform with local refinement near the stabilizer lips (see below).

Flow parameters : Mass flow rate : $\dot{m} = 75$ g/s
Inlet flow velocity : $V = 15.75$ m/s
Inlet kinetic energy : $k = 1\%$ of $V^2 = 2.5$ m^2/s^2
Inlet turbulence lenght scale : $l_T = 20$ mm
Inlet turbulence dissipation : $\varepsilon = 77$ m^2/s^3
Inlet temperature : $T = 300$ K
Equivalence ratio : $\Phi = 0.75$
Inlet fuel mass fraction : $Y_{FU} = 0.0458$

Table 1
Conditions selected for the numerical calculations.

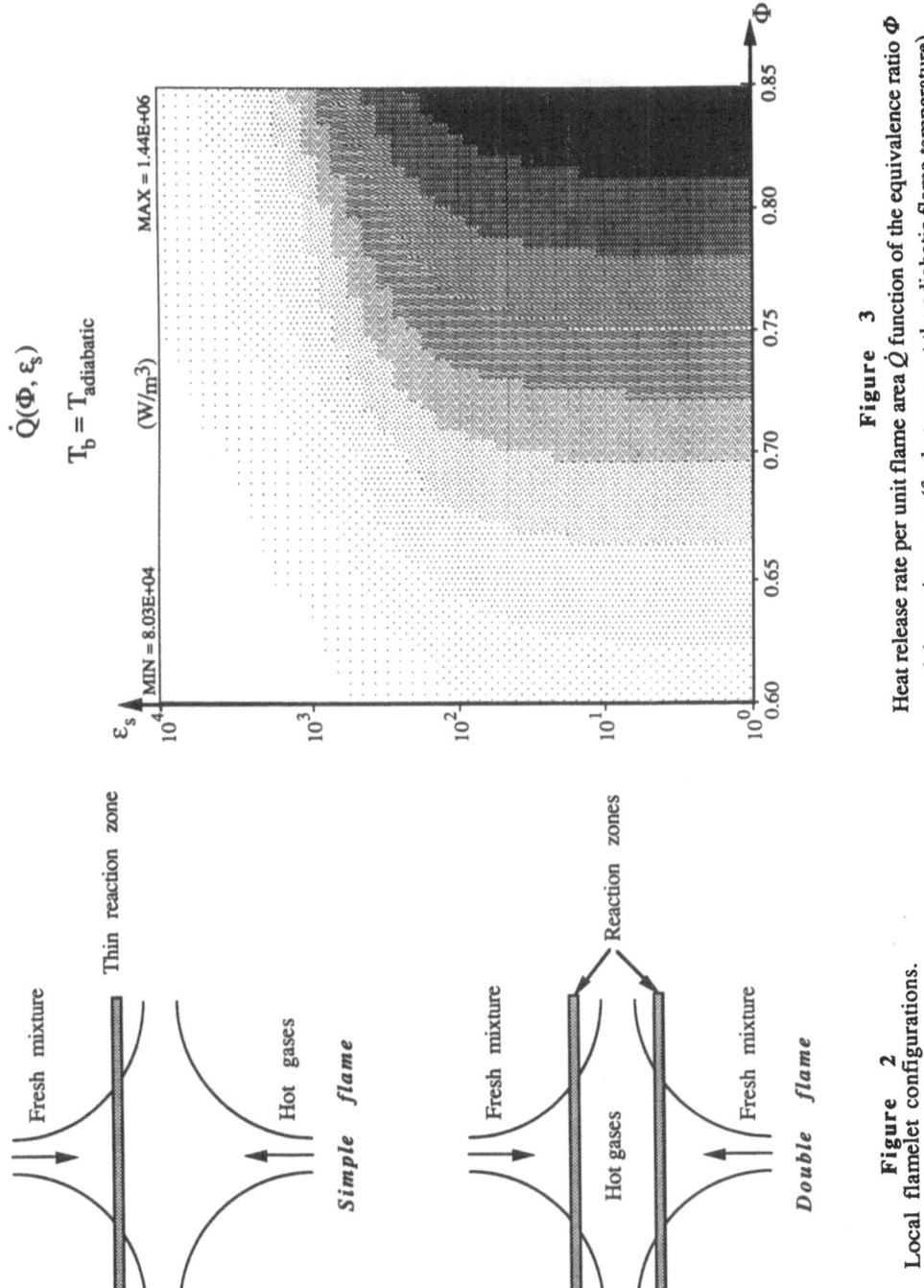

Figure 3

Heat release rate per unit flame area $\dot{Q}$ function of the equivalence ratio Φ and the strain rate ε_s (for hot gases at the adiabatic flame temperature).

Figure 2

Local flamelet configurations.

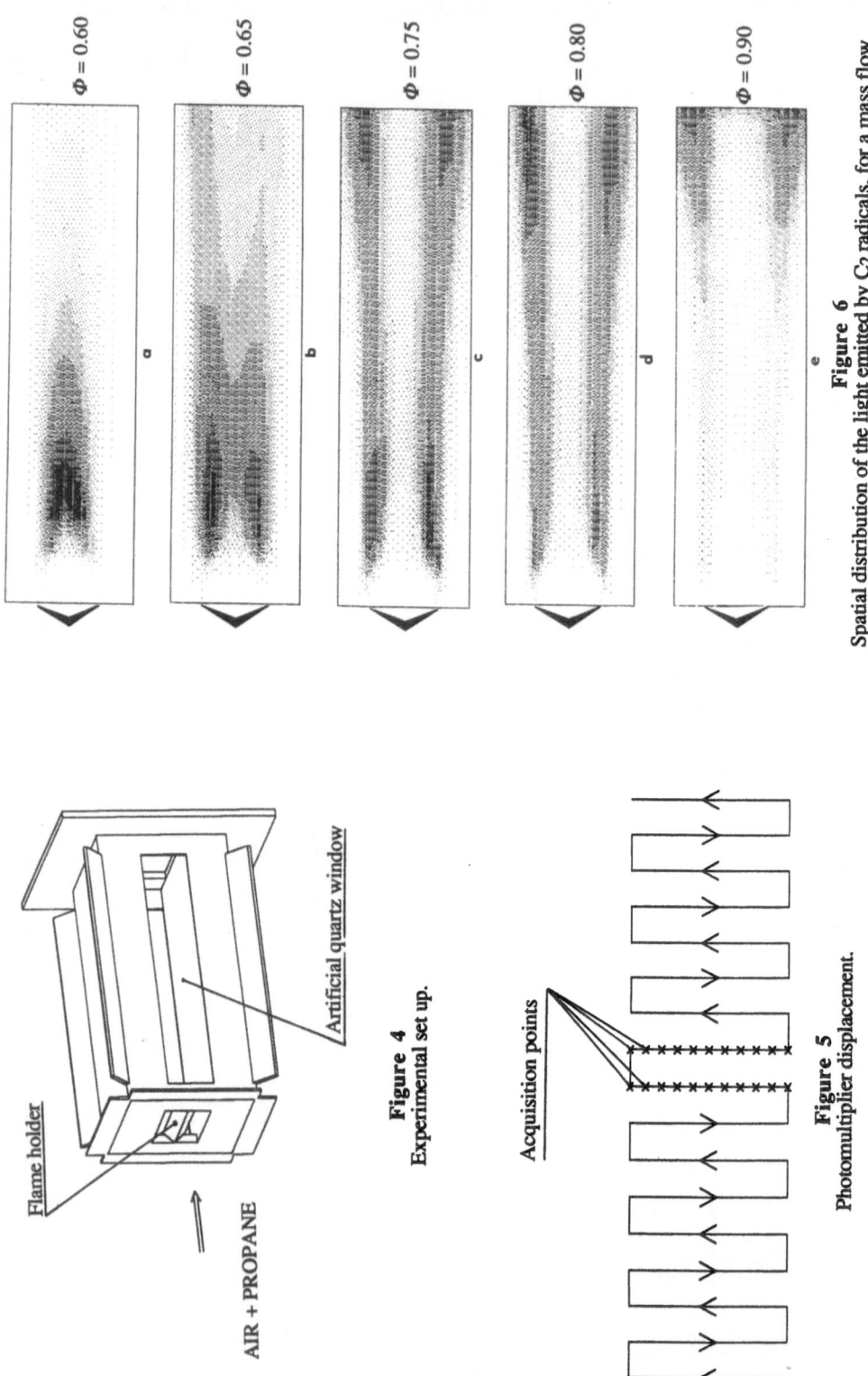

Figure 6
Spatial distribution of the light emitted by C_2 radicals, for a mass flow rate of 75 g/s and different values of the equivalence ratio.

$\Phi = 0.60$

$\Phi = 0.65$

$\Phi = 0.75$

$\Phi = 0.80$

$\Phi = 0.90$

a

b

c

d

e

Flame holder

Artificial quartz window

Figure 4
Experimental set up.

AIR + PROPANE

Acquisition points

Figure 5
Photomultiplier displacement.

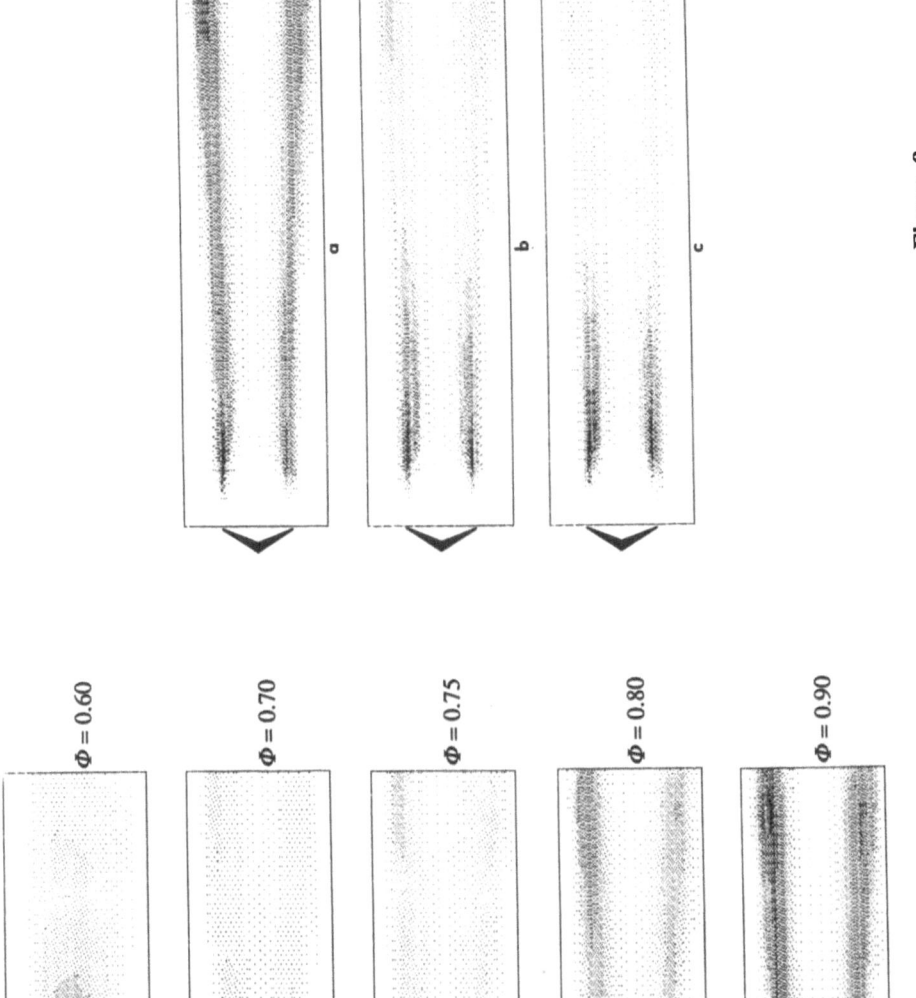

Figure 8
Spatial distribution of the light emitted by CH radicals, for an equivalence ratio of 0.75 and different values of the mass flow rate.

$\dot{m}$ = 50 g/s

$\dot{m}$ = 75 g/s

$\dot{m}$ = 100 g/s

Figure 7
Spatial distribution of the light emitted by CH radicals, for a mass flow rate of 75 g/s and different values of the equivalence ratio.

Φ = 0.60

Φ = 0.70

Φ = 0.75

Φ = 0.80

Φ = 0.90

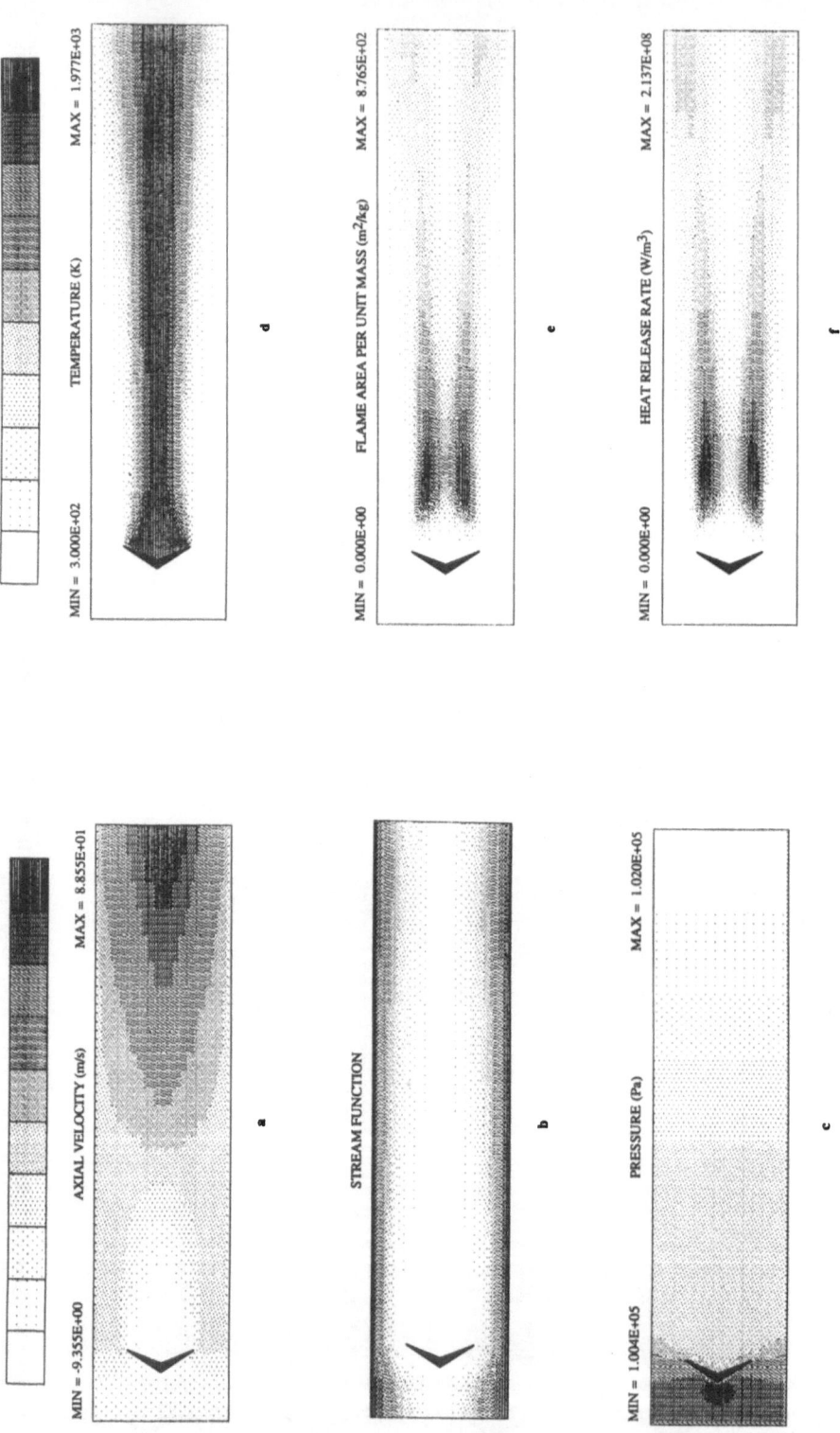

Figure 9

Spatial distributions obtained by numerical calculation for a mass flow rate of 75 g/s and an equivalence ratio of 0.75, using the Coherent Flame Model.

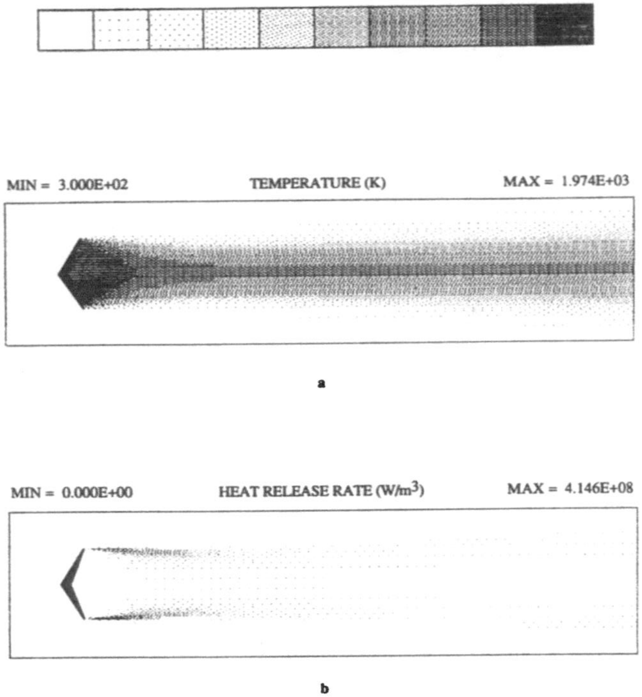

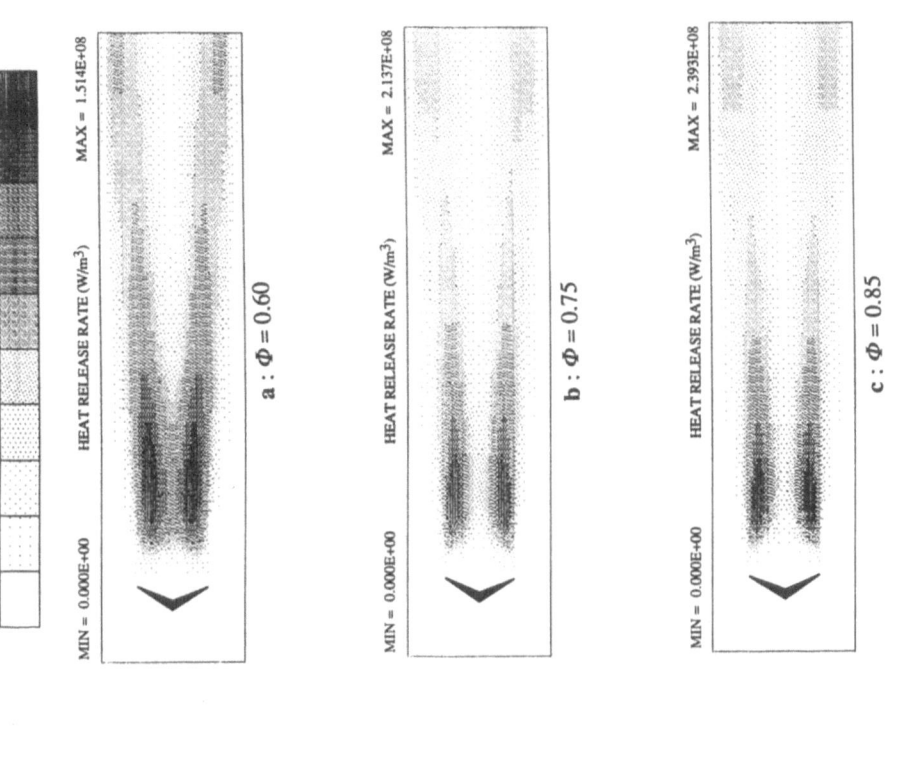

Figure 11
Spatial distributions of heat release rate obtained with the Coherent Flame
Model, for a mass flow rate of 75 g/s, and different values of Φ.

a : Φ = 0.60

b : Φ = 0.75

c : Φ = 0.85

Figure 10
Spatial distributions of the temperature and heat release rate
obtained with the classical "Eddy Break Up" model.

IMPROVEMENTS OF THE KIVA-II COMPUTER PROGRAM FOR NUMERICAL COMBUSTION

P. J. O'Rourke, A. A. Amsden, T. D. Butler
Group T-3, Mail Stop B216
Los Alamos National Laboratory
Los Alamos, NM 87545
and
T. L. McKinley
Cummins Engine Company
Columbus, IN 47202

ABSTRACT

This paper describes and illustrates the principal differences between the newly-released KIVA-II and the KIVA computer programs. Both programs are for the numerical calculation of two- and three-dimensional fluid flows with chemical reactions and sprays. Because of improvements to KIVA-II, it is faster, more accurate, and applicable to a wider variety of problems involving combustion and two-phase flow.

I. INTRODUCTION

The KIVA-II computer program[1] has been developed for the numerical solution of transient two- and three-dimensional fluid flows with chemical reactions and sprays. KIVA-II extends and enhances the capabilities of the earlier KIVA code,[2,3] improving its numerical efficiency and accuracy and its ease-of-use and versatility. In this paper we describe the principal differences between KIVA-II and KIVA and illustrate these with computational examples.

KIVA-II retains several of the basic features of the KIVA code, and users of KIVA should find the transition to KIVA-II to be straightforward. Finite differencing is based on a finite volume method called the ALE (Arbitrary Lagrangian-Eulerian) method.[4,5] Spatial differences are formed with respect to a computational mesh composed of arbitrary hexahedrons. Spatial differencing is made conservative whenever possible by differencing integral forms of the conservation equations. Cartesian components of the velocity field are stored at cell vertices, and the linear momentum equations are differenced in a strictly conservative fashion. All other fluid variables are stored at cell centers. As in KIVA,[3] KIVA-II uses cell-face velocities during a portion of the computational cycle to reduce the susceptibility of computed solutions to parasitic velocity modes and thereby largely eliminate the need for alternate node coupling.[1,2]

Temporal differencing is based on a computational cycle that is divided into two phases - a Lagrangian phase and a rezone phase. In the rezone phase the vertices, which have moved with the fluid in the Lagrangian phase, are moved to new user-specified positions, and the flow field is remapped onto the new computational mesh. The formulation allows a Lagrangian, Eulerian, or a mixed description. The arbitrary mesh can move to follow changes in combustion chamber geometry, such as occur in internal combustion engines.

Equations for an evaporating liquid fuel spray are solved using the stochastic particle method.[6,7] Computational particles - which represent droplets of specified size, velocity, and temperature - interact with the fluid by exchanging mass, momentum, and energy in a conservative fashion. Probability distributions often govern the assignment of particle properties at injectors or the evolution of particle properties at downstream locations. When this is the case, the distributions are randomly sampled to determine the particle properties or their evolution.

The differences between KIVA-II and KIVA are summarized in Table 1.

TABLE I

DIFFERENCES BETWEEN KIVA-II AND KIVA

1. Computational Efficiency Improvements
    a. Coupled, implicit differencing of diffusion and pressure wave propagation terms
    b. Subcycled calculation of convection
    c. 2-D to 3-D converter
2. Numerical Accuracy Improvements
    a. Optional quasi-second-order upwind convection scheme
    b. Generalized mesh diffusion algorithm
    c. Method for computing droplet turbulent dispersion when the time step Δt exceeds the droplet turbulence correlation time
3. New or Improved Physical Submodels
    a. $k - \varepsilon$ turbulence model
    b. Revised subgrid scale turbulence model
    c. Model for droplet aerodynamic breakup
4. Improvements in Ease-of-Use and Versatility
    a. More flexible mesh generation for internal combustion engines
    b. Inflow/Outflow boundaries
    c. Simplified velocity boundary conditions
    d. Gravitational terms
    e. Library of thermophysical properties of common hydrocarbons
    f. Alphabetized epilogue giving FORTRAN variables and their definitions

All of these differences are fully documented in the KIVA-II report. In this paper we describe and illustrate the most important changes, which are 1a, 1b, 2a, 3a, 3b, and 3c of Table I.

II. NUMERICAL EFFICIENCY IMPROVEMENTS

The main criticism of the KIVA code was its slowness. This was largely due to the fact that explicit difference schemes were used for diffusion and convection terms, and a subcycled explicit scheme was used to calculate pressure wave propagation.[2] Each of these schemes had stability conditions that limited the size of the computational timestep. The stability conditions for the convection and diffusion terms usually controlled the main computational timestep, and often limited it to very small values. The diffusional stability criterion was particularly restrictive because the timestep was proportional to the square of the computational cell dimension. In low Mach number problems, the subcycle timestep for calculating pressure waves could become very small and result in further inefficiency.

To improve computational efficiency we have implemented two major changes in KIVA-II. First, implicit differencing is used to calculate the diffusion terms and terms associated with pressure wave propagation. Second, although it still uses explicit schemes, the convection calculation can be repeated or subcycled an arbitrary number of times using timesteps that satisfy the explicit stability criterion for convection and that are submultiples of the main computational timesteps. Because of these changes, the main computational timestep in KIVA-II is no longer controlled by stability criteria, but by accuracy criteria associated with fluid accelerations and cell distortions during the Lagrangian phase and with coupling spray and chemical source terms to the fluid flow.[1]

The implicit equations for diffusion and pressure wave propagation are solved during the Lagrangian phase by a modified SIMPLE method[8] with individual equations solved by the conjugate residual method.[9] It should be noted that we first tried a noniterative alternative to the SIMPLE method that was similar to the PISO[10] method, but we found this to be unsatisfactory. Although the numerical truncation errors of the PISO-like method could be shown to be formally of second-order in the timestep Δt, in practice these errors were sometimes unacceptably large. The conjugate residual method[9] was chosen because of its rapid convergence properties and its low computer storage requirements and because it is vectorizable on computers that offer this feature.

A novel feature of the implicit differencing in KIVA-II is that the amount of implicitness is variable. Weighted averages of the old- and new-time values of pressures and diffused quantities are automatically calculated. The weighting factor for the pressures is based on the Courant number $c\Delta t/\Delta x$, where c is the speed of sound, and the weighting factor for diffused quantities is based on the local diffusion Courant number $D\Delta t/\Delta x^2$, where D is the diffusivity. The weighting factors are chosen to ensure numerical stability and also to reduce the amount of implicitness and thereby to improve computational efficiency. When the Courant number is small enough, KIVA-II automatically uses explicit schemes for which no costly iterative solution is required. When Courant numbers are large enough that implicitness is required, KIVA-II uses partially-implicit schemes that can be solved more quickly than fully-implicit schemes by iterative methods such as the conjugate residual method.

The idea of convective subcyling is that we calculate convection using an explicit scheme and timestep Δt_c that is an integral submultiple of the main computational timestep Δt. The timestep Δt_c must satisfy the Courant condition $u_r \Delta t_c/\Delta x < 1$, based on the fluid velocity u_r relative to the computational grid, but because there is no limit on the number of subcycles, we can have $u_r \Delta t/\Delta x > 1$. Convective subcycling was fairly easy to implement in KIVA-II because the calculation of convection, which is performed in the rezone phase, is split from the calculation of the remaining fluid flow terms. The method saves computational time because the rezone calculation typically takes only ten percent of the time of the Lagrangian phase calculation. Assuming the rezone phase calculation takes a fraction f of the computational time of the Lagrangian phase and that we subcycle N times per computational cycle, one can shown that a calculation with convective subcycling takes a fraction

$$\frac{1 + Nf}{N(1 + f)}$$

of the time of a calculation without subcycling.

Convective subcycling has advantages and a disadvantage relative to the alternative of implicit differencing of convection terms. Generally, it is more accurate because if we subcycle, the timesteps we use to calculate convection are smaller and thus temporal truncation errors are smaller. Also because the calculation is explicit it is easier to use convective subcycling in conjunction with nonlinear monotone difference schemes. Such a scheme has been incorporated in KIVA-II and will be described in the next section. The disadvantage is that convective subcycling is generally slower than implicit differencing. With a little extra computational work convection could be calculated implicitly at the same time that the diffusion terms are calculated (the description "Lagrangian phase" would then no longer apply), but we believe the advantages of convective subcycling outweigh this one disadvantage.

Table II gives computational time comparisons between KIVA and KIVA-II for some two- and three-dimensional internal combustion engine calculations. The two codes were equally fast on a two-dimensional coarse mesh calculation,

TABLE II

COMPUTATIONAL TIME COMPARISONS FOR
2-D AND 3-D ENGINE CALCULATIONS

	KIVA	KIVA-II
2-D Calculations (6.25×10^{-3} s problem time)		
Course Mesh (440 cells)		
Unrestricted Δt	49.5"	53.5"
$\Delta t_{max} = 1 \times 10^{-5}$ s	...	130.8"
$\Delta t_{max} = 1 \times 10^{-5}$ s, fully-implicit diffusion	...	176.8"
Fine Mesh (1760 cells)	5.13'	3.93'
3-D UPS Engine Calculations[11] ($-118°$ to $+92°$, 8712 cells)	3.0 hrs	1.44 hrs

but KIVA-II was faster on two- and three-dimensional calculations in which the meshes were more refined and the timestep used by KIVA was limited by either convective or diffusional stability criteria. Generally, the more refined one's mesh is, the faster KIVA-II will be in comparison to KIVA. We also performed two coarse mesh calculations using KIVA-II in which the timestep was required to be less than 1.0×10^{-5} s. With this small timestep KIVA-II automatically differenced the diffusion terms explicitly, but we also performed a calculation in which we overrode this automatic feature and differenced the diffusion terms fully implicitly. The fully-implicit calculation was thirty-five percent slower, showing the benefit of using variable implicitness.

III. NUMERICAL ACCURACY IMPROVEMENT

The major improvement in numerical accuracy of KIVA-II results from the implementation of a monotone scheme for convection that we call quasi-second-order upwind (QSOU) differencing. For a definition of monotone schemes and a detailed discussion of the QSOU method, the reader is referred to Appendix M of Ref. 1. Here we simply discuss and illustrate some of the general features of QSOU.

The QSOU method is a combination of Leith's method[12] or interpolated donor cell differencing, which is second-order accurate in space, and upwind or donor cell differencing, which is first-order accurate and monotone. QSOU aims to maximize the proportion of Leith's method that is used, while preserving

123

monotonicity. There have been many schemes based on this same general concept,[13,14,15] but QSOU is based primarily on the ideas of van Leer.[16] By using a slightly more strict gradient-limiter than van Leer, QSOU ensures monotonicity in one-dimensional calculations. We have implemented a simple multidimensional extension in which fluxes in each coordinate direction depend only on gradients in that direction. This extension has given monotone results in all test cases, as evidenced by the following calculational results.

Figure 1 gives results from five calculations in which we convect a scalar function through a mesh of square cells and at a 45 degree angle to the mesh directions. The function has initial values equal to one in a square-shaped region and zero outside this region. The calculations differ in the convection schemes used; one uses QSOU and two others use explicit donor cell and inter-polated donor cell. Only the explicit donor cell and interpolated donor cell schemes are available as options when using the KIVA code. Two calculations use the fully-implicit donor cell and QUICK[28] schemes, which are not available

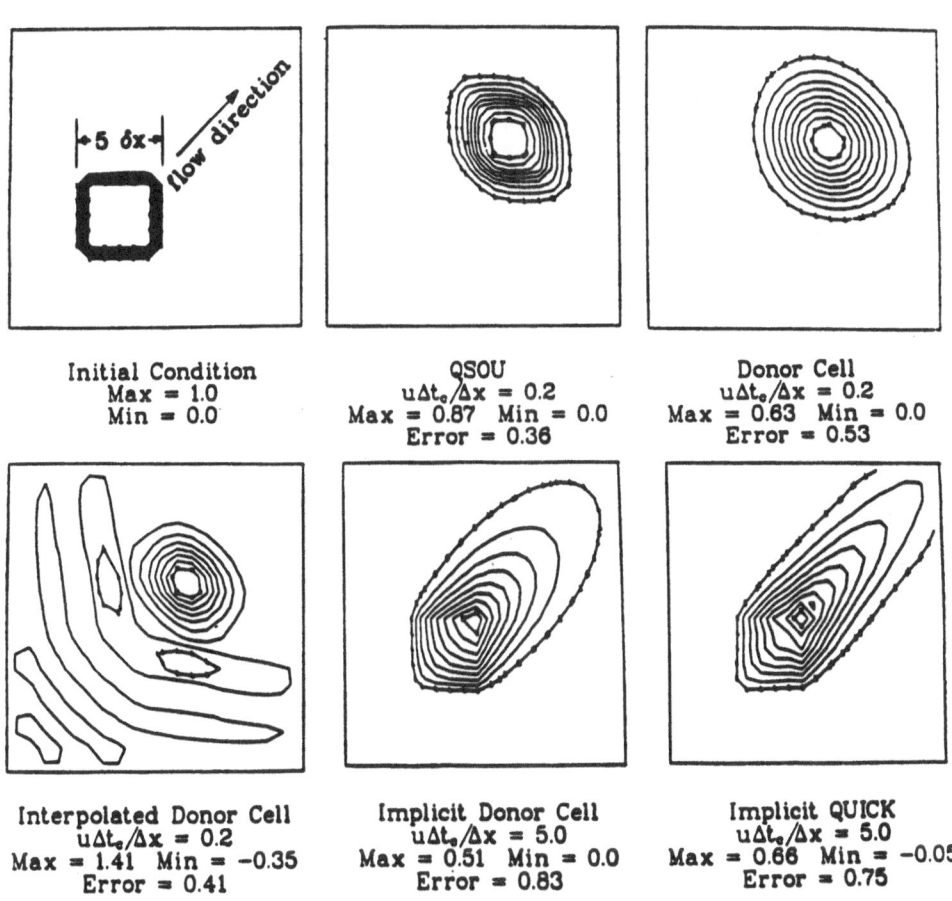

Fig. 1. *Isopleths from calculations of convection of a square-shaped scalar profile.*

in KIVA-II or KIVA, and timesteps that are twenty-five times larger than the timestep used for the explicit schemes. These larger timesteps are chosen to show that although they are unconditionally stable, the fully-implicit methods can be much less accurate than subcycled explicit methods when $u\,\Delta t/\Delta x > 1$. The results are plots of isopleths of the computed solution at a time when the exact solution is a translation of the initially square-shaped region a distance equal to the side of the square in each mesh direction. The reported error is the root-mean-square error between the computed and exact solutions, averaged over the region in which the exact solution is unity. By this error criterion the QSOU solution is most accurate. The interpolated donor cell solution is more accurate than donor cell, but because interpolated donor cell is not monotone its solution assumes values that lie outside the range of the initial conditions. The implicit calculations using the large timestep are least accurate.

IV. NEW OR IMPROVED PHYSICAL SUBMODELS

A.   Turbulence Models

The user of KIVA-II has the option of using one of two turbulence models: a two equation $k-\varepsilon$ model that is not available in KIVA and a subgrid scale (SGS) model that is a modification of the one-equation model of KIVA.[2] In this section we give the equations of both models and illustrate some differences between them with a computational example.

The $k-\varepsilon$ equations are the standard ones for incompressible flow,[17] modified to include volumetric expansion effects[18] and spray/turbulence interactions.[19]

$$\frac{\partial \rho k}{\partial t} + \nabla \cdot (\rho \mathbf{u} k) + \tfrac{2}{3} \rho k \nabla \cdot \mathbf{u} = \sigma : \nabla \mathbf{u} + \nabla \cdot (\mu_T/\sigma_k \nabla k) - \rho \varepsilon - W_s \tag{1}$$

and

$$\frac{\partial \rho \varepsilon}{\partial t} + \nabla \cdot (\rho \mathbf{u} \varepsilon) + (\tfrac{2}{3} c_{\varepsilon_1} - c_{\varepsilon_3}) \rho \varepsilon \nabla \cdot \mathbf{u} = \frac{\varepsilon}{k} c_{\varepsilon_1} (\sigma : \nabla \mathbf{u}) + \nabla \cdot (\mu_T/\sigma_\varepsilon \nabla \varepsilon) - c_{\varepsilon_2} \rho \frac{\varepsilon^2}{k} - c_s \frac{\varepsilon}{k} W_s . \tag{2}$$

In these equations, ρ is the fluid mass density and $\mathbf{u}$ the fluid velocity. The stress tensor σ is given by

$$\sigma = \mu_T [\nabla \mathbf{u} + (\nabla \mathbf{u})^T] - \tfrac{2}{3} \mu_T \nabla \cdot \mathbf{u}\, I , \tag{3}$$

where the turbulent viscosity is obtained from

$$\mu_T = \rho c_\mu \frac{k^2}{\varepsilon} + \mu_\ell \tag{4}$$

and μ_ℓ is the laminar viscosity. The spray interaction term W_s accounts for the work done by the turbulence in dispersing the spray droplets. KIVA-II uses standard values of the constants $c_{\varepsilon 1}$, $c_{\varepsilon 2}$, $c_{\varepsilon 3}$, σ_k, σ_ε, and c_μ, and these are given in Table III. A value of c_s equal to 1.50 has been suggested[20] based on the postulate of length scale conservation, and has been found to give good agreement in experimental comparisons.[19]

The wall boundary conditions for these equations are

$$\nabla k \cdot \mathbf{n} = 0 \tag{5}$$

and

$$\varepsilon = \left[\frac{c_\mu}{\sigma_\varepsilon (c_{\varepsilon_2} - c_{\varepsilon_1})} \right]^{1/2} \frac{k^{3/2}}{L} \; , \tag{6}$$

where $\mathbf{n}$ is the unit normal to the wall and (6) is prescribed at the centers of cells that lie next to the wall and whose centers are distance L from the wall. In addition, when the $k - \varepsilon$ model is used, wall momentum and energy losses are calculated by matching to turbulent law-of-the-wall profiles. The details of this calculation are given in Appendix B of Ref. 1.

The SGS model of KIVA-II is implemented by solving Eqs. (1) and (2) and requiring that the local length scales computed from Eq. (6) be less than or equal to $m\Delta x$, where Δx is a typical computational cell dimension and m equals 3 or 4. More precisely, we take ε to be the maximum of its value computed from Eq. (2) and

$$\varepsilon' = \left[\frac{c_\mu}{\sigma_\varepsilon (c_{\varepsilon_2} - c_{\varepsilon_1})} \right]^{1/2} \frac{k^{3/2}}{m\Delta x} \; . \tag{7}$$

In regions where $\varepsilon = \varepsilon'$ (or $L = m\Delta x$), the model reduces to a one-equation model that is identical to the SGS model of KIVA,[2] except some of the constants are slightly different. In regions where $\varepsilon > \varepsilon'$ (or $L < m\Delta x$) the model reduces to the full $k - \varepsilon$ model. The old SGS model in KIVA is in error in taking $L = m\Delta x$ because there are often regions of the flow field, for example near walls, where all length scales become less than $m\Delta x$. Near walls the velocity solution to

TABLE III

$k - \varepsilon$ MODEL CONSTANTS

$c_{\varepsilon 1} = 1.44$	$\sigma_k = 1.0$
$c_{\varepsilon 2} = 1.92$	$\sigma_\varepsilon = 1.3$
$c_{\varepsilon 3} = -1.0$	$c_\mu = 0.09$

one's turbulence equations should, under certain ideal conditions,[1] be the logarithmic velocity profile. This is true for the $k - \varepsilon$ model, but is not true for a model that assumes a constant length scale.

Although the $k - \varepsilon$ and SGS models of KIVA-II are formally very similar, the results they give can be very different because the SGS turbulent diffusivity can be much smaller than the $k - \varepsilon$ diffusivity. This is demonstrated by recent three-dimensional calculations we have performed of the decay of a swirling flow in a cylindrical vessel. The computational mesh is shown in Fig. 2. We initially specified an axisymmetric flow field with only an azimuthal or swirl velocity component. In experiments[21, 22] such flow fields are observed to be unstable and to develop a precessional motion of the swirl center. We performed calculations using both $k - \varepsilon$ and SGS, and plots of the swirl center location in the midplane of the cylinder are shown in Fig. 3. In the SGS calculation a precessional motion is obtained, but in the $k - \varepsilon$ model the flow remains nearly axisymmetric. Apparently, the $k - \varepsilon$ diffusivity is large enough to suppress the instability.

B.   Droplet Aerodynamic Breakup Model

Recently it has been found[23] that droplet aerodynamic breakup is important in certain engine spays. This finding is consistent with an earlier study[24] that it is primarily an aerodynamic mechanism that causes atomization of liquid jets. In response to these findings, we have developed and incorporated in KIVA-II a model for droplet aerodynamic breakup. We now summarize the basis for the model and give additional evidence for the importance of droplet breakup in diesel-type sprays.

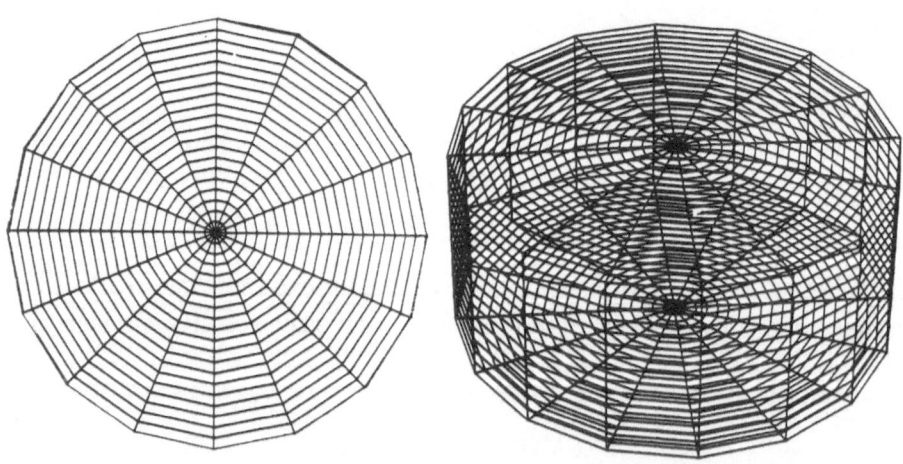

Fig. 2.   Top and perspective views of computational mesh for swirling flow calculations.

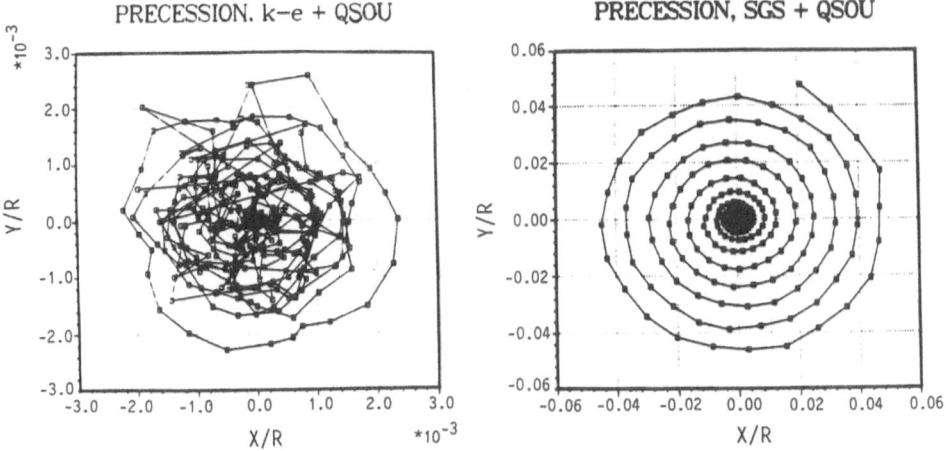

Fig. 3. *Swirl center location in $k-\varepsilon$ [left, $(x/R)_{max} \approx .0025$] and SGS [right, $(x/R)_{max} \approx .05$] calculations. R is the cylinder radius.*

We call our model the Taylor-Analogy Breakup (TAB) method because it is based on an analogy proposed by G. I. Taylor[25] between an oscillating droplet and a spring/mass system. The restoring force of the spring is analogous to surface tension forces, and external forces on the spring/mass system are analogous to aerodynamic forces on the droplet. To the analogy we have added damping forces due to liquid viscosity. A droplet breaks up when its distortion exceeds a certain critical value. Details concerning the equations solved and their solution procedure can be found in Refs. 1 and 26.

The model has been used in comparison with experimental measurements of diesel sprays,[26] and we give here a result that was not previously reported. Good agreement was obtained between computed and measured spray penetrations and drop sizes for an axisymmetric, nonvaporizing spray. We wished to use the computer program to assess the importance of aerodynamic breakup in the experimental sprays. One measure of how well the spray is broken up is the spray surface area S per unit length along the axis of the spray. An equation for S can be derived from the spray equation,[27] and for a steady, nonvaporizing spray this is

$$S \frac{\partial u_s}{\partial z} + u_s \frac{\partial S}{\partial z} = \left. \frac{dS}{dt} \right|_{coll} + \left. \frac{dS}{dt} \right|_{breakup} . \tag{8}$$

The term $dS/dt|_{coll}$ is the time-rate-of-change of S due to collisions and will be negative since collisions frequently result in coalescences, which reduce spray surface area. The term $dS/dt|_{breakup}$ is the time-rate-of-change of S due to breakups and will be positive. The quantity u_s is a surface-area averaged velocity. We calculated S, u_s, $dS/dt|_{coll}$, and $dS/dt|_{breakup}$ by time-averaging the computed steady solution for 500 computational cycles. The values of $\partial S/\partial z$ and $\partial u_s/\partial z$ were then found by simple numerical differentiation.

Figure 4 shows plots of the terms $S\,\partial u_s/\partial z$, $u_s\,\partial S/\partial z$, $-dS/dt|_{coll}$, and $-dS/dt|_{breakup}$ as functions of distance along the axis of the spray. It is seen that at all axial locations the collision and breakup terms are much larger than the other two and that they nearly balance each other, with differences between them being comparable in magnitude to the other two terms. If collisions or breakups, individually or in combination, were ignored, the rate of change of S for an observer moving with velocity u_s down the spray axis, which is $u_s\,\partial S/\partial z$, would differ from the computational result. This supports the contention that to calculate diesel-type sprays accurately, it is necessary to include the effects of both droplet collisions and breakups.

ACKNOWLEDGMENT

This work was supported by the U.S. Department of Energy, Office of Energy Utilization Research, Energy Conversion and Utilization Technologies Program.

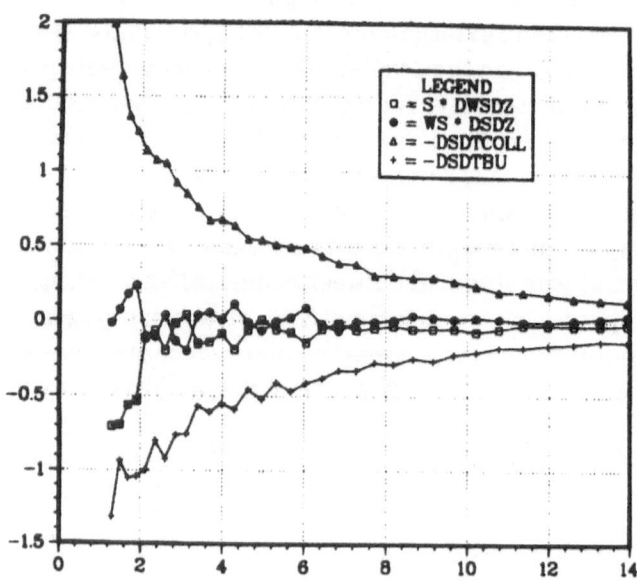

Fig. 4. Terms in S-equation (cm^2/cm-s) vs. distance from the injector (cm).

REFERENCES

1. A. A. Amsden, P. J. O'Rourke, and T. D. Butler, "KIVA-II: A Computer Program for Chemically Reactive Flows with Sprays," Los Alamos National laboratory report LA-11560-MS (May 1989).

2. A. A. Amsden, J. D. Ramshaw, P. J. O'Rourke, and J. K. Dukowicz, "KIVA: A Computer Program for Two- and Three-Dimensional Fluid Flows with Chemical Reactions and Fuel Sprays," Los Alamos National Laboratory report LA-10245-MS (February 1985).

3. A. A. Amsden, J. D. Ramshaw, L. D. Cloutman, and P. J. O'Rourke, "Improvements and Extensions to the KIVA Computer Program," Los Alamos National Laboratory report LA-10534-MS (October 1985).

4. C. W. Hirt, A. A. Amsden, and J. L. Cook, *J. Comput. Phys.* **14**, 227 (1974).

5. W. E. Pracht, *J. Comput. Phys.* **17**, 132 (1975).

6. J. K. Dukowicz, *J. Comput. Phys.* **35**, 229 (1980).

7. P. J. O'Rourke, "Collective Drop Effects in Vaporizing Liquid Sprays," Ph.D. Thesis 1532-T, Princeton University (August 1981).

8. S. V. Patankar, *Numerical Heat Transfer and Fluid Flow* (Hemisphere Publishing Corporation, Washington D.C., 1980).

9. P. J. O'Rourke and A. A. Amsden, "Implementation of a Conjugate Residual Iteration in the KIVA Computer Program," Los Alamos National Laboratory report LA-10849-MS (October 1986).

10. R. I. Issa, *J. Comput. Phys.* **62**, 1, 40 (1986).

11. P. J. O'Rourke and A. A. Amsden, "Three-Dimensional Numerical Simulations of the UPS-292 Stratified Charge Engine," SAE Technical Paper 870597 (1987).

12. P. J. Roache, *Computational Fluid Dynamics* (Hermosa Publishers, Albuquerque, New Mexico, 1982).

13. S. T. Zalesak, *J. Comp. Phys.* **31**, 335 (1979).

14. M. Chapman, *J. Comp. Phys.* **44**, 84 (1981).

15. L. D. Cloutman, "Numerical Experiments with CONCHAS-SPRAY: Installation of the FLOE Algorithm," Los Alamos National Laboratory report LA-10626-MS (February 1986).

16. B. van Leer, *J. Comput. Phys.* **32**, 101 (1979).

17. B. E. Launder and D. B. Spalding, *Mathematical Models of Turbulence* (Academic Press, New York, 1972).

18. W. C. Reynolds, "Modeling of Fluid Motions in Engines - An Introductory Overview," in *Combustion Modeling in Reciprocating Engines*, eds. J. N. Mattavi and C. A. Amann (Plenum Press, New York, 1981).

19. R. D. Reitz and R. Diwakar, "Structure of High Pressure Fuel Sprays," SAE Technical Paper 870598 (1987).

20. P. J. O'Rourke, oral presentation to the Twenty-third Direct Injected Stratified Charge Working Group Meeting, Los Alamos National Laboratory, March 1986.

21. T. M. Dyer, "Characterization of One- and Two-Dimensional Homogeneous Combustion Phenomena in a Constant Volume Bomb," SAE paper 790353 (1979).

22. S. Wahiduzzaman and C. R. Ferguson, "Convective Heat Transfer from a Decaying Swirling Flow within a Cylinder," 8th International Heat Transfer Conference, Paper #86-IHTC-253 (August 1986).

23. R. D. Reitz and R. Diwakar, "Effect of Drop Breakup on Fuel Sprays," SAE Paper 860469 (1986).

24. R. D. Reitz and F. V. Bracco, "On the Dependence of Spray Angle on Nozzle Design and Operating Conditions," SAE Paper 790494 (1979).

25. G. I. Taylor, "The Shape and Acceleration of a Drop in a High Speed Air Stream" in *The Scientific Papers of G. I. Taylor*, ed. G. K. Batchelor Vol. III (University Press, Cambridge, 1963).

26. P. J. O'Rourke and A. A. Amsden, "The TAB Method for Numerical Calculation of Spray Droplet Breakup," SAE Technical Paper 872089 (1987).

27. F. A. Williams, *Combustion Theory* (Benjamin/Cummings, Menlo Park, California, 1985).

28. B. P. Leonard, *Comput. Meths. Appl. Mech. Engrg.* **19**, 59 (1979).

RENORMALIZATION CONCEPT OF TURBULENT FLAME SPEED

Gregory I. Sivashinsky[*]

The Levich Institute for Physico-Chemical Hydrodynamics
The City College of New York, New York, N.Y. 10031

1. Introduction

It is well known that the main difficulty of the theoretical description of flame propagation through the turbulent flow field is the wide range of spatio-temporal scales involved. However, it transpires that the premixed flame is the system ideally suitable for the so-called renormalization group method (Yakhot ,1988a; Sivashinsky, 1988).

The basic idea of the renormalization approach is the successive averaging over gradually increasing scales. After its remarkable success in the theory of phase transition (Wilson, 1971) this idea proved to be quite effective in the evaluation of turbulent transport coefficients (Rose, 1977; Moffatt, 1981; Zeldovich, 1982; Yakhot and Orszag, 1986) and now, we hope, also in solving the problems of turbulent combustion.

There are a number of ways in which the renormalization group idea may be realized in the study of turbulent flame propagation. One way, adopted by Yakhot (1988a) is the successive averaging by perturbative elimination of small spherical shells of high-wavenumber modes and the "ε-expansion" (by which the properties of a $E(k) \sim k^{1-2\varepsilon/3}$ spectrum range are examined through an expansion in powers of ε about the properties of a spectrum $E(k) \sim k$).

We propose an alternative implementation of the renormalization group approach. The basic idea of the new method, which is akin in spirit to Wilson's (1965) block-renormalization scheme, is the replacement of the actual turbulent flow field of continuous spectrum by a cascade of effectively isotropic self-similar flows with widely separated scales. The relations established for the discrete cascade are then extrapolated to the original continuous system. As a result an equation for turbulent flame speed as a function of the turbulent field intensity is obtained.

[*] On leave from Department of Mathematical Sciences, Tel-Aviv University, Ramat-Aviv, Tel-Aviv 69978, ISRAEL.

Compared to the mode elimination procedure, the cascade approach seems to have a wider range of validity (as far as the turbulence intensity is concerned) and thus, to be capable of covering certain nontrivial effects unfeasible by more conventional methods. Of course, this is provided that the cascade model reflects correctly the basic features of the actual flame-flow interaction. This question is still to be answered.

This paper presents a refined, and we believe, more natural approach to the cascade-renormalization concept of turbulent flame speed introduced in our first communication on this topic, recently published in CST (1988).

2. The passive front model of flame propagation

Assuming turbulence to be large scaled, the premixed flame may be regarded as a geometrical surface moving relative to the gas with some constant normal speed U_L. Any change in the flow field immediately affects the configuration of the flame front. In turn, due to heat expansion, the flow field is influenced by the flame.

Due to the difficulty of the corresponding mathematical problem, it is instructive, as a preliminary step, to elucidate the partial features of the phenomenon by studying idealized formal models.

In this respect, it is useful to have a detailed description of the dynamics of the front moving relative to the prescribed divergenceless flow $\vec{w}$.

The corresponding passive front model may be formulated in terms of the single evolution equation

$$F_t + \vec{w} \cdot \nabla F = U_L |\nabla F| \ ,$$

$$\operatorname{div} \vec{w} = 0 \tag{2.1}$$

where $F(\vec{x},t)$ is a scalar field whose zero-level surface $F = 0$ represents the flame front configuration.

For all its seeming simplicity the passive front (or constant density) model is still rich enough to provide important insights relevant to more realistic models (cf. Williams, 1985; Kerstein, Ashurst and Williams, 1988; Osher and Sethian, 1988).

If the flame propagates through a motionless media ($\vec{w} = 0$) Eq. (2.1) admits the solution

$$F = U_L t - \vec{n} \cdot \vec{x} \qquad (|\vec{n}| = 1) \tag{2.2}$$

corresponding to an undisturbed planar flame front (Figure 1(a)).

If $\vec{w}$ is a homogeneous and isotropic turbulent flow field with zero spatio-temporal mean ($\langle \vec{w} \rangle_{\vec{x},t} = 0$) the solution (2.2) is modified to become:

$$F = U_T t - \vec{n} \cdot \vec{x} + G(\vec{x},t) \tag{2.3}$$

Where U_T is the effective (turbulent) flame speed and G is the fluctuation, i.e., $\langle G \rangle_{\vec{x},t} = 0$ (Figure 1(b)).

Due to the isotropy of the flow field, U_T does not depend on the orientation of $\vec{n}$.

For U_T to be $\vec{n}$-independent, $\vec{w}$ may in fact be only effectively isotropic and possibly even nonrandom (e.g. an appropriately arranged quasi-periodic flow).

3. Splitting of the flow field $\vec{w}$

Let the turbulent flow field $\vec{w}$ be characterized by a power law energy and frequency spectra E(k) and ω(k).

To guarantee the absence of a typical length we set (cf. Zeldovich, 1982)

$$\omega(k) = \beta k W_{rms}(k) \tag{3.1}$$

where β is a parameter of the order of unity, and $W_{rms}(k)$ is the k-dependent turbulence intensity defined as

$$W_{rms}^2(k) = 2 \int_k^\infty E(k') dk' \tag{3.2}$$

Condition (3.1) holds, for example in Kolmogorov turbulence, where

$$\omega(k) \sim \bar{\varepsilon}^{1/3} k^{2/3} \; , \; W_{rms}(k) \sim \bar{\varepsilon}^{1/3} k^{-1/3} \tag{3.3}$$

($\bar{\varepsilon}$ is the mean dissipative rate).

Consider the identity

$$W_{rms}^2(K) \equiv \int_0^{W_{rms}^2(K)} W_{rms}^2(k) d\left[\ln W_{rms}^2(k)\right] \tag{3.4}$$

where K^{-1} may be regarded as an integral scale of turbulence.

Let us replace the continuous range of wave numbers k ($K < k < \infty$) by a discrete one

$$k \to k_n \qquad (n = 0, 1, ..., N)$$
$$(K = k_N < k_{N-1} < ... < k_1 < k_0) \tag{3.5}$$

and replace the integral in (3.4) by a finite sum

$$\int_0^{W_{rms}^2(k)} W_{rms}^2(k) d\left[\ln W_{rms}^2(k)\right] \to \sum_{n=0}^{N-1} W_{rms}^2(k_n) \tag{3.6}$$

where the differential is replaced by unity, i.e.,

$$d\left[\ln W_{rms}^2(k)\right] \to \ln\left[W_{rms}^2(k_{n+1})\right] - \ln\left[W_{rms}^2(k_n)\right] = 1 \tag{3.7}$$

The latter relation yields

$$W^2_{rms}(k_{n+1}) = e \, W^2_{rms}(k_n) \tag{3.8}$$

Hence

$$W^2_{rms}(k_n) = e^n W^2_{rms}(k_o) \tag{3.9}$$

Introduce the system of N effectively isotropic self-similar flow fields

$$\vec{w}_n = \vec{w}_n(k_n\vec{x}, \omega_n t) \tag{3.10}$$

where

$$\langle \vec{w}_n \cdot \vec{w}_n \rangle = W^2_{rms}(k_n) \tag{3.11}$$

$$\omega_n = \beta k_n \, W_{rms}(k_n)$$

Following the basic assumptions of the Kolmogorov theory that large scale eddies convect small scale eddies without causing their significant distortion we introduce the superposition $\vec{w}$, defined as

$$\vec{w} = \sum_{n=0}^{N-1} \vec{w}_n \left[k_n(\vec{x} - \vec{v}_n t), \omega_n t \right] \tag{3.12}$$

where

$$\vec{v}_n = \sum_{m=n+1}^{N-1} \vec{w}_m[k_m(\vec{x} - \vec{v}_m t), \omega_m t] \; ,$$

$n = 0, 1, ..., N - 2$ and $\vec{v}_{N-1} = 0$.

Since k_n/k_{n-1} and ω_n/ω_{n-1} are of the order of unity, the scales of the flow field (3.12), being distinct are not, strictly speaking, strongly separated. Thus, despite the undertaken discretization, the whole problem is still not readily accessible to analytical treatment.

To make the problem more pliable, introduce into the system an artificial small parameter ε:

$$k_n \to \varepsilon^n k_n \; , \quad \omega_n \to \varepsilon^n \omega_n \tag{3.13}$$

The modified flow field (3.12) then becomes

$$\vec{w} = \sum_{n=0}^{N-1} \vec{w}_n \left[\varepsilon^n k_n(\vec{x} - \vec{v}_n t), \varepsilon^n \omega_n t \right] \tag{3.14}$$

with

$$\vec{v}_n = \sum_{m=n+1}^{N-1} \vec{w}_m \left[\varepsilon^m k_m(\vec{x} - \vec{v}_m t), \varepsilon^m \omega_m t \right]$$

$n = 0, 1, ..., N - 2$ and $\vec{v}_{N-1} = 0$.

Thus, one obtains the multiple-scale flow field with strongly separated spatio-temporal scales. The associated problem of flame propagation is much simpler than that of the flow field (3.12) and may well be treated perturbatively by appropriately carried out multiple-scale asymptotic analysis.

As a result, one obtains the following recursive relation for the effective (turbulent) flame speed (Sivashinsky, 1988),

$$U_T^{(n)} = U_T^{(n-1)} S\left(\frac{W_{rms}(k_{n-1})}{U_T^{(n-1)}}\right)$$

(3.15)

where $U_T^{(n)}$ is the effective speed of the wrinkled flame propagating through a n-scale flow field and S is the governing function determined by a one-scale problem. In the latter case,

$$U_T^{(1)} = U_L \; S\left(\frac{W_{rms}(k_o)}{U_L}\right)$$

(3.16)

Recursive relation (3.15) admits to a simple geometrical interpretation: the turbulent (wrinkled) flame averaged over the scales k_o^{-1}, k_1^{-1}, ..., k_{n-2}^{-1} may be regarded as an effectively laminar (smooth) flame moving with a renormalized speed $U_T^{(n-1)}$ through a one-scale flow field $\vec{w}_{n-1}(k_{n-1}\vec{x}, \omega_{n-1}t)$.

We would like to emphasize that the special structure of the flow (3.14) imposing advection of small scale motions by large scale ones is quite cruical for the relation (3.15) to hold true.

4. One scale flow field

To determine the governing function S one has to consider a one scale problem for which

$$F_t + \vec{w}_o(k_o\vec{x}, \omega_o t)\nabla F = U_L\left|\nabla F\right|$$

(4.1)

$$\langle \vec{w}_o \cdot \vec{w}_o \rangle = W_{rms}^2(k_o) \;, \; \omega_o = \beta k_o W_{rms}(k_o)$$

(4.2)

$$F = U_T^{(1)}t - \vec{n} \cdot \vec{x} + G(\vec{x}, t)$$

(4.3)

$$U_T^{(1)} = U_L \, S\!\left(\frac{W_{rms}(k_o)}{U_L}, \beta\right) \tag{4.4}$$

For low level isotropic turbulence

$$S \approx 1 + \frac{1}{2} S''(0)\left|\frac{W_{rms}^2(k_o)}{U_L}\right|^2 \tag{4.5}$$

where for the three dimensional case

$$S''(0) = \frac{2}{3} \quad \text{(Clavin and Williams, 1979)} \tag{4.6}$$

Understanding of the global behavior of the function S is not a simple problem. The scanty data available from some concrete examples suggest the qualitative picture shown on Figure 2 (cf. Sivashinsky, 1988).

The emergence of the resonance points (Q_{res}) may be explained as follows: fluctuations of the flame front propagating through the flow field $\vec{w}(k_o\vec{x}, \omega_o t)$ are characterized by two typical frequencies ω_o and $U_T^{(1)}k_o$. The first one (ω_o) corresponds to the flow field oscillations, while the second ($U_T^{(1)}k_o$) is induced by the flow field inhomogeneity. The flame front undergoes maximal distortion when both frequencies coincide. This is clearly accompanied by a marked increase in the effective propagation speed $U_T^{(1)}$ (4.4).

The saturation effect ($S(\infty) < \infty$, Figure 2) is due to the damping influence of the flow field oscillations (Ashurst, Sivashinsky and Yakhot, 1988). This may be understood through consideration of the behavior of the interface between two unmixible fluids situated in the flow field of oscillating vorticity:

$$\vec{w} = 4W_{rms}(\sin k_o x \cos k_o y \cos \omega_o t \, , \, -\cos k_o x \sin k_o y \cos \omega_o t) \tag{4.7}$$

$$(\langle \vec{w} \cdot \vec{w} \rangle = W_{rms}^2)$$

If the flow is time independent ($\omega_0 = 0$) the interface wraps up spirally undergoing unlimited growth (Figure 3(b)). At $\omega_0 \neq 0$ the velocity field periodically changes its sign. This clearly produces a restraining influence on the interface growth, and at $\omega_0 \gg k_0 W_{rms}$ the interface remains nearly unaffected by the flow field oscillations (Figure 3(c)).

The essence of this effect certainly survives if the impenetrable interface is replaced by the flame front. As a result, as $\omega_0/k_0 W_{rms} \to \infty$ the wrinkled front gradually smooths out and its effective speed U_T (conditioned by the increase of the interface) drops down to U_L.

When $\omega_0/k_0 W_{rms} (k_0) = \beta$ is finite, ω_0 and $W_{rms}(k_0)$ grow simultaneously. As a result, $U_T^{(1)}$ approaches a certain finite level corresponding to an equilibrium between the smoothing effect of oscillations and the wrinkling effect of flow field inhomogeneity.

5. Asymptotic solution of the recursive equation (3.15)

The qualitative features of the function S being established, one can obtain the asymptotic solution of the recursive equation (3.15).

There are two basic alternatives:

i) At $N \gg 1$ and $S_{max} \geq \sqrt{e}$

$$U_T \simeq A \, W_{rms} \tag{5.1}$$

where the coefficient A is determined from the relation

$$S\left(\frac{1}{A}\right) = \sqrt{e} \tag{5.2}$$

ii) At $N \gg 1$ and $S_{max} < \sqrt{e}$

$$U_T \sim W_{rms}^{\alpha} U_L^{1-\alpha} \qquad (0 < \alpha < 1) \tag{5.3}$$

where

$$\alpha = 2\ln S(\infty) < 1 \tag{5.4}$$

In both relations (5.1) and (5.3) $U_T = U_T^{(N)}$, $W_{rms} = W_{rms}(k_N)$.

It is interesting that despite the invariance of the equation (3.5) with respect to the simple rescaling

$$U_T \rightarrow CU_T \, , \, W_{rms} \rightarrow CW_{rms} \tag{5.5}$$

the corresponding scale invariant solution (5.1) exists only for sufficiently small β (i.e., at $S_{max} \geq \sqrt{e}$)

At large β (i.e., at $S_{max} < \sqrt{e}$), a completely new solution (5.3) emerges, whose scaling properties are not universal, but are quite sensitive to the details of the underlying flow field.

One, therefore, has to be cautious enough in carrying over the scaling features of the equation (2.1) onto its possible solutions, which actually may not exist at all (cf. Yakhot, 1988b).

6. Differential equation for turbulent flame speed

The recursive relation (3.15) may be associated with a differential equation, which is clearly a much more convenient object to deal with.

As a first step rewrite (3.15) as

$$U_T^{(1)} = U_L \, S\!\left(\frac{W_{rms}(k_0)}{U_L}\right)$$

$$U_T^{(2)} = U_T^{(1)} \, S\!\left(\frac{W_{rms}(k_1)}{U_T^{(1)}}\right)$$

$$\dots$$

$$U_T^{(N)} = U_T^{(N-1)} \, S\!\left(\frac{W_{rms}(k_{N-1})}{U_T^{(N-1)}}\right) \tag{6.1}$$

The system (6.1) implies that

$$U_T^{(N)} = U_L \, S\!\left(\frac{W_{rms}(k_0)}{U_L}\right) \dots S\left(\frac{W_{rms}(k_{N-1})}{U_T^{(N-1)}}\right) \tag{6.2}$$

or

$$\ln\left(\frac{U_T^{(N)}}{U_L}\right) = \sum_{n=0}^{N-1} \ln\left[S\left(\frac{W_{rms}(k_n)}{U_T^{(n)}}\right)\right]$$

(6.3)

Since

$$\ln(W_{rms}^2(k_{n+1})) - \ln(W_{rms}^2(k_n)) = 1$$

(6.4)

the relation (6.3) may be rewritten as

$$\ln\left(\frac{U_T^{(N)}}{U_L}\right) = \sum_{n=0}^{N-1} \ln\left[S\left(\frac{W_{rms}(k_n)}{U_T^{(n)}}\right)\right]\left[\ln\left(W_{rms}^2(k_{n+1})\right) - \ln\left(W_{rms}^2(k_n)\right)\right]$$

(6.5)

We now take the bold step of returning to the continuous spectrum, replacing the summation by integration (cf. (3.6), (3.7)).

Thus we write

$$\ln\left(\frac{U_T(\bar{K})}{U_L}\right) = \int_0^{W_{rms}(K)} \ln\left[S\left(\frac{W_{rms}(k)}{U_T(k)}\right)\right] d\ln\left(W_{rms}^2(k)\right)$$

(6.6)

The integral equation (6.6) may be more conveniently written as an initial value problem for a first order differential equation

$$\frac{dU_T}{dW_{rms}} = 2\frac{U_T}{W_{rms}}\ln\left[S\left(\frac{W_{rms}}{U_T}\right)\right] = 0$$

$$U_T = U_L \text{ at } W_{rms} = 0$$

(6.7)

Eq. (6.7) being homogeneous is easily simplified by the introduction of the auxilliary independent variable

$$Q = \frac{W_{rms}}{U_T(Q)}$$

(6.8)

which transforms it into

$$\frac{dU_T}{dQ} = \frac{2U_T \ln S(Q)}{Q[1 - 2 \ln S(Q)]}$$

$$U_T = U_L \text{ at } Q = 0$$

(6.9)

There are a number of possibilities regarding the asymptotic solution of Eq. (6.7) at $W_{rms} \gg U_L$.

(i) If $S_{max} < \sqrt{e}$ the denominator of (6.9) never vanishes. In this case the straightforward asymptotic analysis yields

$$U_T \sim U_L^{1-\alpha} W_{rms}^{\alpha} \qquad (\text{cf. (5.3)})$$

(6.10)

$$\alpha = 2 \ln S(\infty) < 1 \text{ , since } S(\infty) < S_{max} < \sqrt{e}$$

Thus, $\dfrac{d^2 U_T}{dW_{rms}^2} < 0$ (power law bending, Figure 4(3)).

(ii) If $S_{max} = \sqrt{e}$, the asymptotic solution of Eq. (6.7) may be written as

$$U_T \simeq \left(\frac{W_{rms}}{Q_{res}}\right) + 0\left\{ W_{rms}\left[\ln\left(\frac{W_{rms}}{U_L}\right)\right]^{-1}\right\}$$

$$S(Q_{res}) = S_{max}$$

(6.11)

Thus, again $d^2U_T/dW_{rms}^2 < 0$ (quasi-linear bending, Figure 4(2)).

(iii) If $S_{max} > \sqrt{e}$, there is a point $Q = Q_o$ such that $S(Q_o) = \sqrt{e}$. In this case Eq. (6.7) yields

$$U_T \simeq \left(\frac{W_{rms}}{Q_o}\right) + O\left(W_{rms}^{\gamma} U_L^{1-\gamma}\right)$$

(6.12)

where

$$\gamma = 1 - \frac{2Q_o S'(Q_o)}{\sqrt{e}} < 1$$

If Q_o is close enough to Q_{res}, the exponent γ is positive, thus once again $d^2U_T/dW_{rms}^2 < 0$ (quasi-linear bending, Figure 4(2)).

Note that $U_T - W_{rms}/Q_o$ is unbounded as $W_{rms}/U_L \to \infty$.

For S_{max} considerably larger than $\sqrt{e}$, within the interval $0 < W_{rms}/U_L < Q_o$, S is likely to be well approximated by the Clavin-Williams relation (4.5)(4.6). In this case, as is readily seen, $\gamma < 0$ and $d^2U_T/dW_{rms}^2 > 0$ (no bending, Figure 4(1)).

7. Discussion

When other conditions are fixed the "bending" ($d^2U_T/dW_{rms}^2 < 0$) always emerges at sufficiently high values of β, i.e., when the turbulent eddies are markedly nonsteady. In Taylor's frozen turbulence with $\beta = 0$ and $S_{max} = \infty$ the bending is unlikely.

The bending effect found resembles that observed in some experimental studies (e.g. Abdel-Gayed, Al-Khishali and Bradley, 1984). However, in the absence of reliable estimates on the parameter β and the function S one may not be certain that the β-effect (i.e., temporal activity of the flow) is indeed the basic mechanism controlling the actual phenomenon. Furthermore, there are strong experimental and theoretical indications that the actual bending may at least be partially (if not entirely) induced by the drop in the burning speed, caused by the deformation of the thermo-diffusive flame structure as the flame propagates through the turbulent flow field. Due to this deformation, at high W_{rms} the dynamic equilibrium between the heat released in the reaction zone and the heat extracted from the flame may be violated causing the flame to slow down and even to quench entirely. There are some experimental indications of such a possibility. Studies of

Sokolik, et al. (1967) of confined turbulent flames have shown that there is a certain level of turbulence at which the speed of a turbulent flame approaches its maximal value. Further increase in the intensity of turbulence causes a sharp drop in the flame speed and eventual extinction of the flame. This phenomenon has been reconfirmed in a recent work of Chomiak and Jarosinsky (1982).

Our recent analytical study (Sivashinsky and Berestycki, 1988) carried out with the unidirectional <u>time-independent</u> periodic flow field has shown that the periodically corrugated premixed flame may indeed be quenched provided the intensity of the flow fluctuations is high enough. To capture this effect one clearly has to take into account diffusive-thermal and Lewis number effects which control variations of the flame speed along the front.

It is plausible that if such effects may be modeled in one scale flows they will survive in multiple scale flows as well.

The systematic extension of the cascade-renormalization scheme to the thermo-diffusive flame model is one of the most intriguing and challenging problems of the theory.

Acknowledgement

These studies were supported by the U.S. Department of Energy under Grant No. DE-FG02-88ER13822.

References

Abdel-Gayed, R. G., Al-Khishali, K. J., and Bradley, D. (1984). Turbulent burning velocities and flame straining in explosions. Proc. Royal Soc. London <u>A391</u>, 393.

Ashurst, Wm. T., Sivashinsky, G. I., and Yakhot, V. (1988). Flame front propagation in nonsteady hydrodynamic field. Combust. Sci. and Tech. (1988), to appear.

Chomiak, J., and Jarosinsky, J. (1982). Flame quenching by turbulence. Comb. Flame <u>48</u>, 241.

Clavin, P., and Williams, F. A. (1979). Theory of premixed flame propagating in large-scale turbulence. J. Fluid Mech. <u>90</u>, 589.

Kerstein, A., Ashurst, Wm. T., and Williams, F. A. (1988). Field equation for Huygens propagation in an unsteady homogeneous flow field. Phys. Rev. <u>A37</u>, 2728.

Moffatt, K. K. (1981). Some developments in the theory of turbulence. J. Fluid Mech. 106, 27.

Osher, S., and Sethian, T. A. (1988). Front propagation with curvature dependent speed: algorithms based on Hamilton-Jacobi formulations. J. Comp. Physics 31, 3571.

Rose, H. A. (1977). Eddy diffusivity, eddy noise and subgrid-scale modelling. J. Fluid Mech. 81, 719.

Sivashinsky, G. I. (1988). Cascade-renormalization theory of turbulent flame speed. Combust. Sci. and Tech. 62, 77.

Sivashinsky, G. I., and Berestycki, H. (1988). Flame extinction by periodic flow field, preprint.

Sokolik, A. S., Karpov, W. P., and Semenov, E. S. (1967). Turbulent combustion of gases. Combust. Expl. Shock Waves 3(1), 36.

Williams, F. A. (1985). Turbulent combustion, in Mathematics of Combustion, J. D. Buckmaster Ed., SIAM, Philadelphia, 97.

Wilson, K. G. (1965). Model Hamiltonians for local quantum field theories. Phys. Rev. 2B, 140, 445.

Wilson, K. G. (1971). Renormalization group and critical phenomena. I. Renormalization group and the Kadanoff scaling picture. Phys. Rev. B4, 3174.

Yakhot, V., and Orszag, S. A. (1986). Renormalization group analysis of turbulence. I. Basic Theory. J. Scientific Computing 1(1), 3.

Yakhot, V. (1988). Propagation velocity of premixed turbulent flame. Combust. Sci. and Tech. 60, 191.

Yakhot, V. (1988). Scale invariant solutions of the theory of thin turbulent flame propagation. Combust. Sci. and Tech., to appear.

Zeldovich, Y. B. (1982). Exact solution of the problem of diffusion in a periodic velocity field and a turbulent diffusion. Soviet Physics, Doklady, 27(10), 797.

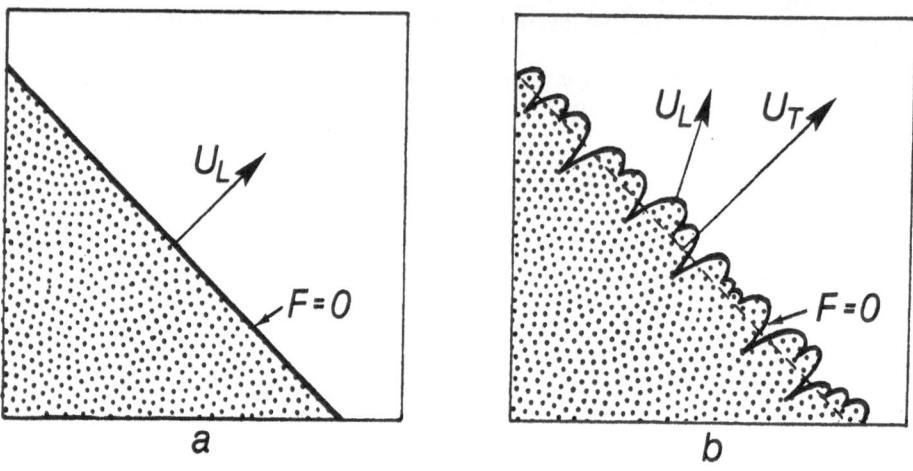

Figure 1 Diagrams showing: (a) - a planar laminar flame moving through a motionless gas at a normal speed U_L; (b) - a "planar" turbulent flame moving through an isotropic homogeneous turbulent flow field at an effective (turbulent) speed U_T.

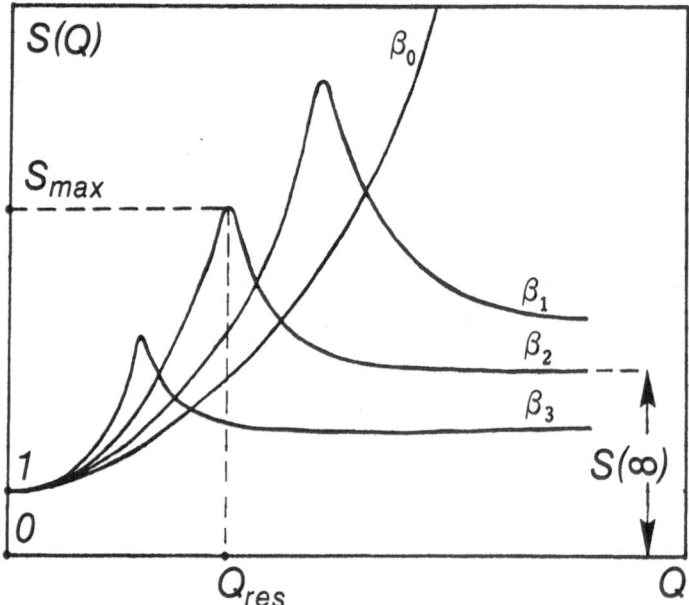

Figure 2 The governing function $S(Q)$ for different values of the parameter β.

$$(0 = \beta_0 < \beta_1 < \beta_2 < \beta_3)$$

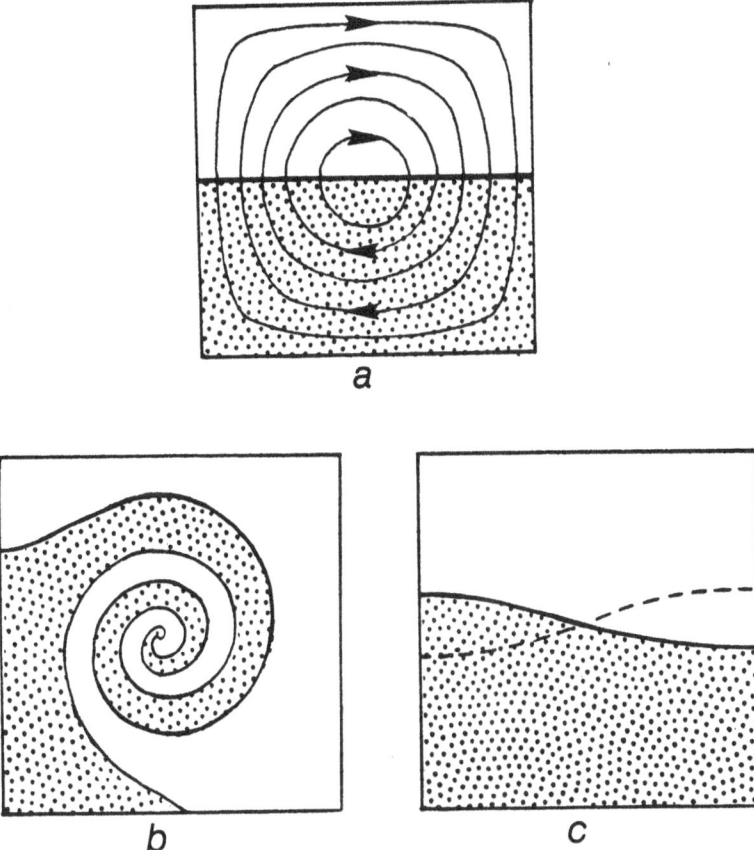

a

b

c

Figure 3 Diagrams showing the effect of oscillations on the evolution of the impenetrable interface between two fluids situated in the flow field of vorticity. (a) - the initial (planar) configuration of the itnerface (t = 0); (b) - the interface configuration at $\omega = 0$ and $t \gg (k_0 W_{rms})^{-1}$, (c) - the interface configuration at $\omega_0 \gg k_0 W_{rms}$ and $t \gg (k_0 W_{rms})^{-1}$.

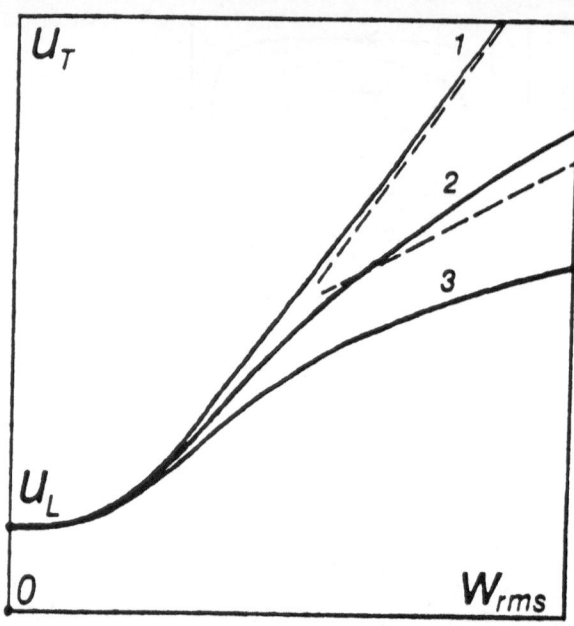

Figure 4 Wrinkled (turbulent) flame speed as a function of intensity of a multiple-scale flow field. (1) - no bending; (2) - quasi-linear bending; (3) - power law bending.

NUMERICAL SIMULATION OF IGNITION PROCESSES

J. Warnatz

Institute of Combustion Technology , Stuttgart University

1.   INTRODUCTION

Combustion processes result from a complex interaction of chemical reaction, transport, and convection. Thus, the simulation needs very much computational efforts and has to be restricted to relatively simple cases. Simple flame fronts can be described by time-dependent or stationary one-dimensional partial differential equation systems with temperature, species concentrations, and flow velocity as dependent variables. Even in the case of combustion of simple hydrocarbons, about 50 species involved in 500 reactions have to be considered, resulting in large non-linear stiff partial differential equation systems. The solution of those equations is considered for three different cases:

1.   Chemistry of Auto-Ignition Processes (e. g. in shock tubes or Otto engines): The simulation of homogeneous 0D problems leads to large non-linear, stiff ordinary differential equation systems which are solved by implicit methods. For the 1D case, much smaller mechanisms have to be used at the present.

2.   Induced Ignition and Instationary Flame Propagation: In this case, spatial discretization using finite differences leads to systems of differential/algebraic equations which are solved using both backward differencing formula (BDF) and extrapolation codes. The main problem (for both 1D and 2D problems) is the construction of adaptive, moving grid point systems in order to treat the flame front as well as the acoustic wave propagation occuring in ignition processes.

2.   SIMULATION OF THE CHEMISTRY OF AUTO-IGNITION

2.1 Background

Engine knock is a decades old problem connected with the operation of Otto engines, in special since the development of high performance versions (e. g. for airplanes). Very early it became apparent, that knocking is kinetically controlled, and must be closely connected with the chemistry of combustion, as can be seen for instance from the strong effect of addition of traces of inhibitors like lead tetraethyl (for reference see [1]). Extension of reaction schemes of the high temperature oxidation of hydrocarbons [2-6] to lower temperatures now allows the explanation of ignition delay up to octane using sophisticated mechanisms for the alkyl radical decomposition of higher hydrocarbons [7-9], as will be demonstrated in the following.

2.2 Calculation Method

Calculation of ignition delay times is performed by solving the conservation equations of the reaction system, considering the respective experimental conditions of pressure, volume, and temperature which, in some cases (e. g. end-gas of Otto engines), are time dependent. Treating homogeneous mixtures, those equations represent a system of ordinary differential equations with the time t as independent variable and with the species concentrations c_i as dependent variables. The equation system may be written as

$$\dot{c}_i = \dot{\omega}_i + c_i \cdot \left\{ \frac{\dot{p}}{p} - \frac{RT \sum \dot{\omega}_i}{p} - \frac{\dot{T}}{T} \right\}$$

if the temperature T and the pressure P are given, or as

$$\dot{c}_i = \omega_i - c_i \cdot \frac{\dot{V}}{V} \qquad i = 1, \ldots, n$$

if the temperature T and the volume V are given, with ω_i = molar rate of formation of species i, R = gas constant, n_s = number of species (typically ~50); $c_i(t=t_0)=c_{io}$ are the initial conditions.

For the given system of ordinary differential equations for the concentrations and their derivatives

$$\vec{c} = (c_1, \ldots, c_n)^T, \quad \vec{c}\,'(t) = f(t, \vec{c}\,'(t), \vec{p})^T$$

with the system parameters p_j (i. e. for example the rate coefficients), differentation with respect to the parameter vector $\underline{p}$ yields then the linear equation system for the (n x m) matrix S of the sensitivity coefficients $S_{ij}(t) = \partial c_i(t)/\partial p_j$

$$S'(t) = J(t) \cdot S(t) + f_p(t)$$

where $J(t)$ is the Jacobian matrix and $f_p(t)$ the n x m matrix of local parametric derivatives

$$f_{p_{i,j}}(t) = \left(\frac{\partial f_i}{\partial p_j}\right)_{p_{k\neq j},c}$$

with initial values which are given by the expressions at the start time of integration

$$S(t = t_0) = \frac{\partial c_0(p)}{\partial p}$$

Solution of this initial value problem is obtained by use of the programs DASSL [10] or LIMEX [11, 12], both of them designed to solve stiff differential/algebraic systems. Solution of the sensitivity equations is performed simultaneously with the solution of the model equations [13, 14].

2.3 Reaction Mechanism

The reaction mechanism used for the description of the oxidation of molecules up to C_8-species is very similar to that discussed in [2-6]. This reaction mechanism contains a complete oxidation scheme for hydrocarbons up to C_4-species [3]. Nearly all the rate coefficients for the higher species have to be estimated or deduced from data of analogous reactions (discussion in [15]).

2.4 Results at Isothermal Conditions

By solving the conservation equations for a 0D homogeneous isothermal system with chemical reaction, ignition delay times are calculated for several higher hydrocarbons up to octane and compared to experimental shock tube results.

The differences in ignition delay result from the pattern of alkyl radical decomposition that changes from one fuel to another. Fig. 1 shows a comparison between experimental data and results of the simulations.

Fig. 2 gives an example of a sensitivity analysis showing the rate-limiting role of H_2O_2 chemistry, OH + alkene reactions, C_8H_{17} decomposition and reaction with O_2, and initial attack on the parent hydrocarbon.

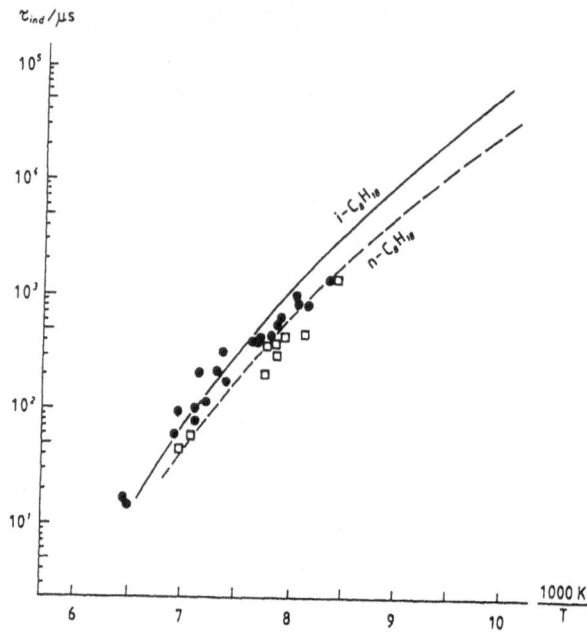

Fig. 1: Experimental results (points) and simulation (line) of ignition delay time τ_{ind} of i-C_8H_{18} and n-C_8H_{18} in O_2/Ar [16]

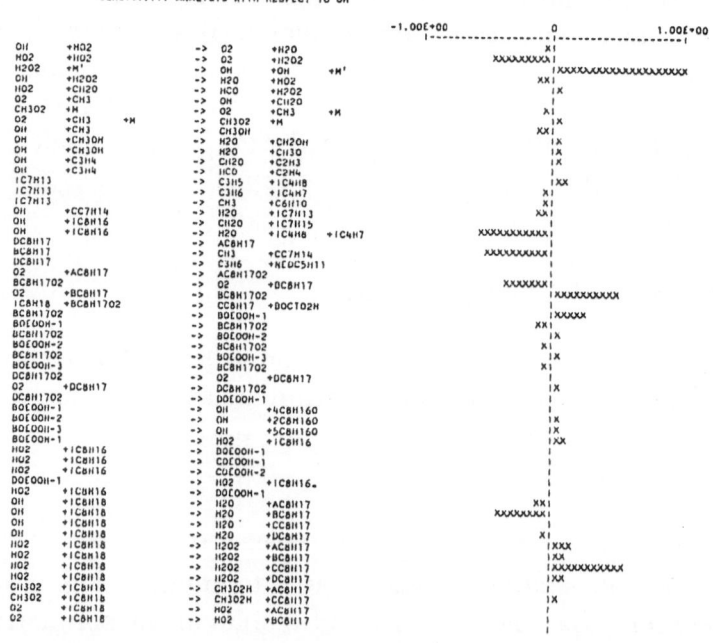

Fig. 2: Sensitivity analysis with respect to the OH concentration for a i-C_8H_{18}-O_2-Ar mixture, T = 800 K

2.5 Results at Changing Temperature and Pressure

The example considered here (Fig. 3) is knock in an engine driven with n-C$_8$H$_{18}$ [17]. Both knocking and non-knocking case can be predicted with a mechanism described above.

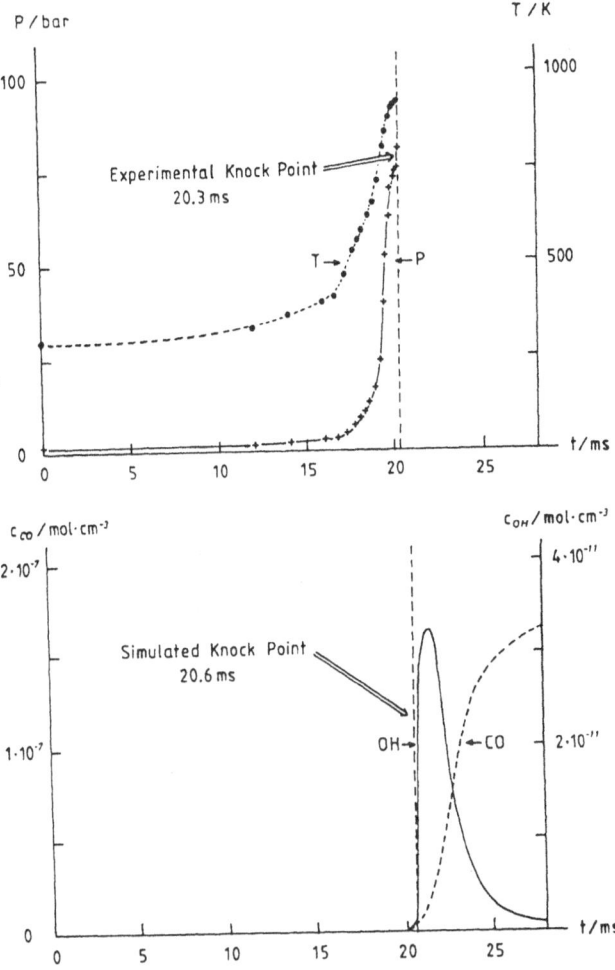

Fig. 3: Measured (upper drawing [17]) and calculated (lower drawing) knocking behaviour in an engine fueled with n-octane

3.   SIMULATION OF INSTATIONARY 1D-COMBUSTION

3.1 Background

A detailed knowledge of instationary combustion processes is required for the understanding of phenomena like ignition or extinction which are of importance in many practical applications (Otto engines, safety considerations). Time-dependent one-dimensional simulations have to be used to treat such instationary combustion processes. They are performed by solving the corresponding conservation equations (i. e. conservation of mass, energy, momentum and species mass) without restriction to systems with spatially uniform pressure, therefore allowing to study the interaction of pressure waves with the flame front and the influence of the uniform pressure assumption.

Spatial discretization using finite differences and an adaptive grid point system lead to a differential/algebraic equation system which is solved numerically by both extrapolation and backward differencing codes [10-12]. Computer simulations are used to study the thermal ignition of ozone-oxygen as well as hydrogen-oxygen mixtures. As comparisons with corresponding experiments show [18-20], minimum ignition energies in O_3-O_2 mixtures can be simulated within the experimental error limits.

3.2 Calculation Method

Mathematical simulation of the ignition process is performed by solving the corresponding system of conservation equations. After transformation into Lagrangian coordinates the instationary one-dimensional conservation equations may be written as

$$\frac{\partial r}{\partial \psi} - \frac{1}{\rho r^{\alpha}} = 0$$

$$\frac{\partial \rho}{\partial t} + \rho^2 \frac{\partial}{\partial \psi} (v r^{\alpha}) = 0$$

$$\frac{\partial v}{\partial t} + r^{\alpha} \frac{\partial p}{\partial \psi} - \frac{4}{3} r^{\alpha} \frac{\partial}{\partial \psi} \left(\rho \mu \frac{\partial}{\partial \psi} (v r^{\alpha}) \right) + 2\alpha r^{\alpha} \frac{v}{r} \frac{\partial \mu}{\partial \psi} = 0$$

$$\frac{\partial T}{\partial t} - \frac{1}{\rho C_p} \frac{\partial P}{\partial t} - \frac{1}{C_p} \frac{\partial}{\partial \psi} \left(\rho r^{2\alpha} \lambda \frac{\partial T}{\partial \psi} \right) + \frac{r^{\alpha}}{C_p} \sum_{i=1}^{n_s} \rho w_i V_i C_{pi} \frac{\partial T}{\partial \psi}$$

$$+ \frac{1}{\rho C_p} \sum_{i=1}^{n_s} \dot{w}_i h_i M_i - \frac{4\rho\mu}{3C_p} \left(\frac{\partial v r^{\alpha}}{\partial \psi} \right)^2 + \frac{2\alpha\mu}{C_p} \frac{\partial}{\partial \psi} (v^2 r^{\alpha-1}) = \dot{q}$$

$$\frac{\partial w_i}{\partial t} + \frac{\partial}{\partial \psi} \left(\rho r^{\alpha} w_i V_i \right) - \frac{\dot{w}_i M_i}{\rho} = 0$$

$$P - \frac{\rho R T}{M} = 0$$

with t = time, α = 0 for infinite slab, α = 1 for infinite cylinder, α = 2 for sphere, n_s = number of species, r = radius, P = pressure, T = temperature, ρ = density, v = velocity, w_i = mass fraction of species i, M_i = molar mass of species i, ω_i = mass scale rate of formation of species i, q = source term for deposition of energy, C_{pi} = constant pressure heat capacity of species i, C_p = constant pressure heat capacity of the mixture, h_i = specific enthalpy of species i, V_i = diffusion velocity of species i, λ = thermal conductivity of the mixture. The independent variables are: t and Ψ; the dependent variables are: r, T, P, w_i, v, and ρ.

The calculations are simplified to a rather great extent, if the pressure in the reaction volume is assumed to be uniform in space. In this case, the momentum equation then is replaced by the equation $\partial P/\partial r = 0$, and the continuity equation (automatically fulfilled after transformation into Lagrangian coordinates) is no longer needed.

The source term in the energy conservation equation is used to simulate ignition by external energy sources (corresponding to the heating by a laser beam in the experiments) [18-20]. For spherical and cylindrical geometries at the center of the reaction vessel symmetry conditions are used. Outer boundary conditions for the temperature and the species mass fractions in the simulations of ignition by external energy sources are simplified by assuming zero gradients of temperature and mass fractions; see [21] for details.

Dealing with auto-ignition phenomena in closed vessels, one has to take into consideration reactions occuring at the vessel surface, like surface recombination of atoms or surface destruction of reactive molecules as well as energy transfer to the vessel. The temperature at the outer boundary is assumed to be constant (thermostated in the experiments), and the system is considered at constant volume. Considering the reactions

$$a_{lk} A_l \xrightarrow{\gamma_k} \tilde{a}_{1k} A_1 + \tilde{a}_{2k} A_2 + \quad + \tilde{a}_{n,k} A_{n,}$$

occuring at the surface, the outer boundary conditions for species mass conservation are given by

$$0 = j_i'' = \rho w_i V_i + \dot{\omega}_i^s \qquad \dot{\omega}_i^s = \sum_{k=1}^{n,} \gamma_k Z_l M_i \{\tilde{a}_{ik} - \delta_{li} a_{lk}\}$$

with γ_k = surface destruction efficiency, Z_l = surface collision number of species l, a_{ik} = stoichiometric coefficient of species i in reaction k, n_s =

number of surface reactions, $\omega_i{}^s$ = mass scale rate of formation of species i per surface unit.

A system of coupled ordinary differential and algebraic equations which can be solved numerically is obtained by spatial discretization using finite differences. For the simulation of ignition by artificial energy sources, adaptive gridding has to be used due to the large ratio of vessel diameter to flame front thickness and to the diameter of the artificial energy source. A static approach is used which controls the grid point system after each successful step of the time integration by computing a new mesh which is compared with the old one. The grid point density is determined by equipartitioning the integral of a mesh function and inverse interpolation, the mesh function F given by a weighted norm of gradients and curvature of the dependent variables f_m

$$F(\psi) = \sum_{m=1}^{M} a_m \frac{\int_0^\psi \left| \frac{\partial f_m}{\partial \psi} \right| d\psi}{\int_0^{\psi_0} \left| \frac{\partial f_m}{\partial \psi} \right| d\psi} + b_m \frac{\int_0^\psi \left| \frac{\partial^2 f_m}{\partial \psi^2} \right| d\psi}{\int_0^{\psi_0} \left| \frac{\partial^2 f_m}{\partial \psi^2} \right| d\psi}$$

with a_m, b_m = weighting factors. If necessary, the adaption of the mesh is performed by insertion and deletion of grid points in the old mesh.

This method is chosen in order to minimize the error associated with the interpolation of the solution onto the new mesh. Piecewise monotonic cubic hermite interpolation [22] is used for the interpolation to avoid overshoots known from spline interpolation.

Simulations without the uniform pressure assumption are complicated by the problem of resolving the shock fronts. In order to avoid numerical instabilities resulting if shocks are not resolved by the grid point system, an artificial viscosity term (smart numerical diffusion), as it is proposed by *Richtmyer and Morton* [23] is introduced, which spreads the shocks over some grid points. The system of ordinary differential/algebraic equations is solved using the packages DASSL [10] or LIMEX [11, 12].

The block-tridiagonal structure allows an efficient numerical evaluation of the Jacobian. As the integration has to be restarted after each adaption of the grid point system, the one-step method LIMEX needs less computing time. Typical cpu-times on a CRAY-1 computer range from less than 1 minute (simulation without regridding and using the uniform pressure assumption) to more than 3 hours (simulation of ignition and flame propagation without the uniform pressure assumption).

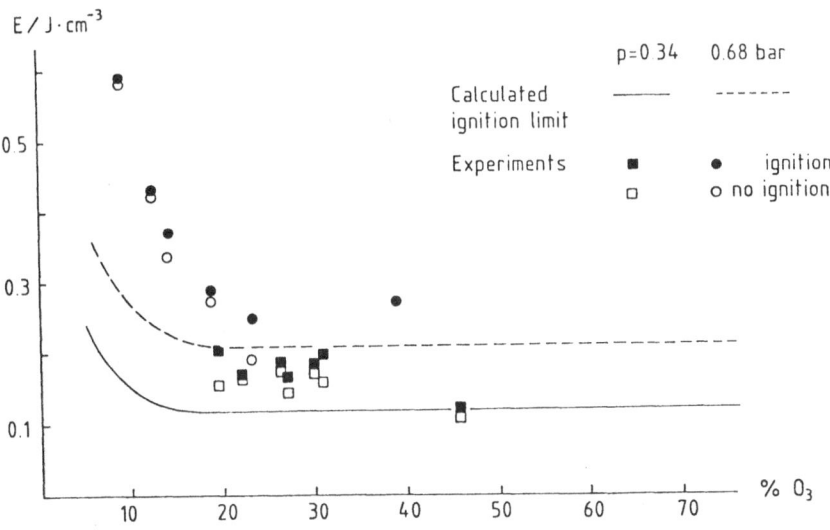

Fig. 4: Measured (points) and calculated (lines) minimum thermal ignition energies in O_3-O_2 mixtures at 0.34 and 0.68 bar; unburnt gas temperature T_u = 298 K, ignition source time $t_s = 1\ \mu s$

3.3 Results on Induced Ignition and Minimum Ignition Energies

As a comparison between experimental [18-20] and computational results shows (Fig. 4), the mathematical model described above is able to calculate minimum ignition energies in ozone-oxygen mixtures. For hydrogen-oxygen mixtures, only computational results are presented. An example of a simulation of an igniting hydrogen-oxygen mixture (without using the uniform pressure assumption) is shown in Figs. 5 and 6 for typical values of ignition source time t_s, ignition source diameter r_s, and ignition source energy E_s.

For short ignition times (<10 μs), the ignition process may be characterized as follows: In the heating period there is a quick rise of temperature and pressure in the source volume, the time being too short for the pressure to distribute over the whole volume of the mixture. Then a diverging pressure wave and a converging rarefaction wave are formed. The rarefaction wave is reflected at the vessel centre forming a diverging wave, the shock wave (moving in direction of the outer boundary) is reflected, forming a converging shock. The pressure waves perturb the propagation of the flame front, and the overall movement of the flame may be described as a superposition of the "normal" flame propagation and the oscillatory movements caused by the crossing shocks.

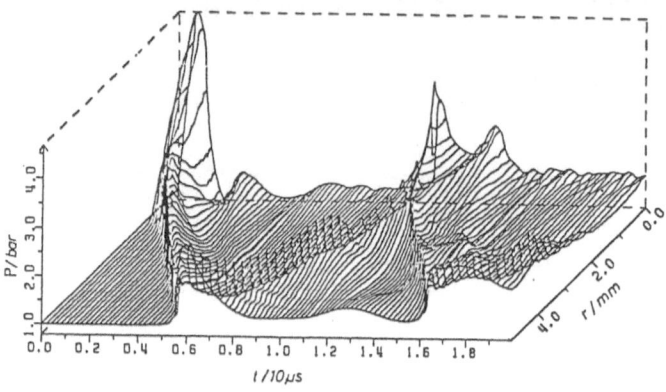

Fig. 5: Calculated pressure profiles in an igniting hydrogen-oxygen mixture (stoichiome-tric); cylindrical geometry, P = 1 bar, t_s = 1 µs, r_s = 1 mm, E_s = 4 J

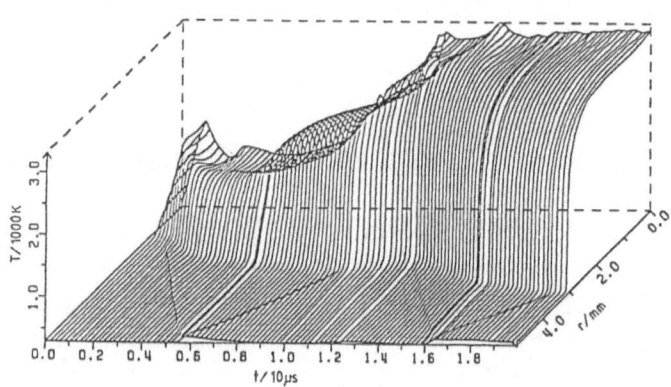

Fig. 6: Calculated temperature profiles in an igniting stoichiometric hydrogen-oxygen mixture; cylindrical geometry, P = 1 bar, t_s = 1 µs, r_s = 1 mm, E_s = 4 J

If ignition times are long in comparison to the time scale of the gas-dynamic processes, there is enough time for the pressure to equilibrate in the reaction system during the heating of the source volume, and the pressure is nearly uniform in space, which means that the uniform pres-sure assumption should be a good approximation in those cases. In fact there are really no differences in the computed minimum ignition ener-gies for an ignition time of 100 µs (for further reference on this topic see [21]).

Fig. 7 shows the dependence of minimum ignition energies on the radius of the external energy source for cylindrical geometry in a hydrogen-oxygen-mixture at an initial pressure of 1 bar at two different ignition times. The slope of the broken line indicates a proportionality between the minimum ignition energy and the source volume, i. e. the minimum energy density for ignition (or ignition temperature) is nearly constant. For smaller radii, diffusion and heat conduction cause the energy densities necessary for ignition to increase.

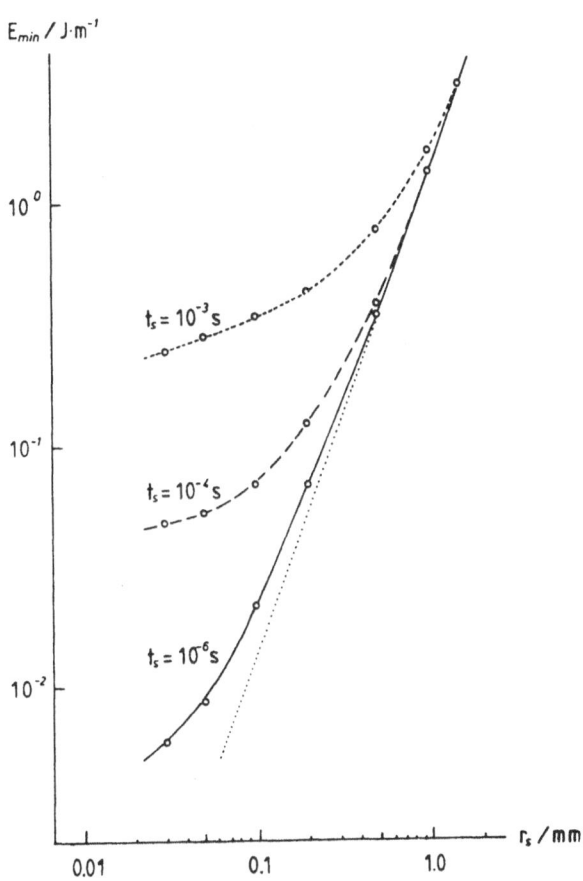

Fig. 7: Calculated minimum ignition energies in 2:1 hydrogen-oxygen mixtures for different radii of the external energy source (P = 1 bar)

The variation of minimum ignition energies with the mixture composition is shown in Fig. 8. There is almost no dependence of the minimum ignition energies on the mixture composition for a source radius of 1 mm between the flammability limits. For small ignition radii (calculation performed here for 0.2 mm) the minimum ignition energies become dependent on the mixture composition because rich mixtures (with a higher thermal conductivity) need higher minimum ignition energies than lean mixtures.

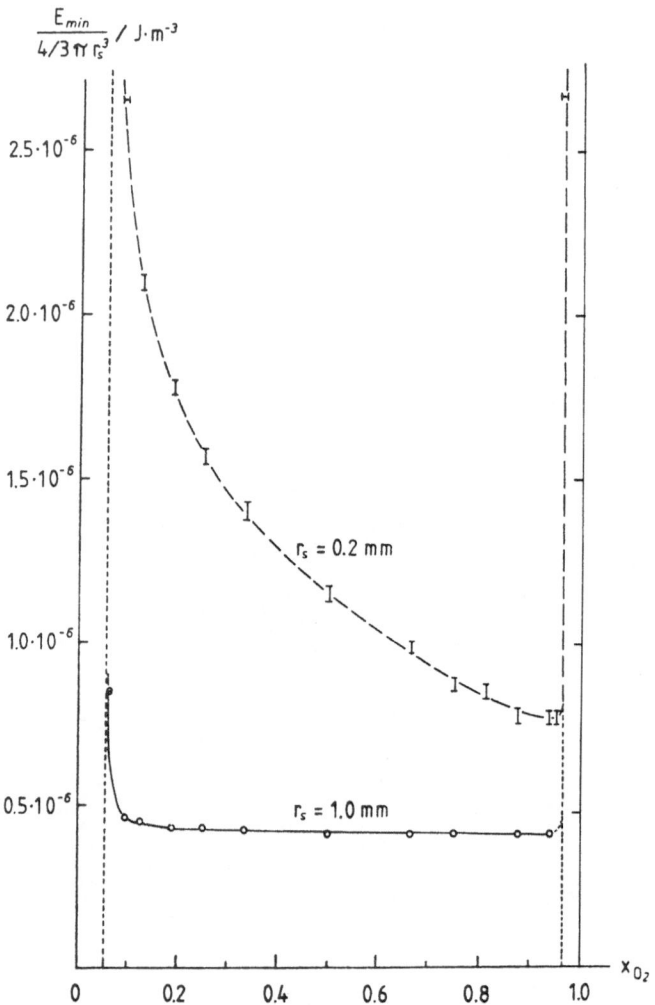

Fig. 8: Calculated minimum ignition energies in hydrogen-oxygen mixtures for different mixture compositions (uniform pressure assumption), spherical geometry, $\tau_s = 0.1$ ms, P = 1 bar

3.4. Results on Auto-Ignition in Hydrogen-Oxygen Mixtures

Since the detection of P-T explosion limits in H_2-O_2 mixtures in the thirtieth and fourtieth, many efforts have been made to explain this phenomemenon quantitatively. All of these attempts included severe simplifications as e. g. steady state assumptions, truncation of the mechanism or restriction to homogeneous (zero-dimensional) systems.

The mathematical model described above is able to simulate all the explosion limits of the hydrogen-oxygen system without those simplifications, using detailed chemistry, a multi-species transport model and realistic surface chemistry. Ignition limits are determined by simulating the reaction in closed vessels for various initial pressures and temperatures.

At the explosion limits, there is a quite sharp transition from slow reaction (moderate temperature rise over a long time, sometimes more than 100 s) to ignition (with an immediate temperature rise in the vessel centre and propagation of the flame).

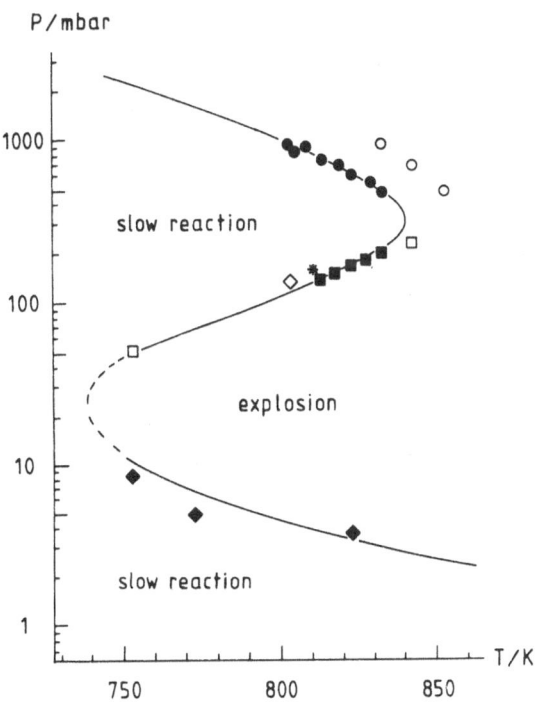

Fig. 9: Calculated (lines) and experimental (points) ignition limits for 2:1 mixtures of hydrogen and oxygen in vessels of different diameters; see [24] for further reference

Calculated P-T ignition limits for hydrogen-oxygen mixtures are shown in Fig. 9. As can be seen, calculated and experimental values are in quite good agreement.

4. INSTATIONARY 2D PROBLEMS

As another example for some typical methods used to solve stiff problems, an instationary 2D reactive flow (full chemistry in terms of sets of elementary reactions) shall be considered. Due to the large computational effort, only recently calculation of 2D Navier-Stokes equations including detailed chemistry in terms of sets of elementary reactions has become possible. An attempt for stationary cases has been reported [25]; the present paper demonstrates an extension to instationary cases.

4.1 Problem and Solution Method

Mathematical simulation of ignition processes have been performed by solving the corresponding system of conservation equations. The equation system [26] is simplified by using the ideal gas law and restricting the problem to 2D geometries (infinite rectangular column, finite cylinder). The finite cylinder geometry is considered here (see [27, 28] for further details).

The partial differential equation system, which consists of one algebraic (ideal gas law) and $n_s + 4$ partial differential equations (continuity, momentum, energy and species conservation equations; n_s = number of species), form an initial-boundary value problem for the dependent variables density ρ, velocities v_r and v_z, temperature T, and pressure P.

Again, spatial discretization of all derivatives different from that of the convective terms is performed by a standard central difference approximation for non-uniform grids, using nine-point stencils for the dissipative terms (to express mixed second derivatives) and five-point stencils in all other cases. The convective terms of the conservation equations have to be treated differently, because standard central difference approximations can cause severe numerical instabilities in regions of high gradients and curvatures. These instabilities can be avoided by using upwind schemes. However a major disadvantage of upwind schemes is the low order of approximation, leading to serious numerical diffusion.

In order to avoid this numerical diffusion, the approach of this work uses central difference approximations whenever possible and a modified upwind scheme whenever necessary (i. e. if overshoots might take place). In this case, locally monotonic interpolants were constructed as developed by *Fritsch and Butland* [22].

One further problem arising in the numerical solution of the partial differential equation system considered is the resolution of shock fronts. The
thickness of shocks is about 1 µm, and severe numerical instabilities
result if the shock is not resolved by the mesh. Because it is not necessary
to resolve the shock front for the problems considered in this work, an
artificial viscosity term ("numerical diffusion") proposed by *Richtmyer
and Morton* (see [23]) is applied, which spreads the shocks over a certain
number of grid points.

The system of ordinary differential/algebraic equations (resulting from
spatial discretization) again can be solved using the extrapolation code
LIMEX [11, 12]. This globally implicit method requires the evaluation of
the Jacobian matrix. Using nine-point stencils for the spatial discretization, the Jacobian has a block-nonadiagonal structure.

The dimension of the Jacobian is given by $n_{pde} \cdot n_l \cdot n_m$ where n_l and n_m
are the numbers of grid points in the different directions and n_{pde} the
number of partial differential equations. Even for the ozone-oxygen system
considered, the dimension of the Jacobian is 16000 (using a 40 x 50
spatial mesh). The solution of the linear equation systems $A \cdot x = z$ (required by the implicit method) is performed by a block-iterative Gauss-Seidel
method. The algorithm used for the iteration is given in detail in [29].

The simulation of the ozone-oxygen ignition (see below) takes about 15
hours on an IBM 3090; the code contains about 30 000 lines written in
FORTRAN.

4.2 Results and Discussion

The model described above is tested simulating two-dimensional ignition
processes in the ozone-oxygen system. The mixture is deposited in a cylindrical reaction vessel. Thermal ignition occurs in a pocket with a
diameter $r_s = 1.37$ mm and an ignition energy $E_s = 10.9$ mJ along the axis
with a decreasing energy density (in order to simulate absorption of
energy).

Figure 10 shows spatial profiles of temperature and pressure in the reaction vessel at different times, beginning at $t = 1$ µs, i. e. just after the external energy source has been turned off. Since the heating period is very
short, there is no possibility for the pressure to equilibrate all over the
reaction volume, and the pressure increase is approximately proportional
to the temperature increase.

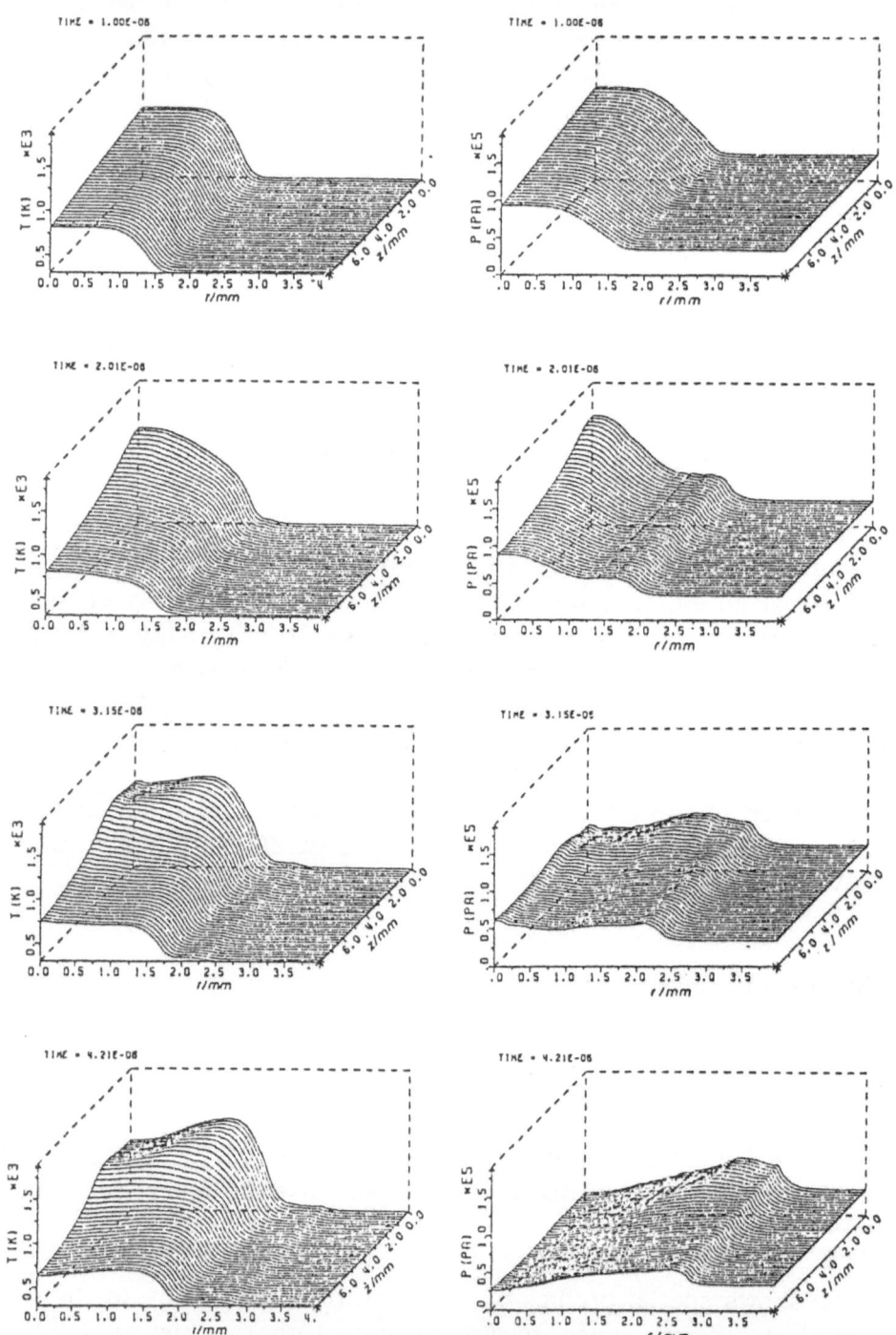

Fig. 10a: Calculated time dependent pressure and temperature profiles in an igniting ozone-oxygen mixture (28% O_3). Cylindrical geometry, P_0 = 0.34 bar, r_s = 1.37 mm, E_s = 10.9 mJ; time = 1 µs - 4 µs. Corresponding experiments are described in [19, 30]

165

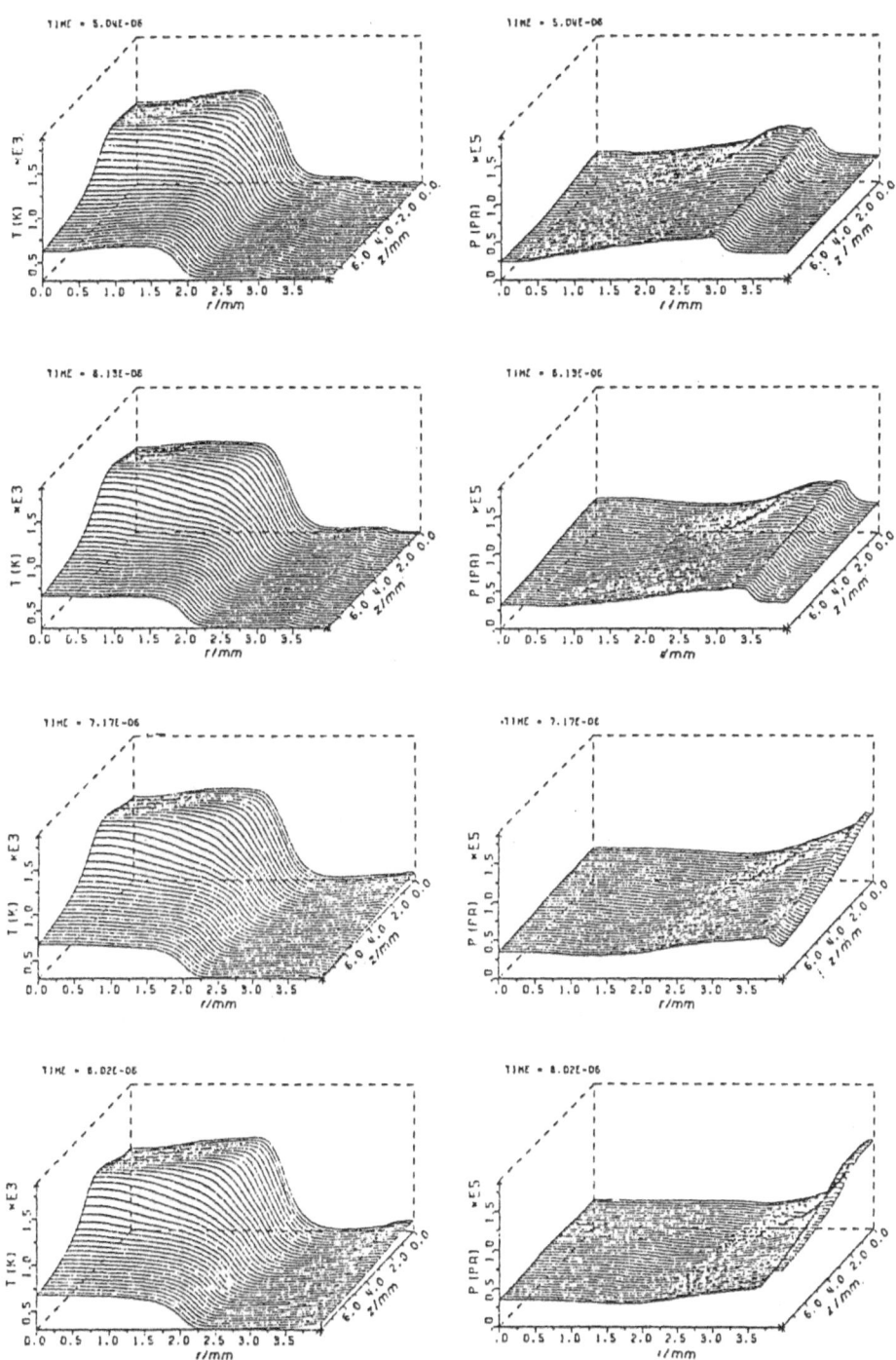

Fig. 10b: Calculated time dependent pressure and temperature profiles in an igniting ozone-
oxygen mixture (28% O_3). Cylindrical geometry, P_0 = 0.34 bar, r_s = 1.37 mm, E_s =
10.9 mJ; time = 5 μs - 8 μs. Corresponding experiments are described in [19, 30]

166

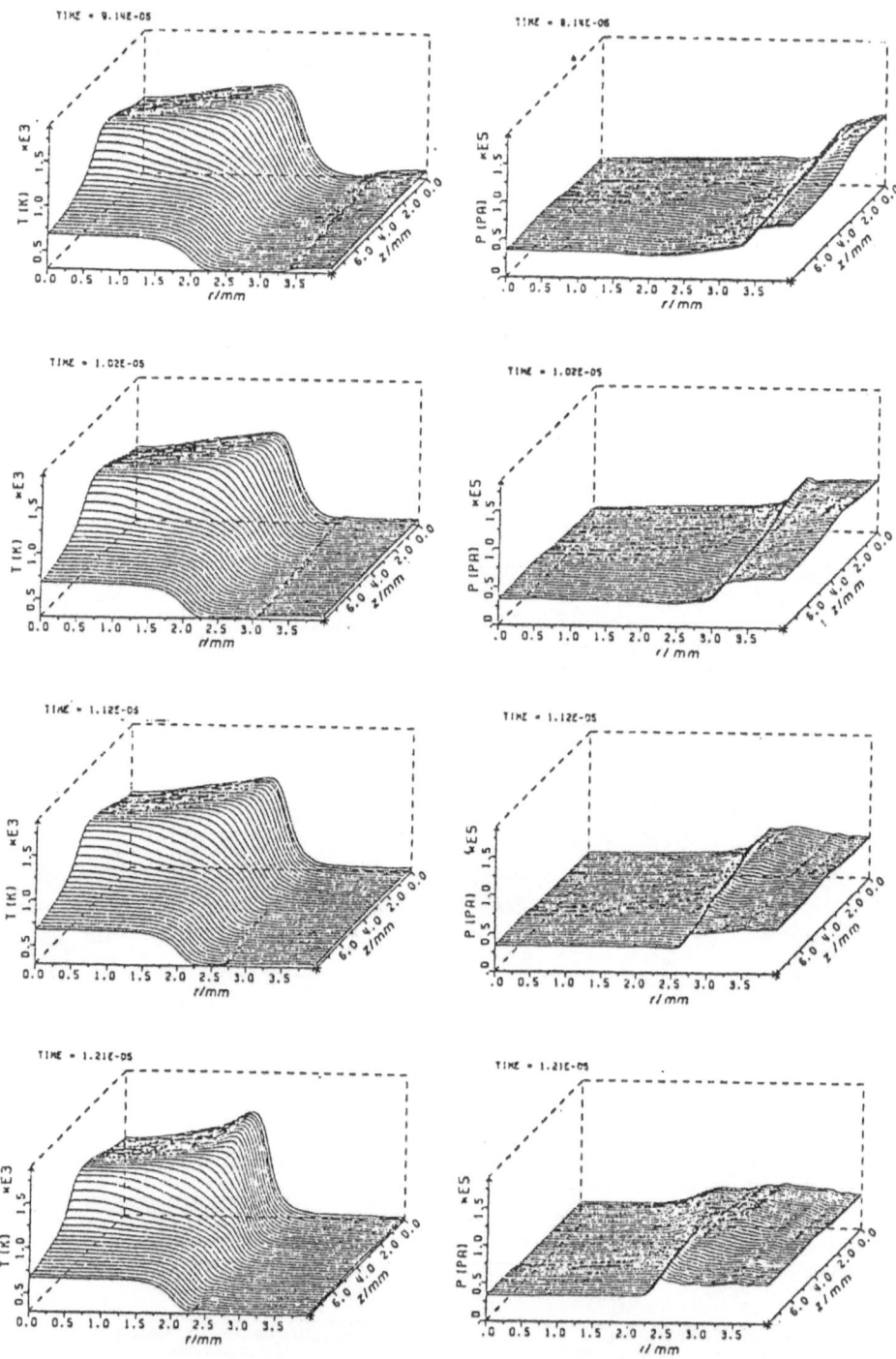

Fig. 10c: Calculated time dependent pressure and temperature profiles in an igniting ozone-oxygen mixture (28% O_3). Cylindrical geometry. P_0 = 0.34 bar, r_s = 1.37 mm, E_s = 10.9 mJ; time = 9 μs - 12 μs. Corresponding experiments are described in [19, 30]

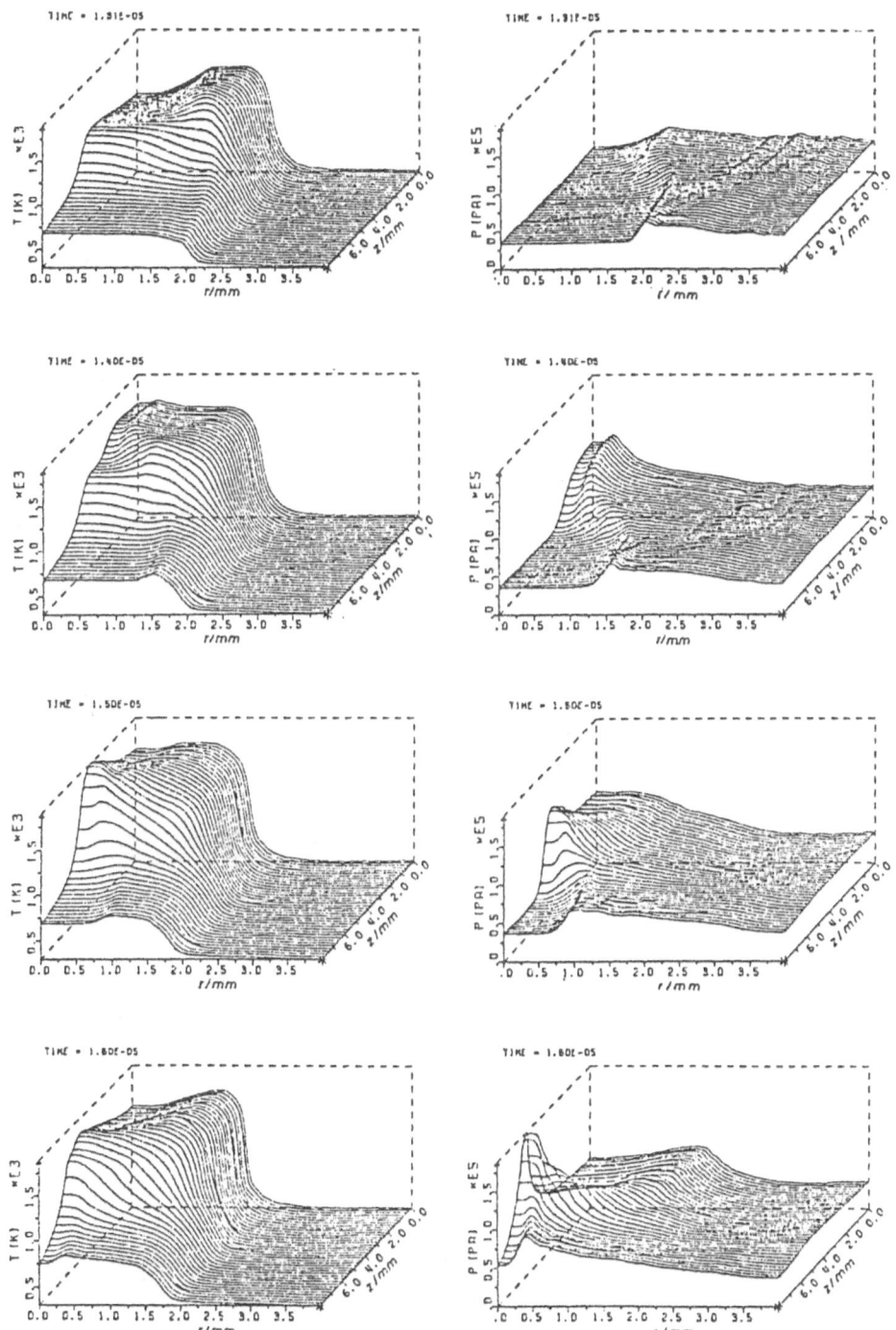

Fig. 10d: Calculated time dependent pressure and temperature profiles in an igniting ozone-oxygen mixture (28% O_3). Cylindrical geometry, P_0 = 0.34 bar, r_s = 1.37 mm, E_s = 10.9 mJ; time = 13 μs - 16 μs. Corresponding experiments are described in [19, 30]

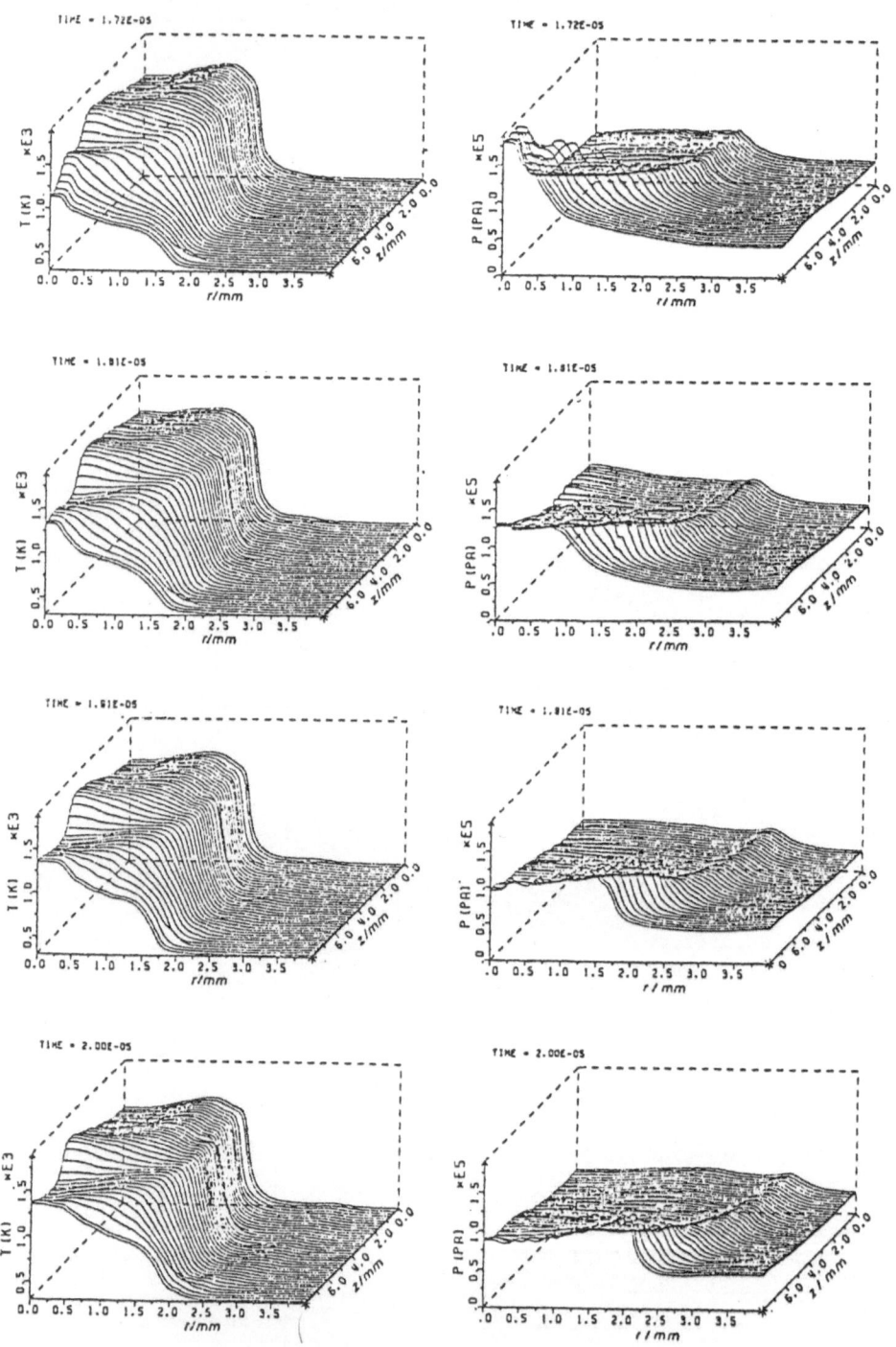

Fig. 10e: Calculated time dependent pressure and temperature profiles in an igniting ozone-oxygen mixture (28% O_3). Cylindrical geometry, P_0 = 0.34 bar, r_s = 1.37 mm, E_s = 10.9 mJ; time = 17 μs - 20 μs. Corresponding experiments are described in [19, 30]

The pressure gradient at the outer boundary of the ignition volume leads to the formation of a shock wave moving in the radial direction and to a rarefaction wave moving towards the cylinder axis. At locations where the amount of energy deposited during the heating period was high enough, ignition occurs after a short induction period, which can be seen from the rapid temperature rise. The flame front formed is moving in radial direction towards the outer boundary. Simultaneously, the flame propagates in the axial direction in accordance with experimental results [19, 30].

5.   ACKNOWLEDGEMENTS

The author is grateful to the "*Fonds der Chemischen Industrie*" and the "*Max-Buchner-Stiftung*" for financial support of this work and thanks *Dr. U. Maas* for valuable help.

6.   REFERENCES

[1]   B. *Lewis*, G. *v. Elbe*: Combustion, Flames, and Explosions in Gases. Academic Press, New York (1961)

[2]   J. *Warnatz*: 18th Symposium (International) on Combustion, p. 369. The Combustion Institute, Pittsburgh (1981)

[3]   J. *Warnatz*, in: W. C. Gardiner (ed.), Combustion Chemistry. Springer, New York (1984)

[4]   J. *Warnatz*: 20th Symposium (International) on Combustion, p. 845. The Combustion Institute, Pittsburgh (1984)

[5]   J. *Warnatz*, H. *Bockhorn*, A. *Möser*, H. W. *Wenz*: 19th Symposium (International) on Combustion, p. 197. The Combustion Institute, Pittsburgh (1983)

[6]   J. *Warnatz*: Ber. Bunsenges. Phys. Chem. 87, 1008 (1983)

[7]   F. *Dryer*, I. *Glassman*, in: C. T. Bowman, J. Birkeland (eds.), Alternative Hydrocarbon Fuels, Combustion and Kinetics. AIAA, New York (1979)

[8]   F. O. *Rice*: JACS 55, 3035 (1933)

[9]   C. *Esser*, U. *Maas*, J. *Warnatz*: Proc. Intl. Symp. on Diagnostics and Modelling of Combustion in Reciprocating Engines, p. 335. The Japanese Society of Mechanical Engineers, Tokyo (1985)

[10] L. R. *Petzold*: A Description of DASSL: A Differential/Algebraic System Solver, Sandia Report SAND82-8637. Sandia National Laboratories, Livermore (1982); IMACS World Congres, Montreal 1982

[11] P. *Deuflhard*, E. *Hairer*, J. *Zugck*: One-Step and Extrapolation Methods for Differential/Algebraic Systems. Univ. Heidelberg, SFB 123: Tech. Rep. 318, (1985)

[12] P. *Deuflhard*, U. *Nowak*: Extrapolation Integrators for Quasilinear Implicit ODEs. Univ. Heidelberg, SFB 123: Tech. Rep. 332, (1985)

[13] J. R. *Leis*, M. A. *Kramer*: Computers and Chemical Engineering 9, 93-96 (1985)

[14] M. *Caracotsius*, W. E. *Stewart*: Computers and Chemical Engineering 9, 359-365 (1985)

[15] W. J. Pitz, J. Warnatz, C. K. Westbrook: 22nd Symposium (International) on Combustion, in press

[16] R. Zellner, K. Niemitz: publication in preparation

[17] P. Pinchon: personal communication

[18] B. Raffel, J. Warnatz, H. Wolff, J. Wolfrum, R. J. Kee, in: J. R. Bowen, J.-C. Leyer, R. I. Soloukhin (Eds.), Dynamics of Reactive Systems, Part II, p. 335. AIAA, New York (1986)

[19] U. Maas, B. Raffel, J. Warnatz, J. Wolfrum: 21th Symp. (Intl.) Comb. The Combustion Institute, Pittsburgh (1987), in press

[20] B. Raffel, J. Warnatz, J. Wolfrum: Appl. Phys. B37, 189 (1985)

[21] U. Maas, J. Warnatz: presented at the 12th ICDERS, Warsaw 1987. AIAA (1989), in press

[22] F. N. Fritsch, J. Butland: SIAM J. Sci. Stat. Comput. 5, 300 (1984)

[23] R. Richtmyer, K. Morton, in: L. Bers, R. Courant, J. Stoker (eds.), Interscience Tracts in Pure and Applied Mathematics No.4, 2nd edition (1967)

[24] U. Maas, J. Warnatz: Comb. Flame 74, 53 (1988)

[25] M. D. Smooke: 3rd Workshop on Numerical Problems in Flame Propagation. Leeds (1988), publication in preparation

[26] R. B. Bird, W. E. Stewart, E. N. Lightfoot: Transport Phenomena. John Wiley & Sons, New York (1960)

[27] U. Maas, J. Warnatz: Zeitschrift für Physikalische Chemie NF (1988), in press

[28] U. Maas, J. Warnatz: Impact of Computing on Science and Engineering (1989), submitted

[29] D. Marsal: Die numerische Lösung partieller Differentialgleichungen in Wissenschaft und Technik, Bibliographisches Institut Mannheim (1976)

[30] B. Raffel, J. Wolfrum: Zeitschrift für Physikalische Chemie NF (1988), in press

PART 2: CONTRIBUTED LECTURES

NUMERICAL SIMULATION OF INERT AND REACTIVE FLOWS BY THE FINITE ELEMENT METHOD

S.AITA*, A. FORESTIER**, J.F. LEVY*, N. MONTMAYEUR*, A. TABBAL*, H. BUNG**

ABSTRACT

This paper relates different numerical and modeling aspects necessary to develop a Computational Fluid Dynamics Code, including combustion. Industrial needs give a real advantage to the Finite Element Method, by its potentialities and versatility.

The paper discusses then two F.E. approaches developed to realize such a code. Choices for these approaches are discussed, mainly for what concerns the basic numerical scheme. Typical applications with these approaches are also indicated.

RESUME

Ce papier relate différents aspects nécessaires au développement d'un logiciel de simulation en Mécanique des Fluides avec combustion. Les besoins industriels donnent un avantage certain à la méthode des Eléments Finis, par ses potentialités et sa versatilité.

Le papier montre ensuite deux approches qui ont été développées pour la réalisation d'un tel code. Il discute les solutions choisies dans ces deux approches, notamment au niveau du schéma numérique. Des applications typiques avec les deux approches sont également données.

****

1. INTRODUCTION

The development of an industrial reactive CFD code relies on different fields, such as fluid mechanics, numerical analysis, computer science, combustion and chemistry. All these fields have undergone lately an explosive growth.

The Industrial needs for high performance engines and the advent of super-computers have made the numerical simulation approach to reactive flow analysis feasible. Such a feasibility, has been demonstrated earlier by different authors (see for example ref.[1 and 2] for automotive applications). Nevertheless, the use of such codes at an industrial level for design optimization, trouble shooting or pollutants reduction, necessitates the solution of several key points, involving both numerical and modeling aspects.

Engineering Systems International, Silic 270, 94578 Rungis Cedex
**Commissariat à l'Energie Atomique, DEMT/SMTS, CEN Saclay, 91191 Gif-sur-Yvette*

Taking benefit of that growth, two independant teams have undertaken important developments these last years, in order to elaborate a reactive CFD code. The first team, at ESI, developed a 3D code named PAM-FLUID™ [3], with a special focus on automotive engine combustion chamber applications. The second team, at the CEA/DEMT, developed a 2D Euler version in Plexus code, focused on Nuclear Power Plant safety analysis. Both codes are explicit.

This paper discusses several issues in modeling and numerical implementation which appear to both teams. It then describes similarities and differences in implementing the same basic choice, i.e. the Finite Element Method. Finally, typical applications obtained with the developed codes are shown.

2. KEY ISSUES IN A REACTIVE CFD CODE DEVELOPMENT

2.1. Finite Element versus Finite Volume Techniques

Major growth in CFD codes, both for inert and reactive flows, has been realized using Finite Difference/Finite Volume methods. Simplicity of programming allowed early developments, enabling extensive testing of numerical techniques and fluid modeling.

The Finite Element technique is best suited for realistic industrial configurations, where typically the code user wishes to get automatically his meshing from one of the CAD/CAM systems or Finite Elements meshers available in the market.

Comparatively to Finite Volume technique, programming effort is more important for a Finite Element code, but in the end, this latter can have the same CPU performance.

In addition, the finite element method offers the versatility of having locally refined mesh, when needed by geometric or modeling considerations, without increasing the total number of mesh points. Both teams adopted the F.E. technique as a basic choice.

2.2. Numerical diffusion

Even in the context of inert fluid flow simulation, numerical diffusion appears to be a key issue. Such a diffusion reduces the validity of results for several thousands of time steps of total calculation. In addition, it competes also with other "physical" reasons of diffusion such as turbulence or species diffusion.

When chemical reactions are to be simulated, the question is more acute. In fact, high order spatial schemes requested for accuracy, give rise to density oscillations involving negative densities !! In that case, several numerical ad-hoc treatments are used to handle rare species during combustion phase calculations.

The two teams adopted two different schemes which eliminate these oscillations and related treatments, insuring conservation and positivity of thermodynamic quantities (see § 3 below).

2.3. Other topics

Several other topics are of interest for the reactive fluid flow modeling [4].

Turbulence modeling is first necessary, as direct turbulence simulation is not realistic even on present supercomputers. The state of the art of turbulence modeling for compressible flows is subject to many discussions concerning averaging techniques, turbulence transport and closure models.

Combustion numerical treatment is also of interest, especially for chemical reactions with characteristic times very short comparatively to flow and diffusion (Chemical Equilibrium reactions). A specific implicit scheme with convergence acceleration has been implemented in PAM-FLUID, in addition to an explicit scheme for chemical kinetic reactions (Reaction time ~ Diffusion time). A basic validation example, i.e. Burke-Schumann flame, is shown in figure 1.

Turbulence combustion interaction models have also to be considered, as the resolved quantities are only space averaged ones. Several such models have been developed recently ; they are mainly phenomenological and adapted to typical types of flames (premixed or diffusion). Accounting for complex chemical reactions in this context, i.e. for pollution prediction for example, is of major complexity.

Finally, most engines are multiphase systems, where reactants are injected in the form of liquid jets or droplets. Specific numerical and modeling techniques are requested, per example to track and describe mean droplets interaction with surrounding gas and boundaries.

3. FINITE ELEMENT FORMULATION

For treatment of complex geometries, unstructured mesh generation is fundamental. In fact, obstacles in Fluid domain lead to great difficulties when treated by an "i,j,k" spatial description.

Several finite element approaches has been developed to deal with unstructured grids. Note that their formulation has to insure the three conservation laws of mass, momentum and total energy. It is also necessary to insure second thermodynamic law concerning entropy inequality.

One advanced formulation is the "SUPG" developed by T.J.R. Hughes and M. Mallet associated to discontinuous Galerkin elements. Note that this scheme is non-linear and the choice of entropy variables can assume the discrete Clausius-Dukeim inequality.

Another approach, initiated by Boris and Book, is the FCT (Flux Corrected Transport) method. The principal idea is to associate to a conservative first order scheme an antidiffusion phase. This purpose has been developed in a Finite Element formulation by R. Lohner [5] with an unstructured formulation. Another idea which assumes to give positive function for mass and energy has been included in PAM-FLUID code, with a multispecies ALE (moving mesh) general formulation in 3D. It is important (in view of combustion treatment) to assure positivity of thermodynamic variables ; which are constant by elements and where velocity unknowns are Q1 or P1 (see figure 2, for basic 2D convection test [6])*.

Another branch of development is found with Riemann solver according to the first Godunov approach. It wants to associate TVD (Total Variation Diminishing) and second order formulation. The principal difficulty lies in the resolution of Riemann problem and we know that an analytic solution can be obtained in only a γ-law state equation. The Van Leer's scheme wants to use this possibility [7]. The conservative variables are assumed to be linear and discontinuous by elements and the slope of each fundamental variable insures TVD notion.

4. TVD SCHEMES

To explain the basic principles of both schemes, let the following be an hyperbolic convection equation :

$$\frac{\partial U}{\partial t} + \frac{\partial}{\partial X} F(U) = 0$$

Here U can be a vector insuring global conservations like : $U = (\rho, \rho V, \rho E)^T$, where ρ is the density, V the velocity and E the total energy.

An explicit finite element procedure is adopted, and the time maching scheme can be written in the form :

$$U_j^{n+1} = U_j^n - \frac{\Delta t}{\Delta X} \left[F(U_{j+1}^n) - F(U_j^n) \right]$$

In the FCT method, the fluxes in the right hand are simply written as a combination of fluxes obtained from a high order scheme (precise but oscillating) and low order scheme (monotone), as following :

$$U_j^{n+1} = U_j^n + \Delta U^L + C (\Delta U^H - \Delta U^L)$$

where $\qquad 0 \leq C \leq 1$

Results are smoothed by post-procosser

The choice of coefficient C is made localy, in a way such, that the solution has no new extrema, comparatively to the low order scheme TVD).

The Reimann solver approach consists on predicting the discontinuity of the solution at elements interfaces, using the slopes Δ_j :

$$U_{j+}^{n+1/2} = U_j^n - \frac{1}{2} \Delta_j^n - \frac{\Delta t}{\Delta X} \left[F(U_{j+1}^n) - F(U_j^n) \right]$$

$$U_{j-}^{n+1/2} = U_j^n + \frac{1}{2} \Delta_j^n - \frac{\Delta t}{\Delta X} \left[F(U_{j+1}^n) - F(U_j^n) \right]$$

The solution of this discontinuity is calculated exactly and is denoted

$$W_{j+}^{n+1/2} (U_{j-}^{n+1/2}, U_{j+}^{n+1/2})$$

Then, the prediction of U is recalculated using the Riemann solutions W :

$$U_{j+}^{n+1} = U_j^n - \frac{\Delta t}{\Delta X} \left[F(W_{j+1}^{n+1/2}) - F(W_j^{n+1/2}) \right]$$

Using this solution, the slopes Δ_j^{n+1} are calculated again to insure TVD and stability.

The 1D approximation is now often used and several versions exist : third order 1D extension by Collela, Finite Difference 2D scheme for an arbitrary thermodynamic law by CEA/DAM (J. Ovadia & al.) and another Finite Difference 3D formulation by ONERA (J.L. Montagne). The CEA/DEMT team made a 2D Finite Element generalization of the method with arbitrary meshes in the new module of Plexus code [8].

5. EXAMPLES OF APPLICATIONS

5.1. Shock-jet Interaction
A model problem in a channel is treated. A fluid at two different states (same pressure but different temperatures) is still. At t = 0, we inject an amount of fluid in subsonic conditions. Two interaction types are observed. The evolution of the pressure and the velocity field are shown in figure 3. It is important to notice that the isopressures are normal to the wall.

5.2. Interaction between two jets
In a motionless gas, the same gas (γ = 1.23) is injected as two jets at different thermodynamic conditions. First jet is axial, and the second jet is at 45°.

In a first calculation, both jets are supersonic (Mach 1.7 and Mach 1.1 respectively). Transient and stationary velocity fields are shown in figure 4. The stationary result of a calculation in a reduced domain is also shown, in order to verify correctness of the solution and boundary treatment.

In a second calculation, the 45° angle jet is brought to subsonic conditions (stagnation pressure reduced by factor 10). Notice that jet interaction features are completely different (figure 5).

Finally, a third calculation is made, bringing also the axial jet to subsonic conditions (corresponding stagnation pressure reduced by a factor 0.6). Weak interaction can be observed as well a double vortexes near the wall.

5.3. 3D flow in an automotive engine

A 3 dimensional engine is considered. Mesh includes cylinder and intake port. Piston and valve have motorized motion. Air motion to the cylinder through valve is studied, with stagnation inlet conditions at intake port limit.

Figure 6 shows instaneous streamlines in the engine, for 3 successive crank angles (i.e. respectively 30, 60 and 120°). Notice at 60° crank angle, the swirling motion of the fluid. This typical motion is destroyed in simulation conditions at 120°. Notice also at this angle, the formation of recirculation region between at its seat, reducing volumetric efficiency.

Figure 7 shows velocity contours and pressure fields at 120° crank angle. Velocity fields are very different from earlier 2D simulation [8]. Pressure contours show the upstream propagation of depression between the valve and its seat. It demonstrates also the presence of a 3D shock in the other side of the intake.

6. CONCLUSION

Finite element second order method has been implemented and successfully used, by the two teams, in industrial codes. Two different schemes have been implemented insuring conservation of hydrodynamic phase and positivity of thermodynamic variable. In addition, the FCT scheme allows immediate correct treatment of ALE problems in multispecies flow with combustion.

Other developments are requested, many of them are related to research problems on physical modelization. One can hope that the explosive growth in research activities in the concerned domains will help coping with the industrial needs.

REFERENCES

[1] Amsden, A.A., Ramshaw, J.D., O'Rourke, P.J. and Dukowicz, J.K. : "KIVA : A computer program for two- and three-dimensional fluid flows with chemical reactions and fuel sprays". Los Alamos National Laboratory (1985).

[2] Cloutman, L. : "CONCHAS-SPRAY : A computer code for reactive flows with fuel sprays". Los Alamos National Laboratory (1982).

[3] E.S.I. "PAM-FLUID Technical Presentation". 10/1988.

[4] Oran, E.S. and Boris, J.P. : "Numerical simulation of reactive flows". (1987).

[5] Lohner, R. : "Finite element flux corrected transport". (1988).

[6] Williams, F.A. : "Combustion Theory. The Fundamental Theory of Chemically Reacting Flow Systems", (1985).

[7] Van Leer, B. : "Flux-vector splitting for the Euler equations". Lecture Notes in Physics. 170, Springer-Berlin, pp.502-512.

[8] Forestier, A., Gaudy, C. & Bung, H. : "Second order scheme for arbitrary mesh in Compressible Fluid Dynamic". I.C9MD, Williamsburg, July 1988.

[9] Aïta, S. and Haug, E., Ulrich, D. : "Supercomputer Simulations in Applied Mechanics and Perspectives in Composite Damage, Internal Fluid Dynamics and Crashworthiness". Presented at ATA 2nd Intl. Symposium : "Use of Supercomputers in the European Automotive Industry", Torino (Italy), April 26-27, 1988.

T 1   *T 2*

T 3   *T 4*

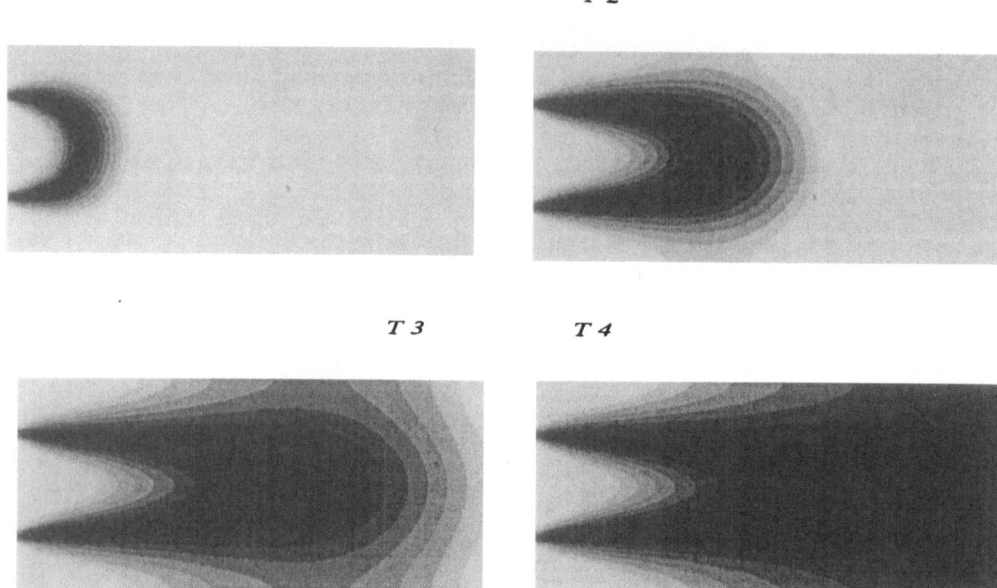

Density contours of product

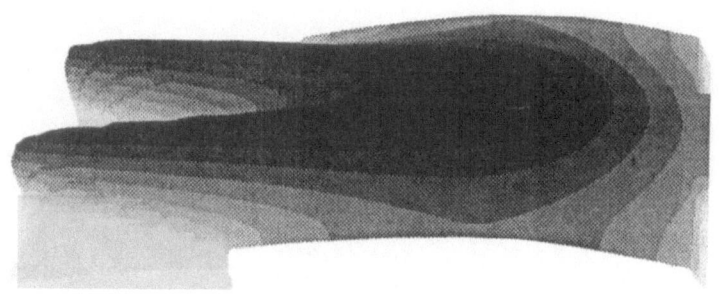

Z-axis warped density contours of product

Figure 1 : Burke-Schumann Flame

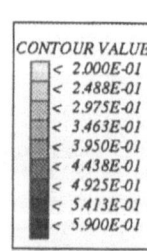

CONTOUR VALUES
< 2.000E-01
< 2.488E-01
< 2.975E-01
< 3.463E-01
< 3.950E-01
< 4.438E-01
< 4.925E-01
< 5.413E-01
< 5.900E-01

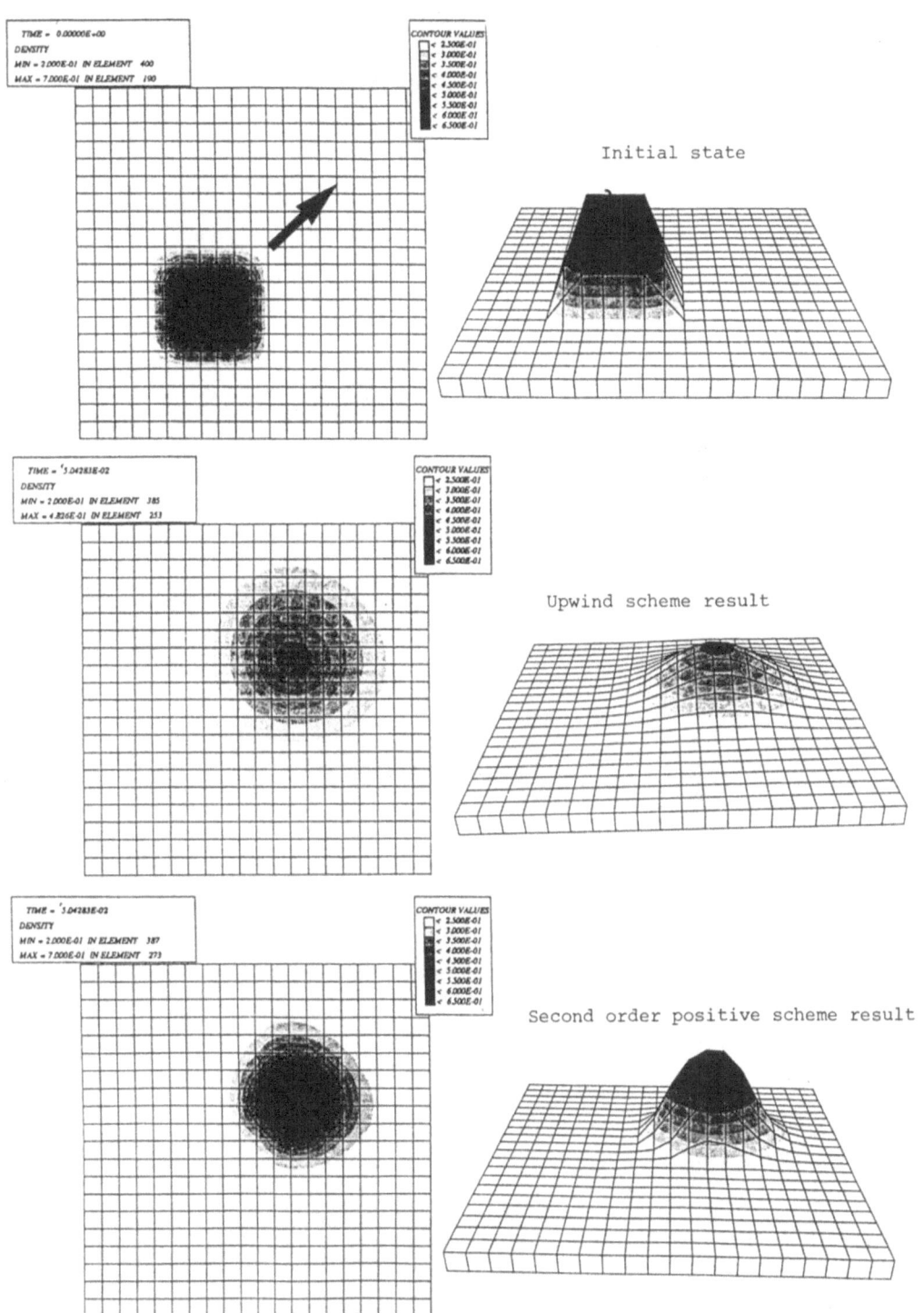

Figure 2 : 2D Convection of a square shaped density jump

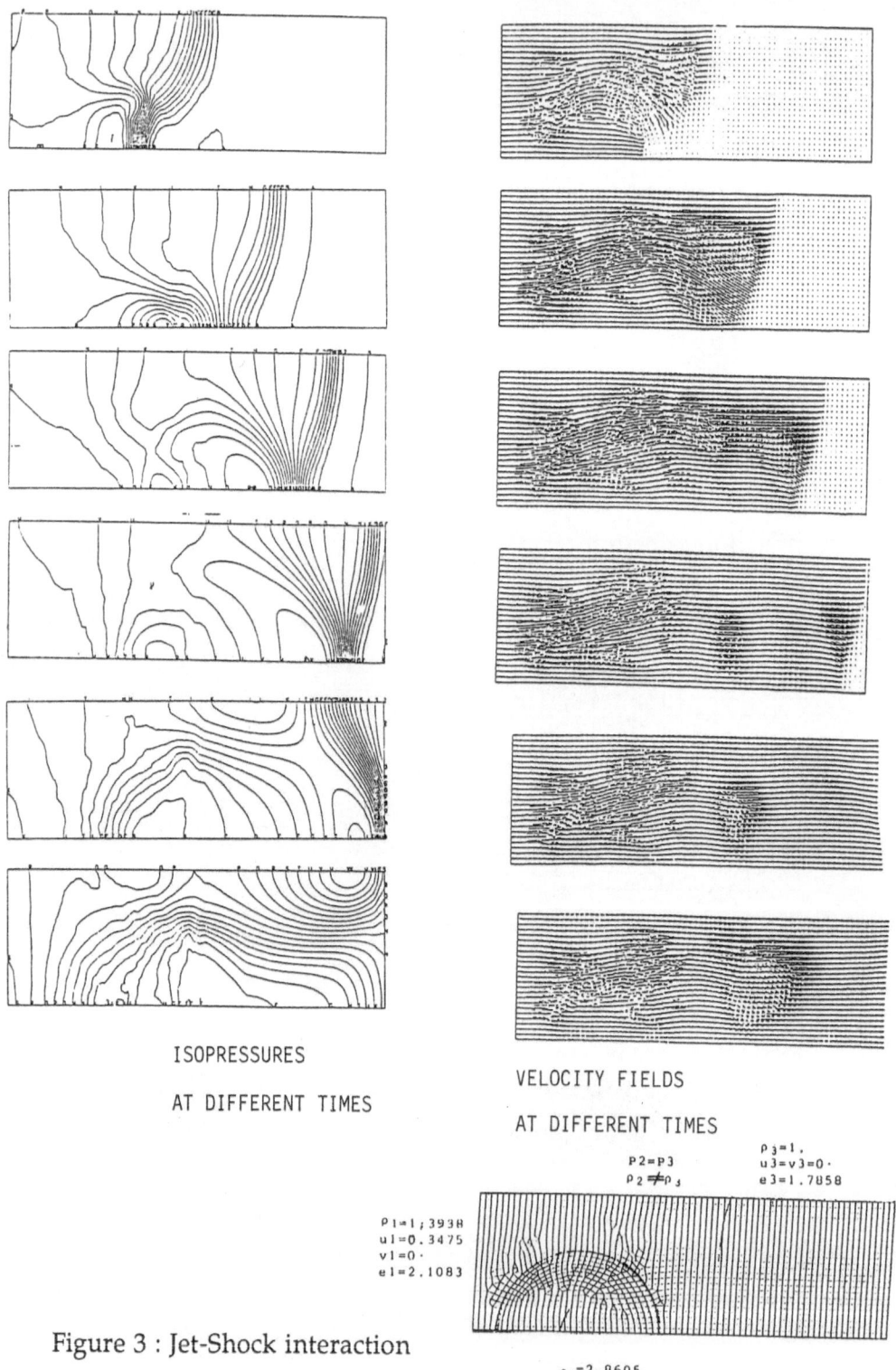

ISOPRESSURES

AT DIFFERENT TIMES

VELOCITY FIELDS

AT DIFFERENT TIMES

Figure 3 : Jet-Shock interaction

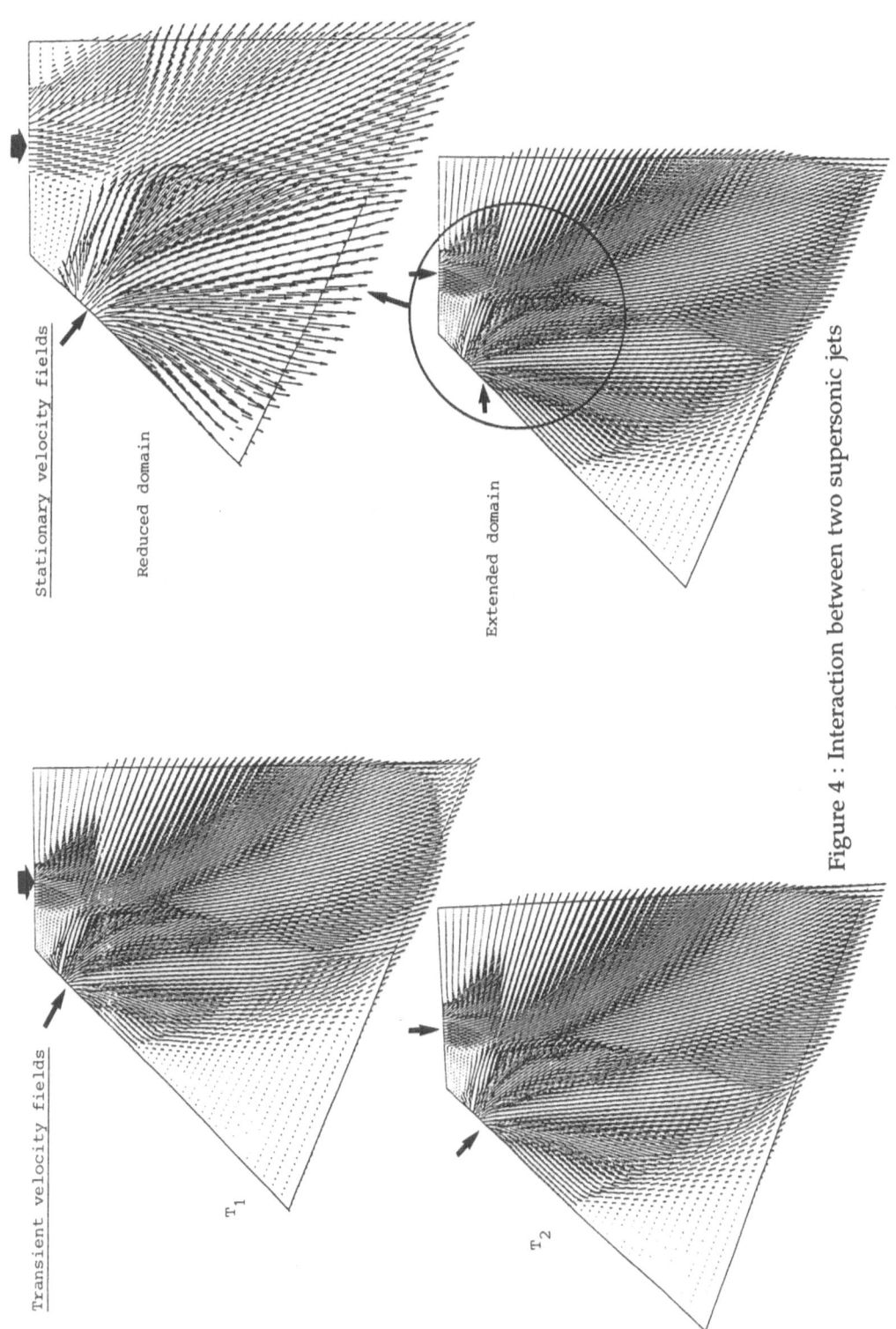

Stationary velocity fields

Reduced domain

Extended domain

Transient velocity fields

T_1

T_2

Figure 4 : Interaction between two supersonic jets

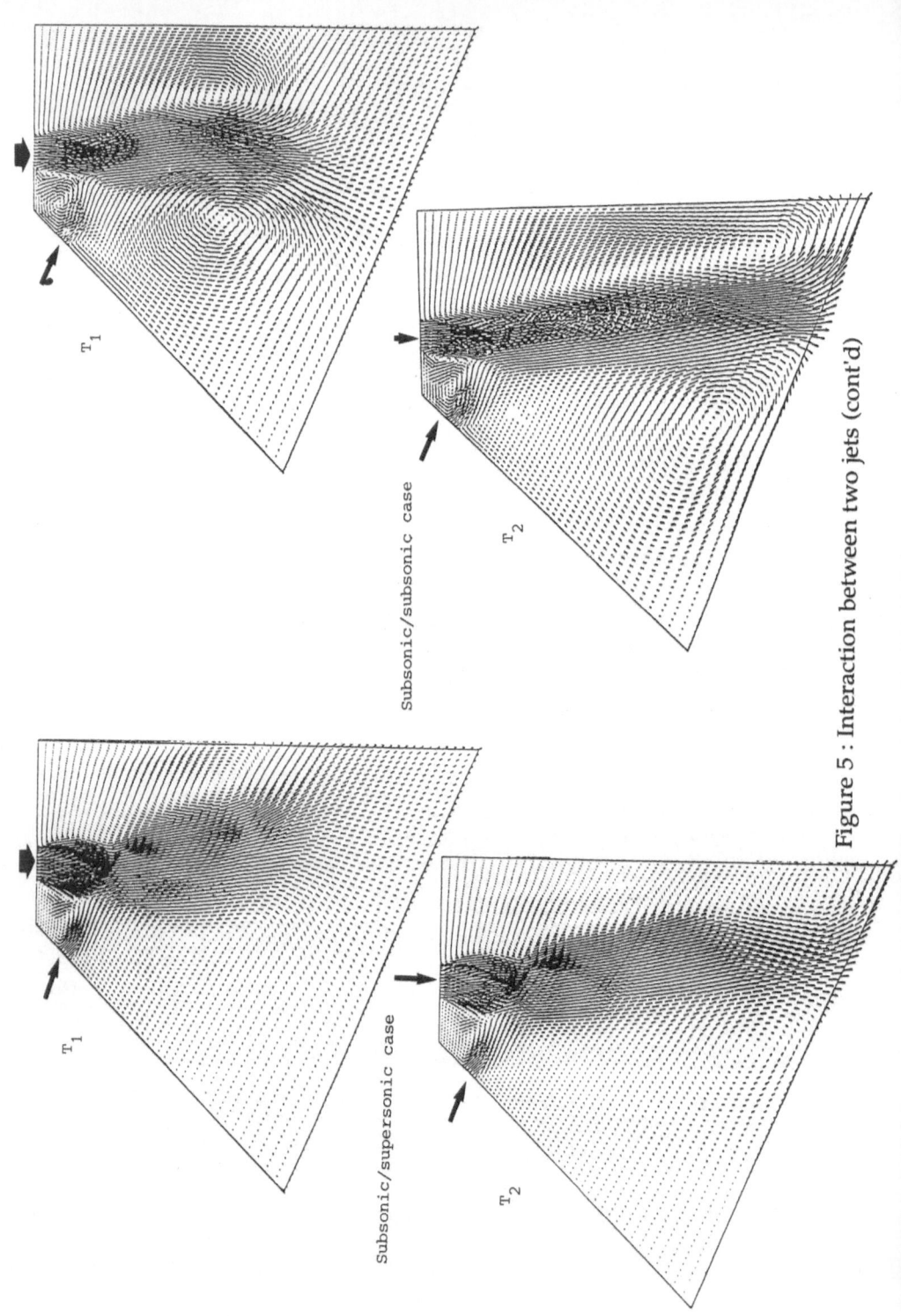

T₁

T₂

Subsonic/subsonic case

T₁

T₂

Subsonic/supersonic case

Figure 5 : Interaction between two jets (cont'd)

Figure 6 : 3D flow in idealized automative engine

STREAMLINES

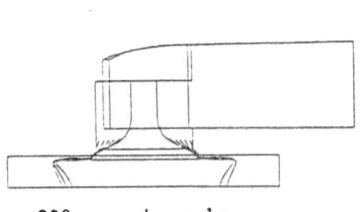

30° cranck angle

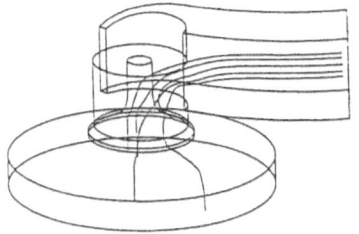

30° cranck angle

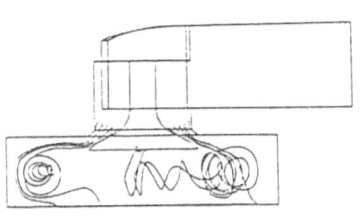

60° cranck angle

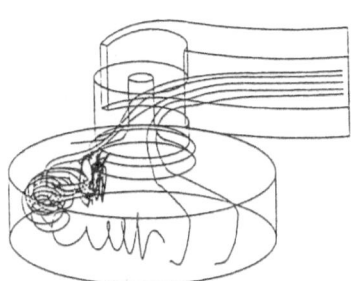

60° cranck angle

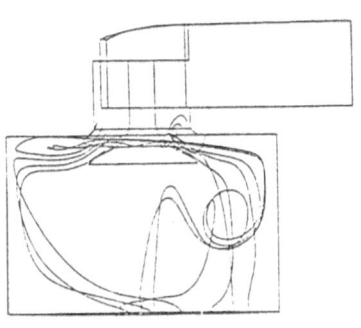

120° cranck angle

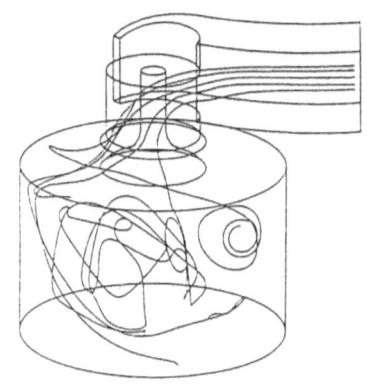

120° cranck angle

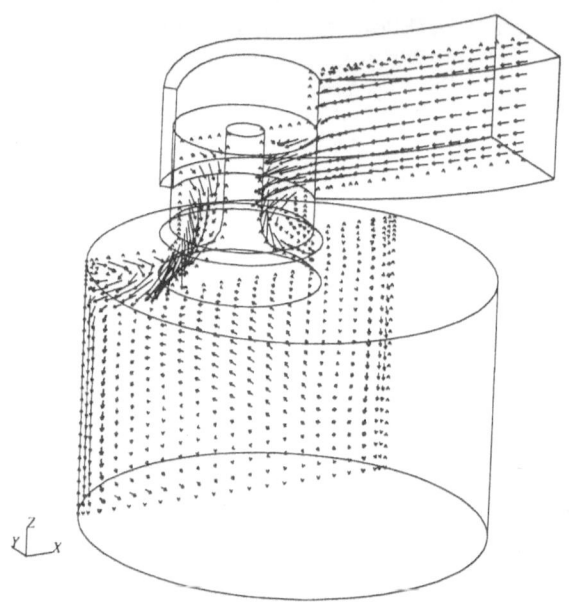

Velocity Field

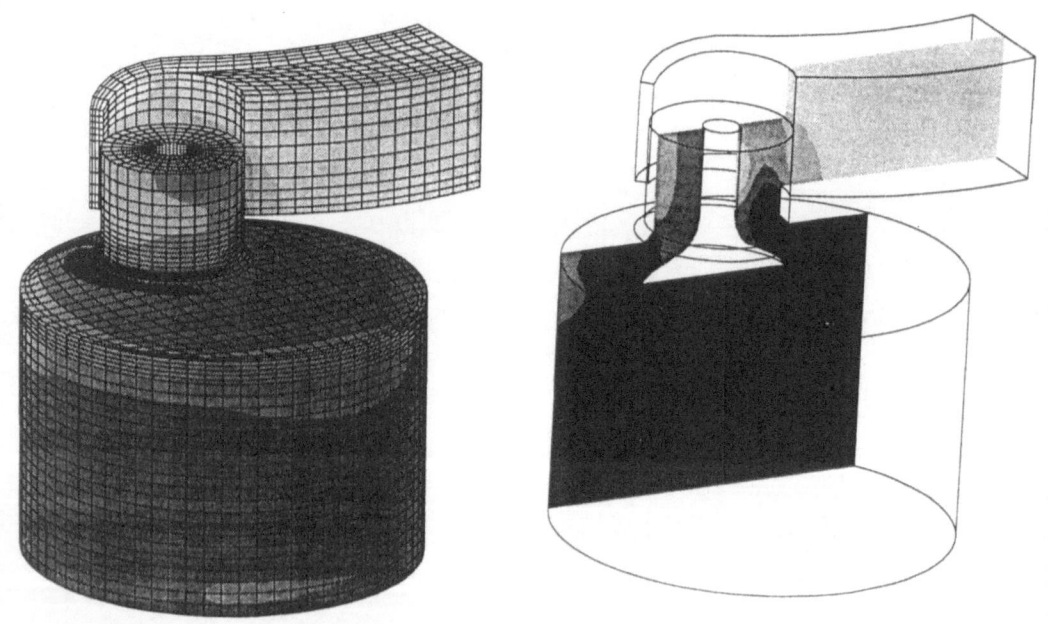

Pressure contours

Figure 7 : 3D flow in idealized automotive engine

NUMERICAL COMPUTATION OF BIFURCATION PHENOMENA AND PATTERN FORMATION
IN COMBUSTION

A. Bayliss and B. J. Matkowsky
Department of Engineering Sciences and Applied Mathematics
Northwestern University
Evanston, IL 60208 USA

M. Minkoff
Mathematics and Computer Science Division
Argonne National Laboratory
Argonne, IL 60439 USA

INTRODUCTION

We develop and employ a new numerical method to study problems in both gaseous
and solid fuel (condensed phase) combustion. The numerical method is an extension
of the adaptive pseudo-spectral method previously introduced [1-6], in that two
dimensional, non-product coordinate transformations are introduced so as to efficiently compute fronts which have a strong transverse variation.

In condensed phase combustion we consider a reaction front (so-called solid
flame) propagating through a cylindrical sample. Such problems describe the process
of self propagating high temperature synthesis (SHS), which is a new and innovative
method for the fabrication of high tech ceramic and metallic materials. A cylindrical sample is ignited at one end and a thermal wave propagates through the sample,
converting unburned reactants to solid products. We describe various modes of
propagation, which have been experimentally observed [9,10,15] as bifurcations from
a basic solution corresponding to a uniformly propagating planar front. Experimental observations include the case when burning occurs throughout the sample as
well as the case when burning occurs on the surface of the sample, but not in its
interior. The additional modes of propagation include (i) oscillatory combustion -
in which a planar propagates with an oscillatory velocity, (ii) spinning combustion
- in which one or more hot spots (luminous points) are observed to move in a helical
fashion along the surface of the sample, and (iii) multiple point combustion - in
which the hot spots appear, disappear, and reappear repeatedly. More specifically,
a basic solution is obtained which describes a uniformly propagating planar reaction
wave. By varying critical parameters of the problem, we construct additional solutions on branches which bifurcate from the basic solution. These solutions exhibit
both spatial and temporal patterns, which become more and more complex as the distance from the bifurcation point is increased. Thus planar oscillatory combustion
has been analyzed as a Hopf bifurcation from the basic solution [14], while spinning
and multiple point combustion have been described traveling and standing wave patterns obtained as bifurcations to pulsating non-axisymmetric solutions [12].

Bifurcation theory is a local theory valid in a neighborhood of the bifurcation
point. To study the behavior of the system in the fully nonlinear regime, far from

the bifurcation point, numerical computations are employed. These computations reveal new and interesting behavior, not previously predicted by analysis. As an example, in planar condensed phase combustion a nearly sinusoidal (in time) solution is computed for parameters near the bifurcation point. As the parameters are varied away from the bifurcation point, extremely severe relaxation oscillations develop and a period doubling secondary bifurcation occurs. On this branch the relaxation oscillations become yet sharper. Beyond a certain point the period doubled branch becomes unstable and stability returns to the singly periodic branch [1,5].

In this paper we consider non-planar pulsating patterns. We first describe the model and the numerical method.

Model

We employ the model of [14], which was extended in [11] to account for the effects of melting, to study non-axisymmetric pulsating solutions. The model assumes that the reactant melts on some surface ahead of the reaction, at which the temperature $\tilde{T}$ reaches the melting temperature $\tilde{T}_m$. We consider the case that the burning occurs on the surface of a cylindrical sample and not in its interior. The independent variables are thus the angle ψ and the axial coordinate $\tilde{x}_3$. The melting surface is given by

$$\tilde{x}_3 = \tilde{\phi}(\psi, t) \quad ,$$

on which $\tilde{T} = \tilde{T}_m$.

We assume one-step, irreversible, Arrhenius kinetics for the reaction rate, and assume that the heat of fusion $\tilde{\gamma}$ is reabsorbed during the reaction, as the product is in the solid phase. In addition, melting causes the reaction rate to be increased by a factor $\alpha > 1$, due to the increased surface to surface contact in the liquid phase. The unknowns are temperature $\tilde{T}$, and concentration of a deficient component $\tilde{C}$. Dimensional constants are the rate constant $\tilde{A}$, the thermal diffusivity $\tilde{\lambda}$, the activation energy $\tilde{E}$, the gas constant $\tilde{R}$, and the heat of reaction $\tilde{\beta}$.

The equations of the model are

$$\frac{\partial \tilde{T}}{\partial \tilde{t}} = \tilde{\lambda}\nabla^2\tilde{T} + \begin{bmatrix} \tilde{\beta} \\ \alpha(\tilde{\beta}+\tilde{\gamma}) \end{bmatrix} \tilde{A} \exp(-\tilde{E}/\tilde{R}\tilde{T})$$

$$\frac{\partial \tilde{C}}{\partial \tilde{t}} = - \begin{bmatrix} 1 \\ \alpha \end{bmatrix} \tilde{A}\tilde{C} \exp(-\tilde{E}/\tilde{R}\tilde{T}) \quad ,$$

where

$$\begin{bmatrix} a \\ b \end{bmatrix} = \begin{cases} a \,, & \tilde{x}_3 < \tilde{\phi}(\psi,\tilde{t}) \\ b \,, & \tilde{x}_3 > \tilde{\phi}(\psi,\tilde{t}) \end{cases} \quad .$$

The boundary conditions are

$$\tilde{T}(x_3,\psi,t) = \begin{cases} \tilde{T}_a & \text{as} \quad \tilde{x}_3 \to \infty \\ \tilde{T}_u & \text{as} \quad \tilde{x}_3 \to -\infty \\ \tilde{T}_m & \text{at} \quad \tilde{x}_3 = \tilde{\phi}(\psi,t) \end{cases}$$

where $\tilde{T}_a$ is the adiabatic temperature, $\tilde{T}_u$ the temperature of the unburned reactant

and $\tilde{T}_m$ the temperature at which melting occurs.

We assume that the reaction occurs on the surface of the cylinder
$\tilde{x}_1^2 + \tilde{x}_2^2 = \tilde{R}_c^2$. We introduce the velocity $\tilde{U}$ of the uniformly propagating planar

front as a reference velocity, and nondimensionalize by defining

$$C = \frac{\tilde{C}}{\tilde{C}_u} \quad , \quad \theta = \frac{\tilde{T}-\tilde{T}_u}{\tilde{T}_a-\tilde{T}_u} \quad , \quad \phi = \tilde{\phi}\tilde{U}/\tilde{\lambda} \quad , \quad t = \tilde{t}\tilde{U}^2/\tilde{\lambda} \quad , \quad \gamma = \tilde{\gamma}/\tilde{\beta} \quad ,$$

$$N = \tilde{E}/\tilde{R}\tilde{T} \quad , \quad \sigma = \frac{\tilde{T}_u}{\tilde{T}_a} \quad , \quad x_i = \frac{\tilde{x}_i\tilde{U}}{\tilde{\lambda}} \quad , \quad R_c = \frac{\tilde{R}_c\tilde{U}}{\tilde{\lambda}} \quad ,$$

where $\tilde{C}_u$ is the initial concentration of the reactant. The axial coordinate is

transformed by
$$z = x_3 - \phi(\psi,t) \quad ,$$

so that the melting surface is located at $z = 0$. Note that the Laplacian is no

longer separable in this coordinate system.

The equations of the model then become

$$C_t = \phi_t C_z - \Lambda C \begin{bmatrix} 1 \\ \alpha \end{bmatrix} \exp\left[\frac{N(1-\sigma)(\theta-1)}{\sigma+(1-\sigma)}\right]$$

$$\theta_t = \phi_t \theta_z + \nabla^2\theta + \Lambda C\left[\frac{1}{\alpha(1+\gamma)}\right] \exp\left[\frac{(N(1-\sigma)(\theta-1)}{\sigma+(1-s)\theta)}\right] \quad ,$$

where

$$\Lambda = \Delta\left[\frac{(1+\gamma)}{\alpha(1-M)} + O(\frac{1}{\Delta})\right] \quad , \quad M = (1 - \frac{(1+\gamma)}{\alpha})e^{\Delta(\theta_m-1)} \quad , \quad \Delta = N(1-\sigma), \text{ and}$$

$$\nabla^2 = \frac{1}{R_c^2}\frac{\partial^2}{\partial\psi^2} + \left[1 + \frac{1}{R_c^2}\left(\frac{\partial\phi}{\partial\psi}\right)^2\right]\frac{\partial^2}{\partial z^2} - \frac{1}{R_c^2}\frac{\partial^2\phi}{\partial\psi^2}\frac{\partial}{\partial z} + \frac{2}{R_c^2}\frac{\partial\phi}{\partial\psi}\frac{\partial^2}{\partial\psi\partial z}$$

The reaction term is multiplied by a cutoff function which vanishes far ahead of the

front, as described in [1,5].

At the melting surface, $z = 0$, the heat of fusion is absorbed by the reactant.

Thus

$$\phi_t = \frac{-1}{\gamma C(0)} [\theta_z]\left[1 + \frac{(\phi_\psi)^2}{R_c^2}\right]$$

where $[\theta_z]$ denotes the jump in θ_z at $z = 0$. The bifurcation parameter is

$\mu = \frac{\Delta}{2(1-M)}$. In [11] a uniformly propagating planar solution was identified as a basic solution, for the related problem with δ-function kinetics. It is shown in [12] that the basic solutions can be unstable to angular disturbance of the form $e^{in\psi}$ when $\mu \geq 4$. The most unstable mode occurs when $n/R_c = 1/2$. In the computations below we take $R_c = 2$ so that the mode $n = 1$ is the only one which becomes unstable.

Numerical Method

The numerical method is based on a Chebyshev pseudo-spectral representation in z and a Fourier pseudo-spectral representation in ψ. However our method is not merely a direct product of the two one dimensional representations, i.e. the two coordinates are not treated independently, as described below. The solutions to these problems have localized regions of rapid spatial variation. In order to improve the accuracy of the Chebyshev pseudo-spectral method in approximating these regions of rapid variation, an adaptive pseudo-spectral method was introduced in [1,2]. In this method the axial coordinate is adaptively transformed by a mapping of the form

$$z' = q(z,\alpha)$$

where for each value of α, $q(\cdot,\alpha)$ is a given mapping. The value of α is determined by minimizing a Sobolev type integral in z, which measures the pseudo-spectral interpolation error of the solution [1].

The parameter α, obtained from this minimization procedure, depends on the strength and location of the regions of rapid variation of the solution, i.e. the reaction zone. For problems with significant transverse variation, the strength and location of the reaction zone has a strong angular dependence. A one-dimensional transformation, i.e. α independent of ψ, may be inefficient and require an excessive number of axial collocation points to accurately resolve the reaction zone in all angles.

In the present computation we employ the two-dimensional transformation

$$z' = q(z,\alpha(\psi))$$
$$\psi' = \psi \ .$$

The function $\alpha(\psi)$ is chosen by minimizing the integral separately for each angle ψ. The resulting mapping is Fourier transformed and may be smoothed by removing the highest Fourier modes. We observe that the additional computational effort required is very small.

The time integration is carried out by a semi-implicit predictor-corrector method in which the axial diffusion term is treated implicitly [2]. A preconditioned Richardson iterative method is used to update the implicit terms. The preconditioning is obtained from a finite difference approximation as described in [7]. This requires the inversion of a tridiagonal matrix at each stage of the iteration. Most of the computational cost is in the tridiagonal solution and the Fast Fourier Transform. The overall cost of the adaptive procedure is minimal.

Numerical Results

We have computed both standing wave and traveling wave solutions to the model, which describe spinning combustion and multiple point combustion respectively. We have found a standing wave solution branch with the parameters $N = 50$, $\gamma = 0.18$, $\alpha = 2.3$, $\theta_m = 0.8$. We increase μ by decreasing σ. In figure 1 we consider the case $\mu = 4.07$ and plot $\theta(t, z=0.4, \psi)$ as a function of t for the angles $\psi = 0$, $\pi/2$, π, $3\pi/2$. The pulsation is nearly sinusoidal. At the null of the standing wave, a small half period pulsation can be observed due to the effect of nonlinearity.

A progressive sharpening of the pulsation occurs as μ is increased along this branch. We illustrate the case $\mu = 4.16$ in figure 2, where we plot $\theta(t, z=0.4, \psi)$. In figures 3a and 3b we plot the temperature field at a time corresponding to formation of the luminous point (figure 3a) and a time where there is a relatively small angular temperature variation. In figure 4 we plot the melting surface $\phi(\psi, t)$ at a succession of time values. The standing wave pattern together with a nonlinear distortion is clearly evident.

Spinning wave patterns have been found for the parameters $\gamma = 0.05$, $\alpha = 2.0$, $\theta_m = 0.8$, $N = 50$. An example of such a solution is shown in figure 5 where the melting surface $\phi(\psi, t)$ is shown for various values of t.

Gaseous Combustion

We consider the problem of a flame stabilized by a line source of fuel of strength $2\pi\kappa$. A basic solution, in the limit of infinite activation energy, describes a smooth stationary cylindrical flame front, separating burned from unburned gases. By varying critical parameters of the problem, we construct additional solutions on branches which bifurcate from the basic solution. These solutions exhibit both spatial and temporal patterns, which become more and more complex as the distance from the bifurcation point is increased. They describe e.g. (i) cylindrical flames which oscillate (both sinusoidal and relaxation oscillations) about a mean position given by the stationary cylindrical flame front, (ii) stationary cellular flames, and (iii) oscillatory cellular flames, which describe both traveling waves about, and standing waves on the cylindrical front.

Assuming an appropriate non-dimensionalization, and finite activation energy, the equations of the diffusional thermal model are

$$\theta_t = \nabla^2\theta - \underset{\sim}{U}\cdot\nabla\theta + \Lambda C \exp\left[\frac{N(1-\sigma)(\theta-1)}{\sigma + (1-\sigma)\theta}\right]$$

$$C_t = \frac{1}{L}\nabla^2 C - \underset{\sim}{U}\cdot\nabla C - \Lambda C \exp\left[\frac{N(1-\sigma)(\theta-1)}{\sigma + (1-\sigma)\theta}\right]$$

Here N and σ are defined above, L is the Lewis number and the flow field

$$\underset{\sim}{U} = \frac{\kappa}{r}\hat{r}.$$ The flame front eigenvalue $\Lambda = \frac{\Delta^2}{2L}$, where $\Delta = N(1-\sigma)$ [13].

This problem has been analyzed in the limit $\Delta \to \infty$. There exist critical values $L_c < 1$ and κ_c such that for $L < L_c$ and $\kappa > \kappa_c$, stationary cellular flame patterns exist [13]. These patterns were studied numerically in [3,4,6]. The value of L was fixed and solution branches were computed by varying κ. A sequence of transitions to stationary cellular flame patterns of increasing mode number was found.

The problem has also been analyzed for $L > 1$ and κ large [8]. The analysis indicates that stable pulsating axisymmetric and pulsating cellular flames may exist for sufficiently large values of κ. Computations of axisymmetric pulsating solutions are presented in [2]. In this paper we illustrate the computation of pulsating cellular flames. Specifically, we have computed a standing wave pattern for the parameters $N = 40$, $\sigma = 0.5$, $L = 4.4$, $\kappa = 20$. In figures 6a,b,c we plot the temperature at three different times. The standing wave pattern is clearly evident. We note that figure 6b corresponds to a time at which the standing wave perturbation of the axisymmetric solution, is almost zero. This is one of the features that distinguishes a standing wave from a traveling wave. We note that cellular flames have also been computed [16] by a Fourier expansion in the direction parallel to the front, coupled to an adaptive finite difference approximation in the direction normal to the front.

This research was supported in part by D.O.E. grant DEFG02-87ER25027, and contract W31-109ENG38, and N.S.F. grant DMS87-01543.

References

1. A. Bayliss and B. J. Matkowsky, "Fronts, Relaxation Oscillations, and Period Doubling in Solid Fuel Combustion", J. Comp. Phys., 71 (1987), 147.
2. A. Bayliss, D. Gottlieb, B. J. Matkowsky and M. Minkoff, "An Adaptive Pseudo-Spectral Method for Reaction Diffusion Problems", J. Comp. Phys. 81 (1989), 421
3. A. Bayliss, M. Minkoff and B. J. Matkowsky, "Adaptive Pseudo-Spectral Computation of Cellular Flames Stabilized by a Point Source", Appl. Math. Lett., 1 (1988), 19.
4. A. Bayliss, B. J. Matkowsky and M. Minkoff, "Bistable Cellular Flames", Proc. Symp. honoring C. C. Lin, World Sci. Publ. (1988), 108.
5. A. Bayliss, M. Minkoff and B. J. Matkowsky, "Period Doubling Gained, Period Doubling Lost", SIAM J. Appl. Math., to appear.
6. A. Bayliss, B. J. Matkowsky and M. Minkoff, "Cascading Cellular Flames", SIAM J. Appl. Math., to appear.
7. C. Canuto and A. Quarteroni, "Preconditioned Minimal Residual Methods for Chebyshev Spectral Calculations", J. Comp. Phys., 60 (1985), 315.
8. M. Garbey, H. G. Kaper, G. K. Leaf and B. J. Matkowsky,, "Linear Stability Analysis of Cylindrical Flames", (to appear), Quar. Appl. Math.
9. J. B. Holt, "The Use of Exothermic Reactions in the Synthesis and Densification of Ceramic Materials", Materials Res. Soc. Bulletin 12 (1982), 60.
10. Y. A. Maksimov, A. T. Pak, G. B. Lavrenchuk, Y. S. Naiborodenko and A. G. Merzhanov, "Spin Combustion of Gasless Systems", Combustion, Explosion and Shock Waves, 15 (1979), 415.
11. S. B. Margolis, "An Asymptotic Theory of Condensed Two-Phase Flame Propagation", SIAM J. Appl. Math., 43 (1983), 331.
12. S. B. Margolis, H. G. Kaper, G. K. Leaf and B. J. Matkowsky, "Bifurcation of Pulsating and Spinning Reaction Fronts in Condensed Two-Phase Combustion", Combust. Sci. and Tech., 43 (1985), 127.
13. B. J. Matkowsky, L. J. Putnick and G. I. Sivashinsky, "A Nonlinear Theory of Cellular Flames", SIAM J. Appl. Math., 38 (1980), 489.
14. B. J. Matkowsky and G. I. Sivashinsky, "Propagation of a Pulsating Reaction Front in Solid Fuel Combustion", SIAM J. Appl. Math., 35 (1978), 465.
15. A. G. Merzhanov, "SHS Processes: Combustion Theory and Practice", Archivum Combustionis 1 (1981), 23.
16. H. Guillard, B. Larrouturou and N. Maman", Etude Numerique des Instabilities Cellulaires d'un front de flamme par une methode pseudo-spectrale", INRIA Report 721, 1987.

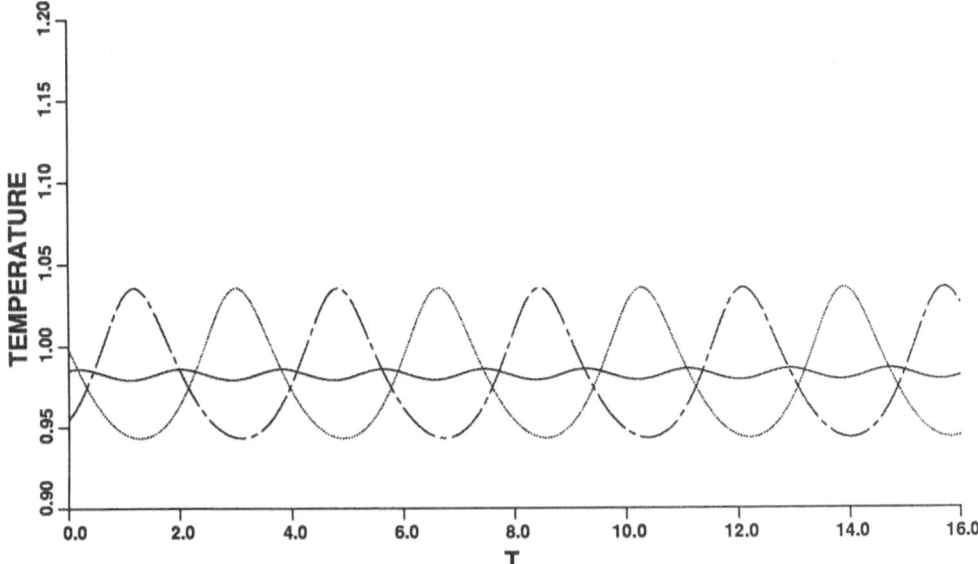

Fig. 1. $\theta(t,z=0.4,\psi)$ at angles $\psi=0$ (——), $\psi = \pi/2$ (••••). $\psi = \pi$ (----),
$\psi = 3\pi/2$ (—— - —— -). Standing wave branch, $\mu = 4.06951$.

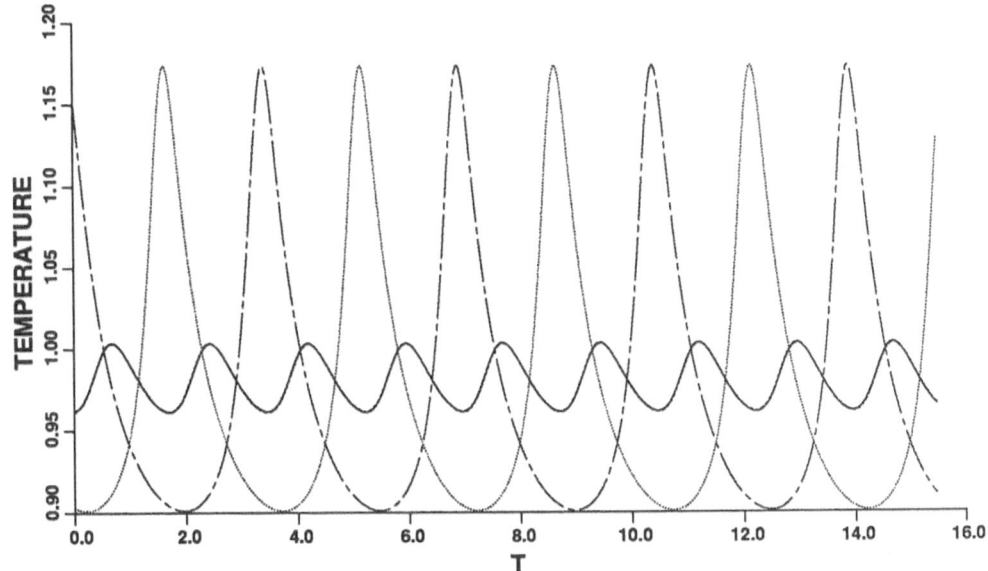

Fig. 2. $\theta(t,z=0.4,\psi)$ at angles $\psi=0$ (——), $\psi = \pi/2$ (••••). $\psi = \pi$ (----),
$\psi = 3\pi/2$ (—— - —— -). Standing wave branch, $\mu = 4.16128$.

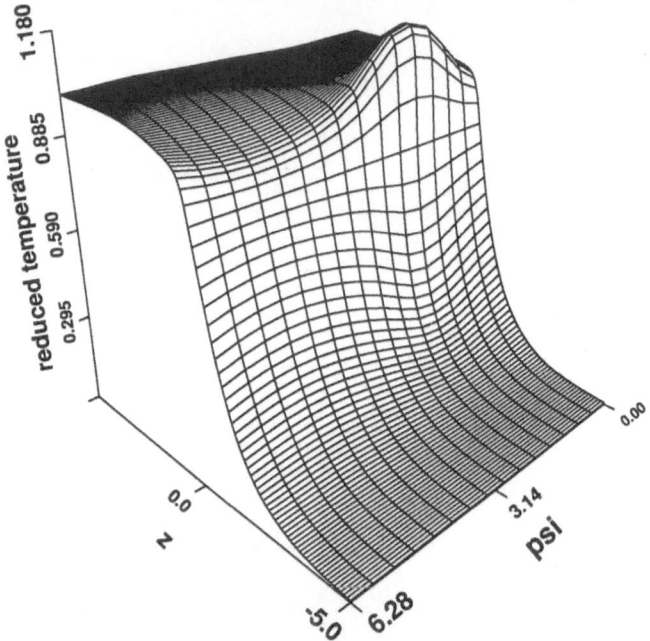

Fig. 3a: Temperature field at time corresponding to a luminous point, $\mu = 4.16128$.

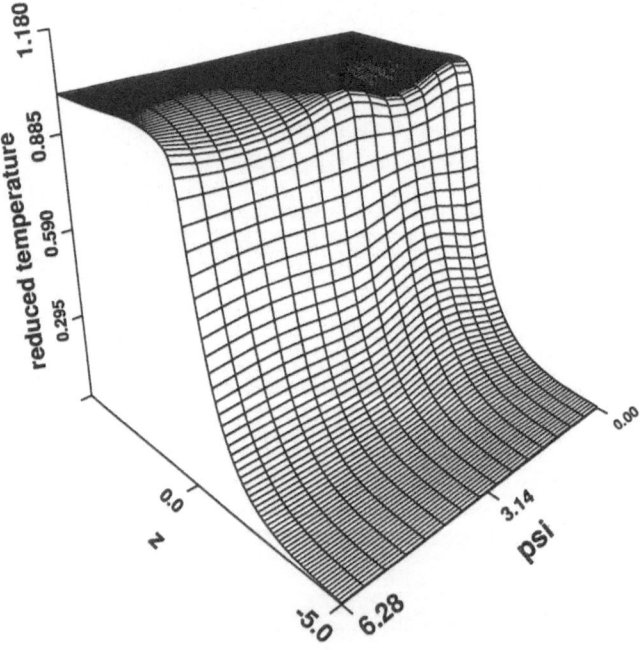

Fig. 3b: Temperature field at a time of small angular variation, $\mu = 4.16128$.

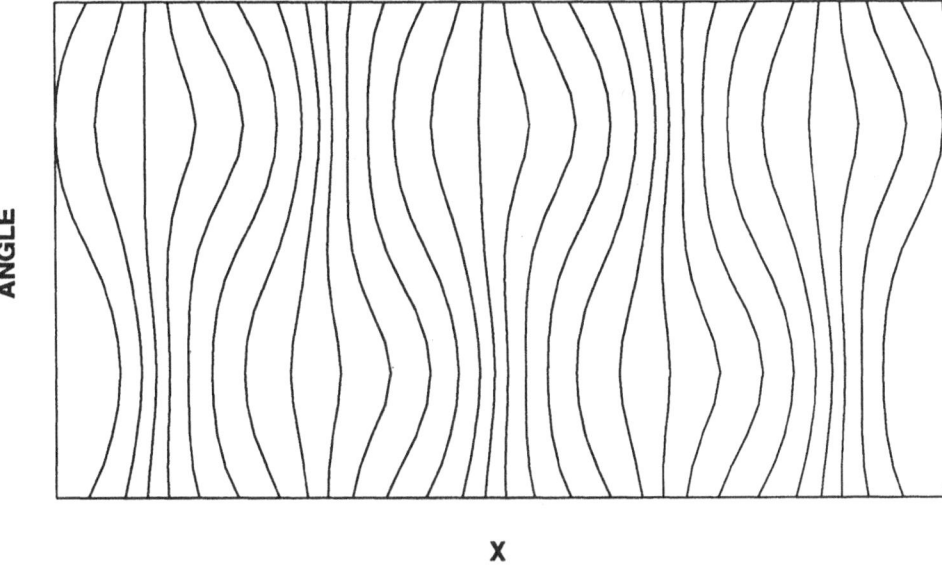

Fig. 4:   Melting surface at different values of t, standing wave branch, μ=4.16128.

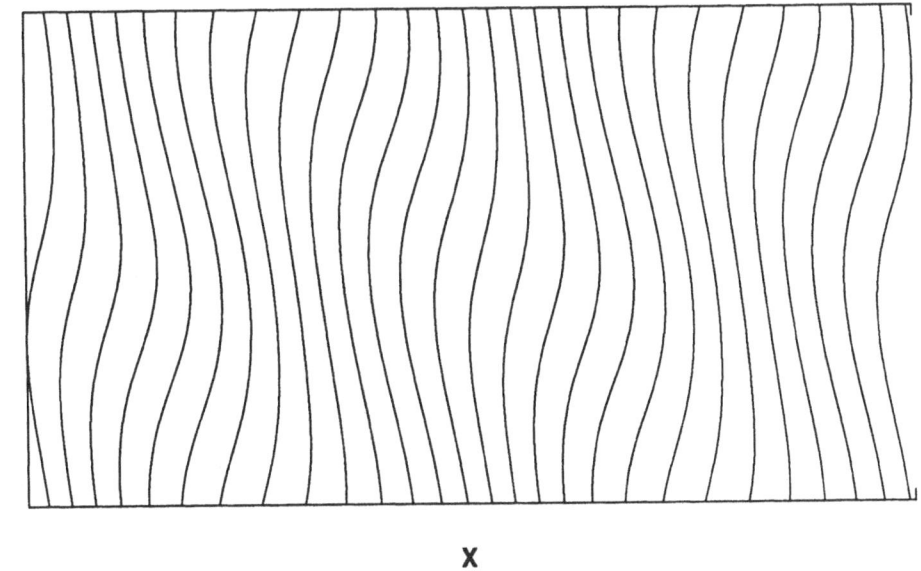

Fig. 5:   Melting surface at different values of t, spinning wave branch, μ=4.11407.

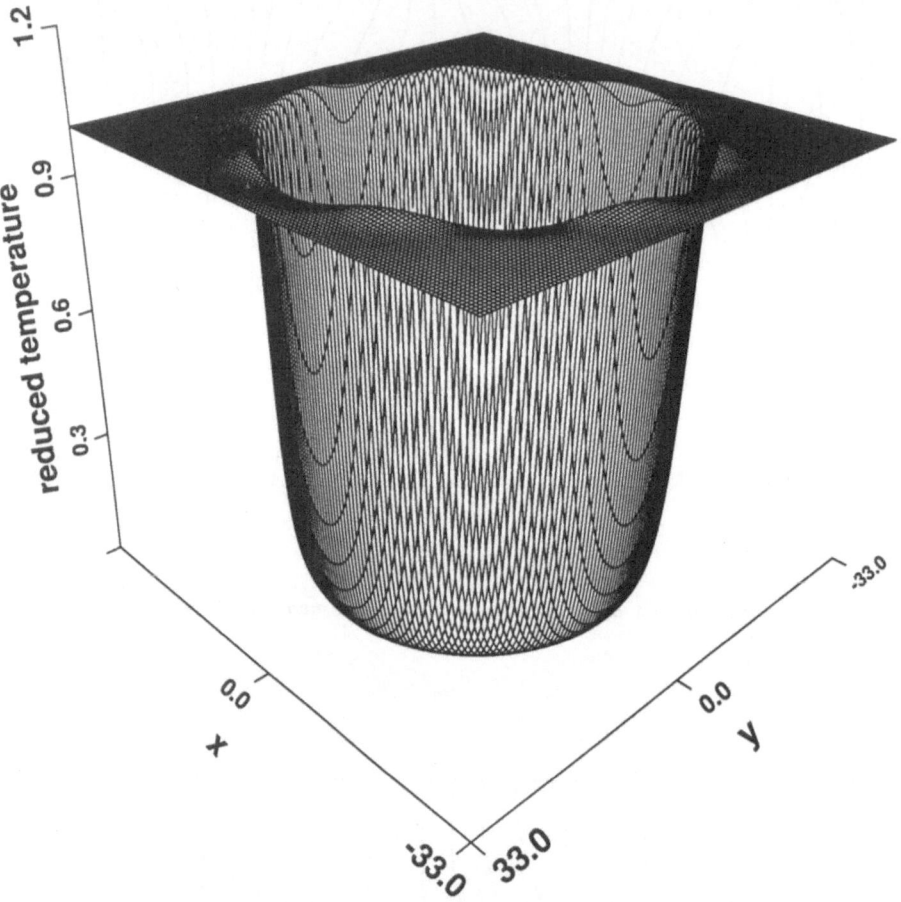

Fig. 6a: Temperature field for gaseous combustion problem, $L = 4.4$, $\kappa = 20$, $t = t_1$.

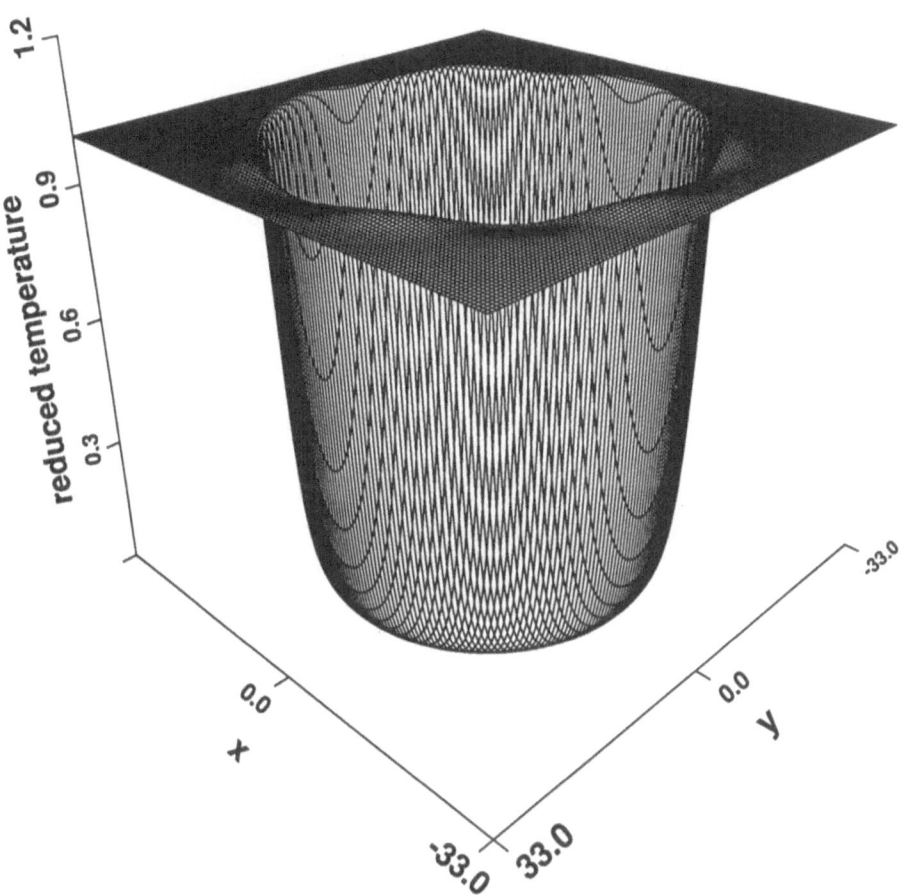

Fig. 6b: Temperature field for gaseous combustion problem, $L = 4.4$, $\kappa = 20$, $t = t_2 > t_1$.

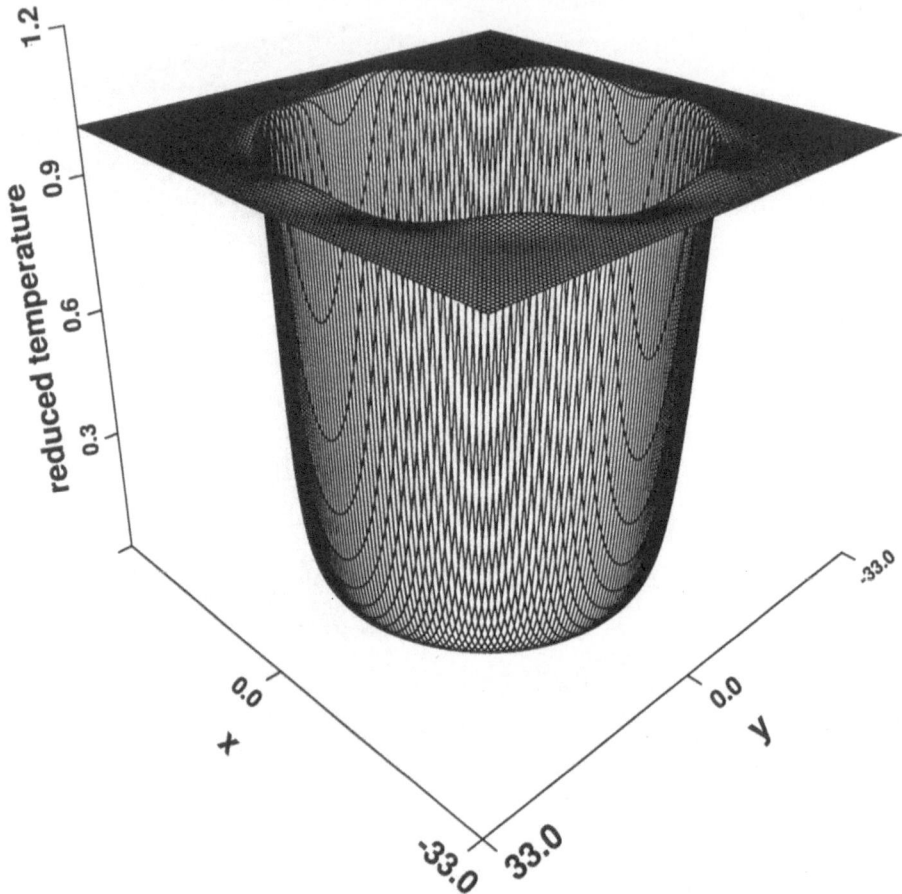

Fig. 6c: Temperature field for gaseous combustion problem, L = 4.4, κ = 20, $t = t_3 > t_2 > t_1$.

AN EXPLICIT RUNGE-KUTTA METHOD FOR
TURBULENT REACTING FLOWS CALCULATIONS

A.A. Boretti^

Dipartimento di Energetica, Universita' di Firenze,
Via S.Marta, 3 50139 Florence, ITALY
and Internal Fluid Mechanics Division
NASA Lewis Research Center, Cleveland, OHIO, USA

Abstract

The paper deals with a numerical method for the solution of the conservation equations governing steady, reacting, turbulent viscous flows in two dimensional geometries, in both cartesian and axisymmetric coordinates. These equations are written in Favre-averaged form and closed with a first order model. A two-equation $K-\varepsilon$ model, where low Reynolds number and compressibility effects are included, and a modified eddy-breack up model are used to simulate fluid mechanics turbulence, chemistry and turbulence-combustion interaction. The solution is obtained by using a pseudo-unsteady method with improved perturbation propagation properties. The equations are discretized in space by using a finite volume formulation. An explicit, multi-stage, dissipative Runge-Kutta algorithm is then used to advance the flow equations in the pseudo-time. The method is applied to the computation of both diffusion and premixed turbulent reacting flows. The computed temperature distributions compare favorably with experimental data.

List of symbols

AE = Arrhenius activation energy
c   = concentrations fluctuations parameter
C   = constant
CFL = Courant number
D   = diffusion vector
e   = specific internal energy
E   = total specific energy
f = function
f   = unknown vector
F   = flux tensor
H   = total enthalpy
H_p = optimization matrix
K   = turbulence kinetic energy
I   = identity matrix
j   = index of the spatial discretization
k   = index of the multistage algorithm
L   = length scale of turbulent motions
m   = index of the temporal discretization
M   = Mach number
n   = wall normal
N   = unit outward normal
NV = updating rate
p   = pressure

---

^ Actually at Sistemi Elettronici Tecniche di Controllo, Centro Ricerche FIAT, Strada Torino 50, 10043 Orbassano, Italy

P = production of turbulence kinetic energy
PF = Arrhenius prefactor
Pr = Prandtl number
q = heat flux vector
Q = work due to turbulence
R = low Reynolds number term
Re = Reynolds number
s = stoichiometric ratio
S = source vector
Sc = Schmidt number
t = time
T = residual
tu = turbulence intensity
v = volume
V = velocity
x = axial coordinate
y = radial coordinate
Y = mass fraction

β = perturbation speed
γ = specific heat ratio
δr = characteristic volume dimension
δt = time step
$\delta \Sigma$ = surface area
ΔH = chemical heat release
χ = thermal conductivity
θ = multistage scheme coefficient
Σ = boundary of the fixed volume
ρ = density
μ = viscosity coefficient
Γ = diffusion coefficient
τ = viscous stress tensor
Ω = artificial viscosity coefficient
ε = turbulence kinetic energy dissipation rate
ϕ = Schwab-Zeldovich function
ϕ = reaction rate

subscript

i = inviscid
fu = fuel
l = laminar
ox = oxidizer
t = turbulent
v = viscous

Introduction

The design of future combustion concepts requires accurate numerical methods to simulate the behaviour of liquid or gaseous fuels in a highly turbulent air stream [1]. These chemical reacting turbulent flows involve fluid mechanics, chemistry and turbulence-combustion interaction. The present approach is devoted to determine the solution of the full turbulent reacting flow problem according to an engineering accuracy. The reactive gas is described as a mixture of three species, fuel, oxidizer and products. The combustion is supposed to be controlled by a single step irreversible chemical reaction.

The reliability of a numerical method depends on both the mathematical modeling of the physical process and the solution algorithm. The governing equations are written here in

Favre-averaged form. No simplification is used in deriving the energy conservation equation, as usually performed [5,8]. The closure is achieved by using a first order model. The K-ε turbulence model adopted here includes low Reynolds number terms, so that the equations are valid all over the laminar, transition and turbulent region, as described in [2]. Furthermore, the model includes a density gradient term to better simulate variable density effects. The modified "eddy break-up","mixing controlled" reaction rate expression adopted here, introduces a concentration gradients dependence in addition to the classical concentration dependence [8]. This model has been developed for imperfectly premixed flow [2], but it is also useful in cases when the combustion is neither fully premixed nor entirely diffusion controlled. This kind of flow occurs under many circumstances even if the fuel and the oxidizer enter the combustion chamber in separate streams. These two models allow to simulate fluid mechanics turbulence, chemistry and turbulence-combustion interaction.

The numerical solution is obtained by using a pseudo-unsteady method with improved perturbation propagation properties [3]. Pseudo-unsteady methods with artificial equations are quite common in turbomachinery applications [9, 10], but up to now they have received little use in computing reacting flows, despite they are faster and simpler than classical pressure iteration methods. At low speeds, the maximum allowable time step for the proposed artificial equations [12] is indeed very close to the one obtained in the pressure iteration method of [11] even by using the simple Lax-Wendroff algorithm. The equations are discretized in space by using a finite volume formulation, and integrated in the pseudo-time with an explicit multi-stage dissipative Runge-Kutta algorithm. Multi-stage schemes for the numerical solution of ordinary differential equations are usually designed to give a high order of accuracy, but in a pseudo-unsteady solution these schemes are selected only for their properties of stability and damping. The four stage scheme adopted here allows a CFL number of about 2.6 although it is only marginally more complex than the simple Lax-Wendroff algorithm, and it introduces only a minimum of artificial viscosity.

Governing equations

The unknown vector f is the solution of a system of conservation equations. This system is written in Favre-averaged, dimensionless, vector, integral form as follows

$$\iint_\Sigma N \cdot (F_i + F_v)\, d\Sigma = \iiint_V S\, dv$$

The basic dependent variables are the density, the velocity and the energy. The reactive variables are the fuel mass fraction and a Shvab-Zeldovich function

$$\phi = Y_{fu} - Y_{ox}/s$$

The turbulence variables are the turbulence kinetic energy and its dissipation rate. Their conservation equations are the following

$$f = \begin{Bmatrix} \rho \\ \rho\cdot V \\ \rho\cdot E \\ \rho\cdot Y_{fu} \\ \rho\cdot \phi \\ \rho\cdot K \\ \rho\cdot \varepsilon \end{Bmatrix} \qquad F_i = \begin{Bmatrix} \rho\cdot V \\ \rho\cdot V\,V + p\cdot I \\ \rho\cdot E\cdot V + p\cdot I\cdot V \\ \rho\cdot V\cdot Y_{fu} \\ \rho\cdot V\cdot \phi \\ \rho\cdot V\cdot K \\ \rho\cdot V\cdot \varepsilon \end{Bmatrix} \qquad F_v = \begin{Bmatrix} 0 \\ -\tau \\ -\tau\cdot V + q \\ D_Y \\ D_\phi \\ D_K \\ D_\varepsilon \end{Bmatrix}$$

$$
S = \begin{bmatrix}
0 \\
0 \\
\rho \bullet \varepsilon - P + Q + \Delta H \bullet \phi \ \text{fu} \\
- \ \phi \ \text{fu} \\
0 \\
P - Q - \rho \bullet \varepsilon + R_K \\
(P - Q) \bullet C_{\varepsilon 1} \bullet f_{\varepsilon 1} \bullet \varepsilon / K - C_{\varepsilon 2} \bullet f_{\varepsilon 2} \bullet \rho \bullet \varepsilon / K + R_\varepsilon
\end{bmatrix}
$$

The equation of state of a perfect gas and the constitutive relations are written as follows

$$
p = (\gamma - 1) \bullet \rho \bullet e = (\gamma - 1) \bullet \rho \bullet (E - V^2/2) = (\gamma - 1) \bullet \rho \bullet (H - V^2/2) / \gamma
$$

$$
\tau = -2/3 \bullet (\ \mu / Re \bullet div(V) + \rho \bullet K) \bullet I + 2 \bullet \mu / Re \bullet def(V)
$$

$$
q = - \ \gamma \bullet x \bullet grad(e) / (Re \bullet Pr)
$$

$$
D_Y = - \ \Gamma_Y \bullet grad(Y) / (Re \bullet Sc_Y)
$$

The stress tensor, the heat flux vector and the diffusion vector are the sum of a laminar and a turbulent part. The latter is expressed according to a classical eddy viscosity concept, i.e. by assuming

$$
\mu = \mu_\ell + \mu_t \qquad x = x_\ell + x_t \qquad \Gamma = \Gamma_\ell + \Gamma_t
$$

The turbulent viscosity coefficient is expressed according to the Prandtl-Kolmogorov formulation

$$
\mu_t = C_\mu \bullet f_\mu \bullet \rho \bullet Re \bullet K^2 / \varepsilon
$$

The production of turbulent energy from the mean flow energy and the work due to turbulence are given by

$$
P = \mu_t \bullet (def(V) : def(V)) / Re
$$

$$
Q = C_\rho \bullet \mu_t \bullet (grad(\rho) \bullet grad(p) / \rho^2) / Re
$$

while the low Reynolds number functions are [2]

$$
f_\mu = exp(-3.4/(1. + Re_t /50.))
$$

$$
f_{\varepsilon 1} = 1
$$

$$
f_{\varepsilon 2} = 1.0 - 0.33 \bullet exp(-Re_t^2)
$$

$$
R_\varepsilon = -2. \bullet \mu_\ell \bullet (grad(\varepsilon))^2 / Re
$$

$$
R_K = -2. \bullet \mu_\ell \bullet (grad(K))^2 / Re
$$

where

$$\text{Re}_t = \rho \cdot K^2 \cdot \text{Re} / (\mu_\ell \cdot \varepsilon)$$

and the model constant are taken as follows

$$C_{\varepsilon 1} = 1.43 \qquad C_{\varepsilon 2} = 1.92 \qquad \text{Sc}_\varepsilon = 1.3 \qquad \text{Sc}_K = 1.0 \qquad C_\mu = 0.09 \qquad C_\rho = 1.0$$

The above two equation turbulence model is valid all over the laminar, transition and turbulent regions. The model adopts the assumptions and approximations which are normally used for constant density flows, and the gradient diffusion model is supposed to be rewritten in the density weighted form without explicitly taking into account the density fluctuations. However, the compressibility term Q allows a partial consideration of variable density effects.

The reaction rate is finally expressed according to a modified "eddy-break-up" model. The mixing controlled reaction rate is given as follows

$$\phi_{fu,mix} = C_{\phi 1} \cdot \rho \cdot c^{1/2} \cdot K / \varepsilon$$

where for imperfect premixing we use [3]

$$c = \min (Y_{fu}^2 , (Y_{ox}/s)^2 , K^3/\varepsilon^2 \cdot \text{grad}(Y_{fu})^2 \cdot C_{\phi 2})$$

The "kinetics controlled" reaction rate is then evaluated as follows

$$\phi_{fu,kin} = \rho^2 \cdot Y_{fu} \cdot Y_{ox} \cdot PF \cdot \exp(-AE / e)$$

The actual reaction rate is taken to be the smaller between the values obtained from the previous expressions.

Along the inflow boundary, all the flow variables except the static pressure are specified. Along the outflow boundary, only the static pressure is specified. Along the solid boundaries, the no slip condition requires the vanishing of velocity, turbulence kinetic energy and turbulence kinetic energy dissipation rate, the latter being obviously to be intended as the modified quantity used in the conservation equations. The specific internal energy is then set equal to the value resulting from the wall temperature for constant temperature walls, or the energy gradient normal to the wall is set equal to zero for adiabatic walls. Furthermore, the species concentration gradient normal to the wall is set equal to zero. Along a symmetry plane (axis) the normal derivatives of all the flow parameters vanish, with the exception of the normal velocity component, to be set equal to zero. Better details of the adopted boundary conditions are given in the following applications of the method, and a discussion about these boundary conditions can be easily found elsewhere [14].

Numerical Solution

The proposed equations are solved in two dimensional geometries. The derivation of the two dimensional equations from the previous general form is straightforward and not presented here. The equations are written in a form useful in both cartesian and axysimmetric geometries by introducing a switch parameter distinguishing between these two coordinates [2].

The discretization adopted is a pseudo unsteady, finite volume, dissipative, explicit one. In the proposed pseudo-unsteady method, the solution of the steady equations is obtained as the asymptotic solution of the following artificial unsteady equations

$$\iiint_V (\frac{\partial f}{\partial t}) \bullet H_p^{-1} \, dv + \iint_\Sigma N \bullet (F_i + F_v) \, d\Sigma = \iiint_V S \, dv$$

These unsteady equations are generally constructed in order to obtain the better convergence rate, obviously providing that the steady state solution is not altered.

From the identity between the convergence process and the elimination process of the initial perturbations to the steady solution, the convergence parameters are determined in order to improve the perturbation propagation or damping [3]. We use

$$H_p = \begin{vmatrix} 1 & 0 & 0 & 0 & .. & 0 \\ 0 & 1 & 0 & 0 & .. & 0 \\ 0 & 0 & 1 & 0 & .. & 0 \\ f_1 \bullet V^2/2 & -f_1 \bullet V_x & -f_1 \bullet V_y & f_2 & .. & 0 \\ \cdots\cdots\cdots\cdots\cdots & & & & & \\ 0 & 0 & 0 & 0 & .. & 1 \end{vmatrix}$$

where

$$f_1 = \min(M^2 - 1., 0)$$

$$f_2 = \max(f_1 + 1., C_p)$$

$$M = V \bullet (\gamma \bullet p / \rho)^{-1/2}$$

and C_p is a small positive number.

If the diffusion of signals is neglected when compared with the signal convection, the maximum allowable time step for an explicit discretization is given by the classical CFL stability condition

$$\delta t = CFL \bullet \delta r / \beta_{max}$$

where δt is the time step, δr is the mesh size, CFL is the Courant number and β_{max} is the maximum perturbation speed. The proposed artificial time dependent equations [3] give

$$\beta_{max} = V \bullet \min\{ (1. + M^2)/2. + [((1. - M^2)/2.)^2 + 1.]^{1/2} , (1. + 1./M)\}$$

while the physical time dependent equations (H_p is the identity matrix) give

$$\beta_{max} = V \bullet (1. + 1./M)$$

As a result of the previous expression for H_p, an improved perturbation propagation follows in subsonic flows.

The equations are then discretized in space by using a finite volume discretization. The mesh is nonorthogonal and curvilinear, conforming to the boundaries of the domain, with lines intersecting at arbitrary angles, properly refined where high gradients are expected to occur. The discretization nodes, located at the intersection of these lines, are the centers of hexagonal control volumes, obtained by connecting the six surrounding nodes. A sample computational domain and the hexagonal control volume are shown in figure 1. The discretized equations are written as follow [2,4]

$$\frac{\partial f}{\partial t} = H_p \cdot \{ -[\sum_{j=1}^{6} (F_{i,j} + F_{v,j}) \cdot N_j \cdot \delta \Sigma_j]/v + S \}$$

where the subscript j refers to every face of the finite volume.

Finally, the equations are discretized in time by using an explicit, dissipative discretization. Adding a dissipative term, the previous equation can be rewritten as follows

$$\frac{\partial f}{\partial t} = T(f) + D(f)$$

where T represent the residual and D is the dissipative term. The explicit k-stage Runge-Kutta algorithm is written as follows

$$f^{(0)} = f^m$$

$$f^{(1)} = f^{(0)} - \theta_1 \cdot \delta t \cdot [T_i (f^{(0)}) + T_v (f^{m*}) + D(f^{(0)}) - \Omega \cdot D(f^{m*})]$$

....

$$f^{(k-1)} = f^{(0)} - \theta_{k-1} \cdot \delta t \cdot [T_i (f^{(k-2)}) + T_v (f^{m*}) + D(f^{(0)}) - \Omega \cdot D(f^{m*})]$$

$$f^{(k)} = f^{(0)} - \delta t \cdot [T_i (f^{(k-1)}) + T_v (f^{m*}) + D(f^{(0)}) - \Omega \cdot D(f^{m*})]$$

$$f^{m+1} = f^{(k)}$$

The dissipative terms are given as follows [2]

$$D^m = 1/(6 \cdot \delta t) \cdot \sum_{j=1}^{6} (f_j^m - f^m)$$

where

$$\Omega = C_{\Omega 1} \cdot |1. - C_{\Omega 2} \cdot 1/6 \cdot \sum_{j=1}^{6} (\rho_j^m - \rho^m) |$$

and $C_{\Omega 1}$ and $C_{\Omega 2}$ are vectors of coefficients of the order of unity [2].

The subscript j refers now to every surrounding node involved in the finite volume approximation, and the superscript m refers to local time $m \cdot \delta t$. The terms referring to time $m* \cdot \delta t$ are updated only at specific iterations and assumed constant between two updatings. The updating rate is taken equal to NV iterations, to be determined according to a numerical optimization.

A four-stages scheme, with the standard coefficients

$$\theta_1 = 1/4 \qquad \theta_2 = 1/3 \qquad \theta_3 = 1/2$$

has a Courant limit of about CFL = 2.6.

The time step is evaluated according to the classical CFL stability limit all over the computational domain. A safety coefficient smaller than unity is introduced in order to take into account the neglected stability limits due to the viscous terms.

Results

First calculations have been performed for a partially premixed turbulent reacting flow. Figure 2 shows schematically the flow domain. Experimental and theoretical results are presented in [13]. In this configuration, mixing of two parallel streams, one of hot gases and the other of a fresh mixture of air and methane, in a constant area duct is considered. The hot jet causes a flame to be ignited and stabilized in the duct. The inlet duct, with a cross section of 100x100 mm^2, is split into two parts. The upper section (80x100 mm^2) is assigned to the fresh air and methane mixture, with a velocity of 65 m/s, a temperature of 580 K, and a mixture ratio of 0.8; the lower one (20x100 mm^2) is assigned to the pilot flame made up of hot gases, with a velocity of 130 m/s and a temperature of 2000 K; the walls are insulated.

Calculations have been performed on a 35x45 computational grid, nonuniform but orthogonal. Conventional steady solutions have been obtained in about $2 \cdot 10^3$ iterations with a total CPU time of about 1 1/2 h on a VAX 8800. Figures 3,4 show a comparison of predicted and measured transverse temperature profiles at x=42 mm and at x=122 mm. The agreement in both sections lies within engineering accuracy. The introduction of the influence of kinetics in the evaluation of the reaction rate and a better calibration of the model constants lead to a better accuracy than the one previously obtained [2], even if at x=122 mm the temperatures in the mixing layer are still underestimated.

The method has then be applied to a turbulent diffusion reacting flow. Figure 5 shows schematically the flow domain. Experimental and theoretical results are presented in [8]. In this configuration, there is a central jet of fuel, substantially methane, with a velocity of 21.3 m/s and a temperature of 300 K, and a concentric jet of oxidizer, with a velocity of 34.3 m/s and a temperature of 589 K. The outer radius of the fuel jet is 8 mm, the inner and outer radius of the air jet are respectively 11.1 and 28.6 mm. The radius of the combustor is 101.6 mm. The wall temperature is 1140 K.

Calculations have been performed on a 45x45 computational grid, nonuniform but orthogonal. Conventional steady solutions have been obtained in less than $3 \cdot 10^3$ iterations with a total CPU time of about 3 h on a VAX 8800. Figures 6,7 show a comparison of predicted and measured radial temperature profiles at x=95 mm and x=398 mm.

Even if the accuracy appears to be not optimal, the proposed method allows a better accuracy than the method presented in [8], despite of the single step chemical reaction adopted here. This is substantially due to a better modeling of the reaction rate. The mixing controlled reaction rate of [8] is simply taken proportional to the smaller between the fuel and the oxidizer concentration. The present mixing controlled reaction rate is taken proportional to the fuel concentration gradients with limitations arising from the availability of fuel and oxidizer.

In a laminar diffusion flame, little oxidizer is detectable within the reaction zone envelope, and little fuel is detectable outside of it. The reaction zone is a very thin envelope, and fuel and oxidizer concentrations show a small overlapping. In a turbulent diffusion flame, the reaction zone is thicker, and the averaged values of fuel and oxidizer overlap significantly. The reaction zone is now a finite volume, even if always relatively small. The introduction of the gradient concentration dependence other than the concentration dependence appears to lead to a better simulation of the thickness of the reaction zone.

Conclusions

The paper has presented a numerical method for the study of steady, reacting, turbulent viscous flows in two dimensional geometries, both cartesian and axisimmetric.

The proposed two-equation K—ε model, where low Reynolds number and compressibility effects are included, and the modified "eddy breack-up" model give a satisfactory description of fluid mechanics turbulence, chemistry and turbulence-combustion interaction within an engineering accuracy.

The pseudo-unsteady method with improved perturbation propagation properties and the explicit multi stage dissipative Runge-Kutta algorithm appear to be fast and reliable.

The application of the method to the computation of the temperature distributions for a diffusion and a premixed turbulent reacting flow shows a satisfactory agreement between experimental and computational results.

Acknowledgements

The author would like to express his indebtedness to Dr. Enrica Giuntoli for the encouragement and many helpful suggestions she has provided during this study.

References

[1]   Mularz, E.J., and Sockol, P.M., "Chemical Reacting Flows," in "Aeropropulsion '87. Session 3 - Internal Fluid Mechanics Research", NASA CP 10003, 1987.
[2]   Boretti, A.A. and Martelli, F., "Accuracy and Efficiency of Time Marching Approach for Combustor Modelling," paper presented at the "AGARD Fluid Dynamics Panel Simposium on Validation of Computational Fluid Dynamics", Lisbon, Portugal, May 1988.
[3]   Boretti, A.A. and Martelli, F., "Application of a Fast Pseudo Unsteady Method to Steady Compressible Turbulent Reacting Flows," proceedings of the "First National Fluid Dynamic Congress," Cincinnati, Ohio, July 1988.
[4]   Arts, A., "Cascade Flow Calculations Using a Finite Volume Method," in "Numerical Methods in Turbomachinery Bladings", VKI LS 1982-05, Bruxelles, Belgium, 1982.
[5]   Bray, K.N.C., "Turbulent Flows with Premixed Reactants," in Libby, P.A., and Williams, F.A., Turbulent Reacting Flows, Springer-Werlag, New-York, 1981.
[6]   Jameson, A.,"Transonic Flow Calculations," Princeton University MAE Report 1651, 1983.
[7]   Jorgenson, P.C.E., and Chima,R.V.,"An Explicit Runge-Kutta Method for Unsteady Rotor-Stator Interaction," NASA TM 100787, 1988.
[8]   Lin, C.-S., "Numerical Calculations of Turbulent Reacting Flow in a Gas-Turbine Combustor," NASA TM 89842, 1987.
[9]   Essers, J.A., "Artificial Evolution Techniques for Transonic Flows," in "Numerical Methods in Turbomachinery Bladings", VKI LS 1982-05, Bruxelles, Belgium, 1982.
[10]   Viviand, H.,and Veuillot, J.P.,"Methodes pseudo-instationnaires pour le calcul d'ecoulements transoniques," ONERA TP no.1984-69, 1984.
[11]   Hirt, C.W., Cloutman, L.D., and Romero, N.C., "SOLA-ICE : A Numerical Solution

208

Algorithm for Transient Compressible Fluid Flows," Los Alamos Laboratory Report LA-6236, 1976.

[12]   Boretti, A.A., and Martelli, F.,"Numerical Modeling of Turbulent Combustion in Premixed Flows," proceedings of the "1988 National Heat Transfer Conference", Houston, Texas, July 1988.
[13]   Dupoirieux, F., "Numerical Calculations of Turbulent Reactive Flows and Comparison with Experimental Results," ONERA T.P. 1986 -80.
[14]   Cline, M.C., "VNAP2: A Computer Program for Computation of Two-Dimensional, Time-Dependent, Compressible, Turbulent Flow," Los Alamos Report LA-8872, 1981.

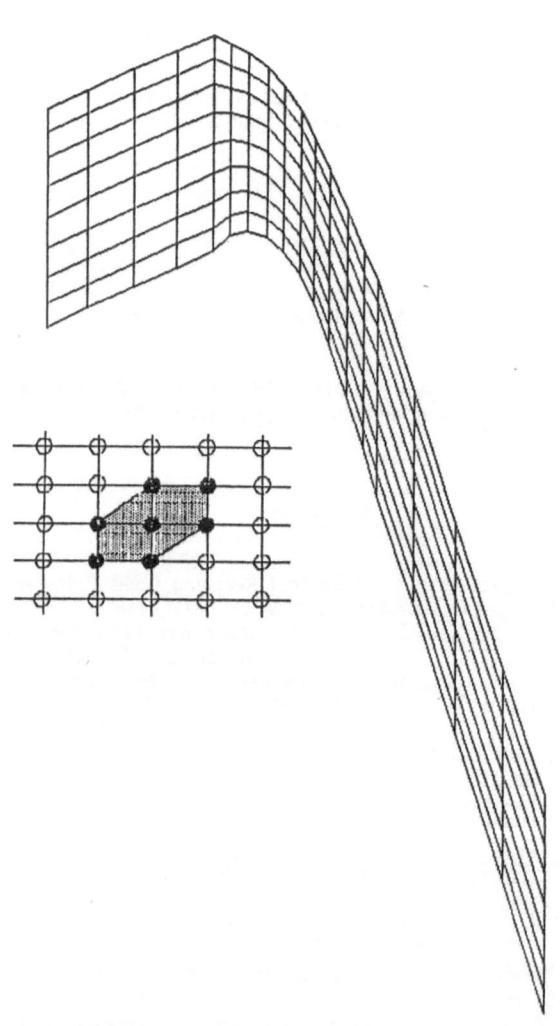

Figure 1 - Sample computational domain and hexagonal control volume.

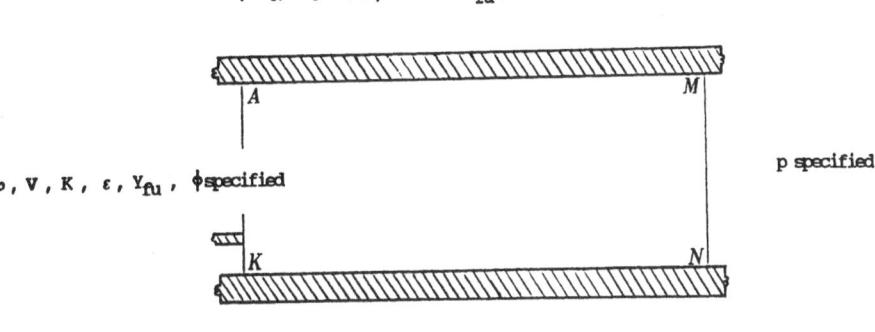

$$v = K = \varepsilon = \partial e / \partial n = \partial Y_{fu} / \partial n = \partial \phi / \partial n = 0$$

ρ , V , K , ε , Y_{fu} , ϕ specified

p specified

$$v = K = \varepsilon = \partial e / \partial n = \partial Y_{fu} / \partial n = \partial \phi / \partial n = 0$$

Figure 2 - Premixed turbulent reacting
flow: physical domain.

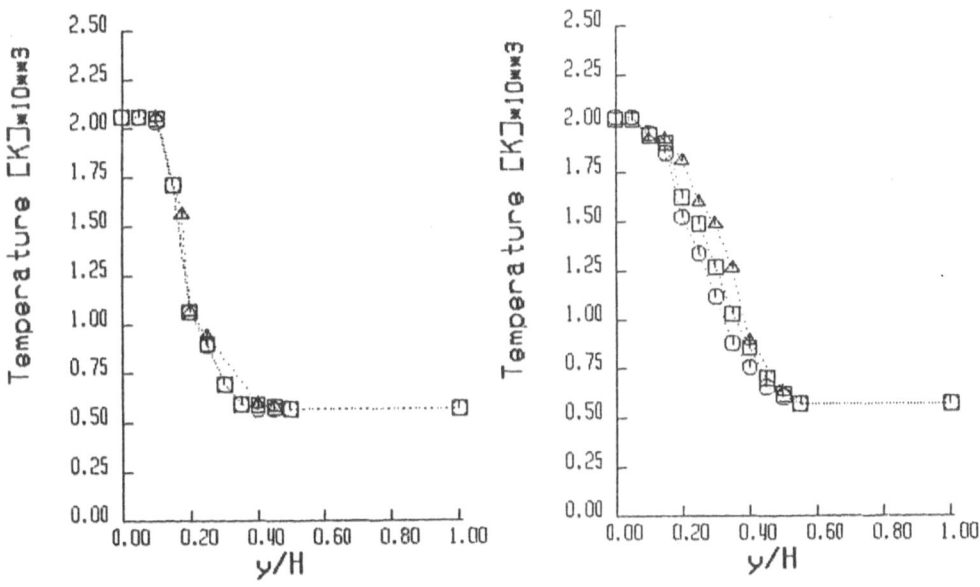

□ theoretical, present method
⊕ theoretical, reference [2]
△ experimental, reference [13]

Figure 3 - Premixed turbulent reacting
flow: comparison of predicted and
measured temperature distributions.

Figure 4 - Premixed turbulent reacting
flow: comparison of predicted and
measured temperature distributions.

$V = K = \varepsilon = \partial Y_{fu} / \partial n = \partial \phi / \partial n = 0$, e specified

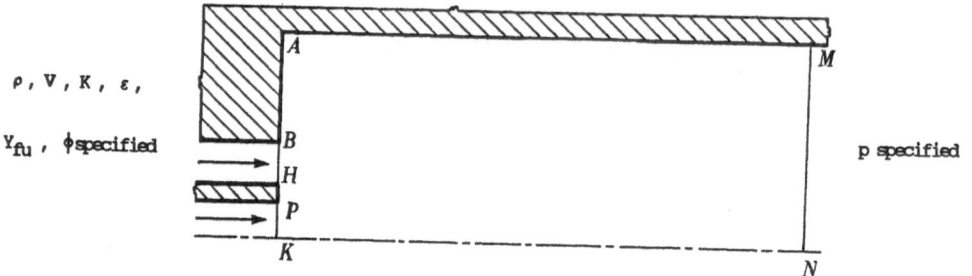

$\rho, V, K, \varepsilon,$

Y_{fu}, ϕ specified

p specified

$\partial V_{tangential} / \partial n = \partial K / \partial n = \partial \varepsilon / \partial n = \partial e / \partial n = \partial Y_{fu} / \partial n = \partial \phi / \partial n = 0, \quad V_{normal} = 0$

Figure 5 - Diffusion turbulent reacting
flow: physical domain.

⬛ theoretical, present method
⊕ theoretical, reference [8]
△ experimental, reference [8]

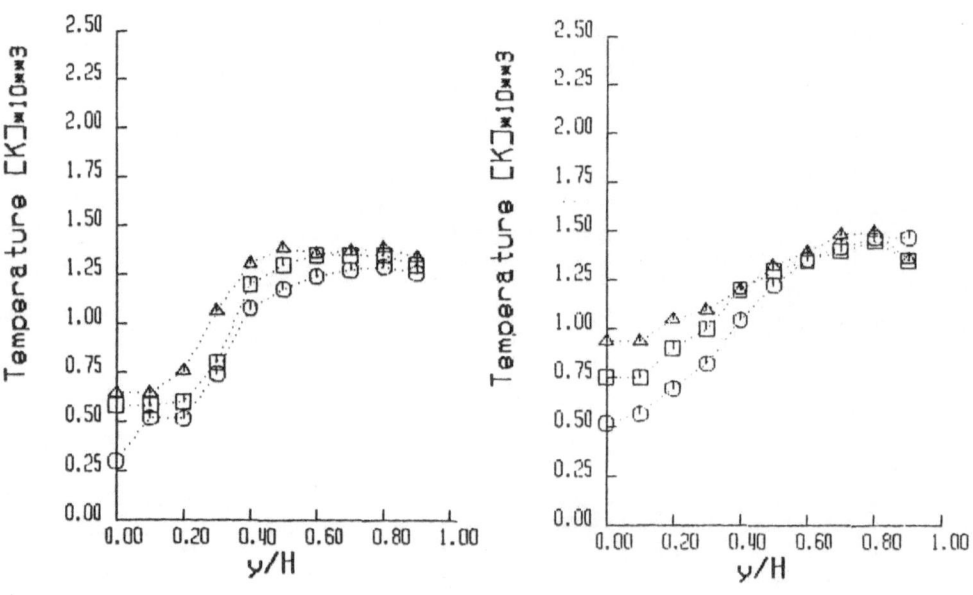

Figure 6 - Diffusion turbulent reacting
flow: comparison of predicted and
measured temperature distributions.

Figure 7 - Diffusion turbulent reacting
flow: comparison of predicted and
measured temperature distributions.

Knock Prediction in
Spark Ignition Engines

by

R. A. Cox, R. G. Myhill and W. P. Sweetenham

Harwell Laboratory, Didcot, Oxon, OX11 0RA, United Kingdom

ABSTRACT

A computer simulation of the thermodynamic processes within the cylinder of a spark ignition engine, together with a detailed chemical model of autoignition, is used to predict experimental data for gas temperatures and pressures taken from a production engine. Values for parameters which determine the rates of burning and heat loss are fitted to experimental data for temperature and pressure and then both the thermodynamic and the chemical simulations agree well with the experimental results. Moreover, after fitting to one set of temperature and pressure data, the autoignition model's sensitivity to spark advance is similar to knock sensitivity of the real engine.

1 Introduction

This work is being carried out at Harwell on behalf of the Harwell Petrol-Engine Working Party, whose members comprise Ford, Jaguar Cars, Elf and the Lucas Group, together with support from the Vehicle Division of the Department of Trade and Industry.

The work described in this paper aims at simulating the conditions in the mixture of air and unburnt fuel sufficiently accurately to allow a simulation of the chemical reactions governing engine knock. For this purpose a two zone thermodynamic simulation of the gas mixture within the cylinder is used. The model is deliberately made simple by assuming that the only effect of fluid flow is to ensure that the two gas zones are each homogeneous. The intention is to produce a model that is sufficiently sophisticated for the simulation of autoignition while including as few free parameters as possible.

The simulation covers the compression and expansion strokes. It incorporates a turbulent combustion model where the combustion rate depends on the flame geometry which in turn depends on the cylinder geometry. The heat release rate is based on the fuel composition and the rate of combustion. The simulation includes the temperatures and pressures in the gas allowing for the temperature dependent heat capacity of the gas and also incorporates a detailed autoignition reaction scheme for the unburnt gas.

The paper describes the experimental data, the analysis used to allow for cycle-to-cycle variation and examples of simulations whose temperature and pressure predictions have been fitted to the experimental data. These simulations predicted knock under conditions for which the experimental engine knocked. So, fitting some of the parameters in the thermodynamic model to thermodynamic data was sufficient to provide good experimental agreement in the chemical modelling as well. Moreover, when the ignition was retarded, the experimental engine stopped knocking and also the simulation no longer predicted knock.

The paper also shows how the knock predictions vary with different engine operating conditions such as engine speed, engine load and air/fuel ratio.

2 The Experimental Data

The simulation is based on data from experiments[1] carried out at Harwell for the Harwell Petrol-Engine Working Party. The data came from a production engine which was run on the laboratory bench under carefully measured conditions. Hence values needed by the simulation, such as ignition angle and fuel RON number, are known accurately. (The Research Octane Number of a fuel is defined by comparing the knock behaveour of the fuel against the behaveour of mixtures of N-heptane and iso-octane in a particular test engine. The RON value is the percentage of iso-octane in the mixture of N-heptane and iso-octane that exhibits knock behaveour similar to that of the fuel under test.) Temperature measurements were made using the C.A.R.S. technique to measure the gas temperature at a certain point in the cylinder at a predetermined crank angle. The gas pressure was measured at each degree crank angle during the engine cycle.

Each experiment consisted of many engine cycles with the same engine operating conditions and the same crank angle for temperature measurement. The experiment was repeated with the temperature measurement at several different crank angles. A clear pattern emerged from these experiments. As the crank angle, at which the temperature measurement occurred, was moved through the engine cycle, the temperature first rose slowly as the gas was compressed. Then there was a period during which no measurement could be made because the flame front distorted the measurement volume. After the flame front, the gas temperature was much greater than before. Then the gas temperature slowly fell as the piston moved away from the cylinder head.

This suggested that a two-zone model for the gas thermodynamics would be appropriate.

3 The Thermodynamic Model

Initially there is one homogeneous zone of unburnt gas in the cylinder. The composition of the gas is calculated from the cylinder dimensions, the air/fuel mass ratio, the initial cylinder pressure and the fuel RON number. It is assumed that the fuel is completely vapourised and that it is composed of n-heptane and iso-octane. During the simulation, the heat capacity of the gas is calculated from the tabulated heat capacities of the various gas components as functions of temperature. All the gases are assumed to be ideal.

The cylinder is assumed to be a geometric cylinder with the spark plug at a given distance below the top of the cylinder and at a certain distance from the axis of the cylinder. During the burning process, it is assumed that the gas is in two homogeneous zones. One zone is of burnt gas. It is spherical in shape and centred on the spark plug. The zone of unburnt gas occupies the volume of the cylinder outside the spherical zone of burnt gas; see Figure 1. The volumes and surface areas of the two gas zones can be found at any given time by calculations based on the position of the spark plug, the radius of the zone of burnt gas and the position of the piston. Details of these geometry calculations are given in Sweetenham et al., in preparation.

In practice the flame front, which can be assumed to be of negligible thickness, has a very complex geometry which may even have a disconnected structure. Pockets of unburnt gas may be entrained in the region of burnt gas. This complicated geometry affects the rate at which gas is burnt and thus the rate at which the burnt gas zone expands. These effects are covered by the flame model which is described later in the paper.

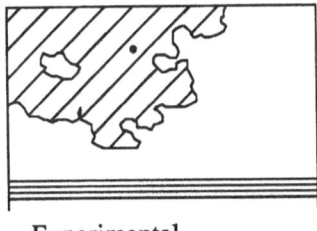   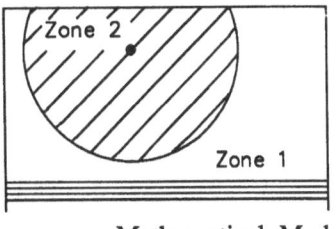

Experimental   Mathematical Model

Fig 1. Comparison of actual and model flame geometries.

The thermodynamic model is evaluated by solving a set of six simultaneous ordinary differential equations. The independent variable is time and the six quantities to be solved for are:

the radius of the zone of burnt gas,

the pressure in the cylinder,

the temperature in each of the two gas zones,

the fraction of gas that has been burned and

the fraction of unburnt gas entrained within the burnt gas zone.

These quantities define the thermodynamic state of the two gas zones. It is assumed that a quasi-static approximation holds and so pressure and temperature of each zone is related to its volume and number of moles by the perfect gas law.

The processes simulated in the thermodynamic model are:

the movement of the piston, which is determined by the engine speed, and structure geometry,

burning at the flame front, simulated using a turbulent flame model, which heats up the zone of burnt gas,

heat loss from each gas zone to the cylinder walls and
the heat gained from the autoignition chemistry.

The thermodynamics result in the following three differential equations:

$$\frac{dT_1}{dt} = \frac{1}{\tilde{C}_p^1}\left(k(T_1 - T_E)A_1 + H_1 + V_1\frac{dP}{dt}\right).$$

$$\frac{dT_2}{dt} = \frac{1}{\tilde{C}_p^2}(K(T_2 - T_E)A_2 - \frac{1}{M}\frac{dM_2}{dt}\Delta H + V_2\frac{dP}{dt}).$$

$$\frac{dP}{dt} = \frac{-\tilde{C}_p^1\tilde{C}_p^2 P}{\tilde{C}_v^2\tilde{C}_p^1 V_2 + \tilde{C}_v^1\tilde{C}_p^2 V_1}$$
$$\left\{\frac{dV}{dt} - \frac{RT_1}{P}\frac{dn_1}{dt} - \frac{RT_2}{P}\frac{dn_2}{dt} + \frac{R\Delta H}{\tilde{C}_p^2 PM}\frac{dM_2}{dt} - \frac{R}{\tilde{C}_p^1 P}\frac{dQ_1}{dt} - \frac{R}{\tilde{C}_p^2 P}\frac{dQ_2}{dt}\right\}$$

P is the gas pressure, M is the total mass of gas R is the molar gas constant. The zones of unburnt and burnt gas, are numbered 1 and 2 respectively. For gas zone i, T_i is the temperature, $\tilde{C}_v^i$ and $\tilde{C}_p^i$ are the heat capacities at constant volume and constant pressure, V_i is the volume, A_i is the wall area in contact with the gas zone, M_i is the mass of gas, Q_i is the thermal energy and n_i is the number of gas molecules. ΔH is the enthalpy of combustion, k is the heat loss coefficient and T_E is the wall temperature.

3.1 The four stages of combustion

The combustion process can be divided into four distinct phases:

Adiabatic compression: The first stage of combustion consists of the compression of the initial charge of air-fuel mixture up until the spark plug is fired. During this first stage there is only one zone of unburnt gas and no burning.

Ignition and homogeneous burn: This is an intermediate stage that occurs between the ignition of the air-fuel mixture by the spark and the development of a spherical flame front. At the commencement of this phase a small spherical region of gas, centred on the spark plug, is singled out to be burnt during the phase. The contents of this region are assumed to burn homogeneously at a constant rate. It is a necessary step as the next stage does not allow for an initial flame development.

Development and propagation of a flame front: This stage describes the part of the cycle when there is a propagating flame front in the cylinder. It starts with a spherical flame whose radius is equal to that of the burnt gas zone after the homogeneous burning has finished. The flame then expands rapidly and turbulently. This propagation is described by the turbulent flame model described below. The phase ceases when the flame area becomes negligible.

Final burning of entrapped gas: This stage covers the final burning of the entrapped gas and the subsequent expansion and cooling of the burnt gas. At the end of the previous stage there is still some unburnt unentrapped gas left as the imaginary flame front reaches the cylinder walls before all the unburnt gas has been either burnt or entrapped. At the commencement of the phase any such residual gas is instantly transferred to the region of entrapped gas so that it is finally burnt.

3.2 The Turbulent Flame Model

The turbulent flame model is based on a paper by Keck[2]. It includes a number of simplifying assumptions about the flame geometry. There are two processes by which the gas is burnt.

Gas is burnt at the flame front at a rate proportional to its area, A_f, the density of the unburnt gas, ρ_1, and the laminar flame speed, s_l. The radius of the flame is a function of the radius of the zone of burnt gas. When the radius of the zone of burnt gas is small, the radius of the flame front is the same as that of the burnt zone. As the radius of the burnt zone grows, the radius of the flame front grows faster. This is to simulate the increase in turbulence as burning proceeds. The empirical relation given by Keck for the flame radius is

$$r_f = r_2 + u_T \tau_b \left(1 - e^{-\left(\frac{\rho_2 r_2}{\rho_1 3 s_l \tau_b} \right)^2} \right),$$

where ρ_1, ρ_2 are the densities of the unburnt and burnt gas zones respectively and r_2 is the radius of the burnt gas zone.

Unburnt gas is entrained within the zone of burnt gas at a rate proportional to the area of the flame front, the density of the unburnt gas and a characteristic speed, u_T, and then burns at a rate proportional to the mass of entrained gas, μ, over a characteristic time, τ_b. Within the model this gas is still considered to be in the region of unburnt gas. The equation governing the mass of the entrained unburnt gas, μ, is

$$\frac{d\mu}{dt} = \rho_1 A_f u_T - \frac{\mu}{\tau_b}.$$

The combination of the two processes gives the following rate of increase in the mass of burnt gas, M_2,

$$\frac{dM_2}{dt} = \rho_1 A_f s_l + \frac{\mu}{\tau_b},$$

There are three parameters in the turbulent flame model:

s_l: a laminar flame speed which determines the rate of burning at the flame front,

u_T: a characteristic speed which determines the rate at which unburnt gas is entrained within the zone of burnt gas and

τ_b: a characteristic time which determines the rate at which the entrained gas burns.

The rate of change of the cylinder volume is known *a priori* from the engine speed. The burning rate and temperature changes determine the rate of change of the volume of burnt gas and this is solved as a differential equation for the radius of the zone of burnt gas:

$$\frac{dr_2}{dt} = \frac{1}{A_2}\left(\frac{dV_2}{dt} - \frac{A_p}{A_c}\frac{dV}{dt}\right).$$

A_p is the area of the piston head in contact with the zone of burnt gas and A_c is the total area of the piston head.

3.3 The Heat Loss Model

The rate of heat loss from each gas zone is proportional to the product of the area of the gas zone in contact with the cylinder wall and the temperature difference between that gas zone and the cylinder wall. The heat transfer coefficient is taken to be the same for each gas zone. This empirical model brings in two further parameters whose values are not known exactly· *a priori* ,

   k: the heat transfer coefficient and

   T_E: the temperature of the engine wall.

Radiative heat transfer is assumed to be small[3] and any that does occur is adsorbed, in the model, into the heat transfer coefficient by suitably adjusting it to describe the experimental data. It is assumed that there is no heat transfer across the flame front.

The coding of the model allows for easy replacement of the heat loss model by a more detailed model.

4 The Autoignition Model

The autoignition model is intended to simulate the chemical reactions which can lead to engine knock. The thermodynamic model provides the temperature and pressure of the unburnt gas as functions of time and the chemical model is used to simulate the progress of autoignition under these conditions. The autoignition model acts as a source of heat for the thermodynamic model but the corresponding changes in the composition of the unburnt gas are *not* included in the thermodynamic model; *i.e.* the composition of the unburnt gas is frozen with respect to the thermodynamic model.

The chemical model is based on the Shell Thornton[4] model which describes autoignition in terms of a hydrocarbon oxidation mechanism appropriate for alkane-type fuels with more than five carbon atoms, in the temperature region 700-850K. The model assumes that oxidation of the fuel proceeds by a degenerately branched free radical chain mechanism. Heat is released from the chemical reaction at a rate proportional to the reaction rate, resulting in temperature increases in the gas additional to those derived in the thermodynamic model. The thermokinetic feedback from this provides a satisfactory explanation of the occurrence of cool flames, 2-stage ignition and other phenomenological features of hydrocarbon oxidation.

The original Shell model containing eight generalised reaction steps was refined and extended in the light of the expanding body of kinetics information on the type of elementary reactions involved[5]. The resulting fifteen step model was able to predict satisfactorily the temperature, fuel and oxygen dependence of ignition delay time for selected primary reference fuels in a rapid compression machine.

During the development of this model, the fuel structure dependent reaction steps were identified and the rate parameters of these could be adjusted to give appropriate ignition delay times for specific primary reference fuels, whilst keeping those of common reaction steps, e.g. those involving HO_2 radicals, constant. Parameters for the structure dependent reactions for pure iso-octane and n-heptane (primary reference fuels of 100 RON and 0 RON respectively) were determined in this way and a combined model for the two fuels was formulated with 31 chemical reactions involving 24 chemical species.

The combined model, when run with appropriate initial concentrations of the primary reference fuels, was capable of predicting experimental ignition delay in a rapid compression machine for fuels with RON values in the range 70-100 without further parameter adjustment. Further details of the chemical basis for this model have been given elsewhere[6].

5 Solving the Equations

The thermodynamic model and the chemical model together result in a set of 31 simultaneous, initial valued, ordinary differential equations in time. To solve the equations, the computer program FACSIMILE[7] was used. The thermodynamic and chemical models were described in the FACSIMILE language. That is, code was written to tell FACSIMILE how to calculate the initial values of all the variables, how to evaluate the time derivatives for the variables and how to alter the model when the equations changed (for example at spark ignition and when there was no unburnt gas left). The FACSIMILE program took the model and used a numerical integrator for stiff ordinary differential equations[8] to calculate the values of the variables as functions of time.

The use of FACSIMILE in coding the mathematical model means that it is a relatively easy task to change any of the sub-models, such as,

autoignition model,

flame model,

heat loss model,

without altering the underlying thermodynamic model. Thus the model can be updated as new theoretical models and experimental evidence becomes available.

6 Using the Experimental Data

The experimental data showed that under the same engine operating conditions, there was great variation in pressure and temperature from cycle to cycle. To get accurate temperature values, the raw C.A.R.S. data from engine cycles with a temperature measurement at the same crank angle were grouped together so that in each group the gas pressures at the time of the temperature measurement were in a range whose width was 0·2MPa. The average of the raw C.A.R.S. data from all the engine cycles in each group was used to get a temperature value. Thus the data to be fitted consists of temperature values throughout the engine cycle with associated pressures. At some crank angles through the cycle there is one than one pressure value, because the gas pressures recorded at that angle fell into more than one group. The crank angles with several pressure values have several temperature values, one for each pressure value.

From looking at typical pressure curves, a pressure value was chosen for each point through the engine cycle, so that the chosen set of pressures were a good representation of a commonly occurring engine cycle. The data used in the fitting process consisted of those chosen pressure values together with their associated temperature values.

6.1 Fitting to the Experimental Data

FACSIMILE has a parameter fitting facility which was used to fit values of the burning parameters S_l, U_T and τ_b and values of the wall temperature and heat transfer coefficient to the temperature and pressure data. At each of the times for which a temperature value was available, FACSIMILE was given a value for the pressure and and either a value for the temperature of the burnt gas or a value for the temperature of the unburnt gas.

FACSIMILE then performed the simulation many times, improving its estimates of the parameters until it found a set of parameter values that fitted the data within a pre-selected error.

6.2 Engine Running at 1500 r/min, Ignition 26° B.T.D.C.

The first set of data used were from an engine running at 1500 r/min with spark ignition at 26° before top dead centre (B.T.D.C.). Under these conditions, many of the engine cycles knocked and the parameters were fitted to measurements selected from those cycles which knocked.

Figures 2 and 3 show the comparison between the experimental data and the predictions for the parameter values which gave the best fit. Figure 2 is the comparison for pressure values and figure 3 is the comparison for temperature values. The first five temperature values come from the zone of unburnt gas and the last temperature value comes from after the flame front has passed across the measurement position.

The fit to the pressure curve is reasonable from 180° B.T.D.C., where the simulation starts, up to 20° after top dead centre (A.T.D.C.), where the simulation ends. The fit to the temperatures is reasonable for the unburnt gas temperatures, but the estimate of the burnt gas temperature at 20° A.T.D.C. is far too high. The calculation of the burnt gas temperature soon after ignition, at 26° B.T.D.C., is based on the heat of combustion of the air/fuel mixture and is probably fairly accurate. The over-estimate of the burnt gas temperature at 26° A.T.D.C. may well be due to under-estimating the rate of heat loss from the burnt zone after a period of turbulent burning. The heat transfer coefficient, expressed in $W K^{-1} cm^{-2}$ is assumed to be the same for each gas zone and to be constant in time.

The fitting provides a simulation covering the thermodynamic and chemical models for which some of the parameters of the thermodynamic model have been fitted to experimental temperature and pressure data. The model with the fitted parameter values follows the progress of the autoignition reactions. It is found that these reactions progress very quickly at about 16° of crank angle A.T.D.C., which is a few degrees after peak pressure is attained. Thus, after fitting the thermodynamic model, the chemical model predicts knock. The real engine knocks under these conditions and the knock perturbations on the pressure curve are first seen soon after peak pressure.

6.3 Engine Running at 1500 r/min, Ignition 22° B.T.D.C.

The next set of data came from the same engine with spark ignition retarded to 22° B.T.D.C.. Under these conditions the engine did not knock. The same comments apply to this fit as for the previous case.

Under these conditions, the chemical model does not predict fast autoignition before 20° A.T.D.C., when the simulation ended. After 26° A.T.D.C., the engine temperatures and pressure have fallen sufficiently to preclude fast autoignition on the time scale of an engine cycle.

7 Qualitative Predictions from the Model

To test the model, we have observed its behaviour under varying engine conditions. One test was to change the engine load and mark the resulting change in the model's predictions. A reduction in engine load was simulated by reducing the initial pressure of the air/fuel mixture at the start of the compression stroke while maintaining the same air/fuel ratio.

Starting from the simulation for 1500 r/min with ignition at 26° B.T.D.C. and an initial cylinder pressure of 1 atmosphere, 0·1MPa, the initial cylinder pressure was reduced on successive runs. As the initial cylinder pressure was reduced, knock was predicted later and later in the cycle, until at a pressure of 0·08MPa, knock was not predicted.

Next, FACSIMILE's parameter fitting facility was used to discover how much the ignition angle had to be advanced for the model to predict fast autoignition at 16° A.T.D.C.. Figure 4 plots initial cylinder pressure against the spark ignition angle which causes knock.

Figure 5 shows the effect of altering the air/fuel ratio. The graph plots air/fuel ratio against the spark ignition angle required to predict fast autoignition at 16° A.T.D.C.. An air/fuel ratio of 1 means a stoichiometric mixture.

Figure 6 shows the effect of altering the engine speed while keeping the parameter values for burning and heat transfer fixed. In agreement with experiment, the model predicts that a greater ignition advance is needed to achieve the knock at a higher engine speed.

8 Conclusions

From the fitting to experimental data it is clear that the model is capable of giving a reasonable simulation of the combustion process. However, the model would benefit from an improved description of the heat transfer from the cylinder. The qualitative results show that the model predicts trends in the correct direction, but the values of the fitted burning and heat transfer parameter need to be altered when the model is used to simulate conditions well away from those at which the parameter values were found.

The model should prove to be a useful tool in investigating cycle to cycle variation and changes in engine behaviour under varying conditions. The next stage will be to fit the burning parameters to different engine cycles under the same conditions and to a range of different engine cycles under different operating conditions. The behaviour of the fitted parameters should give useful insights into engine behaviour and could well be valuable in improving engine design through mathematical modelling.

9 Acknowledgements

The authors wish to express their gratitude to the Harwell Petrol-Engine Working Party for permission to publish this work. Due credit must go to Dr M. J. Norgett who originally started the thermodynamic model and to Miss J. Cole who worked both with Dr Norgett on the thermodynamic model and with Dr Cox on the autoignition model.

10 References

1. Greenhalgh, D.A., Williams R.W. and Baker, C.A. (1987) *The Development and Application of the C.A.R.S. Technique for In-cylinder Thermometry.* I.S.A.T.A., Florence.

2. Keck, J.C., *Turbulent Flame Structure and Speed in Spark-Ignition Engines,* Proceedings of the Nineteenth International Symposium on Combustion, Sponsored by the the Combustion Institute, Haifa, Israel, 1982, pp1451-1466.

3. Annand, W.J.D.: *Heat transfer in the cylinder and porting,* The Thermodynamics and Gas Dynamics of Internal Combustion Engines. pp773-805. Ed. J.H. Horlock and D.E.Winterborne. Oxford University Press, 1986.

4. Halstead, M.P., Kirsch, L.J., Prothero, A. and Quinn, C.P., **Proc. Roy. Soc. Lond. A,** 1975, 346-515

5. Cox, R.A. and Cole, J.A.: **Comb. Flame** 1985, 60, 109.

6. Cox, R.A., Cole, J.A. and Cavanagh, J.: *Computer Modelling Studies of Autoignition Chemistry*, 6th Task Leaders' Meeting of the I.E.A. Working Party on Conservation in Combustion, Livermore, California.

7. Curtis, A.R. and Sweetenham, W.P.: FACSIMILE/CHEKMAT User's Manual, Harwell Report AERE R.12805, February 1988.

8. Curtis A.R., I.M.A. Conference on The State of the Art in Numerical Analysis, Birmingham, 1986

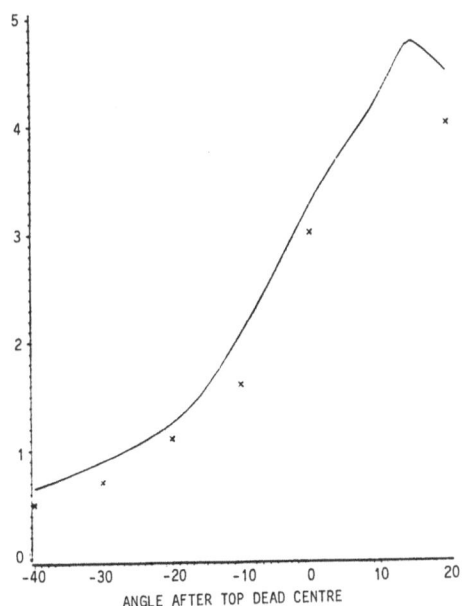

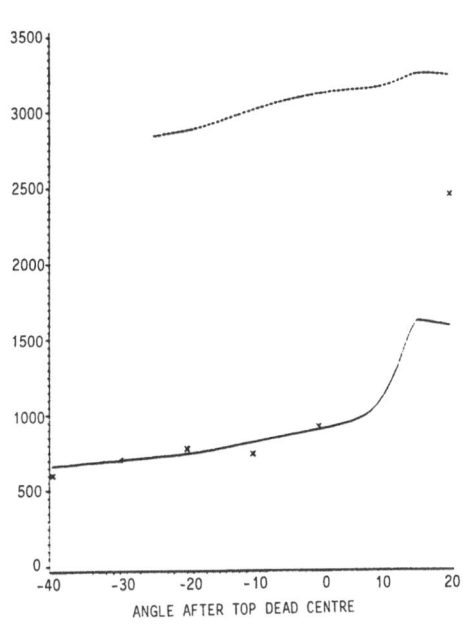

Fig 2. Predicted and experimental pressure curves at 1500 r/min with ignition at 26° B.T.D.C.. The crosses are the experimental values, the line shows the predicted values.

Fig 3. Predicted and experimental temperature curves at 1500 r/min with ignition at 26° B.T.D.C.. The crosses are the experimental values, the first five before flame front arrival and the last in the burnt gas. The solid line is the predicted unburnt gas temperature and the dashed line is the predicted burnt gas temperature.

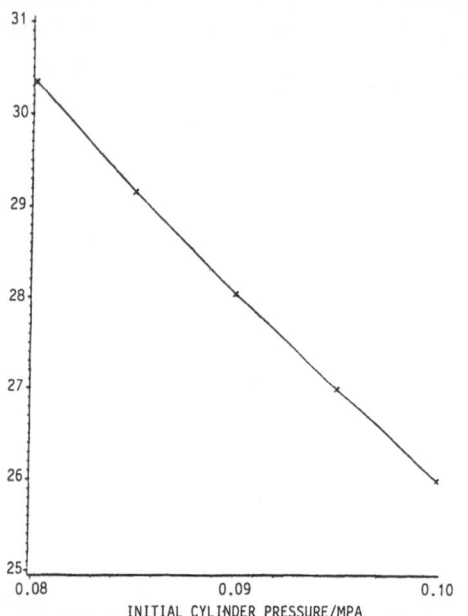

Fig 4. Ignition advance against engine load for knocking. The crosses are predicted values, joined by straight lines.

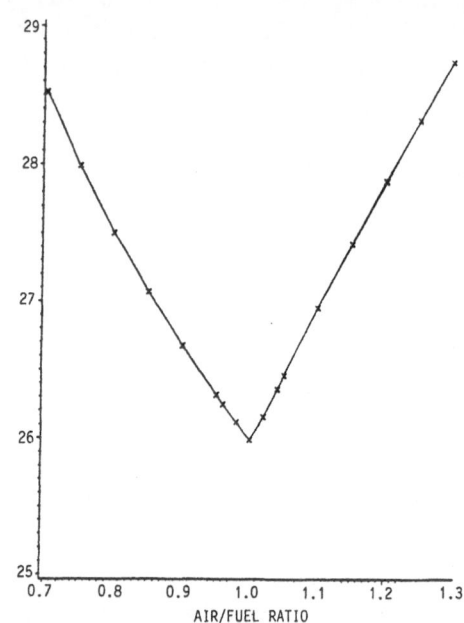

Fig 5. Ignition advance against air/fuel ratio for knocking. The crosses are predicted values, joined by straight lines.

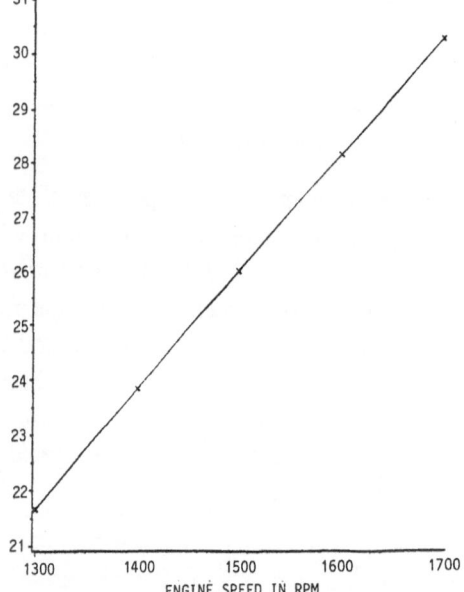

Fig 6. Ignition advance against engine speed for knocking. The crosses are predicted values, joined by straight lines.

A local extinction of the thermo-diffusive premixed flame at low Lewis number

Bruno DENET and **Pierre HALDENWANG**
Laboratoire de Recherche en Combustion
Université de Provence - Centre de Saint Jérôme (S 252)
13397 Marseille Cedex 13, France

Abstract

We solve the 2-D thermo-diffusive model of premixed flames in the framework of Fourier Spectral Methods. Although the temperature and concentration fields are not periodic in the direction perpendicular to the flame, we suggest a particular treatment, simple to implement, that is applied to these quantities in order to transform them into the sum of a known profile and a periodic unknown. This process also takes advantage of the fact that the physics of flames allows us to consider as periodic the higher derivatives. "Infinite" order convergence of Spectral Methods is thus recovered. This algorithm being very efficient, we can perform numerical simulation concerning the diffusive-thermal instability, far in the non-linear domain. Thus, at low Lewis number, we numerically observed, for the first time to our knowledge, a phenomenon of local extinction. This brings a plausible explanation to the presence of unburnt combustible in the lean hydrogen flame.

I./ Introduction

The numerical simulation of flame propagation is usually supposed to involve very sophisticated algorithms because two reasons prevail : the first one considers the fact that we have to track a free surface corresponding to the very thin region of space where the chemical reaction takes place. The second reason is that this discontinuity is very corrugated and its profile rapidly develops in time. The latter point stems from the fact that plane flames rarely exist. The interface between (cold) fresh gases and (hot) combustion products is indeed subject to instabilities leading to well-known patterns of wrinkled flames - for recent reviews see Ref.[1] to [3].

Two different types of mechanism are usually considered : the thermo-diffusive instability and the Darrieus-Landau instability. The latter one is of hydrodynamic origin and is the most often put forward because it appears as soon as the density changes across the flame. The first one is a consequence of diffusive effects through the finite thickness of the real flame, while the second one considers the flame as an infinitely steep interface. As observed in experiments [4-6], the flames in the unstable regime are wrinkled, exhibiting cell patterns which become more and more cusped as the non-linearity increases. Further in the non-linear domain, this interface is subjected to turbulent fluctuations. Cusped forms continuously appear or merge in a chaotic way, exhibiting the so-called self-turbulization.

The numerical approach that we shall describe in this paper, has been previously employed for studying this self-turbulization [7]. This study allowed us to confirm that the competition between the diffusive effects is the source of this time-dependant behaviour. Moreover, we qualitatively compared the results of our simulation with those corresponding to the Kuramoto-Sivashinsky equation which was considered as a model equation of flame dynamics [8]. It has been concluded [7] that the Kuramoto-Sivashinsky equation is not fictitious for flame front dynamics - it has a finite (and non-vanishing) domain of validity. Furthermore, those results have been obtained in a parameter range for which standard criteria of any weakly non-linear derivation are near their limits. This indicates that the domain of validity -in a qualitative sense- is larger than the one that is usually recognized [1].

The aim of the present numerical work, in solving the "field equations" for large non-linearities, is to propose an efficient way to bring theoretical approaches nearer to experiment. Up to now, most of the numerical studies in combustion consider the flame as an infinitely steep interface propagating with the normal velocity usually derived from asymptotics. In very important cases such as the combustion of hydrogen, this simplified model can neglect essential phenomena. The study we present in this contribution shows a new effect which requires an internal treatment of the flame to be pointed out. Obviously this effect of local quenching due to the intrinsic dynamic of the flame introduces a new flamability limit. As for the model to solve, our approach is restricted to the thermo-diffusive model. Nevertheless, we are fully aware of the fact that we neglect some leading contribution to the flame wrinkling. From an experimental point of view, it seems [9] that diffusive effects play a leading role when the non-linear regime of the thermo-diffusive instability is reached. The present numerical approach can be viewed as a first step towards a complete study of hydrogen-oxygen combustion.

Except for very recent contributions [10-13], the direct simulation of flame propagation with finite thickness was limited to 1-D computations of the reaction-diffusion equations (see e.g. [14]). The 2-D field equations are rarely solved because the tracking of 2-D largely curved fronts requires a multi-dimensional fully adaptive method [15,16]. The more the front is wrinkled, the more the computational effort in adaptation increases. In the present contribution we avoid paying the cost of self-adaptation, expecting that the efficiency of Fourier Spectral Methods will allow us to adequately represent the small scales wherever they are located.

Part II briefly recalls the constant-density model. In the third paragraph we present the numerical scheme, the originality of which lies in the periodification process. This feature is believed to be extensible to more complex flames. Part IV is devoted to the presentation of the numerical results. For a large control parameter an extinction phenomenon is locally observed. This brings a plausible explanation to the presence of unburnt combustible in the lean hydrogen flame which corresponds to very non-linear conditions.

II./ **The physical model**

The simplest 2-D system of premixed flame dynamics is the thermo-diffusive model. It assumes that the gas expansion plays a negligible role. Moreover, to clearly put forward the phenomenum of local extinction, we choose the simplest chemistry : a single one-step chemical reaction is assumed. Non-dimensional quantities are obtained using a classical approach : the length scale is the flame thickness obtained from asymptotics, likewise the velocity unit is the asymptotic flame speed. The use of the normalized variables allows us (see e.g. [3], [17]) to write the model, in a frame moving with the flame front, as follows :

$$\frac{\partial T}{\partial t} + U\frac{\partial T}{\partial x} = \Delta T + \Omega \qquad \text{II.1.a}$$

$$\frac{\partial C}{\partial t} + U\frac{\partial C}{\partial x} = \frac{1}{Le}\Delta C - \Omega \qquad \text{II.1.b}$$

with

$$\Omega = \frac{\beta^2}{2\,Le}\, C\, \exp\left(\frac{\beta(T-1)}{1+\gamma(T-1)}\right) \qquad \text{II.2}$$

where T and C correspond to the reduced temperature of the gas mixture and concentration of a reactant, (the other reactant being in excess), L_e, β and γ are respectively the Lewis number of the reactant, the reduced activation energy (or Zeldovich number) and the heat release parameter. U is the reduced flame speed which is an unknown of the problem. U is supposed to be parallel to the x-direction.

Because we are interested in the resolution of unstable flame fronts, we shall assume that the flat front is unstable with respect to patterns, periodic in y the direction perpendicular to the flame speed. The linear stability analysis leads [18] to the following threshold of the diffusive thermal instability :

$$\beta(L_e-1) < -2 \qquad \text{II.3}$$

The maximum growth rate σ_{max} is obtained for the wave-number k_{max}

$$\sigma_{max} = \varepsilon^2/16 \qquad \text{at} \qquad k_{max} = \frac{1}{2}\left(\frac{\varepsilon}{2}\right)^{1/2} \qquad \text{II.4.(a-b)}$$

where $\varepsilon = -1 - \frac{\beta}{2}(L_e - 1)$ is the control parameter of the instability

III/ **The numerical algorithm**

The numerical description of a largely wrinkled front in its smallest length scales, as it is here the purpose, generally requires [15,16] special ingredients such as a fully 2D self-adaptive gridding. This is due to the fact that three different length scales are present in the computational domain. The smallest one is related to the reaction zone, the second being the length scale of the pre-heating region. The largest one is scaled by the typical amplitude of the wrinkles which are, in the present problem, of one order of magnitude larger than the pre-heating zone.

The 2-D tracking of a zone of large gradients requires a large computational effort in order to adapt the grid to the position of small length scales. For the present problem, we suggest a different approach, easily implemented and certainly competitive in term of CPU cost. Since the contribution by Babuska [19], it is well known that two different options can be chosen in order to increase the accuracy : the "h" version consists of increasing the number of grid points where needed, as it is generally done in self-adapting processes. On the other hand, the "p" version locally extends the order of the scheme when a lack of precision is observed. By using Spectral Methods we have chosen to globally extend the order of the scheme. Numerical calculations of flames using Chebyshev expansion have already been performed [12], but because the expansion in finite Fourier series is known to be the fastest Spectral Method [20-21], we have attempted to transform a non-periodic problem into a periodic one in all directions. As it will be recalled later, this method is very efficient if periodicity at all order can be attained.

It is well known [20], that the direct application of Fourier Spectral Methods to a problem with non-periodic boundary conditions leads to the well-known Gibbs phenomenon. The method therefore loses its interest. A classical way to increase the convergence properties is to substract [20] a simple form (generally a polynomial) from the unknowns, the first non-periodicity being reported to higher derivatives. Here, we take advantage of specific properties of flames in order to extend this principle and to recover the feature of exponential convergence.

We are interested in the computation on the $(-X_0, X_0)*(-Y_0, Y_0)$ rectangle with periodic boundary conditions in the y-direction. In the x-direction the boundary conditions are :

$$T(-X_0, y) = 0 \quad , \quad T(X_0, y) = 1 \quad ; \quad C(-X_0, y) = 1 \quad , \quad C(X_0, y) = 0 \qquad \text{III.1.(a-d)}$$

Strictly speaking such boundary conditions should be imposed at infinity. Nevertheless, if X_0 is large enough, these boundary conditions can be satisfied at finite distances without loss of accuracy [7].

Taking into account these non-periodic boundary conditions, let us define θ and ψ as the following intermediate unknowns :

$$T(x,y) = S_0(x) + \theta(x,y)$$
$$C(x,y) = 1 - S_0(x) + \psi(x,y)$$

$$\text{III.2.(a-b)}$$

where $S_0(x)$ is a smooth "step" function satisfying the following boundary conditions:

$$S_0(-X_0) = 0 \quad ; \quad \frac{\partial^n S_0}{\partial x^n}(-X_0) = 0 \quad ; \quad n = 1, 2, \ldots$$

$$S_0(X_0) = 1 \quad ; \quad \frac{\partial^n S_0}{\partial x^n}(X_0) = 0 \quad ; \quad n = 1, 2, \ldots$$

III.3

Among the functions satisfying these conditions we have chosen the following one :

$$S_0(x) \;=\; \frac{1}{2}\left[\, 1 + \tanh\left(\, \Gamma \;\; \tan\left(\frac{z-\pi}{2}\right)\right)\right]$$

III.4

where $z=(x+X_0)\pi/X_0$ and Γ is a parameter that determines the slope of S_0 at the centre of the integration domain.

If we suppose X_0 large enough we can then assume all the x-derivatives of $T(x,y)$ and $C(x,y)$ to be negligible at $x=\pm X_0$. Hence it is straightforward to show that $\theta(x,y)$ and $\psi(x,y)$ are periodic in all directions, likewise their derivatives at every order. The equations governing $\theta(x,y)$ and $\psi(x,y)$ become:

$$\frac{\partial\theta}{\partial t} + U\frac{\partial\theta}{\partial x} - \Delta\theta \;=\; f_\theta \;=\; \Omega + \frac{\partial^2 S_0}{\partial x^2} - U\frac{\partial S_0}{\partial x}$$

III.5.(a-b)

$$\frac{\partial\psi}{\partial t} + U\frac{\partial\psi}{\partial x} - \frac{1}{L_e}\Delta\psi \;=\; f_\psi \;=\; -\Omega - \frac{1}{L_e}\frac{\partial^2 S_0}{\partial x^2} + U\frac{\partial S_0}{\partial x}$$

Because the physics of the flame considers that Ω, the production term, has a small support of order $1/\beta$, then Ω and its further derivatives vanish at $x=\pm X_0$. Moreover it is easy to verify that f_θ and f_ψ (and their further derivatives) are periodic in all directions. The thermo-diffusive model is thus posed in terms of a periodic problem having excellent properties of convergence in the framework of Fourier Spectral Methods : exponential convergence to the exact solution can be achieved because all quantities and their successive derivatives are periodic in all directions.

As usually in Spectral Methods, the time discretization is furthermore achieved using finite differences. Several two or three points schemes can be easily implemented. The results presented here have been carried out with a simple first order scheme, treating implicitly the diffusion terms. The time step is actually limited by the reaction term. A more detailed description of the algorithm is given in reference [7]. Because standard flames have a high activation energy ($\beta\sim10$) the three length scales mentioned in the introduction are separated by an order of magnitude. So that, if the front is strongly wrinkled, at least a hundred Fourier modes are required in the x-direction while the discretization in the y-direction depends on the Y_0 value. Thanks to FTT algorithms the computational cost of our approach increases about linearly with the number of degrees of freedom. Moreover the vectorization of each elementary step of the algorithm is easy to implement on a vectorized computer.

IV/ Local extinction of the thermo-diffusive premixed flame

The result we present in this part is obtained with $\beta=10$, $\gamma=0.8$; for this value of b the threshold of the thermo-diffusive instability is attained at $L_e=0.8$. For a lower value, for instance $L_e=0.6$, the growth rate has a non-vanishing value : $\sigma_{max}=1/16$ obtained for $k_{max}=0.25\sqrt{2}$. The typical wave-length is thus more than 15 times the flame thickness.

Let us define L_y as : $L_y = 2\,Y_0$. This quantity corresponds to the diameter of the tube in which the pre-mixed flame propagates. This image of a pipe is to be interpreted loosely because periodic boundary conditions in the y-direction are not consistent with the presence of duct walls. We intend simply that the picture of propagation in a pipe fixes a lower bound to the wave-numbers allowed to be unstable. This is additionally a source of quantification : i.e. all unstable wave-numbers are integer multiples of the basic quantity given by $k_1 = 2\pi/L_y = \pi/Y_0$ A small tube is then characterized by a diameter allowing a limited number of unstable modes.

We want now to study the flame dynamics with a large control parameter. Considering Eq.(II.5) we have to choose a large value for β and a small Lewis number. By decreasing the Lewis number we increase the fuel mobility compared to the thermal diffusivity. Unburnt fuel thus tends to flee the cusped regions where the production term consequently decays. Local quenching, accompanied by unburnt fuel, is thus expected at low Lewis numbers and has actually been studied in experiments related to lean hydrogen flames [9].

To simulate this phenomenon we have chosen $L_e=0.2$ ($\varepsilon=3$, $1-L_e=8/\beta$) which is close to the Lewis number of hydrogen in the lean hydrogen-oxygen flame. The chosen parameters are $L_y =24$ with 32 y-modes, and $L_x=36$ with 256 x-modes. The integration domain being small in the y-direction, we initiate the computation with a two wave-length solution. The evolution of isotherms is given in Fig.(1) and Fig.(2). The temperature profile clearly develops towards a solution which is locally smooth at the cusps and, the more time goes on, the more the temperature locally decreases in large pockets. We then had to stop the computation because these pockets rapidly reached the boundary of the integration domain. The final profiles are shown on Fig.(3) to Fig.(5a) where thermal, fuel, and production profiles are drawn. We can conclude that the flame is strongly wrinkled with points where local extinction occurs. This development of the flame pattern is accompanied by an important increase of the flame velocity, the time evolution of which is plotted on Fig.(6). We have stopped the time integration because the cold zone of the temperature field was getting ready to leave the computational domain.

However, there is an important issue that we have not yet answered : does this process lead to a steady solution ? (i.e. a flame front profile propagating with a stable velocity). It is not clear that patterns, such as the present iso-production lines on Fig.(5b) in the form of moon crescents, correspond to a steady solution. For instance, one can imagine an oscillating asymptotic behaviour. Obviously, more investigation is needed to conclude on the existence of a dynamical extinction in the cusps.

V/ **Conclusion**

We have presented a numerical algorithm that allowed us to adopt a new approach to flame front dynamics. At first glance, it was not obvious that a problem of front tracking could be efficiently treated with an elementary Fourier Pseudo-Spectral algorithm. This has been made possible thanks to a periodification process which takes into account the physics of flames. We believe that the present method can be easily implemented for studying more complex situations such as hydrodynamic flames or non-adiabatic flames. The efficiency of the present Fourier expansion allows us to treat rather complicated non-linear behaviours. This indeed represents real progress in the study of wrinkled flame dynamics. For low Lewis number we have indeed noticed an effect of local extinction that has already been observed in experiments with lean hydrogen flames. Although the role of the high diffusivity of hydrogen has been suspected by theoretical arguments [9], that is the first time that local quenching has been clearly exhibited. This phenomenon is accompanied by a strong increase of the flame speed. Nevertheless we have to moderate this point because we are not sure of the existence of such a steady curved state. More computational effort is needed to make this issue clear.

We wish to thank Professor Paul Clavin who, through extremely helpful advice and encouragements, stood at the origin of the present study.

This work has received support from the scientific commitee of the "Centre de Calcul Vectoriel pour la Recherche" which provided the computational resources.

References

1. Sivashinsky G.I. : Instabilities, Pattern Formation, and Turbulence in Flames. *Ann. Rev. Fluid Mech. 15* : 179-199 (1983)
2. Williams F.A. : *Combustion Theory*, The Benjamin/Cummings Publishing Company (1985)
3. Clavin P. : Dynamic behavior of premixed flames fronts in laminar and turbulent flows. *Prog. Energy Combust. Sci., 11* : 1-59 (1985)

4. Bregeon B., Gordon A.S., and Williams F.A. : Near-Limit Downward Propagation of Hydrogen and Methane Flames in Oxygen-Nitrogen Mixtures. *Combust. Flame, 33* : 33-45 (1978)

5. Sabathier F., Boyer L. and Clavin P. : Experimental Study of a Weak Turbulent Premixed Flame. *Prog. Aeronaut. Astronaut., 76,* 246-258 (1981)

6. Quinard J., Searby G. and Boyer L. : Cellular structures on premixed flames in a uniform laminar flow. *In "Cellular structures in instabilities",* Lecture Notes in Physics, Weisfreid J.E. and Zaleski S. Ed., Springer Verlag, vol. **210**, 331-341 (1984)

7. Denet B. and Haldenwang P. : A pseudo-spectral scheme for turbulent thermo-diffusive premixed flames. *to appear.*

8. Sivashinsky G.I., Non-linear analysis of hydrodynamic instability in laminar flames. Part I : Derivation of basic equations. *Acta Astronautica. 4:,* 1177-1206 (1977)

9. Mitani T. and Williams F.A., Studies of Cellular Flames in Hydrogen-Oxygen-Nitrogen Mixtures. *Combust. Flame, 39* : 169-190 (1980)

10. Ashurst W.M.T, Peters N. and Smooke M.D. : Numerical Simulation of Turbulent Flame Structure with Non-unity Lewis Number. *Combust. Sci. Tech.,* **53**, 339-375 (1987)

11. Benkhaldoun F. and Larrouturou B. : A finite element adaptive investigation of two-dimensional flame front instabilities, to appear in *Comp. Meth. in Appl. Mech. Eng.*

12. Ehrenstein U., Guillard H and Peyret R. : Flame computations with a Chebyshev multi-domain method, to appear in *Int. J. for Num. Meth. in Fluids.*

13. Guillard H., Larrouturou B. and Maman N. : Numerical investigation of two-dimensional flame front instabilities using a pseudo-spectral method. *INRIA Report 721* (1987)

14. Glowinski R., Larrouturou B. and Temam R. : *Numerical Simulation of Combustion Phenomena ;* Proceedings, Sophia-Antipolis, France 1985, Lecture Notes in Physics 241, Springer-Verlag (1985)

15. Smooke M.D. and Koszykowski M.L. : Two-Dimensional Fully Adaptive Solutions of Solid-Solid Alloying Reactions. *J. Comput. Phys.,* **62**, 1-25 (1986)

16. Matsuno K. and Dwyer H.A. : Adaptive Methods for Elliptic Grid Generation, *J. Comput. Phys.,* **77**, 40-52 (1988)

17. Buckmaster J.D. and Ludford G.S.S, *Lectures on Mathematical Combustion,* (SIAM-CBMS, Philadelphia) (1983)

18. Sivashinsky G.I., Diffusional-thermal theory of cellular flames. *Combust. Sci. Tech., 15* : 137-145 (1977)

19. Babuska I. : The p and h-p Versions of the Finite Element Method. A Survey. *In : Proceedings of the Workshop on Theory and Application of Finite Elements* (R. Voigt, M.Y. Hussaini, ed.) Berlin, Heidelberg, New-York : Springer Verlag (1987)

20. Gottlieb D. and Orszag S.A., *Numerical Analysis of Spectral Methods : Theory and Applications.* (SIAM-CBMS, Philadelphia) (1977)

21. Canuto C., Hussaini M.Y., Quarteroni A. and Zang T.A., *Spectral Methods in Fluid Dynamics.* (Springer Series in Computational Physics, Springer - Verlag) (1988)

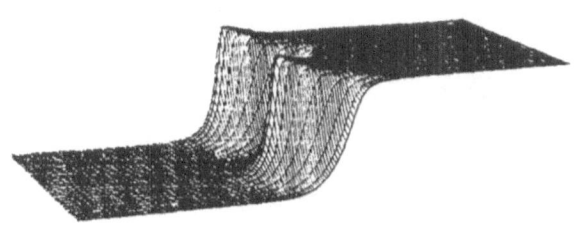

Fig.1 *Beginning of the non-linear behaviour.*
The thermodiffusive instability is rapidly growing thanks to a large control parameter. The temperature profiles are plotted at t=1. The parameters are $\beta=10$, $L_e=0.2$, $L_y=24$, $\varepsilon=3$.

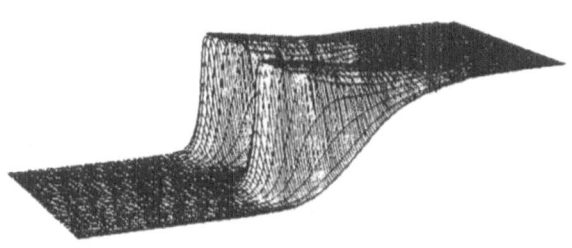

Fig.2 *Creation of fresh mixture pockets.*
At low Lewis number, unburnt fuel rapidly diffuses towards hot regions leading to the creation of cold gas pocket. Temperature profiles are plotted at t=4. The parameters are $\beta=10$, $L_e=0.2$, $L_y=24$, $\varepsilon=3$.

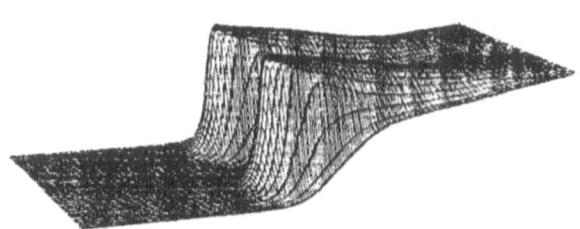

Fig.3 *Local quenching of the flame.*
At low Lewis number, unburnt fuel rapidly diffuses towards hot regions leading to the creation of a local extinction. Temperature profiles are plotted at t=9. (No steady solution is yet attained). The parameters are $\beta=10$, $L_e=0.2$, $L_y=24$, $\varepsilon=3$.

232

Fig.4 *Local quenching of the flame (continued).*
Concentration profiles of dilute fuel are plotted at t=9.

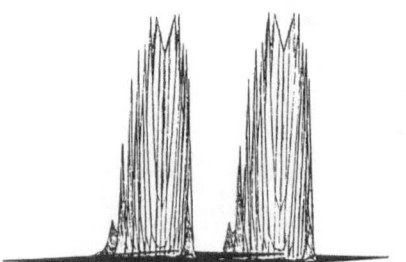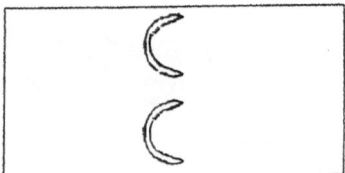

Fig.5 *Local quenching of the flame (continued).*
Production profiles (a) and an iso-production line (b) are plotted at t=9.

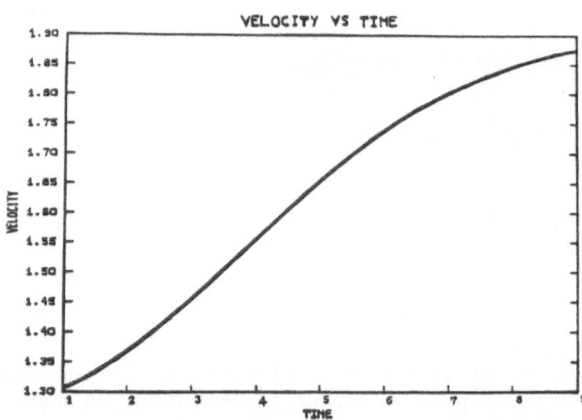

Fig.6 *Time evolution of the flame speed.*
Flame velocity versus time is plotted from t=1 to t=9. The parameters are $\beta=10$, $L_e=0.2$, $L_y=24$, $\varepsilon=3$.

COMPUTER SIMULATION OF DETAILED PROCESSES OCCURRING AT NEAR EXTINCTION CONDITIONS OF FLAME SPREAD OVER SOLID FUELS

Colomba Di Blasi , Silvestro Crescitelli, Gennaro Russo

Dipartimento di Ingegneria Chimica, Università
di Napoli, Piazzale V. Tecchio, 80125 NAPOLI, Italia

ABSTRACT

A computer model of near limit flame spread processes over solid fuels is presented. The unsteady, two-dimensional, fully elliptic mathematical model accounts for finite rate combustion kinetics and for heat, momentum and mass transfer in the gas phase. Gas phase processes are coupled through the boundary conditions at the interface to solid phase processes that are heat conduction and thermal degradation, modeled by a finite rate zero order Arrhenius reaction. Numerical solution is computed by a finite difference formulation of equations and by a semi-implicit procedure to account for the coupling between the two phases.

Detailed simulations of the spread process in a counterflow environment have been used to investigate the detailed flame structure and the mechanisms leading to extinction when the gas flow velocity is increased or the oxygen level of the oxidizing flow is lowered. Although the time evolution of the process is very different for the two cases, extinction can be always explained in terms of a decrease of the Damkohler number at the flame leading edge. This number, defined as the ratio of the residence time to the chemical time, decreases, as the opposed flow velocity is increased or as the oxygen concentration is lowered, causing the retreat of the flame against the forced flow until extinction occurs.

INTRODUCTION

Inflammability limits of solid fuels change with environmental conditions. A better understanding of their dependence on these conditions can give useful information for fire safety and control. To this end, flame spread processes have been extensively investigated from an experimental point of view. However only global parameters, such as flame and pyrolysis spread rates, have been measured. Detailed measurements are not easy to obtain, especially at near extinction conditions when the process is very sensitive to small changes in the environmental conditions. In a similar way, theoretical analyses emphasized predictions of rates of flame propagation and only very simplified descriptions of the spread process and its controlling mechanisms are available. In particular, all numerical studies of opposed flow flame spread and extinction [1-5] have been made with a superimposed flow field. On the other hand, it is well known that, for the chemically controlled regime [5,7], the interaction between the flow field and chemical processes becomes very important and very different spread rates and flame structure are predicted on dependence of the shape of the assigned velocity profile [4,8].

The aim of the present paper is to formulate a more complete mathematical model that includes the description of the flow field in order to predict detailed flame structure and extinction over a thermally thick solid fuel. To have a proper description of the process, the effects of finite rate combustion kinetics are accounted for. Moreover, at near extinction conditions, streamwise diffusion cannot be neglected so that the marching techniques usually applied for boundary layer flow computations are not adopted. Detailed simulations of phenomena occurring at the flame leading edge are performed to give the evolution of the spread process from steady conditions of propagation to extinction, when the flow velocity is increased or the oxygen concentration is lowered. Results are used to explain the controlling role played by mass diffusion and gas phase chemistry.

THE COMPUTER MODEL

The two-dimensional problem to be modeled is schematically represented in Fig. 1. A thick fuel slab is placed horizontally in a channel. An initial external heating (thermal radiation) of a small portion of the solid fuel (left side of the fuel slab) is used to cause ignition, then the flame spreads forward (to the right in the Figure) against the oxidizing flow.

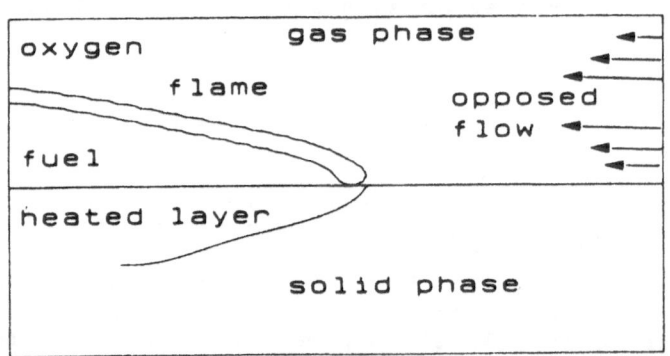

Fig. 1
Schematic of
opposed flow
flame spread

The model assumes that the solid fuel is thermally thick and pyrolyzes only at the surface according to a zero order Arrhenius reaction. Therefore solid phase processes are described by an energy balance equation:

$$c\rho_s \, [\partial T_s /\partial t] = k_s \partial/\partial x[\partial T_s /\partial x] + k_s \partial/\partial y[\partial T_s /\partial y] + I_0 K exp(K(y-\tau)) \qquad (1)$$

where c is the solid heat capacity, ρ_s the density, k_s the thermal diffusivity, T_s the temperature, I_0 the incident radiant flux, K the absorption constant and τ the solid thickness.

Combustion in the gas phase is described by a global second order Arrhenius reaction: $F + \nu_0 O = \nu_p P$. Gas phase properties are assumed constant, but the gas density variations in the buoyancy induced body force are modeled through the Boussinesq approximation. The viscous dissipation term and the compressive work are neglected, due to the low speed problem. Radiation from the flame to the fuel and from the solid surface to ambient are neglected too, given the small scale of the process. In the present formulation, the processes occurring in the gas phase are modeled by the unsteady Navier Stokes equations expressed in vorticity ζ and stream function χ transport form and by energy and chemical species balance equations:

- vorticity

$$\rho(\partial\zeta/\partial t + u\partial\zeta/\partial x + v\partial\zeta/\partial y) = \mu\partial/\partial x(\partial\zeta/\partial x) + \mu\partial/\partial y(\partial\zeta/\partial y) + \rho g\beta\partial T/\partial x \qquad (2)$$

- stream function

$$\partial/\partial x(\partial\chi/\partial x) + \partial/\partial y(\partial\chi/\partial y) = \zeta \qquad (3)$$

where
$\partial\chi/\partial y = u$, $\partial\chi/\partial x = -v$ and $\zeta = \partial u/\partial y - \partial v/\partial x$

- chemical species

$$\rho(\partial Y_i /\partial t + u\partial Y_i /\partial x + v\partial Y_i /\partial y) = w_i + \partial/\partial x[D\rho(\partial Y_i /\partial x)] \\ + \partial/\partial y[D\rho(\partial Y_i /\partial y)] \qquad (4)$$

$\sum_i Y_i = 1 \qquad$ i = I, P, O, F
- energy

$$c_p\rho(\partial T/\partial t + u\partial T/\partial x + v\partial T/\partial y) = q + k\partial/\partial x(\partial T/\partial x) + k\partial/\partial y(\partial T/\partial y) \qquad (5)$$

where
$w_i = - A_g exp(-E_g/RT) Y_O Y_F \nu_i M_i /M_F\rho \ \rho \qquad$ i = I,O,P,F
$q = w_F \, \Delta H$

In the eqns.(2-5), u and v are the longitudinal and normal velocity components, μ the viscosity, Y_i the species mass fraction, ρ the density, D the diffusion coefficient, c_p the specific heat at constant pressure, T the temperature, k the thermal conductivity, A_g and E_g the pre-exponential factor and the activation energy of the combustion reaction, ΔH the heat of combustion and M_i the molecular weight. Prescribed values of temperature and species mass fractions with a uniform velocity profile are imposed at the inlet. Zero gradients conditions at the exit are assumed. At the upper wall zero gradients conditions are assigned for temperature and species mass

fractions and no-slip conditions for velocity. At the interface
between solid and gas phase the following conditions are
applied:

$$\rho D \partial Y_F / \partial y = m(Y_F - 1) \tag{6}$$
$$\rho D \partial Y_i / \partial y = m Y_i \qquad i = O, I, P \tag{7}$$
$$- k \partial T / \partial y = - k_s \partial T_s / \partial y + mL \tag{8}$$
$$m = \rho v \tag{9}$$
$$u = 0, \ T = T_s \tag{10}$$

where the pyrolysis mass flux is given by:
$m = A_s \exp(-E_s / RT) \rho_s$.

As for the initial conditions, after steady spread
conditions are reached, the inlet values for velocity or oxygen
mass fraction are changed and the subsequent transient
processes are described.

Solution of eqns. (1-5) and related initial and boundary
conditions is obtained by means of finite difference
discretization that employs the Euler scheme for the time
derivative, the second upwind formula for convective terms and
the central scheme for diffusive terms on the staggered grid
shown in Fig. 2.

For each time step the solution is computed as follows:
a) the solid phase energy equation is solved by means of an ADI
technique with a known heat flux from the gas phase. The step
gives, besides the solid temperature, the distribution of the
temperature at the surface and the pyrolysis mass flux
necessary to solve the gas phase equations.
b) solution of Navier-Stokes equations that, given the strong
coupling between vorticity and stream function, is computed by
an iterative procedure. The stream function is computed with
the last available values of vorticity; then new values of
velocities and new boundary conditions on vorticity are

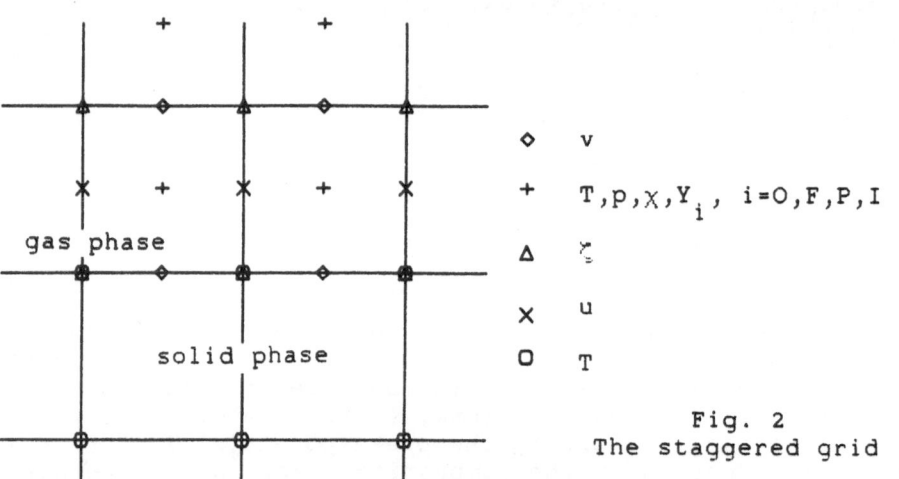

Fig. 2
The staggered grid

computed and a new vorticity field is obtained. Computations
of χ and ζ are iterated until convergence is reached.
c) computation of gas temperature and chemical species

distribution that gives, besides all gas phase variables, the new heat flux from the flame to the fuel.

The three steps a), b) and c) must be repeated for each time step to account for the coupling between the two phases. In particular, step b) must be included too in the iteration procedure since the normal component of velocity at the interface (and the stream function) varies according to the pyrolysis mass flux. Solution of all gas phase equations is computed by a line-by-line method.

d) finally, once the velocity and temperature fields are computed, pressure can be obtained from a Poisson equation derived by the divergence of momentum equations and by the continuity equation:

$$\partial/\partial x(\partial p/\partial x)+\partial/\partial y(\partial p/\partial y)=2[-(\partial u/\partial y)(\partial v/\partial x)+(\partial u/\partial x)(\partial v/\partial y)-g\beta\partial T/\partial y]\rho \quad (11)$$

Neumann boundary conditions for this equation are derived again from momentum equations. Solution of Poisson equation (11), with Neumann boundary conditions, exists only if the compatibility condition, derived from the Green's theorem, which relates the source and the value of derivatives on the boundary, is verified. Because of truncation error this condition is not verified. Thus, to avoid divergence in the solution, this inconsistency is corrected by modifying the source term of eqn. (11) of as suggested in [9].

RESULTS AND DISCUSSIONS

In this section the results of numerical simulations of opposed flow flame spread over thick solid fuels are discussed. The values for the solid properties are chosen in the range of values typical for cellulosic materials and are the same used in [3,4]. The constant gas phase properties are referred to air at ambient temperature. Integration of finite difference equations has been conducted with a grid of 80 nodes along the x direction and 40 along the y direction both for the gas and the solid phase and a time step of .5E-5 s. Further reductions in the time and space steps do not give rise to any change in the computed solution.

Radiative ignition (.025 cm of the solid fuel are heated for .06 s) is simulated to start flame spread. A description of the spread process in air (oxygen mass fraction of .23) with an opposed flow velocity of 10 cm/s, can be inferred from Figs. 3a-3c where temperature, fuel and oxygen mass fraction and reaction rate isolines and the vector velocity field are plotted for t = .07s. Opposed flow flame spread is dominated by the interaction of fluid-dynamics, chemical reactions and heat transfer at the flame leading edge where a combustible mixture is established. This position is marked by a vertical line in Fig. 3c (and in all isotherm plots) and corresponds to the maximum heat flux from the flame to the solid fuel. Given that the simulated conditions are far from extinction, the edge of the heated layer in the gas is very close to that in the solid. Downstream this position, flame structure is that

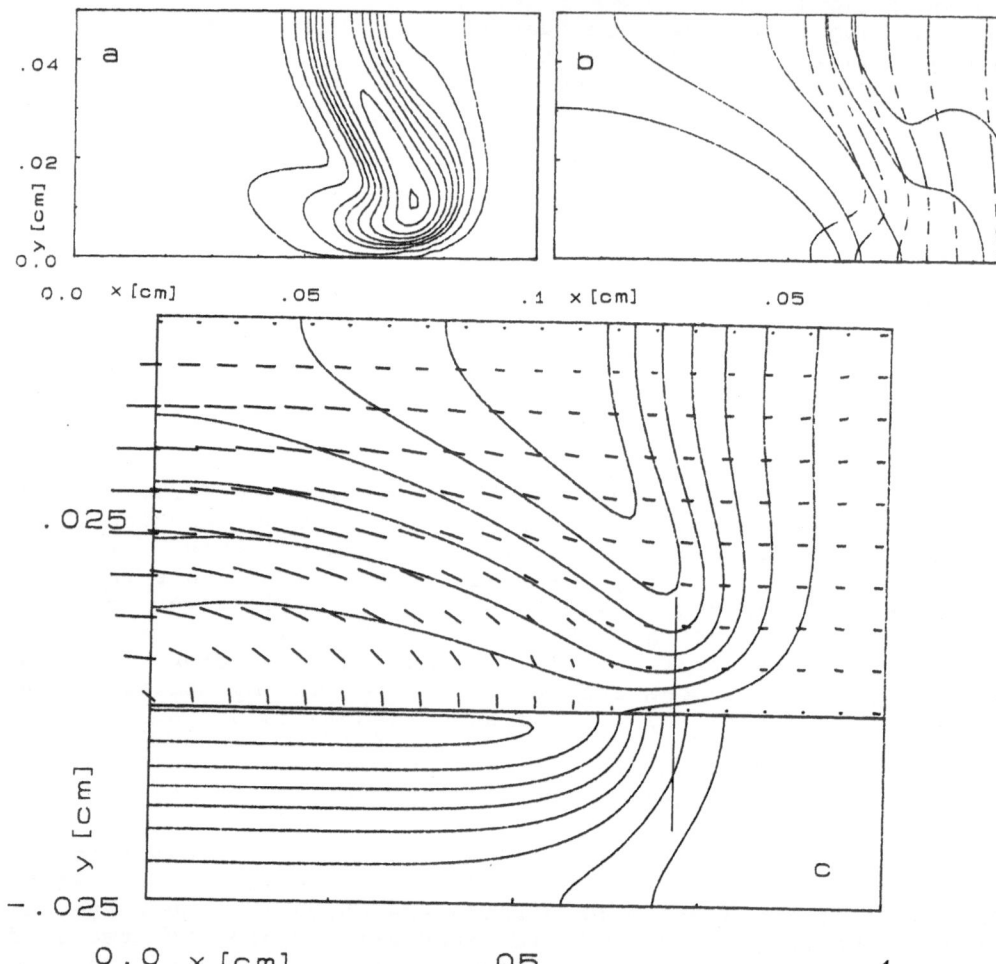

Fig. 3 Flame structure for t=.07 s: a) reaction rate
isolines [g/cm^{3}s]: 3E-4, 3E-3, .0125, .025, .05, .1,
.2, .4, .65; b) (solid lines) isolines of fuel mass
fraction: .5, .3, .1, .01, (dashed lines) isolines of
oxygen mass fraction: .001, .01, .05, .15, .1, .2;
c) solid phase isotherms from T_s=350 K, step 50 K, gas
phase isotherms from 600 K, step 300 K and vector
velocity field (velocity at the inlet 10cm/s).

typical of a diffusion flame in a counterflow environment:
isolines of reaction rate (Fig. 3a) correspond to the flame
position characterized by maximum temperature and minima fuel
and oxygen mass fractions. From the vector velocity field, two
main flows appear, one of vapor fuel from the pyrolyzing
surface and the other of air at ambient conditions from the
inlet. It is interesting to observe that the normal component
of velocity at the interface, proportional to the pyrolyzed
mass flux, reaches the highest values behind the flame leading
edge. Consequently the formation of the upstream combustible

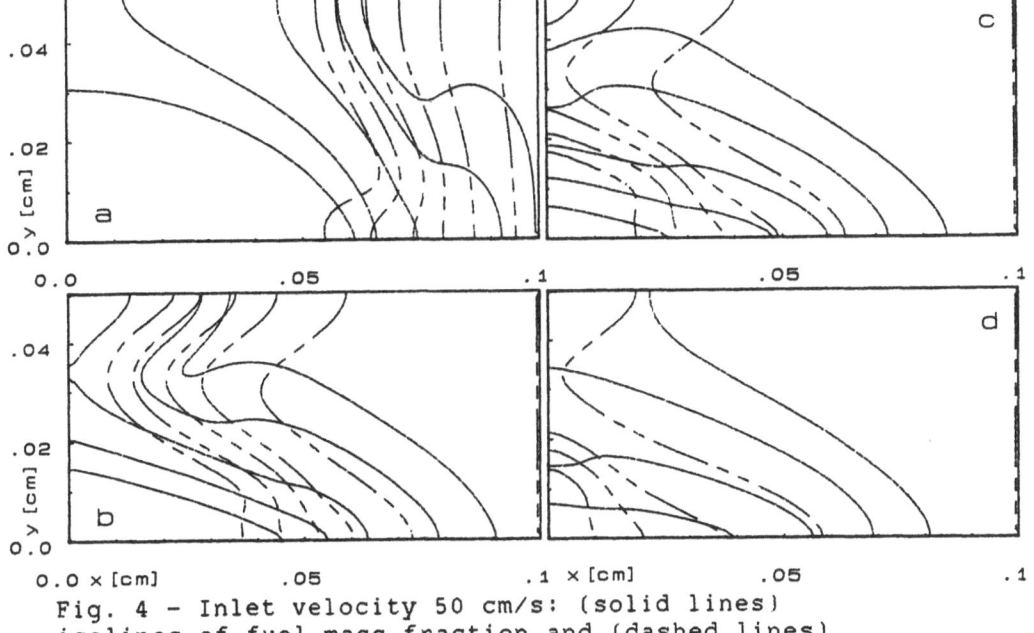

Fig. 4 — Inlet velocity 50 cm/s: (solid lines)
isolines of fuel mass fraction and (dashed lines)
isolines of oxygen mass fraction for t=.07 s, .074 s,
.077 s, .080 s (same values of Fig. 3b)

mixture comes from fuel diffusion, from the flame zone, through
a quenching layer between the solid surface and the flame. The
opposed velocity goes to zero on the surface, allowing this
fuel transport upstream. Also, the process is favoured by the
very steep gradients of fuel mass fraction in this region,
while it does not appear to depend on the surface temperature
gradient. This mechanism was proposed, by means of some
experimental measurements of velocity and surface temperature,
by Glassman and coworkers [7] and is confirmed by the numerical
simulation here presented. The predicted flow field agrees
qualitatively with experiments while the pressure, that
essentially accounts only for hydrodynamics effects, decreases
from the inlet of the channel to the exit.

To analyze the mechanisms leading to extinction and to give a
detailed representation of this phenomenon, simulations have
been made, starting from the flow and flame configuration of
Fig. 3, by increasing the opposed flow velocity or by
decreasing the oxygen concentration at the inlet of the
channel.

Time evolution from spread to extinction, caused by high
opposed flow velocity, is shown through Figs. 4-6 where,
starting from t = .07 s, the opposed velocity at the inlet, is
suddenly changed from 10 cm/s to 50 cm/s. As the opposed
velocity is increased, the flame, that must counteract a higher
velocity, leans towards the surface. Although the maximum in
the gas temperature is not changed, the gradients become
higher. The increase in the heat flux from the flame to the

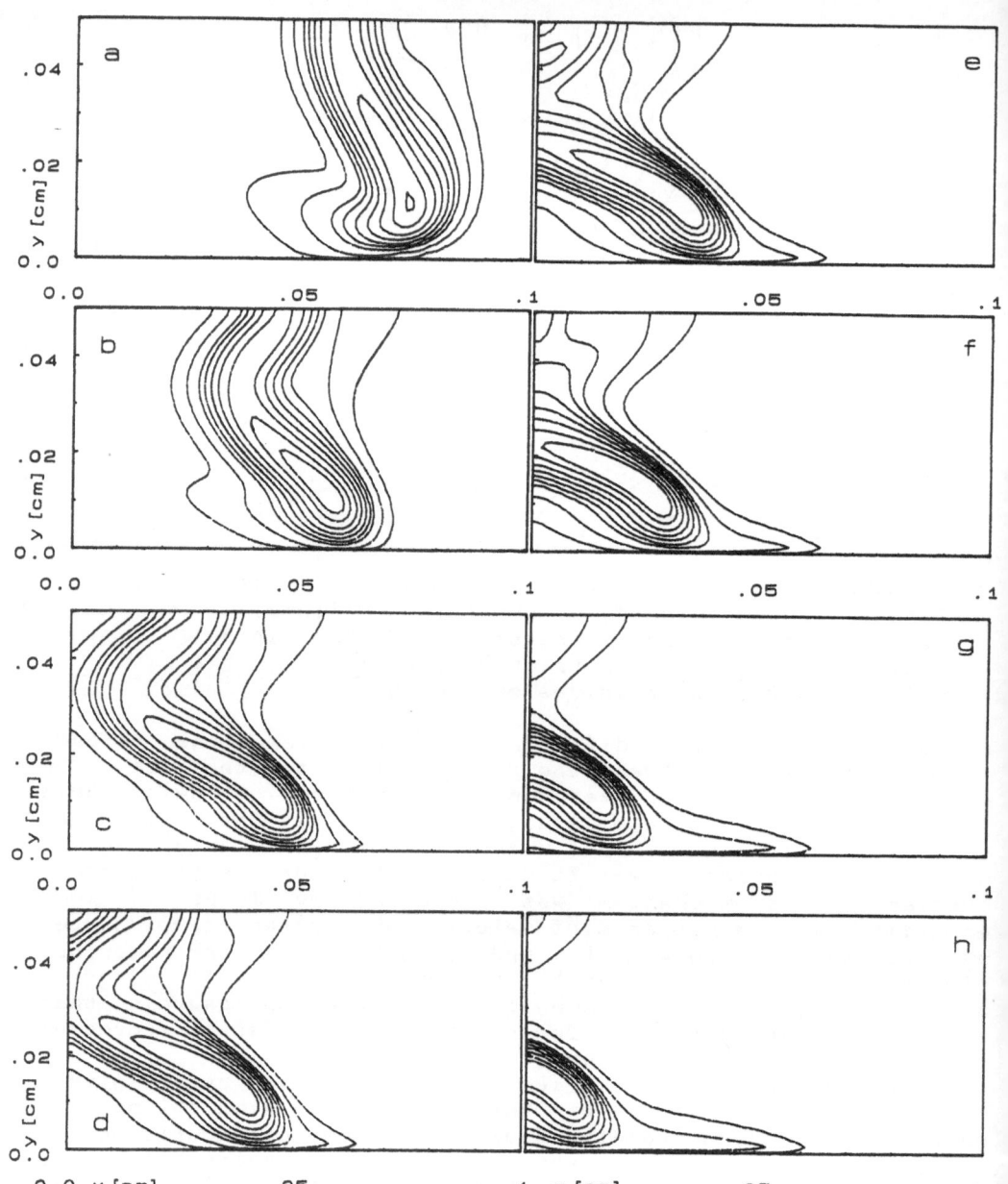

Fig. 5 - Inlet velocity 50 cm/s: reaction rate
isolines for t=.07 s, .072 s, .075 s, .076 s,
.077 s, .079 s, .080 s, .081 s (same values
of Fig. 3a)

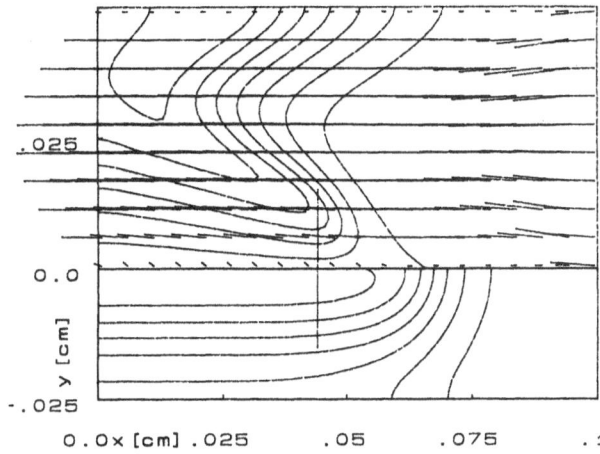

Fig. 6 — Inlet velocity 50 cm/s: temperature and velocity field for t=.074s (same values of Fig. 3b)

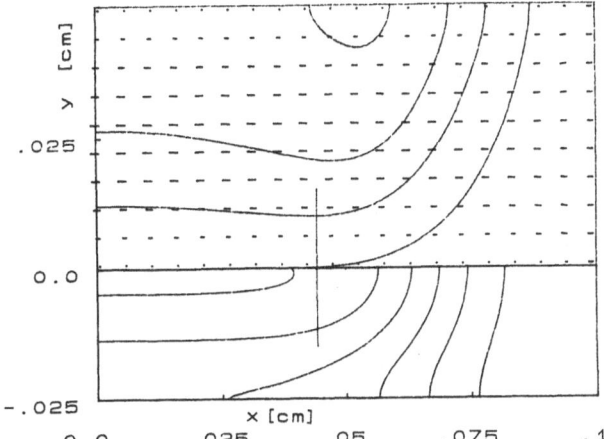

Fig. 7 — Oxygen mass fraction .1: temperature and velocity field for t=.1 s (same values of Fig. 3b)

solid fuel makes the surface temperature increase. This causes an increase of the pyrolysis reaction and of the amount of vaporized fuel. The higher production of vapor fuel gives a slight increase in the maximum combustion reaction rate too. All these processes would enhance the spread process but it is well shown that the flame leading edge recedes behind the edge of the solid heated layer, the oxidizer penetrates the fuel rich side and the size of the flame is reduced until it is blown off.

Time evolution from spread to extinction caused by low oxygen concentrations is shown through Figs 7-9, where, starting from t = .07s s, the oxygen mass fraction at the inlet is suddenly changed from .23 to .1, the value corresponding to the limit oxygen index determined, for the same material, when the flame spreads in the same direction as the oxidizing flow [10]. As the oxygen concentration is decreased, the flame becomes more enlarged, the quenching distance from the solid surface is increased and both the reaction rate and gas phase temperature decrease by a noticeable amount. The decrease in the gas temperature causes a decrease in the heat flux and surface temperature too. Consequently the fuel production is reduced

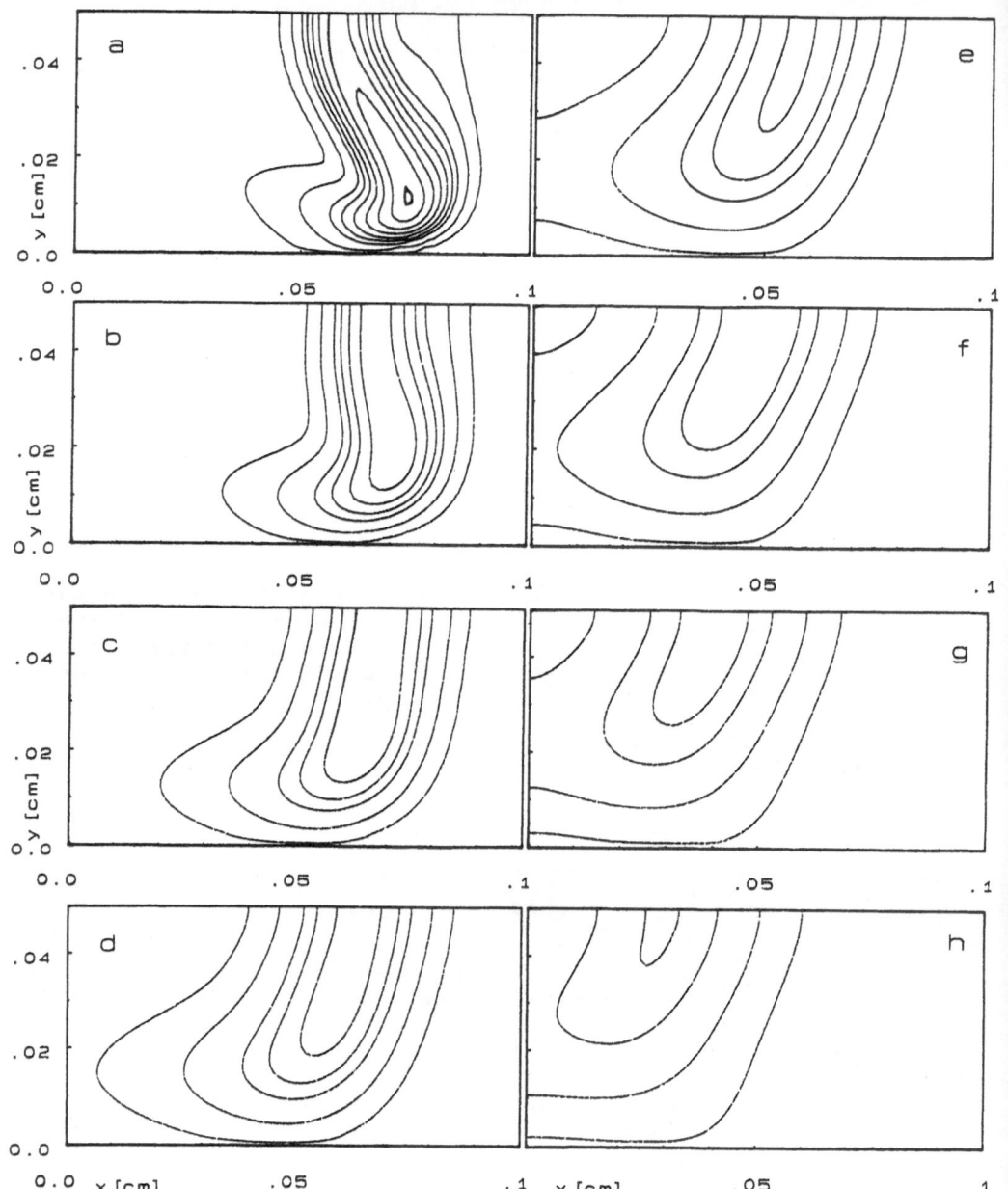

Fig. 8 – Oxygen mass fraction .1: reaction rate
isolines for t=.07 s, .08 s, .09 s, .1 s, .11 s,
.12 s, .13 s, .14 s (same values of Fig. 3a)

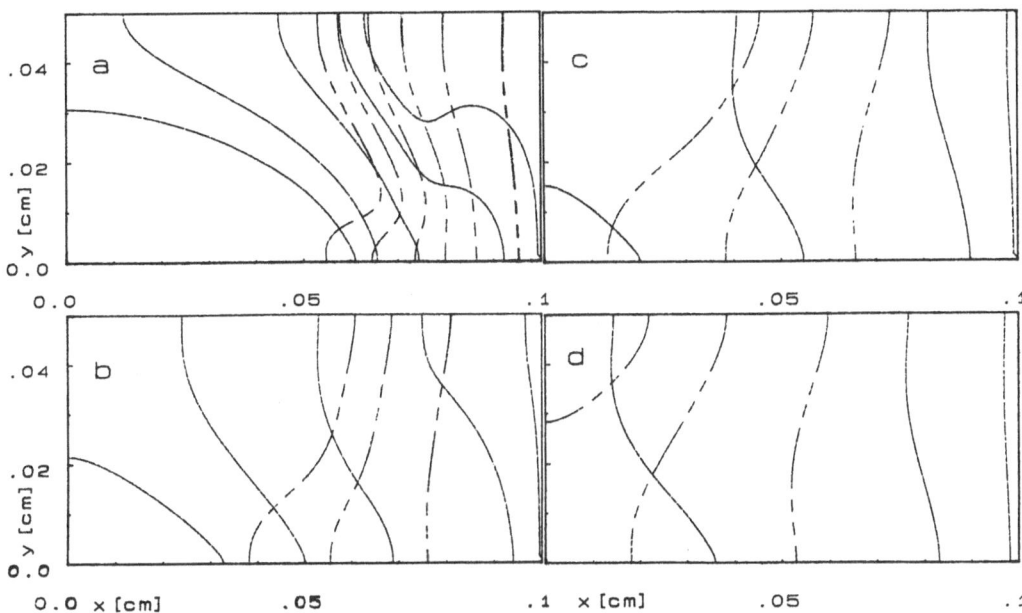

Fig. 9 - Oxygen mass fraction .1 : (solid lines)
isolines of fuel mass fraction and (dashed lines)
isolines of oxygen mass fraction for t=.07 s, .09 s,
.11 s, .13 s (same values of Fig. 3b)

and the flame recedes towards a higher fuel concentration zone, but this is not sufficient to allow spreading and the retreat process goes on until extinction occurs.

Although extinction is approached according to very different time evolution for the two cases here reported, it can be observed that it is related to a continuous decrease of the ratio of the flow time (α/u^2) to the chemical time (ρ/w), that is, the Damkohler number. From both extinction cases simulated, indeed, the flame leading edge recedes with respect to the heated solid fuel, thus the solid heat conduction is not controlling in these conditions. The interactions between fluid-dynamics and chemical processes in the gas phase, which are much faster than solid phase processes, are responsible for extinction. In the former case, when the flow velocity is increased, the slight increase in the chemical time is much lower than that in the flow time, so that the Damkohler number at the flame leading edge continuously decreases, leading to extinction. In the latter case, the flow time is unchanged, but the strong increase in the chemical time due to the decrease of the reaction rate again gives decreasing values of the Damkohler number and extinction occurs.

CONCLUDING REMARKS

A computer model of opposed flow flame spread over solid fuels is presented. Interaction between fluid mechanics and chemistry, both in the gas and solid phase, is properly described by balance equations of momentum, mass and energy for the gas phase and energy for the solid phase. An iterative finite difference method has been used to get detailed simulations of flame structure and extinction. The convective, thermal and chemical structure of the flame agrees qualitatively with experiments while the pressure decreases from the inlet of the channel to the exit. A further improvement of the model will include variable gas phase properties and, in particular, density variations related to pressure and temperature variations through state equation.

REFERENCES

1) Frey A. E. jr. and T'ien J., Combustion and Flame 36, 263 (1979)

2) Borgeson, R., A., T'ien J., S., Combustion Science and Technology, 32, 125 (1983)

3) C., Di Blasi, G., Continillo, Proc. of the Chemical Engineering Fundamental, XVIII Congress (1987), p.361

4) C. Di Blasi, C. Continillo, S. Crescitelli, G. Russo, Combustion Science and Technology 54 (1987),25

5) C. Di Blasi, S. Crescitelli, G. Russo, A. C. Fernandez-Pello, Proc. of the Second International Symposium on Fire Safety Science (1988), in press

6) Fernandez-Pello A. C., Ray S. R. and Glassman I., Eighteenth Symposium (Int.) on Combustion, The Combustion Institute,

7) Ray, S., R. and Glassman, I., Combustion Science and Technology, 32, 33 (1983)

8) C., Di Blasi, S. Crescitelli, G. Russo, A. C. Fernandez-Pello , paper presented at the Fall Technical Meeting of the Eastern State Section of the Combustion Institute, Florida December 5-7 (1988)

9) Briley, W., R., J. Comp. Phys. 14, 8 (1974)

10) C. Di Blasi, S. Crescitelli, G. Russo, Combustion and Flame 72 (1988), 205

Induction Period Generation
of a Supersonic Flame[1]

J.W. Dold

Laboratoire de Recherche en Combustion[2], Université de Provence,
Centre St. Jérôme, Boîte 252, 13397 Marseille Cedex 13, France

Introduction

The way in which detonation waves are initiated has recently been studied theoretically by a number of people. Clarke [1–4] first examined the way in which chemical and gas-dynamic disturbances may interact in an exploding atmosphere. Using the same basic approach, Clarke and Cant [5] and Jackson and Kapila [6] independently examined the problem of reaction initiation by a piston-driven shock wave. The same problem has been addressed using a purely asymptotic method by Crighton and Blythe [7]. In a series of articles [8], Clarke, Kassoy and Riley considered an evolution behind a shock-wave in a reacting atmosphere where the shock-wave is maintained by thermal expansion accompanying an initially existing subsonic (deflagration) flame. Dold [9] looked at the evolution of a single temperature disturbance in an unbounded atmosphere. More recently, Klein and Peters [10] considered the development of weak shock-waves undergoing multiple reflections in a closed vessel containing combustible reactants.

A common feature of all of these studies is that the chemical reaction must, at some stage, be sufficiently widespread for an inertial confinement to take place. This amounts to requiring that the induction time of the chemical reaction should be of a similar magnitude to the time taken for acoustic disturbances to cross the region of strong chemical evolution—a situation that arises if the medium is suitably *preconditioned* by the phenomena examined in each of the studies. Under such circumstances, the chemical reaction feeds energy into both temperature and pressure increases. It is generally found that a localised ignition event takes place in a finite time, involving large increases in both pressure and temperature. Some studies [7,11] have endeavoured to reveal the spatial structure of the local ignition.

More recently, after noticing that the ignition represents not only a local completion of the chemical reaction, but also the first point on a supersonic path of reaction-runaway, Kapila and Dold [12,13] have succeeded in asymptotically analysing the structure of the fast reaction wave in the cases of condensed and gaseous explosives. The first appearance of a shock-wave is described following the moment when the wave slows

---

[1] This work forms part of a larger theoretical study into detonation initiation assisted by a NATO collaborative research grant, held by A.K. Kapila and the author. It was also sponsored by the U.K. Science and Engineering Research Council.

[2] Most of the work was carried out while on leave from: School of Mathematics, University of Bristol, University Walk, Bristol BS8 1TW, U.K. The support of the Université de Provence, and especially of Prof. P. Clavin and all of the staff of the Laboratoire de Recherche en Combustion is warmly acknowledged.

down sufficiently for a sonic point to be produced by the flame. This shock strengthens as it moves forward into less reacted regions, transforming the reaction wave from a weak to a strong detonation [14].

As a complement to this work, it is necessary to be able to calculate the path of the reaction wave if the analysis is to be applied to any particular example. The purpose of this article is to present an approach which appears to be well-suited to the numerical calculation of the path of a supersonic flame as it emerges, after an induction period, from an initial and boundary setup. While the method is presented in terms of solving a set of perturbation equations, it is worth bearing in mind that the same ideas can be applied to the full set of chemically reactive Euler equations, possibly including multiple chemical reactions and changes in the overall number of molecules.

Model

The following set of equations can be used to describe the evolution of a single iso-molar exothermic chemical reaction $F \to P$ in a polytropic medium with adiabatic coefficient γ:

$$\rho T_t = -Q\rho y_t + \frac{\gamma - 1}{\gamma} P_t$$

$$\gamma^{-1} P_t + P\rho u_\psi = -Q\rho y_t$$

$$u_t + \gamma^{-1} P_\psi = 0 \tag{1}$$

$$P = \rho T$$

$$\rho y_t = -\frac{\epsilon}{Q}(\rho y)e^{(1 - 1/T)/\epsilon}$$

where the one dimensional Lagrangean coordinate ψ is defined as

$$\psi = \int_{x_0(t)}^{x} \rho \, dx \qquad \text{for some} \qquad x_0'(t) = u(x_0, t). \tag{2}$$

All variables and constants appearing here have been made dimensionless with respect to a selected dimensional temperature T_c^*, pressure P_c^*, homogeneous induction time $t_c^* = (\epsilon/QA^*)\exp(E^*/R^*T_c^*)$, density $\rho_c^* = W_c^* P_c^*/R^* T_c^*$, sound speed $a_c^* = (\gamma P_c^*/\rho_c^*)^{1/2}$, length $x_c^* = a_c^* t_c^*$ and initial reactant mass-fraction Y_c (used to scale y). We also define the dimensionless heat of reaction Q as $Q^* Y_c/C_P^* T_c^*$, and the dimensionless inverse activation energy ϵ as $R^* T_c^*/E^*$. The constant C_P^* represents the specific heat at constant pressure; R^* is the universal gas constant; W^* is the effective molecular weight; A^* is the Arrhenius pre-exponential factor; Q^* is the heat of reaction; and E^* represents the activation energy of the chemical reaction.

In cases where the activation energy is large, or $\epsilon \ll 1$, a perturbation analysis of these equations about a suitable initial state takes the form

$$\begin{array}{cc} T \sim 1 + \epsilon\phi & P \sim 1 + \epsilon\gamma p \\ u \sim \epsilon v & y \sim 1 - \epsilon w/Q \end{array} \tag{3}$$

leading to the set of equations

$$\phi_t = e^\phi + (\gamma - 1)p_t$$
$$p_t + v_\psi = e^\phi$$
$$v_t + p_\psi = 0 \tag{4}$$
$$w_t = e^\phi.$$

These equations describe the induction period of a thermal explosion which is accompanied by pressure and velocity interactions—the type of process with which this paper is chiefly concerned.

The latter equation simply keeps track of the amount of reactant consumed. Eliminating p and v between the first three equations leads to Clarke's equation [1,4,5,15,16]

$$(\phi_t - \gamma e^\phi)_{tt} = (\phi_t - e^\phi)_{\psi\psi} \tag{5}$$

which has a number of interesting features, each reflecting a possible physical process. Both the adiabatic sound-speed of unity, and the isothermal sound-speed of $\gamma^{-1/2}$ can be observed in relating different terms; if either left or right sides dominate, then the equation models either a constant density or constant pressure reaction runaway respectively—in fact a transition from one to the other is possible [9].

For simplicity, while the method of solution described in this paper can be applied to more general initial conditions, only results arising from the following initial data, at $t = 0$, will be selected for consideration.

$$p \equiv 0, \quad v \equiv 0, \quad w \equiv 0$$
$$\text{and} \quad \phi \equiv \phi_0(\psi) = \frac{\nu}{2}[\cos(\pi\psi/\mu) - 1] \tag{6}$$

In these, ν is the magnitude and μ is the half-wavelength (in terms of ψ) of an initial sinusoidal temperature disturbance. Because the evolution must be symmetric about the locations $\psi = 0$ and $\psi = \mu$, one can equivalently consider values of ψ only in the range $\psi \in [0, \mu]$, along with the conditions of symmetry

$$p_\psi(0,t) = p_\psi(\mu,t) = 0 = v(0,t) = v(\mu,t). \tag{7}$$

It can thus be seen that the conditions (6) describe an initially quiescent setup in a closed one-dimensional container, with simple order one variations in initial reaction rate caused by the temperature nonuniformity.

Flame Path

Under these or any more general, but nevertheless bounded, initial and boundary conditions for ϕ, p, v and w, the value of ϕ is not found to remain bounded. The nonlinear reaction rate e^ϕ causes values of ϕ to grow at a strongly increasing rate. This, ultimately, makes ϕ increase towards infinity in a finite time.

In fact, a path of singularity $\tilde{t}(\psi)$ must exist at which ϕ becomes infinite—that is $\phi(\psi, \tilde{t}) = \infty$ with $\phi(\psi, t) < \infty$ for $t < \tilde{t}(\psi)$. Furthermore, under bounded initial and boundary conditions, the function $\tilde{t}$ must have the property that $|\tilde{t}'(\psi)| \leq 1$.

This condition arises from the fact that finite values of p, v and ϕ at any point (ψ_0, t_0) must depend only on finite values within the range of the adiabatic sound-speed characteristics $t_0 - t = |\psi - \psi_0| > 0$ which converge on the point. In general, the singularity path will be supersonic; that is its dimensionless mass-flux $m(\psi) = 1/\tilde{t}'(\psi)$ will have a magnitude greater than one, $|m| > 1$ or $|\tilde{t}'| < 1$.

Of course, the perturbation expansions (3) break down once ϕ and the other variables become sufficiently large (of order ϵ^{-1}). However, the singularity path $\tilde{t}$ still remains relevant in denoting the first-order location of a flame path which can only be analysed further by returning to the set of full equations (1). Assuming such a flame path to exist, recent analyses [12,13] have succeeded in deducing the nature of the flame, including the possible initiation of shock waves and a subsequent evolution into a strong detonation.

The perturbation equations (4) describe a system with linear acoustic interactions between pressure and velocity, along with a strongly nonlinear forcing effect due to the reaction-rate term e^ϕ. Because of this nonlinearity, it is not a trivial task to calculate the singularity path $\tilde{t}(\psi)$. On the other hand, if the path of the singularity is known, then the asymptotic behaviour of p, v, w and ϕ close to $t = \tilde{t}(\psi)$ is more easily found.

Defining a new coordinate $\tau = \tilde{t} - t$ which measures the time remaining before the singularity arrives at any given value of ψ, the equations (4) can be written in terms of τ and ψ as independent variables.

$$
\begin{aligned}
\phi_\tau + e^\phi &= (\gamma - 1)p_\tau \\
v_\psi + v_\tau \tilde{t}' - p_\tau &= e^\phi \\
p_\psi + p_\tau \tilde{t}' - v_\tau &= 0 \\
w_\tau &= -e^\phi
\end{aligned}
\tag{8}
$$

Close to the singularity path, e^ϕ is large so that derivatives with respect to τ must be large. On the other hand, derivatives with respect to ψ (at fixed values of τ) are taken along paths which lie almost parallel to the singularity path. Thus these derivatives should be relatively much smaller, and can be neglected in obtaining the first-order asymptotic relations

$$
\frac{1 - \tilde{t}'^2}{\gamma - \tilde{t}'^2} \phi_\tau \sim \left(1 - \tilde{t}'^2\right) p_\tau \sim \frac{1 - \tilde{t}'^2}{\tilde{t}'} v_\tau \sim w_\tau = -e^\phi
\tag{9}
$$

from which it can be seen that

$$
\phi + \ln\left(\frac{\gamma - \tilde{t}'^2}{1 - \tilde{t}'^2}\right) \sim -\ln \tau \sim \left(\gamma - \tilde{t}'^2\right) p
$$

$$
\frac{\gamma - \tilde{t}'^2}{1 - \tilde{t}'^2} w \sim -\ln \tau \sim \frac{\gamma - \tilde{t}'^2}{\tilde{t}'} v.
\tag{10}
$$

The expansions can be extended to the order of $\tau \ln \tau$ as follows:

$$\phi \sim -\ln \tau - \ln \left(\frac{\gamma - \tilde{t}'^2}{1 - \tilde{t}'^2} \right) + F\tau \ln \tau$$

$$\phi \sim (\gamma - \tilde{t}'^2)p + a + A\tau \ln \tau \tag{11}$$

$$w \sim (1 - \tilde{t}'^2)p + b + B\tau \ln \tau$$

$$v \sim \tilde{t}'p + c + C\tau \ln \tau$$

in which $F(\psi)$, $A(\psi)$, $B(\psi)$ and $C(\psi)$ are functions of ψ which can be determined by substituting into the equations (8) and equating terms of the order of $\ln \tau$. This gives

$$A = B = \frac{\gamma + 3\tilde{t}'^2}{(\gamma - \tilde{t}'^2)^2} \tilde{t}''$$

$$F = -\frac{\gamma - 1}{2} \frac{A}{1 - \tilde{t}'^2} \tag{12}$$

$$C = -\frac{2\tilde{t}' \tilde{t}''}{(\gamma - \tilde{t}'^2)^2}.$$

The functions $a(\psi)$, $b(\psi)$ and $c(\psi)$, as well as the function $\tilde{t}(\psi)$, will depend on the set of boundary and initial conditions, and need to be determined numerically. Since the equations (8) can be integrated to show that $a = b + \phi_0$, it is sufficient numerically to determine the functions b and c. Supposing these functions are known, the results (11) are then accurate to $O(\tau \ln \tau)$ as $\tau \to 0$.

Numerical Method

a) Characteristic Form

A transformation of equations (4) into characteristic form, using the variables $\xi = \frac{1}{2}(t + \psi)$ and $\eta = \frac{1}{2}(t - \psi)$, allows one to express the problem as follows

$$p + v = \int_{\eta \text{ fixed}} e^{\phi} \, d\xi + g_1(\eta)$$

$$p - v = \int_{\xi \text{ fixed}} e^{\phi} \, d\eta + g_2(\xi) \tag{13}$$

$$w = \int_{\psi \text{ fixed}} e^{\phi} \, dt + g_3(\psi)$$

$$\phi = w + (\gamma - 1)p + g_4(\psi)$$

where g_1, g_2, g_3 and g_4 are functions of integration. Under conditions (6), for example, these equations lead to the more explicit integral formulation

$$p + v = \int_{2\psi}^{\xi} e^{\phi}\, d\xi, \qquad\qquad p - v = \int_{-2\psi}^{\eta} e^{\phi}\, d\eta,$$

$$w = \int_0^t e^{\phi}\, dt \qquad \text{and} \qquad \phi = w + (\gamma - 1)p + \phi_0(\psi). \tag{14}$$

In principle, a numerical solution thus only requires quadrature of the integrals appearing here combined with some incrementing along characteristics. However, for the purposes of accurately obtaining information about the singularity path $\tilde{t}(\psi)$ this formulation is not especially well suited. The singularity path intersects some of the characteristics, and particular care is needed to ensure that the path is accurately captured and that each of the functions $a(\psi)$, $b(\psi)$ and $c(\psi)$ is adequately determined.

b) Parametric Form

It would be more suitable if the integration could be carried out with respect to a variable that grows towards infinity as $t \to \tilde{t}(\psi)$. In this way, by simply taking the integration far enough, the singularity path and its associated properties could be approached arbitrarily closely. Supposing that ς is defined to be such a variable, then the equations (4) can be transformed from the independent coordinates (ψ, t) to the coordinates (ψ, ς), giving

$$p_\varsigma = \frac{t_\varsigma}{1 - t_\psi^2}\left(e^{\phi} - t_\psi p_\psi - v_\psi\right)$$

$$v_\varsigma = \frac{t_\varsigma}{1 - t_\psi^2}\left(t_\psi e^{\phi} - t_\psi v_\psi - p_\psi\right) \tag{15}$$

$$w_\varsigma = t_\varsigma e^{\phi}$$

$$\phi_\varsigma = w_\varsigma + (\gamma - 1)p_\varsigma$$

in which partial derivatives with respect to ψ are taken with ς held fixed. The initial conditions (6) and (7) now adopt the form:

$$\left.\begin{array}{ccc} t \equiv 0, & p \equiv 0, & v \equiv 0, \\ w \equiv 0 & \text{and} & \phi \equiv \phi_0(\psi) \end{array}\right\} \quad \text{at} \quad \varsigma = 0$$

$$\left.\begin{array}{ccccc} t_\psi(0, \varsigma) & = & t_\psi(\mu, \varsigma) & = & 0 \\ \text{with} \quad p_\psi(0, \varsigma) & = & p_\psi(\mu, \varsigma) & = & 0 \\ v(0, \varsigma) & = & v(\mu, \varsigma) & = & 0 \end{array}\right\} \quad \text{for} \quad \varsigma \geq 0. \tag{16}$$

The set of equations (15) needs to be closed by making a suitable choice for the variable ς. There are clearly many possibilities for this.

An extra condition that is chosen in order to achieve closure can be made to fit in with almost any desired manner in which the evolution of the reaction runaway is to be studied. Indeed, there is nothing to prevent one from changing the way in which ς varies at any stage, provided only that $t(\psi, \varsigma)$ remains continuous and, generally, increasing with ς. Equivalently, one can select the function $t_\varsigma(\psi, \varsigma)$ to be any (generally non-negative) function, the definition of which may be changed (if necessary discontinuously) at any stage. For the present purposes, it is useful to note some possible choices, and some properties that may be worth building into a choice for t_ς.

It is clear that the choice $t_\varsigma = 1$ defines changes in ς to be the same as changes in time. If this definition is maintained from the beginning, then conditions (16) show that ς would exactly correspond to t. Similarly, the choices $p_\varsigma = 1$, or $w_\varsigma = 1$ make changes in ς the same as changes in pressure or species concentration, respectively. Maintaining these definitions from the start would, effectively, calculate progressive isobars or iso-concentration paths as ς increases.

At early times, a time-like behaviour for ς ensures that all of the chemical and gasdynamic properties are robustly modelled. Of course, this is no longer the case when the first occurrence of an infinite value for ϕ is approached. One definition that is initially time-like but which approaches the singularity increasingly slowly is obtained by setting

$$ w_\varsigma = e^{\phi_0} \quad \Longleftrightarrow \quad t_\varsigma = e^{\phi_0 - \phi}. \tag{17} $$

Such a definition is only sensitive to pressure and velocity fluctuations in so far as they affect the chemical reaction rate e^ϕ —a property that may prove advantageous where such fluctuations, rather than chemical effects, are initially dominant.

On the other hand, the asymptotic results (11) make it clear that pressure is the most consistently increasing property close to the singularity path $\tilde{t}(\psi)$, especially where $\tilde{t}'$ is near its upper limit of unity. The simple pressure-following choice

$$ p_\varsigma = 1 \quad \Longleftrightarrow \quad t_\varsigma = \frac{1 - t_\psi^2}{e^\phi - t_\psi p_\psi - v_\psi} \tag{18} $$

is therefore more likely to be useful close to the singularity path. However, it is easily seen that this choice is most unsuitable if gasdynamic fluctuations are able to balance the chemical source term e^ϕ in the denominator of the expression for t_ς.

An optimal use of both of the possible expressions above may involve, for instance, starting with the definition (17) and then switching over to the definition (18) when the maximum value of t_ψ exceeds a chosen level. In some circumstances, a more continuous change from one choice to the other may be suitable using an expression of the form

$$ \kappa w_\varsigma + \varsigma p_\varsigma = \kappa e^{\phi_0} + \varsigma \quad \Longleftrightarrow \quad t_\varsigma = \left(\kappa e^{\phi_0} + \varsigma \right) \Big/ \left(\kappa e^\phi + \varsigma \frac{e^\phi - t_\psi p_\psi - v_\psi}{1 - t_\psi^2} \right) \tag{19} $$

which balances both choices—the choice (18) becoming more significant at values of ς which are of the order of the constant κ or greater. In principle, such a combination can be made to avoid any excessive dominance by gasdynamic effects, using an appropriate choice for the constant κ.

c) Algorithm

To test this approach numerically, a fairly straightforward finite difference scheme was used. The equations (15) were solved on a fixed grid covering values of ψ in the range $[0, \mu]$. Using the conditions of symmetry (16) to provide values on an extended grid where necessary, the derivatives t_ψ, p_ψ and v_ψ were calculated from the known values of $t(\psi, \varsigma)$, $p(\psi, \varsigma)$ and $v(\psi, \varsigma)$ at any stage, using a high-order central differencing formula. Knowing these, the derivatives t_ς, p_ς, w_ς, ϕ_ς and v_ς were then calculated using equations (15) and (19), usually with $\kappa = \frac{3}{2}$.

Starting with the conditions (16), the values of t, p, v, w and ϕ were estimated at progressively incremented values of ς using, firstly, an explicit step based on current values of the derivatives with respect to ς. This was followed by iterative refinements of the step using the arithmetic mean of the derivatives at the current and the new values of ς. The size of the step in ς was chosen so as to give an accuracy of about 0.01% in the calculated induction time of $t(\psi, \infty) \equiv \gamma^{-1}$, for the spatially homogeneous case in which $\nu = 0$.

With this method, a problem can arise over the use of equation (19) for stepping forward the values of t. This is most clearly seen in equation (18), which has the property of being able to produce cusps in the values of $t(\psi, \varsigma)$. Starting with the conditions (16), these cusps tend to develop for large enough values of ν or small enough values of μ at the position $\psi = \mu$. In such cases, the flame path $\tilde{t}(\mu)$ reaches this point mainly through the action of pressure-waves propagating from regions of intense reaction close to the path $\tilde{t}(\psi)$ at earlier times, rather than because of local chemical forcing. The cusp causes severe problems with the high-order method used for calculating derivatives and, effectively, destroys the results whenever it arises. In countering this tendency, a simple device proved to be useful; with j measuring the number of mesh points from the position $\psi = \mu$, the right hand side of the formula (19) was multiplied by the function, $1 - \exp(-j^2/6^2)$, as soon as the estimated value of t_ψ at the mesh point $j = 1$ exceeded $\frac{2}{5}$.

Since the system is hyperbolic, it may be noted that cusps in $\tilde{t}(\psi)$ may be quite realistic under the circumstances noted above. Thus, more appropriate means of dealing with them may be possible. However, the device described significantly hindered the progress of $t(\psi, \varsigma)$ at only a few mesh points. By using a non-uniform mesh, these points were made to cover a small range of values of ψ. For the purposes of this paper, and indeed for many practical purposes, this is adequate. In the more complete physical system described by equations (1), shock or detonation waves will almost certainly have destroyed the slow induction growth of the various perturbation quantities long before such cusp-positions are reached [12,13].

Proceeding in this way with the calculation of t, p, w, ϕ and v at progressively increasing values of ς, the path $\tilde{t}(\psi)$ is approached. For a selected case, this process is illustrated in figure 1a where values of time t are plotted at successive stages. As these curves approach $\tilde{t}$, it becomes feasible to use the numerical results to estimate the functions $b(\psi)$ and $c(\psi)$ appearing in the asymptotic relations (11). For this purpose, we may note that

$$w - \left(1 - t_\psi^2\right)p \sim b$$
$$v - t_\psi\, p \sim c. \tag{20}$$

253

The left hand sides of these relations are plotted in figures 1b and 1c where it can be seen that these expressions also converge to discernibly well-defined limits as ς increases.

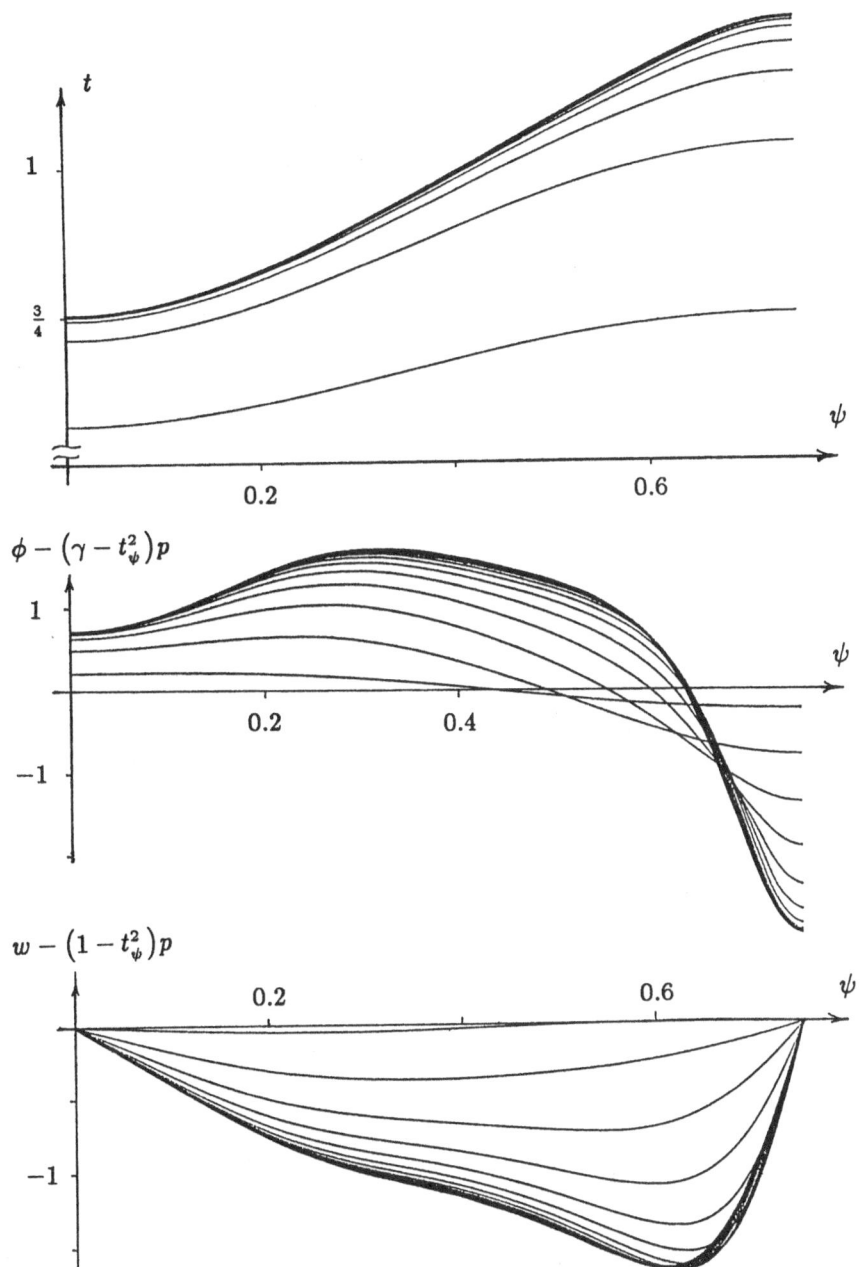

Figure 1. Results for the case $\nu = \mu = \frac{3}{4}$, for values of ς successively incremented by unity, showing **a)** time t, and the relations **b)** $\phi - (\gamma - t_\psi^2)p$, and **c)** $w - (1 - t_\psi^2)p$.

254

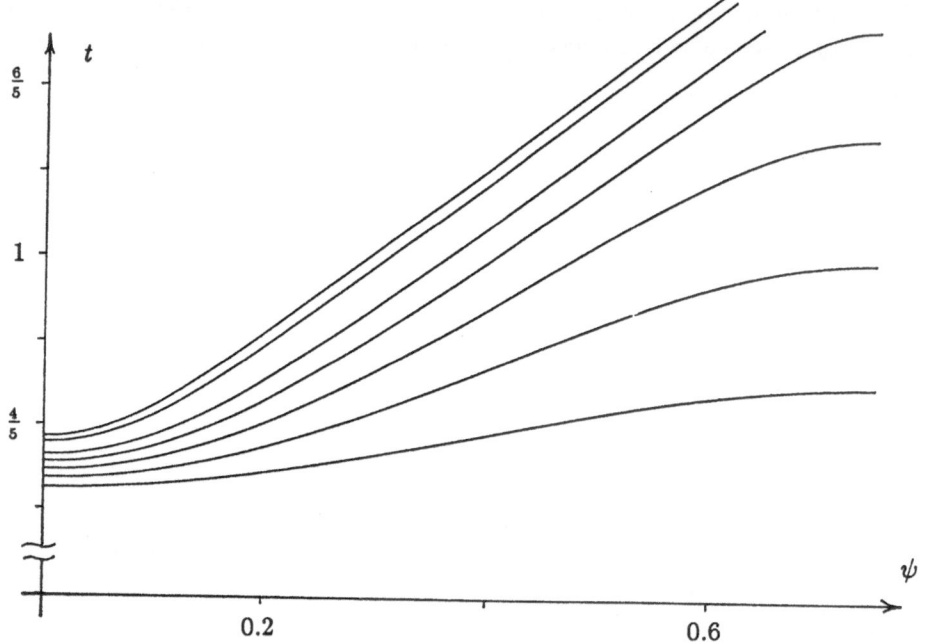

Figure 2. Calculated curves of $\tilde{t}(\psi)$ for $\nu = 0.2, 0.4, \ldots 1.6$, with μ held fixed at $\frac{3}{4}$.

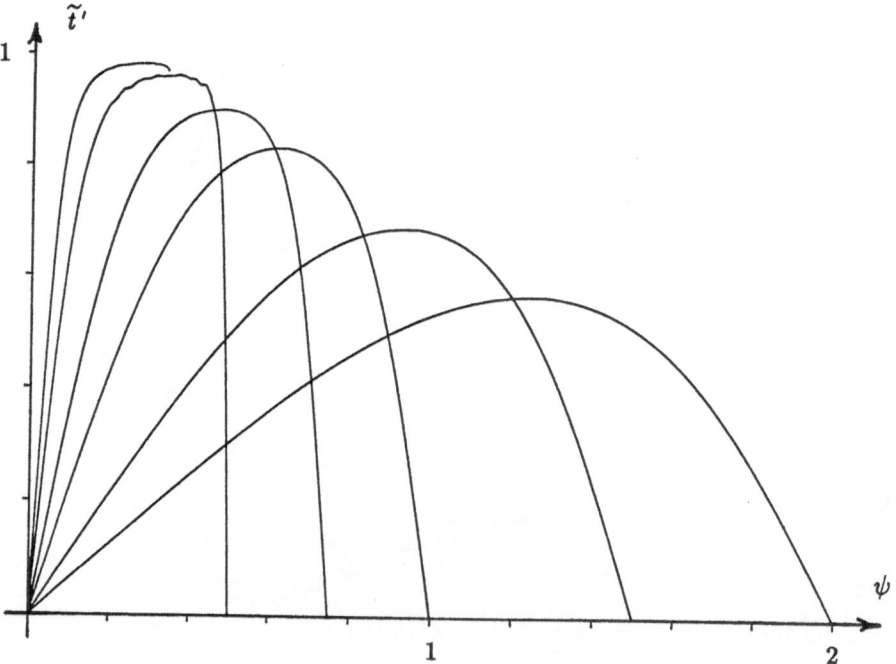

Figure 3. Calculated curves of $\tilde{t}'(\psi)$ for $\mu = \frac{3}{8}, \frac{1}{2}, \frac{3}{4}, 1, \frac{3}{2}$ and 2 with ν held fixed at $\frac{3}{4}$.

Results

The method was used to calculate supersonic flame paths generated by the initial conditions (6) over a range of values of ν and μ. In general, the first occurrence of a singularity takes place at the time $\tilde{t}(0)$ which adopts a value somewhere between the constant volume induction time of γ^{-1} and the constant pressure induction time of one.

$$\gamma^{-1} \leq \tilde{t}(0) \leq 1 \tag{21}$$

The last point at which $\phi \rightarrow \infty$ is the point $\psi = \mu$, which is always reached by the singularity path before the local constant volume induction time of $\gamma^{-1}e^{\nu}$. Moreover, since $|\tilde{t}'(\psi)| \leq 1$, the position $\psi = \mu$ must also explode before the time $\tilde{t}(0) + \mu$. It follows that

$$\gamma^{-1} \leq \tilde{t}(0) \leq \tilde{t}(\psi) \leq \tilde{t}(\mu) \leq \min\{\gamma^{-1}e^{\nu}, \tilde{t}(0) + \mu\}. \tag{22}$$

The latter inequality illustrates that there are essentially two ways in which the reaction can be driven to completion in the initially colder parts of the mixture. If the amplitude of the temperature disturbance ν is initially small, or if the range of the disturbance μ covers a large enough range of values of ψ, then the approach to the singularity is always dominated by chemical effects. In such cases, no cusp is seen to form around the initially coldest point. On the other hand, if the amplitude of the temperature disturbance is initially large, or if it is distributed over a small range (i.e. μ is small), then chemical effects dominate only part of the reaction runaway. Gasdynamic effects tend to dominate the final approaches towards the singularity. Where μ and $e^{\nu} - 1$ are comparable in size, both effects are important.

By looking at examples with either μ fixed or ν fixed, the transition between these two extremes can be observed. In figure 2 the value of μ is held fixed at $\mu = \frac{3}{4}$ and calculated singularity paths $\tilde{t}(\psi)$ are plotted for increasing values of ν. Apart from a region where the singularity first emerges near $\psi = 0$, the paths are seen to approach a form with slope very nearly equal to unity for large enough values of ν. Values at a few final grid points are not plotted in cases where the anti-cusp device, described above, was used.

Figure 3 shows an example with ν held fixed at the value $\nu = \frac{3}{4}$. In this case, values of the inverse mass-flux of the flame $\tilde{t}'(\psi) = 1/m(\psi)$ are plotted for various values of μ. Again it is seen that small values of μ cause the singularity path to travel with a mass-flux value only slightly greater than unity over a significant part of the evolution.

Conclusions

The introducion of a parametric variable ς, that can be chosen appropriately, simplifies the problem of solving for the generation of a supersonic reaction wave as it emerges after an induction period in a non-uniform explosive mixture. Various options for the choice of ς can be selected, and changed at any stage, so as to allow an integration with respect to ς that approaches the reaction wave as slowly as one chooses. This strongly contrasts with (say) a characteristic method, in which the flame-path intersects at least one set of characteristics at finite values.

It is clear that the method can readily be extended to solve for gasdynamic and chemical properties along almost any chosen path in space and time. In the full set of reactive Euler equations, possibly involving more complicated phenomena than those presented in equations (1), one may expect to identify points at which shock-waves would either be produced or where the solution would have lost its reality at an earlier stage because of the intervention of a shock wave.

References

1. **J.F. Clarke**, *A progress report on the theoretical analysis of the interaction between a shock wave and an explosive gas mixture*, College of Aeronautics Report **7801**, Cranfield Institute of Technology, Bedfordshire, U.K. 1978.

2. ———, *Small amplitude disturbances in an exploding atmosphere*, J. Fluid Mech. **89** (1978) 343–356.

3. ———, *On the evolution of compression pulses in an exploding atmosphere; initial behaviour*, J. Fluid Mech. **94** (1979) 195–208.

4. ———, *Propagation of gasdynamic disturbances in an explosive atmosphere*, Prog. Astronautics and Aeronautics **76** (1981) 383–402.

5. **J.F. Clarke and R.S. Cant**, *Non-steady gas dynamic effects in the induction domain behind a strong shock wave*, Progress in Astronautics and Aeronautics **95** (1984) 142–163, AIAA New York.

6. **T.L. Jackson and A.K. Kapila**, *Shock-induced thermal runaway*, SIAM J. Appl. Math. **45** (1985) 130–137.

7. **P.A. Blythe and D.G. Crighton**, *Shock generated ignition*, proceedings of "*Nonlinear waves in active media*," Euromech **241**, Tallinn 1988, to be published by Springer–Verlag.

8. **J.F Clarke, D.R. Kassoy and N. Riley**, Proc. Roy. Soc. Lond. A **393** (1984) 309–329; **393** (1984) 331–351; **408** (1986) 129–148.

9. **J.W. Dold**, *Dynamic transition of a self-igniting region*, in "*Mathematical modelling of combustion and related topics*," Eds. C-M. Brauner and C. Schmidt-Lainé, Martinus Nijhoff, Dordrecht 1988, p.p. 461–470.

10. **R. Klein and N. Peters**, *Cumulative effects of weak pressure waves during the induction period of a thermal explosion*, J. Fluid Mech. **187** (1988) 197–230.

11. **T.L. Jackson, A.K. Kapila and D.S. Stewart**, *Evolution of a reaction center in an explosive material*, Theoretical and Applied Mechanics report No. 484, Department of Theoretical and Applied Mechanics, University of Illinois, 1987.

12. **A.K. Kapila and J.W. Dold**, manuscript, 1989.

13. **J.W. Dold and A.K. Kapila**, manuscript, 1989.

14. **W. Fickett and W.C. Davis**, *Detonation*, University of California Press, Berkeley 1979.

15. **J.F. Clarke**, *Finite amplitude waves in combustible gases*, in "*The mathematics of combustion*," Ed. J.D. Buckmaster, SIAM, Philadelphia 1985

16. ———, *Fast flames, waves and detonation*, College of Aeronautics report, Cranfield Institute of Technology, Bedfordshire, U.K. 1988.

A Numerical Solution for Reacting and Non Reacting Flow

David Elkaim   Mohamed Agouzoul   Ricardo Camarero

Dept. Applied Mathematics Ecole Polytechnique de Montréal

P.O. Box 6079, Stn A, Montréal, Canada H3C 3A7

Abstract

A Numerical technique for the prediction of laminar flows, with and without chemical reaction, through the solution of the compressible time-dependent Navier-Stokes equations is presented. The solution consists of an explicit, time marching, control volume technique which is second order accurate in time and space. The technique has been applied to two cases: Laminar flow through a channel with sudden expansion and laminar channel flow with chemical reaction. The obtained steady state numerical results and their good comparison with experimental data show the accuracy and applicability of the technique.

Nomenclature

a	half the height of the channel
C_p	Coefficient of specific heat (at constant pressure)
D_j	Diffusion coefficient of specie j
$\bar{h}$	Total enthalpy (including enthalpy of formation)
h_{R_j}	Enthalpy of formation of specie j
K	Thermal conductivity
K_R	Constant appearing in source term for m_1 for test case 2
m_j	Mass fraction of specie j
M_j	Molecular weight of specie j
n	Number of species
p	Pressure
$\Re$	Universal gas constant
S	Source term (vector, equation (1))
t	Time
T	Temperature
u	Longitudinal velocity
v	Transverse velocity or volume
V	Volume
x, y	Cartesian coordinates

Greek Symbols

δ	Variation of variable between time level n and $n+1$
Δt	Time step
γ	Ratio of x heats
μ	Coefficient of laminar viscosity
ρ	Mixture density

Superscripts

1	First order variation
2	Second order variation
$n, n+1$	Time level n or $n+1$

Subscripts

1,2,3,4	Calculated at point 1,2,3,4
A, B, C, D	Relative to the cells A, B, C, D
g	Volume delimited by points 7-9-3-5-7
i	Computational point i
in	Conditions at the inlet
j	Relative to the specie j
n	Specie n or time level n

1 Introduction

In this paper we apply a control volume method to laminar flow with and without chemical reaction. Among the different numerical schemes usually employed for a discrete representation of the flow field, a significant amount of work has been done within the finite volume framework. Numerical Simulation of flows following this procedure, requires a special choice of computing cells to yield stable and realistics results. The staggered grid system provides a reliable procedure that meets these requirements; however its implementation demands a particular discretization, additional storage for the dependent variables, and results in some difficulties in implementing the general boundary conditions. With this arrangement it's hard to use techniques that accelerate the convergence like multigrids and this is a great disadvantage especially with explicit methods that usually converge very slowly. Another important point is that the use of dynamic refinement of the grid to optimize the grid size and which is of practical importance in determining the flame front in combustion problems, cannot easily be carried out with the staggered mesh. In order to adopt such techniques, in the future, one must first use a numerical scheme where all variables are calculated at the same computational point. In the present work, a numerical scheme, based on the previous works of Ni(1981) and Davis(1984), is developed. It is an explicit second order accurate time marching control volume scheme in which all the variables are computed at the center of the computational cell. In order to test the method, two different cases are considered:

1. Flow over a backward facing step.

2. Channel flow with chemical reaction. In this case, with appropriate assumptions, an analytical solution is obtained.

2 Governing Equations

The system to be treated is the two-dimensional flow of gas which transports n chemical species. The unsteady Navier-Stokes compressible equations coupled with mass transport equations of the species, govern such flows. Written in conservation form, the governing equations are:

$$\frac{\partial U}{\partial t} + \left(\frac{\partial F}{\partial x} + \frac{\partial G}{\partial y} \right) = S \tag{1}$$

where

$$U = \begin{bmatrix} \rho \\ \rho u \\ \rho v \\ \rho \tilde{h} \\ \rho m_1 \\ \vdots \\ \rho m_n \end{bmatrix} \quad ; \quad F = \begin{bmatrix} \rho u \\ \rho u^2 + p - \sigma_{xx} \\ \rho u v - \tau_{xy} \\ \tilde{h} \rho u - q_x \\ \rho u m_1 - D_1 \frac{\partial}{\partial x} m_1 \\ \vdots \\ \rho u m_n - D_n \frac{\partial}{\partial x} m_n \end{bmatrix}$$

$$
G = \begin{bmatrix} \rho v \\ \rho u v - \tau_{xy} \\ \rho v^2 + p - \sigma_{vv} \\ \tilde{h}\rho v - q_y \\ \rho v m_1 - D_1 \frac{\partial}{\partial y} m_1 \\ \vdots \\ \rho v m_n - D_n \frac{\partial}{\partial y} m_n \end{bmatrix} ; \quad S = \begin{bmatrix} 0 \\ 0 \\ 0 \\ 0 \\ S_1 \\ \vdots \\ S_n \end{bmatrix}
$$

and:

$$
\sigma_{xx} = \frac{2}{3}\mu \left(2\frac{\partial u}{\partial x} - \frac{\partial v}{\partial y} \right) ; \quad \sigma_{vv} = \frac{2}{3}\mu \left(2\frac{\partial v}{\partial y} - \frac{\partial u}{\partial x} \right) ; \quad \tau_{xy} = \mu \left(\frac{\partial u}{\partial y} + \frac{\partial v}{\partial x} \right)
$$

$$
q_x = \left(\frac{K}{c_p} \right) \frac{\partial}{\partial x} \tilde{h} ; \quad q_y = \left(\frac{K}{c_p} \right) \frac{\partial}{\partial y} \tilde{h}
$$

Also, assuming that for each specie the coefficient of specific heat at constant pressure is constant and that the mixture behaves like a perfect gas, the energy, written in terms of the total enthalpy, is given by:

$$
\tilde{h} = T \sum^n c_{p_j} m_j + \sum^n h_{R_j} m_j + \frac{1}{2} \left(u^2 + v^2 \right) \tag{2}
$$

so that the heat released by the chemical reaction is implicitly taken into account by the definition of $\tilde{h}$. Thus we do not have any source terms due to chemical reaction in the equation of the conservation of $\tilde{h}$.

The precedent system is closed by the state equation for a perfect gas which is:

$$
p = \rho \Re T \sum_j \frac{m_j}{M_j} \tag{3}
$$

Finally as a consequence of the definition of the mass fraction, we have:

$$
\sum^n m_j = 1 \tag{4}
$$

The level of complexity of the chemical reaction between the different species and the mass species source term will be given in the appropriate section when dealing with each test case.

3 Boundary Conditions

Solution of differential equations (whether the solution is analytical or numerical) involves the knowledge of the boundary conditions. There are four types of boundaries on which conditions must be specified: inlet, solid walls, symmetry axis and outlet.

Inlet

In order to solve the N-S equations together with the energy equations, we specify three variables. Here the velocities and static temperature are specified (Rudy et al. (1981)).

For the mass transport equations, one has to specify also the concentration. The pressure is extrapolated from the interior of the domain and the remaining variables are calculated using equations (2) and (3).

Solid walls

On solid walls the velocities have to be set to zero. The temperature and mass fraction variables have either specified fluxes or specified values. The conditions used here for solid walls are given in the section dealing with the results of each test case.

Symmetry axis

On symmetry axes normal velocities and normal derivatives of other variables are set to zero.

Outlet

Here, one variable has to be specified and it is given, once again, in the section dealing with the results of each test case. For the rest of the variables, we assume that the flow extends over a sufficiently large spatial region so that at outlet it is fully developed and then the first normal derivative of the dependent variables vanishes.

4 The Numerical Procedure

The numerical procedure is presented in two parts. First a general description of the spatial discretization is given. Then the time discretization is derived.

Grid Configuration and Domain Discretization

The general flowfield is broken into zones and each zone into small polygons. These polygons are quadrilaterals for two-dimensional flows. Within each zone, the computational mesh may be defined through conformal mapping techniques, body fitting scheme, or simple algebraics algorithms. The zones and their meshes used in the calculations herein are all body fitted and are generated completely independent of the flow solver.

Time Discretization

The technique proposed here is an explicit time marching control volume scheme and second order accurate in time. It is based on the Ni scheme (Ni, 1981) for Euler equations.

A portion of the computational grid is presented in figure 1.

Figure 1: Portion of the mesh for the time discretization

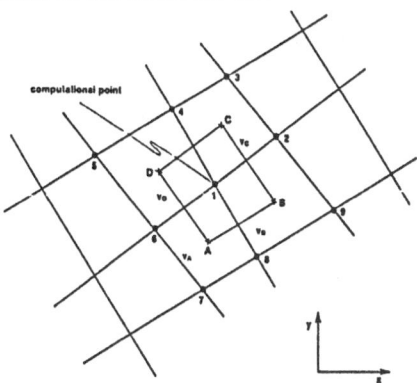

The computational point (point 1), is surrounded by four cells (A, B, C, D) having each a volume v_A, v_B, v_C, v_D. Every cell is delimited by four points (ex: cell C is delimited by points 1,2,3,4) and the points A, B, C, D represent the geometric center of the cell A, B, C, or D respectively.

The procedure of updating the primary variables in time is performed according to a second-order accurate Taylor series expansion in time:

$$U^{n+1} = U^n + \left.\frac{\partial U}{\partial t}\Delta t\right|_n + \left.\frac{\partial^2 U}{\partial t^2}\frac{\Delta t^2}{2!}\right|_n \tag{5}$$

and we denote the first order change at time level n by:

$$\delta U_i^1 = \frac{\partial U}{\partial t} \cdot \Delta t \tag{6}$$

where the upperscript 1 refers to first order while the lowerscript i refers to computational point i.

Similarly the second order change is denoted by:

$$\delta U_i^2 = \frac{\Delta t^2}{2!}\frac{\partial^2 U}{\partial t^2} \tag{7}$$

We shall now calculate first and second order changes without considering the source term of equation (1). The discretization of the source term is given lately.

First Order Change

According to equation (1) without the source term, (6) becomes:

$$\delta U_i^1 = -\left(\frac{\partial F}{\partial x} + \frac{\partial G}{\partial y}\right)\Delta t \tag{8}$$

To get the first order changes, equation (8) is integrated upon the volume V_g (figure 1, points 7-9-3-5-7) by splitting it upon the four volumes v_A, v_B, v_C, v_D and assuming a linear variation of F and G between two points

The first order variation at each computational point is then given by:

$$\delta U_i^1 V_g = v_a \delta U_A^1 + v_B \delta U_B^1 + v_c \delta U_C^1 + v_D \delta U_D^1 \tag{9}$$

Second Order Change

According to equations (1) and (7), the second order change of U at point i is:

$$\delta U_i^2 = -\frac{\Delta t^2}{2!} \frac{\partial}{\partial t} \left(\frac{\partial F}{\partial x} + \frac{\partial G}{\partial y} \right) \tag{10}$$

Changing the order of the derivatives $\frac{\partial}{\partial t}\left(\frac{\partial F}{\partial x}\right)$ and $\frac{\partial}{\partial t}\left(\frac{\partial G}{\partial y}\right)$ by $\frac{\partial}{\partial x}\left(\frac{\partial F}{\partial t}\right)$ and $\frac{\partial}{\partial y}\left(\frac{\partial G}{\partial t}\right)$ and using the chain rule for $\frac{\partial F}{\partial t}$ and $\frac{\partial G}{\partial t}$, equation (10) becomes:

$$\delta U_i^2 = -\frac{\Delta t^2}{2} \left[\frac{\partial}{\partial x}\left(\frac{\partial F}{\partial U}\cdot\frac{\partial U}{\partial t} \right) + \frac{\partial}{\partial y}\left(\frac{\partial G}{\partial U}\cdot\frac{\partial U}{\partial t} \right) \right] \tag{11}$$

Including one Δt into the parenthesis will give:

$$\delta U_i^2 = -\frac{\Delta t}{2} \left[\frac{\partial}{\partial x}\left(\Delta t\frac{\partial U}{\partial t}\cdot\frac{\partial F}{\partial U} \right) + \frac{\partial}{\partial y}\left(\Delta t\frac{\partial U}{\partial t}\cdot\frac{\partial G}{\partial U} \right) \right] \tag{12}$$

By noting that $\Delta t \cdot \frac{\partial U}{\partial t}$ is the first order change of U (equation 6) and using the following notations:

$$\Delta F \equiv \frac{\partial F}{\partial U}\cdot\delta U^1 \quad ; \quad \Delta G = \frac{\partial G}{\partial U}\cdot\delta U^1 \tag{13}$$

where $\frac{\partial F}{\partial U}$ and $\frac{\partial G}{\partial U}$ are the Jacobian matrices, equation (12) becomes:

$$\delta U_i^2 = -\frac{\Delta t}{2} \left\{ \frac{\partial}{\partial x}\Delta F + \frac{\partial}{\partial y}\Delta G \right\} \tag{14}$$

The second order variation is now obtained by integrating equation (10) upon the volume v_p delimited by the centers of the cells (A, B, C, D) and assuming a linear variation of ΔF and ΔG between two points.

Time Step

Time step is calculated according to J. C. Tannehill [1975] in order to satisfy the C.F.L. condition and for cartesian meshes.

Source terms and derivatives

It can be easily shown by means of the same technique that the source term is discretized as follows:

$$\delta U_i|_{\text{source}} = S_i\Delta t + \frac{\Delta t}{2}\delta S_i^1 \tag{15}$$

The velocity or other derivatives are simply approximated by centered differences for non cartesian grids.

Smoothing - Artificial Viscosity

For supersonic or high Reynolds number flows, laminar shear stresses do not produce enough physical viscous smoothing to stabilize the scheme necessitating numerical smoothing.

In the present work, artificial viscosity, as deduced from an upwind difference scheme, is added explicitly. Each primary variable ϕ calculated at time level $n + 1$ is corrected by:

$$
\begin{aligned}
\phi^{n+1}_{corrected} = \phi^{n+1} &+ \frac{|u|}{2}\frac{\Delta t}{\Delta x}\left(\phi^{n+1}_{i+1,j} - 2\phi^{n+1}_{i,j} + \phi^{n+1}_{i-1,j}\right) \\
&+ \frac{|v|}{2}\frac{\Delta t}{\Delta y}\left(\phi^{n+1}_{i,j+1} - 2\phi^{n+1}_{i,j} + \phi^{n+1}_{i,j-1}\right)
\end{aligned}
\tag{16}
$$

Artificial viscosity has been found to be useful for the first test case. For the second one, no artificial viscosity was needed when chemical reaction occured without heat release but artificial viscosity was needed when chemical reaction occured with heat release preventing us from temperature fluctuations.

5   Results and Discussion

The described control volume numerical technique has been tested on two viscous flow problems. The results of the two cases demonstrate the capabilities of the present method.

The Backward Facing Step

The first test case solved numerically with the developed method is the low subsonic flow over a backward facing step. For this flow only one specie is considered and the transport equations of the species concentration do not need to be solved. Some global features of this case are given in figure 2.

Figure 2: Global features for the backward facing step; Geometry and Space discretization

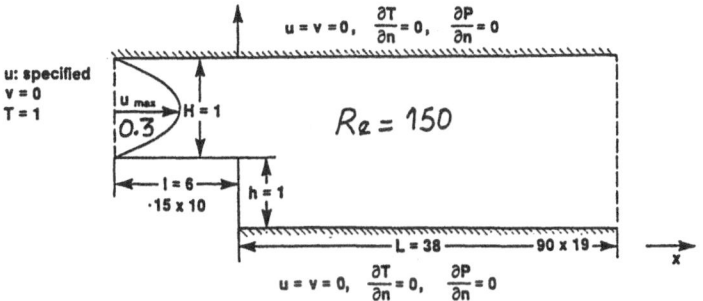

The Reynolds number is based on the maximum inlet velocity and the height of the step. In order to compare our numerical results with the Gamm Workshop experimental data (Gamm, 1984), a non dimensionalized shear stress $\left(\bar{\tau} = \frac{1}{Re_{(\rho u)_{max}}} \times \frac{\partial \rho u}{\partial y}\right)$ along the lower wall is calculated and plotted in figure 3.a.

Figure 3: (a) Non dimensional shear stress (∞) Experimental Gamm (1984), (—) Present numerical results, (b) Recirculation region behind the step and (c) Streamlines

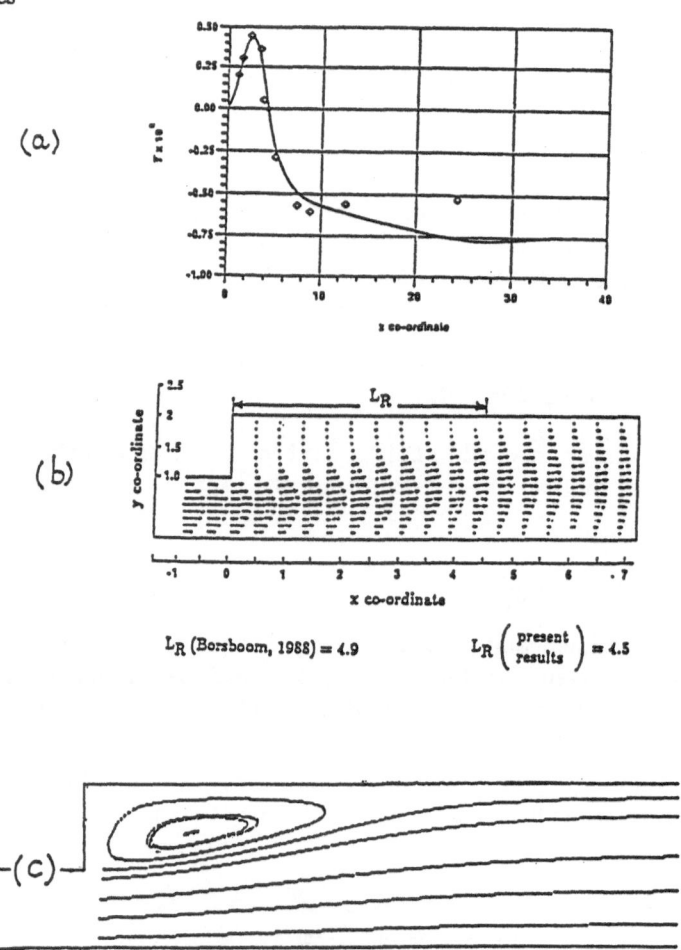

L_R (Borsboom, 1988) = 4.9 $L_R \left(\begin{array}{c} present \\ results \end{array}\right) = 4.5$

The comparison is rather good and even excellent in the recirculation region, however the calculated shear stresses do not tend asymptotically to the experimental downstream values, wich may indicate that the Reynolds number of the experimental flow may have been actually smaller than indicated. In Figure 3.b, the reattachment length as predicted (numerically) by Borsboom (1988) is compared to our predicted value and once again the comparison is good. Finally figure 3.c shows the streamlines for the backward facing step calculation.

Channel Flow with Chemical Reaction

For this test case, we consider the following problem, as suggested by Winters et al. (1981). Two reacting species flow along a two-dimensional channel and the heat released distorts the flow which would have been developed in the absence of heat release. One of the specie (specie 1) is assumed to be very dilute in the mixture so that only the equation for m_1 needs to be solved. Global features for this case are given in figure 4.

Figure 4: Global features for the channel flow; Geometry and Space discretization

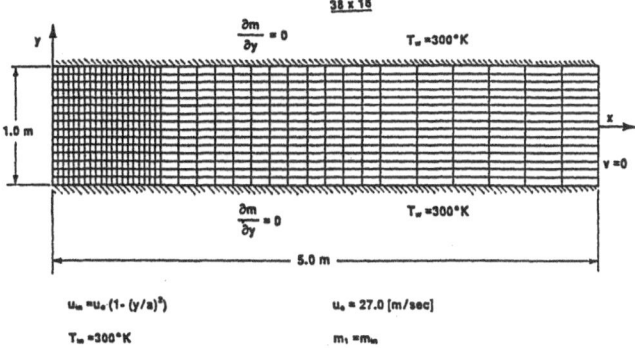

The source term for the mass fraction of the dilute specie is taken to be:

$$S_1 = -K_R m_1$$

which is independent of temperature and K_R is equal to 100. A non uniform refined mesh at the entry, where gradients are expected to be higher because of the choosen value of the reaction coefficient K_R, is used.

For the limiting case of no heat release ($h_{Rj} = 0.0$), the flow remains fully developed, the density is constant, then the flow and chemistry decouple and the equation for m can be solved analytically [Winters et al., 1981].

Results without heat release

Once we have the analytical solution without heat release one can set $x = 0.0$ and get the inlet profile $m_{1_{in}}$ of the specie 1. This inlet profile (displayed in figure 5, $x = 0.0$) is further used to compute the numerical solution of the equations.

Figure 5: Profiles of mass fraction m_1 across the channel at various stations, reaction without heat release, (analytical (—) and present numerical results $(- - -))$

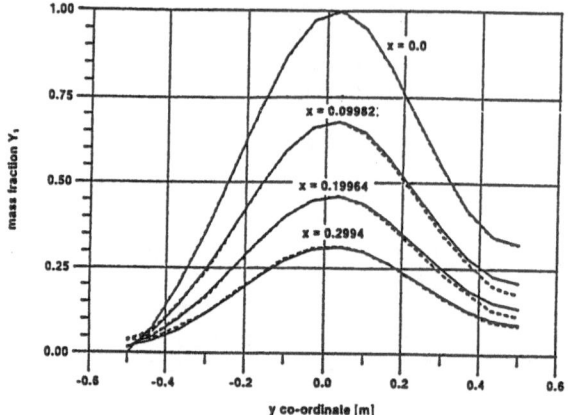

Figure 5 shows the numerical and analytical results of the mass fraction of specie 1 at different stations downstream. The comparison is very good except for nodes near the solid boundaries where the biggest difference was found to be 13%. This discrepancy can be attributed to difficulty in satisfying the zero gradient boundary condition and to the grid spacing in this region where large gradients occur.

Results with Heat Release

In figure 6 we present contours of temperature in the channel when the reaction occurs with heat release (i.e. $h_{R_j} \neq 0$).

Figure 6: Temperature contours for reacting channel flow

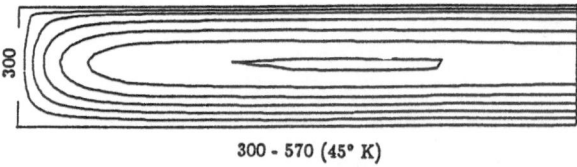

300 - 570 (45° K)

The temperature rises to $570° K$ and then decreases attaining a symmetric parabolic profile at the outlet.
Figure 7 shows our numerical results for the axial velocity across the channel at several stations along the channel.

Figure 7: Profiles of the velocity across the channel at various stations (reaction with heat release)

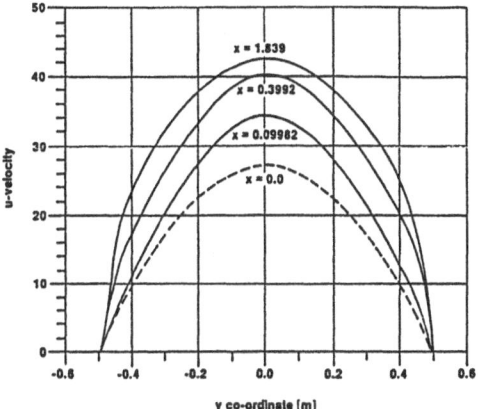

The effect of exothermic reaction is to accelerate the flow as expected (decrease in local density). The acceleration of the fluid has the effect of retarding the rate at which specie 1 is consumed downstream and this is shown in figure 8 where our numerical results both with and without heat release are displayed.

Figure 8: Profiles of mass fraction m_1 across the channel of various stations (reaction without heat release $(---)$, with heat release $(—)$)

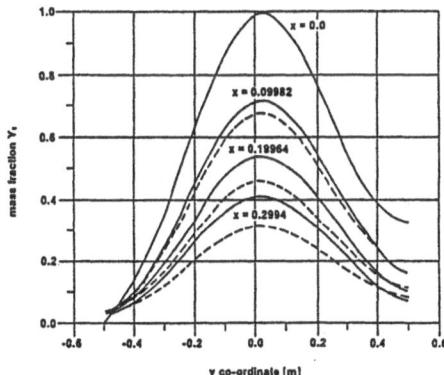

Discussion of the proposed method

There are two major advantages in the method that have been proposed: It's implementation on computers is very easy and straightforward and it could be easily adapted to calculations with dynamic mest refinement. This method has been also widely tested for transonic flow calculations [Davis (1984), Ni (1981)] and the reported results which also include computational costs were very encouraging.

During the computations of the mass fractions, one has to worry that the mass fraction remains positive and less than 1 by correcting it in the computer program.

Artificial viscosity which is needed to stabilize the solution is without any doubt the major draw back of the present proposed method and some further work is to be done in order to keep it to the minimum required without affecting the shape of the flow.

Conclusion

A second order finite volume time marching scheme for solving the compressible Navier-Stokes equations was developped. It was tested with success on two different cases which may suggest to apply it further to turbulent flows. Work in this direction is actually undertaken by the authors.

References

1. Abramowitz, M. and Stegun, I. A., (1965). *Handbook of Mathematical functions*, Dover, New York, 1965.

2. Borsboom, M., (1988). "An Implicit Approximate Factorization Finite Volume Technique with Improved Accuracy for the Compressible Navier-Stokes Equations", Von Karman Institute for Fluid Dynamics, Lecture Series, *Computational Fluid Dynamics*, March 7-11, 1988.

3. Davis, R. L. and Ni, R. H., (1984). "Prediction of Compressible, Laminar Viscous Flows Using a Time-Marching Control Volume and Multiple Grid Technique", *AIAA Journal*, Vol. 22, No. 11, November 1984.

4. Gamm (1984). Workshop on Analysis of Laminar Flow over a Backward Facing Step, eds., K. Morgan, J. Periaux, F. Thomasset, Vieweg, Brannschweig.

5. Mac Cormack, R. W., (1971). "Numerical Solution of the Interaction of a Schock Wave with a Laminar Boundary Layer", Proc. Second Int. Conf. Num. Methods Fluid Dyn., *Lecture Notes in Physics*, Vol. 8, Springer-Verlag, New York, pp. 151-163, 1971.

6. Ni, R. H., (1981). "A Multiple Grid Scheme for Solving the Euler Equations", *AIAA Journal*, Vol. 20, No. 11, 1981.

7. Rudy, D. H. and Strikwerda, J. C. (1981). "Boundary Conditions for Subsonic Compressible Navier-Stokes Calculations", *Computers and Fluids*, Vol. 9, pp. 327-338, 1972.

8. Tannehill, J. C., Holst, T. L. and Rakich, J. V., (1975). "Numerical Computation of Two Dimensional Viscous Blunt Body Flows with an Impinging Schock", *AIAA Paper*, 75-154, Passadena, California, 1975.

9. Winters, K. H., Rae, J., Jackson, C. P. and Cliffe, K. A., (1981). "The Finite Element Method for Laminar Flow with Chemical Reaction", *International Journal for Numerical Methods in Engineering*, Vol. 17, 239-253, 1981.

ON THE EQUATIONS FOR REACTIVE GRANULAR FLOW

Pedro F. Embid
Mathematics Department
University of New Mexico
Albuquerque, New Mexico, 87131, USA

and

Melvin R. Baer
Fluid and Thermal Sciences Department
Sandia National Laboratories
Albuquerque, New Mexico, 87185, USA

Abstract

In this study, we examine a multiphase model proposed by Baer and Nunziato describing nonequilibrium reactive granular flows. The mathematical character of this model is established to provide the foundations for characteristics-based numerical methods. This analysis includes classification of the wave fields, jump conditions and special wave solutions. This system of equations is hyperbolic but degenerates at points where the relative flow is locally sonic. For fixed upstream conditions, multiple solutions of the equations are determined when the relative velocities exceed the sound speed of the gas phase.

Introduction

One of the most complex combustion phenomena is that of multiphase combustion in gas permeable reactive materials. In highly energetic granular materials, such as propellants and explosives, combustion self-accelerates from deflagration to detonation due to the interaction of various thermal, chemical and mechanical processes.

During the accelerated combustion of granular energetic materials, high gas pressure generates flow with large drag forces. In turn, the disparity of solid and gas phase pressures induces compaction of the granular reactant that precedes combustion. Compaction reduces permeability and further increases gas pressurization. This process is inherently unsteady and combustion/compaction eventually produces low amplitude shock wave development whereupon localized heating at 'hot-spots' enhances the growth to detonation. It is now recognized that the mechanical response of the granular reactant is the key to the formation of a sustained shock wave necessary for the compressive stages leading to detonation.

It is clear that the highly coupled thermal-mechanical processes associated with the multiphase combustion of energetic materials can only be described by models that appropriately treat low and high speed reactive flows. In spite of much previous study, multiphase models have only recently been formulated which are capable of describing nonequilibrium reactive granular flows accounting for both the compressibility of all phases and the compaction of the granular reactant. Baer and Nunziato[1] developed a multiphase model using the theory of mixtures and applied the model to several energetic materials. In this description, balance laws are derived allowing the phases to exchange mass, momentum and energy. Constraints are imposed on the phase interactions to assure that the multiphase flows do not violate the entropy inequality. Consistent with this requirement, the solid volume fraction is treated as an independent kinematic variable and an evolutionary equation is used to describe dynamic compaction (all details of this development are given in Reference 1). Most importantly, this mathematical description is hyperbolic and the wave motion described by these multiphase equations is mathematically well-posed.

Much study of this multiphase model has focused on determining appropriate constitutive laws for reactive flow and numerical solutions of the field equations has employed well-established Eulerian methods. In numerical solutions, it is observed that the essential combustion physics accumulates near and within regions of extreme gradients. It is thus desirable to improve numerical methods without introducing numerical artifacts for shock-capture which masks the underlying physics.

As an example of the transient combustion of granular materials, Figure 1 displays the evolution of solid phase pressure obtained by numerical solution of a multiphase mixture model describing a DDT (deflagration-to-detonation transition) event in a column of granular explosive HMX[1]. During the various stages of combustion, the coupled effects of gas flow, convective heat transfer, combustion and shock wave development produces disparate wave fields. Shown in Figure 2 is an overlay of gas and solid phase pressures at a time during deflagration where it is seen that the effects of compaction, gas permeation and two-phase combustion produce regions of extreme gradients. Near the onset to detonation, these waves merge and the mechanism leading to transition is not well defined due to the numerical smearing at shock fronts. A characteristic-based numerical method may provide the insight to the precise mechanisms of accelerated multiphase combustion. Unfortunately, the mathematical structure of multiphase hydrodynamics has not been extensively studied and the appropriate mathematical foundations for the development of improved numerical methods requires additional investigation. This paper provides a brief analysis of the multiphase model.

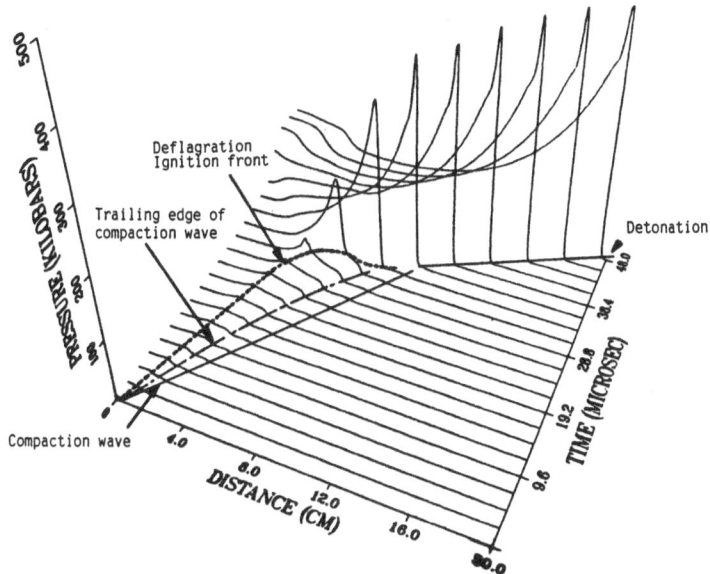

Figure 1. Distance-time evolution of solid pressure field during the accelerated combustion of granular HMX initiated by a thermal pulse at the origin.

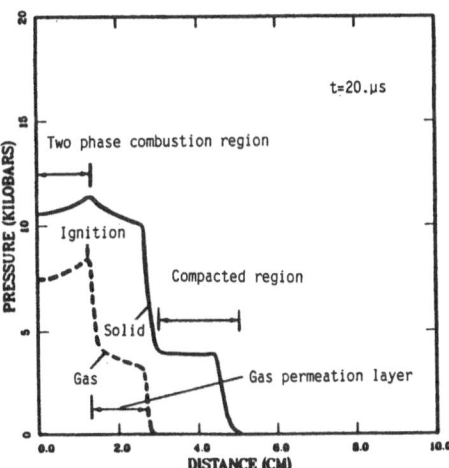

Figure 2. An overlay of gas and solid phase pressures at a time of 20 μs after initiation of accelerated combustion in granular HMX.

Continuum Theory of Multiphase Mixtures

In one dimension, the reactive multiphase equations are written in conservation form as:

Conservation of Mass

$$\frac{\partial}{\partial t}(\phi_a \rho_a) + \frac{\partial}{\partial x}(\phi_a \rho_a v_a) = c_a^+, \tag{1}$$

Conservation of Momentum

$$\frac{\partial}{\partial t}(\phi_a \rho_a v_a) + \frac{\partial}{\partial x}(\phi_a \rho_a v_a^2 + \phi_a p_a) = m_a^+, \tag{2}$$

Conservation of Energy

$$\frac{\partial}{\partial t}(\phi_a \rho_a E_a) + \frac{\partial}{\partial x}((\phi_a \rho_a E_a + \phi_a p_a) v_a) = e_a^+, \tag{3}$$

Compaction Evolutionary Equation

$$\frac{\partial \rho_s}{\partial t} + \frac{\partial}{\partial x}(\rho_s v_s) = -\frac{\rho_s F}{\phi_s}. \tag{4}$$

The subscript $a = s$ denotes solid-phase variables and $a = g$ denotes gas-phase states. Associated with each phase is a set of state variables: phase velocity v_a, material density ρ_a, phase pressure p_a, total phase energy $E_a = e_a + v_a^2/2$, phase internal energy e_a and volume fraction ϕ_a. In this description, independent equations of state for each phase are admissable and are given in the form:

$$p_a = p_a(\rho_a, e_a). \tag{5}$$

Furthermore, the phases occupy all of the total volume, and the mixture obeys the saturation constraint given as:

$$\phi_s + \phi_g = 1. \tag{6}$$

The interaction terms, c_a^+, m_a^+ and e_a^+, represent the exchange of mass, momentum and energy between the phases, respectively. To assure conservation of mass, momentum and energy of the total mixture, the following constraints are required:

$$c_s^+ = -c_g^+, \quad m_s^+ = -m_g^+, \quad e_s^+ = -e_g^+. \tag{7}$$

The forcing function for the compaction equation, F, dictates that compaction is rate dependent and is driven by pressure differences, thus:

$$F = \frac{\phi_s \phi_g}{\mu_c}(p_s - p_g - \beta_s), \tag{8}$$

where β_s is the configuration stress (reflecting distortion of the solid granular material) and μ_c is the compaction viscosity.

Explicit constitutive equations for the interaction terms are given in Reference 1. In the momentum and energy phase interactions it is convenient to separate the volume fraction derivative terms describing geometric coupling between the phases according to:

$$m_s^+ = p_g \frac{\partial \phi_s}{\partial x} + m_s^*, \tag{9}$$

$$e_s^+ = v_s p_g \frac{\partial \phi_s}{\partial x} + e_s^*. \tag{10}$$

The m_s^* and e_s^* terms are now strictly algebraic, however the system is not of a divergence free form.

Finally, the phase entropy, η_a, complements this system by imposing the following restriction:

$$\sum_a \left(\frac{\partial}{\partial t}(\phi_a \rho_a \eta_a) + \frac{\partial}{\partial x}(\phi_a \rho_a v_a \eta_a) \right) \geq 0 \tag{11}$$

Analysis of the Multiphase Equations

In this section, the mathematical structure of the multiphase hydrodynamic equations is studied without the algebraic sources in Equations 1 - 4 (note that the volume fraction gradients of the phase interactions must be retained). Incorporation of highly stiff interactions (*i.e.* chemistry, interphase drag and heat transfer) can be included in a numerical solution using an operator splitting method.[4] For the sake of brevity, mathematical details are not included in this work and the interested reader will find a complete development in Embid and Baer.[2]

The field equations 1-4 constitute a first order system of equations that can be written in matrix form as:

$$\frac{\partial \mathbf{U}}{\partial t} + \mathbf{A}(\mathbf{U}) \frac{\partial \mathbf{U}}{\partial x} = 0, \tag{12}$$

where $\mathbf{U} = (\phi_s \rho_s, \phi_s \rho_s v_s, \phi_s \rho_s E_s, \rho_s, \phi_g \rho_g, \phi_g \rho_g v_g, \phi_g \rho_g E_g)^T$.

The 7x7 matrix $\mathbf{A}(\mathbf{U})$ has eigenvalues $v_s \pm c_s, v_s, v_s, v_g \pm c_g, v_g$. All eigenvalues are real but v_s is a double eigenvalue, hence the system of equations is not strictly hyperbolic. However, this system is hyperbolic provided that $\mathbf{A}(\mathbf{U})$ is diagonalizable. This is permissable except when the eigenvalues v_s and $v_g + c_g$ (resp. $v_g - c_g$) coalesce. At these points v_s becomes a triple eigenvalue and one of the eigenvectors associated with v_s align with the eigenvector associated with $v_g + c_g$ (resp. $v_g - c_g$). As a consequence, at points where $v_s - v_g = \pm c_g$ the matrix $\mathbf{A}(\mathbf{U})$ has a nontrivial 2x2 Jordan block. As a physical interpretation, flow at the pore level is analogous to flow in a moving variable cross-section area duct (due to local variations of porosity) and the choked flow conditions corresponds to a sonic pont of the relative gas flow.

The wave field associated with the eigenvalue λ_k and eigenvector, $\mathbf{r}_k$ is genuinely nonlinear if $\nabla\lambda_k(\mathbf{U})\cdot\mathbf{r}_k(\mathbf{U}) \neq 0$ (or linearly degenerate if $\nabla\lambda_k(\mathbf{U})\cdot\mathbf{r}_k(\mathbf{U}) = 0$) for all $\mathbf{U}$ of interest.[2] The wave fields associated with v_a ($a = s, g$) is linearly degenerate, whereas those associated with $v_a \pm c_a$ are genuinely nonlinear provided that $1/\rho_a\, \partial(\rho_a c_a)/\partial\rho_a$ has a consistent sign. This condition has been verified for the equations of state used in the multiphase model of Baer and Nunziato. Additionally, it is found that:

$$\left.\frac{\partial p_a}{\partial \tau_a}\right|_{\eta_a,\phi_a} < 0\;, \quad \left.\frac{\partial p_a}{\partial \eta_a}\right|_{\tau_a,\phi_a} > 0\;, \quad \left.\frac{\partial^2 p_a}{\partial \tau_a^2}\right|_{\eta_a,\phi_a} > 0, \tag{13}$$

where $\tau_a = \rho_a^{-1}$. These conditions assure that the Rayleigh line intersects the Hugoniot at a maximum of two points.[3]

Additional results concerning the characteristic form of the multiphase flow equations, derivation of Riemann invariants and construction of simple wave solutions such as centered rarefaction waves and contact discontinuities are given in detail in Reference 2.

Jump Conditions and Special Solutions

Although the system represented by Equations 1 - 4 is not in divergence form due to the $\partial\phi_s/\partial x$ terms in m_a^+ and e_a^+, derivation of the jump conditions requires piecewise C^1 generalized solutions with ϕ_s continuous everywhere. Let $[f]$ denote the difference of f at the right (subscript R) and left (subscript L) of a discontinuity, i.e. $[f] = f_L - f_R$, σ the discontinuity speed and $V_a = v_a - \sigma$. The jump conditions are then given as:

$$[\rho_a V_a] = 0 \quad \text{(mass)}, \tag{14}$$

$$[\rho_a V_a^2 + p_a] = 0 \quad \text{(momentum)}, \tag{15}$$

$$[h_a + V_a^2/2] = 0 \quad \text{(energy)}, \tag{16}$$

$$[\phi_a] = 0, \tag{17}$$

where $h_a = e_a + p_a/\rho_a$ is the phase enthalpy. The entropy inequality provides an additional constraint:

$$\sum_a [\phi_a \rho_a \eta_a V_a] \geq 0. \tag{18}$$

Note that Equations 14-16 are also the jump conditions for single-phase flow.

The state variables at the left of the discontinuity are then fixed and we examine the case where the speed of the discontinuity is $\sigma = v_{sL}$. At this condition, $v_{sR} = v_{sL}$ and the jump conditions reduce to:

$$[\phi_s] = 0, \tag{19}$$

$$[v_s] = 0, \tag{20}$$

$$[p_s] = 0, \tag{21}$$

$$[\rho_g(v_g - \sigma)] = 0, \tag{22}$$

$$[\rho_g(v_g - \sigma)^2 + p_g] = 0, \tag{23}$$

$$[\rho_g(v_g - \sigma)(h_g + (v_g - \sigma)^2/2)] = 0, \tag{24}$$

$$[\rho_g \eta_g(v_g - \sigma)] \geq 0. \tag{25}$$

Conditions 20-21 imply that the jump states in the solid phase correspond to a contact discontinuity. Since the gas-phase equation of state satisfies Equation 13, solution of Equations 22-24 follows single-phase hydrodynamic theory[3] given at the intersection of the Rayleigh line

$$p_g - p_{gL} = -m^2(\tau_g - \tau_{gL}), \tag{26}$$

and the Hugoniot curve

$$e_g(p_g, \tau_g) - e_g(p_{gL}, \tau_{gL}) = -(p_g + p_{gL})(\tau_g - \tau_{gL})/2, \tag{27}$$

where $m = \rho_{gL}(v_{gL} - v_{sL})$.

Next, we consider the condition $v_{gL} - v_{sL} > 0$. In the subsonic or sonic case $(v_{gL} - v_{sL} \leq c_{gL})$, only one possible state exists for the right state given as $(\tau_{gR}, p_{gR}) = (\tau_{gL}, p_{gL})$. However, for the supersonic case $(v_{gL} - v_{sL} > c_{gL})$, there exist two solutions of Equations 26 and 27: (τ_{gL}, p_{gL}) and (τ_g^*, p_g^*). For $(\tau_{gR}, p_{gR}) = (\tau_{gL}, p_{gL})$, the entropy flux in Equation 25 is zero and for $(\tau_{gR}, p_{gR}) = (\tau_g^*, p_g^*)$ the entropy flux is positive. The first solution corresponds to a solid-phase contact discontinuity with uniform gas flow and the second solution is a solid-phase contact discontinuity with a gas-phase shock. Both solutions satisfy Lax's geometric shock condition.[5] The issues of stability of these solutions and discontinuous solutions with $[\phi_s] \neq 0$, describing compaction shocks, remain subjects of continued study.

Summary

In this paper, we provide a brief overview of the mathematical characteristics of the multiphase reactive flow equations posed by Baer and Nunziato. Although several flow characteristics parallel those of single-phase flow, coupling of the hydrodynamic fields occurs due to variations of volume fraction. Specifically, we observe that special conditions exist at choked flow points. Having established a mathematical foundation for characteristic-based numerical methods, much of the current developments in numerical computation as applied to single phase compressible flow may be applicable to multiphase reactive flows.

Acknowledgment. — This work was sponsored by Sandia National Laboratories supported by the U. S. Department of Energy under contract number DE-AC04-76DP00789. We gratefully acknowledge fruitful discussions with P. Lax (Courant Inst.) and A. Majda (Princeton University).

References

1. Baer, M. R. and Nunziato, J. W., "A Two-Phase Mixture Theory for Deflagration-to-Detonation Transition (DDT) in Reactive Granular Materials", *Intl. J. Multiphase Flow*, **12**, pp 861-889, 1986.

2. Embid, P. and Baer, M., "Mathematical Analysis of a Two-Phase Model for Reactive Granular Materials", SAND88-3302, Sandia National Laboratories, 1989, in press.

3. Courant, R. and Friedrichs, K., **Supersonic Flow and Shock Waves**, Springer-Verlag, New York, 1976.

4. Baer, M. R., Benner, R. E., Gross, R. J. and Nunziato, J. W., "Modeling and Computation of Deflagration-to-Detonation Transition (DDT) in Reactive Granular Materials", **Lectures in Applied Mathematics**, American Mathematical Society, pp 479-498, 1986.

5. Lax, P., "Hyperbolic Systems of Conservation Laws and the Mathematical Theory of Shock Waves", SIAM, Philadelphia, 1973.

Implicit Schemes for Subsonic Combustion Problems

G. Fernandez, H. Guillard

INRIA, centre de Sophia Antipolis
2004 Av des Lucioles, 06565 Valbonne (France)

1. Introduction: The computations of subsonic laminar reactive flows face some important numerical difficulties related to the simultaneous presence of different physical processes evolving in a large range of disparate length and time scales. To fix the ideas, we consider as an example, the propagation of a laminar flame inside a closed tube one meter long. Relevant time scales for this problem are chemical reaction times of order less than or equal to 10^{-6} s, a characteristic transit time of the acoustic waves in the tube of order 10^{-2} s and the flame transit time of order 1 s. Moreover, one also has to consider the following space scales: the tube length (100 cm), the flame thickness (10^{-2} cm) and the reaction rate thickness (10^{-3} cm). The appearance of multiple space scales is traditionally adressed by the use of adaptive mesh algorithms one example is given in section 5. In this paper we concentrate on the multiple time scales problem and present some results on implicit schemes designed to overcome the difficulties associated with the equations of compressible combustion in the laminar subsonic regime.

This paper is organized as follows: Section 2 deals with the mathematical model and the finite volume/finite element method used for the spatial approximation. Then, in Section 3, a fully linearized implicit non factored scheme is described and analyzed. This scheme is second order accurate for steady-state computations but only first order accurate for unsteady problems. This can be a drawback for some strong transient problems and in Section 4, we present some preliminary results on a second-order accurate (in space and time) semi-implicit scheme specially designed for low Mach number computations. Finally, the paper ends with an example of the computation of a complex reactive flow involving the mixing of combustion processes with aerodynamics phenomena.

2. Governing Equations and Spatial Discretization: A thorough description of the mathematical model and of the spatial approximation used in this study is provided in [1]. These two points are briefly recalled here, to make this paper self contained. The set of equations governing the conservation of mass, momentum, energy and partial density of the reactant is:

$$W_t + \nabla.\mathbf{F}(W) + \nabla.\mathbf{R}(W, \nabla W) + S(W) = 0 \tag{1}$$

where

$$W = \begin{pmatrix} \rho \\ \rho u \\ \rho v \\ E \\ \rho Y \end{pmatrix}, \quad S = \begin{pmatrix} 0 \\ 0 \\ 0 \\ -Q\Omega \\ \Omega \end{pmatrix}, \quad \mathbf{F} = (F_1, F_2) \text{ with } F_1 = \begin{pmatrix} \rho u \\ \rho u^2 + p \\ \rho u v \\ (E+p)u \\ \rho Y u \end{pmatrix}, \quad F_2 = \begin{pmatrix} \rho v \\ \rho u v \\ \rho v^2 + p \\ (E+p)v \\ \rho Y v \end{pmatrix}$$

$$\mathbf{R} = (R_1, R_2) \quad \text{with} \quad R_1 = \begin{pmatrix} 0 \\ 0 \\ 0 \\ -D_T \partial T / \partial x \\ -D_Y \partial Y / \partial x \end{pmatrix}, \quad R_2 = \begin{pmatrix} 0 \\ 0 \\ 0 \\ -D_T \partial T / \partial y \\ -D_Y \partial Y / \partial y \end{pmatrix}$$

In the above equations, classical notations have been used: ρ is the mixture density, u and v are the components of the velocity, p is the pressure, E is the sum of the thermal and kinetic energies per unit volume, D_T is the mixture thermal conductivity, T the temperature, Q the heat released by the chemical reaction and Ω is the rate at which this reaction proceeds. Moreover, Y is the mass fraction of the reactant R and D_Y is its molecular diffusion coefficient. This model makes the assumption of a single overall reaction of the form $R \rightarrow P$ and neglects the viscous terms in the momentum and energy equations. Neverthless, it retains the essential features of the full Navier-Stokes equations. The spatial approximation of this set of equations uses a finite-volume/finite-element approach derived from [2]. First a control volume C_i is constructed around each vertex i of a finite element triangulation $\mathcal{T}$ (see Fig. 1). We denote by ∂C_i the boundary of the cell C_i and by $\vec{\nu_i}$ the outward unit normal on ∂C_i. After integration of the equations on each control volume and use of the Green formula we obtain:

$$\iint_{C_i} W_t \, dxdy + \iint_{C_i} S \, dxdy + \int_{\partial C_i} \mathbf{F}.\vec{\nu}_i \, d\sigma + \int_{\partial C_i} \mathbf{R}.\vec{\nu}_i \, d\sigma = 0 \tag{2}$$

The following quadrature formula are used in (2):
<u>temporal and source terms</u>:

$$\iint_{C_i} W_t \, dxdy = area(C_i)(W_i)_t \qquad \iint_{C_i} S(W) dxdy = area(C_i)S(W_i) \tag{3}$$

<u>diffusive terms</u>:

$$\int_{\partial C_i} \mathbf{R}(W).\vec{\nu} \, d\sigma = \sum_{\tau \in T(i)} \mathbf{R}|_\tau. \int_{\partial C_i \cap \tau} \vec{\nu}_i d\sigma \tag{4}$$

where $T(i)$ is the set of triangles around vertex i and $\mathbf{R}|_\tau$ is the value of $\mathbf{R}$ obtained by a P1 interpolation in each triangle τ.
For the hyperbolic terms the simplest expression:

$$\int_{\partial C_i} \mathbf{F}.\vec{\nu}_i \, d\sigma = \sum_{j \in \mathcal{K}(i)} \mathbf{F}|_{ij}. \int_{\partial C_{ij}} \vec{\nu}_i d\sigma \quad \text{with} \quad \mathbf{F}|_{ij} = \frac{1}{2}(\mathbf{F}_i + \mathbf{F}_j) \tag{5}$$

where $\kappa(i)$ is the set of neighbors of vertex i and $\partial C_{ij} = G_1 I G_2$ (see Fig. 1), leads to a centered scheme. In the absence of viscosity (either physical or numerical) this scheme has an unpleasant oscillatory behavior; a better approximation is provided by an upwind-centered scheme. Thus in this study, we have used Roe's flux function [3]:

$$\Phi(W_i, W_j) = \mathbf{F}|_{ij}. \int_{\partial C_{ij}} \vec{\nu}_i d\sigma = \frac{F(W_i) + F(W_j)}{2}.\vec{\eta}_{ij} + \frac{1}{2}|\tilde{A}(W_i, W_j)|(W_i - W_j) \tag{6}$$

where $\vec{\eta}_{ij}$ is defined by $\vec{\eta}_{ij} = \int_{\partial C_{ij}} \vec{\nu}_i d\sigma$ and where $\tilde{A}(W_i, W_j)$ is a matrix close to the Jacobian $A = \partial(\mathbf{F}.\vec{\eta}_{ij})/\partial W$ that satisfies:

$$\tilde{A}(W_i, W_j)(W_i - W_j) = (\mathbf{F}(W_i) - \mathbf{F}(W_j)).\vec{\eta}_{ij} \tag{7}$$

The preceding scheme is only first order accurate, extension to second order accuracy is done by the MUSCL interpolation technique as in [4].

3. An implicit scheme: The temporal scheme can be written as:

$$\frac{W^{n+1} - W^n}{\delta t} + \nabla.F(W^{n+1}) + \nabla.R(W^{n+1}, \nabla W^{n+1}) + S(W^n) = 0 \tag{8}$$

which represents a nonlinear set of algebraic equations. Linear equations with the same temporal accuracy as (8) can be obtained by using the implicit linearized procedure of Beam & Warming [5].
• For the hyperbolic terms, the form of the flux (6) suggests the following approximate linearization, [6]:

$$
\begin{aligned}
\Phi^{n+1}(W_i^n, W_j^n, W_i^{n+1}, W_j^{n+1}) &= \frac{1}{2}[A(W_i^n)W_i^{n+1} + A(W_j^n)W_j^{n+1}] \\
&\quad + \frac{1}{2}|\tilde{A}(W_i^n, W_j^n)|(W_i^{n+1} - W_j^{n+1})
\end{aligned}
\tag{9}
$$

• The linearization of the diffusive terms make use of a first-order Taylor expansion:

$$R_{1,2}(W^{n+1}, W_x^{n+1}) = R_{1,2}(W^n, W_x^n) + P_{1,2}^n \delta W + Q_{1,2}^n \delta W_x + o(\delta t) \tag{10}$$

with the notations: $P_{1,2} = \partial R_{1,2}/\partial W$; $Q_{1,2} = \partial R_{1,2}/\partial W_x$; $\delta W = W^{n+1} - W^n$ and $\delta W_x = W_x^{n+1} - W_x^n$. In the case of constant diffusion coefficients, we note that $P_{1,2} - (Q_{1,2})_x = 0$ allows us to rewrite the diffusion terms as: $P_{1,2}\delta W + Q_{1,2}\delta W_x = (Q_{1,2}\delta W)_x = (Q\delta W)_x$ Collecting the expression (9) and (10), the implicit scheme (8) can now be written:

$$
\begin{aligned}
\delta W_i + \sigma_i \sum_{j \in K(i)} \Phi^{n+1}(W_i^n, W_j^n, \delta W_i, \delta W_j) &+ \sigma_i \sum_{\tau \in T(i)} \nabla(Q\delta W)|_\tau . \int_{\partial C_i \cap \tau} \vec{v}_i \, d\sigma \\
= -\sigma_i \sum_{j \in K(i)} \Phi(W_i^n, W_j^n) &- \sigma_i \sum_{\tau \in T(i)} R^n|_\tau . \int_{\partial C_i \cap \tau} \vec{v}_i \, d\sigma + \delta t S(W_i)
\end{aligned}
\tag{11}
$$

where σ_i stands for $\delta t/area(C_i)$. This scheme is first-order accurate, a simple way to increase its accuracy is to use in the right hand side of (11) (the explicit phase), the second order extension of [4] in the expression of the hyperbolic fluxes: this results in a second order accurate scheme in the case of steady state computations. Numerical experiments with this scheme are reported in [7].

4. A time split scheme for subsonic computations: The preceding scheme is only first-order accurate due to the spatial approximation of the hyperbolic terms in the implicit matrix. To reach second-order accuracy, one can use the same second-order upwind spatial approximation of the hyperbolic terms in the explicit and implicit phase of (11). However the size of the implicit matrix is larger than for a first order scheme (in 1-D one obtain a pentadiagonal matrix instead of a tridiagonal one) and this results in an increase of computer time. Thus we turned to a semi-implicit strategy. Although this study is only in its beginning, the preliminary results obtained with this type of scheme are very promising and indicate that this approach worths a thorough investigation.

The basic idea of this approach is to recognize that for subsonic flows, the acoustic waves only play a marginal role and the approximation of these phenomena can be done with a first-order accurate scheme without affecting the overall accuracy of the computation. Thus, we split the hyperbolic terms into two groups: an "acoustic part" whose approximation is implicit and first-order accurate and a "convective part" discretized by an explicit second order scheme. In addition, the implicit treatment of the "acoustic part" replace the very restrictive CFL stability criterion by a "convective CFL " stability criterion that also corresponds to a requirement of accuracy. To explain how the hyperbolic fluxes are splitted, let us consider as an example the 1-D Euler equations. Due to the homogeneity property of these equations we have:

$$F(W) = A(W)W \quad \text{with} \quad A = \frac{\partial F}{\partial W} = T\Lambda T^{-1} = T\begin{pmatrix} u & 0 & 0 \\ 0 & u+c & 0 \\ 0 & 0 & u-c \end{pmatrix} T^{-1}$$

and a splitting of the eigenvalue matrix:

$$\Lambda = \Lambda_u + \Lambda_c = \begin{pmatrix} u & 0 & 0 \\ 0 & 0 & 0 \\ 0 & 0 & 0 \end{pmatrix} + \begin{pmatrix} 0 & 0 & 0 \\ 0 & u+c & 0 \\ 0 & 0 & u-c \end{pmatrix} \tag{12}$$

in two matrices isolating the fast waves from the slow ones permits us to define a "convective " and an "acoustic" part of the flux by $F_u(W) = T\Lambda_u T^{-1}W$, $F_c(W) = T\Lambda_c T^{-1}W$:

$$F_u(W) = \frac{\gamma-1}{\gamma}\begin{pmatrix} \rho u \\ \rho u^2 \\ \rho\frac{u^3}{2} \end{pmatrix}, \quad F_c(W) = \begin{pmatrix} \dfrac{\rho u}{\gamma} \\ \dfrac{\rho u^2}{\gamma} + p \\ [\dfrac{p}{\gamma-1} + \dfrac{1}{2}\dfrac{\rho u^2}{\gamma}]u + pu \end{pmatrix}.$$

Using a similar procedure for the two-dimensional equations, a first-order semi-implicit scheme can be defined by:

$$\delta W + \sigma_i \sum_{j\in\mathcal{K}(i)} \Phi_c^{n+1}(W_i^n, W_j^n, \delta W_i, \delta W_j) + \sigma_i \sum_{j\in\mathcal{K}(i)} \Phi(W_i^n, W_j^n) = 0 \tag{13}$$

where Φ is the numerical flux function defined by (6) and Φ_c is defined by a formula similar to (9). A second-order extension of the preceding scheme is then realized by combining the two step explicit Hancock-van Leer (HVL) scheme [8] for the "convective" part of the flux with the implicit upwind scheme for the "acoustic" part. This results in the following two steps semi-implicit scheme:

• explicit centered predictor step (done in primitive variables $\tilde{W}$):

$$\tilde{W}_i^* = \tilde{W}_i - \frac{\Delta t}{2}\tilde{A}_u(\tilde{W}_i)\tilde{W}_{zi} + \tilde{A}_v(\tilde{W}_i)\tilde{W}_{yi} \tag{14}$$

where $\bar{W}_{xi}, \bar{W}_{yi}$ are the components of the averaged gradient at vertex i defined by:

$$\nabla \bar{W}_i = \frac{1}{area[T(i)]} \int \int_{T(i)} \nabla \bar{W} \, dx dy \tag{15}$$

•interpolation : $\bar{W}_{ij}^* = \bar{W}_i^* + \frac{1}{2}\overrightarrow{ij}.\tilde{T}\tilde{T}_u^{-1}\nabla \bar{W}_i$

$$\bar{W}_{ji}^* = \bar{W}_j^* + \frac{1}{2}\overrightarrow{ji}.\tilde{T}\tilde{T}_u^{-1}\nabla \bar{W}_i$$

where the 4×4 matrix $\tilde{T}_u^{-1}$ is obtained from the matrix of left eigenvectors $\tilde{T}^{-1}$ by replacing the last two rows by zeroes. So, the interpolation acts only on the "convective" part of the fluxes.

• semi-implicit corrector step (in conservative variables):

$$\delta W^{n+1} + \sigma_i \sum_{j \in \mathcal{K}(i)} \Phi_c^{n+1}(W_i^n, W_j^n, \delta W_i, \delta W_j) = -\sigma_i \sum_{j \in \mathcal{K}(i)} \Phi(W_{ij}^*, W_{ji}^*, W_i^n, W_j^n) \tag{16}$$

where Φ is now defined by:

$$\begin{aligned} 2\Phi = \ & A_u(W_{ij}^*)W_{ij}^* + A_u(W_{ji}^*)W_{ji}^* + |A_u(W_{ij}^*, W_{ji}^*)|(W_{ij}^* - W_{ji}^*) \\ & A_c(W_{ij}^*)W_i^n + A_c(W_{ji}^*)W_j^n + |A_c(W_{ij}^*, W_{ji}^*)|(W_i^n - W_j^n) \end{aligned} \tag{17}$$

Note that if we linearize the equations by frozing the matrices coefficients, the above scheme exactly results in the HVL scheme applied to the slow characteristic variables and the first order upwind scheme applied to the fast characteristic variables.

We now compare the behavior of this semi-implicit scheme with the one described in Section 3 with respect to both acoustic and convective phenomena. For our first numerical test, we consider the damping of an acoustic wave in a closed square vessel. In Fig. 2 are compared the pressure plots at $t = 3/4$ period obtained by the second-order explicit HVL scheme, the scheme described in section 3 and the semi-implicit scheme. The plots are almost identical for the two implicit schemes indicating that they behave in a similar way with respect to the acoustic phenomena. Next, we investigate the "convective" behavior of each scheme by considering the propagation of a density hill by a constant velocity field. In this case the exact solution of the Euler equations simply corresponds to a translation of the initial solution. Fig. 3 compares the solution obtained with the three schemes. The plots illustrate the increase of accuracy resulting from the use of the semi-implicit scheme: The solution obtained with this scheme really looks like a second-order solution indicating that the splitting (12) is effective in isolating the fast from the slow waves. On the contrary, the result obtained with the implicit scheme is not really second-order accurate due to the first order spatial approximation used in the implicit matrix. One can note that the results obtained with the semi-implicit scheme are even better that those obtained with the second-order explicit HVL scheme (because the semi-implicit scheme allows the use of a "convective" CFL close to one) . Furthermore, the semi-implicit scheme is 25 times faster than the HVL scheme.

5. A case study: The tulip flame: We conclude this study by the computation of an instability of flame front propagation in a closed tube, namely the tulip flame problem. This complex phenomenon has been the subject of several experimental studies and from the numerical standpoint, have been studied in [11] and [9]. At the present time, the physics of this instability is only partially understood and numerous reasons have been invoqued to explain it. Indeed, our numerical experiments reveal that this behavior of the flame front is rather common and appears in very simple situations. Thus reasons like complex chemistry or boundary layer effects due to heat loss at the wall are to be eliminated. In our experiments, a closed rectangular adiabatic box of 100 x 25 (asymptotic) flame thickness was used. The time integration of the system (1) is realized by the implicit scheme described in Section 3. To correctly resolve the flame front, a dynamic mesh adaptation algorithm similar to the one described in [12] is employed except that the "rezone" step is now performed with a second-order upwind ALE procedure inspired from [13]. First we compute the propagation of the flame inside the tube from its ignition at the center of the right hand side of the vessel. The developpement of the tulip flame is illustrated Fig. 4. (a), (b) and (c) where a 41 x 51 mesh was used. First, the flame develops from a semi-circular shape. After interacting with the wall, it flattens and propagates at an almost constant speed under the form of an approximately plane flame front. Then a cusp appears at the center of the flame front. Figures 4 (b) shows the change of sign of the vorticity when the curvature of the flame front inverts. Solid lines are used to represent positive isocontours of vorticity and dotted lines for negative ones. The corresponding velocity fields are shown by figure 4 (c). Next we use a 31 x 21 mesh and begin the computation by chosing as initial condition a plane flame front established at the right hand side of the tube. As can be seen in Fig. 5, during its propagation towards the left, the front develops quite soon a cusp turned towards the fresh gases. Thus this experiment suggests that the development of a tulip flame does not depend on the initial stage of formation of a flame but rather involves a fundamental instability mecanism of the plane flame front. Concerning the adaptive algorithm, observe that the mesh (see Fig. 5 b) follows closely the flame front in its displacement and that the mesh is rather irregular. However, as shows by the smoothness of the contour plots the spatial approximation is robust enough to treat this case. Moreover although the adaptive algorithm is done line by line, we have found that no averaging was necessary between the different horizontal lines.

References

[1] A. HABBAL, A. DERVIEUX, H. GUILLARD, B. LARROUTUROU, "Explicit calculation of reactive flows with an upwind finite-element hydrodynamical code", Rapport INRIA n° 690, (1987).

[2] G. VIJAYASUNDARAM, "Transonic flow simulations using an upstream-centered scheme of Godunov in Finite-Element", J.C.P., 63, 1986.

[3] P. L. ROE, "Approximate Riemann solvers, parameter vectors and difference schemes", J.C.P., 43, p. 357, (1981)..

[4] F. FEZOUI, "Résolution des équations d'Euler par un schéma de Van Leer en éléments finis", INRIA Report n° 358, (1985).

[5] R. F. WARMING, R. M. BEAM, "On the construction and application of implicit factored schemes for conservation laws", SIAM AMS Proceedings, Volume 11, (1978).

[6] B. STOUFFLET, "Résolution numérique des équations d'Euler des fluides parfaits compressibles par des schémas implicites en éléments finis", Thèse, Univ. de Paris, (1984).

[7] G. FERNANDEZ, H. GUILLARD, "An Implicit Method for the Computation of Reactive Flows", 12th IMACS WORLD CONGRESS, July 1988.

[8] B. VAN LEER, "Computational methods for ideal compressible flows", von Karman Institute for Fluids Dynamics, Lecture series 1983-04 , 1983

[9] D. DUNN-RANKIN, P.K. BARR, R.F. SAWYER, "Numerical and Experimental Study of 'Tulip' Flame Formation in a Closed Vessel", Twenty-first Symposium (International) on Combustion, The Combustion Institute, pp 1291-1301, 1986.

[10] F.H. HARLOW, A.A. AMSDEN, "A Numerical Dynamics Calculation Method for All Flow Speeds", J.C.P., 8 pp 197-213, 1971.

[11] L.D. CLOUTMAN, "Numerical Simulation of Turbulent Premixed Combustion and Decay of Turbulent Swirling Flow", Lawrence Livermore National laboratory report L-035, 1988.

[12] B. LARROUTUROU, "Etude Mathématique et Modélisation Numérique de Phénomènes de Combustion", Thesis Paris XIII, 1987.

[13] B. PALMERIO, " A Consistant ALE Rezoned Mesh Adaption Algorithm for Compressible Finite Element Calculations", INRIA Report n° 829, 1988.

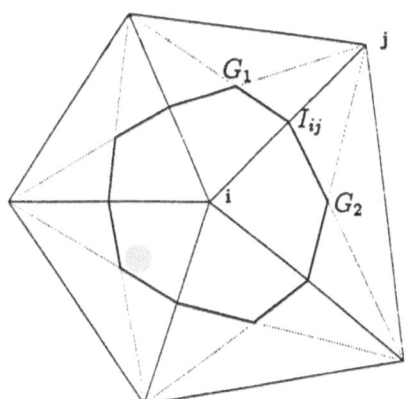

Figure 1: Cell C_i

Initialisation:

$(x, y) \in [0, 1] \times [0, 1]$

$\rho(x, y, 0) = 1.$
$u(x, y, 0) = \sin(\pi x)$
$v(x, y, 0) = \sin(\pi y)$
$p(x, y, 0) = 14285.70$

Explicit $O(\Delta x^2)$, $O(\Delta t^2)$, cfl=0.8

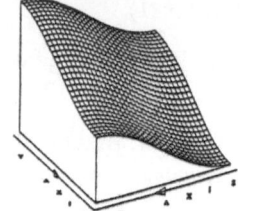

min=14004.44 , max=14572.63

Implicit $O(\Delta x)$, $O(\Delta t)$

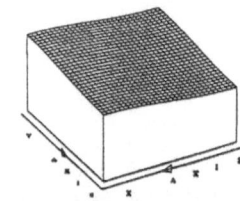

min=14237.66 , max=14334.50

Semi-implicit $O(\Delta x^2)$, $O(\Delta t^2)$

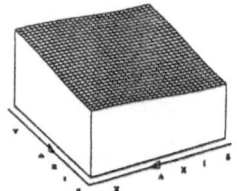

min=14237.62 , max=14334.46

Figure 2 : Acoustics, pressure surface $t = \dfrac{3}{4}$ period, cfl=10.

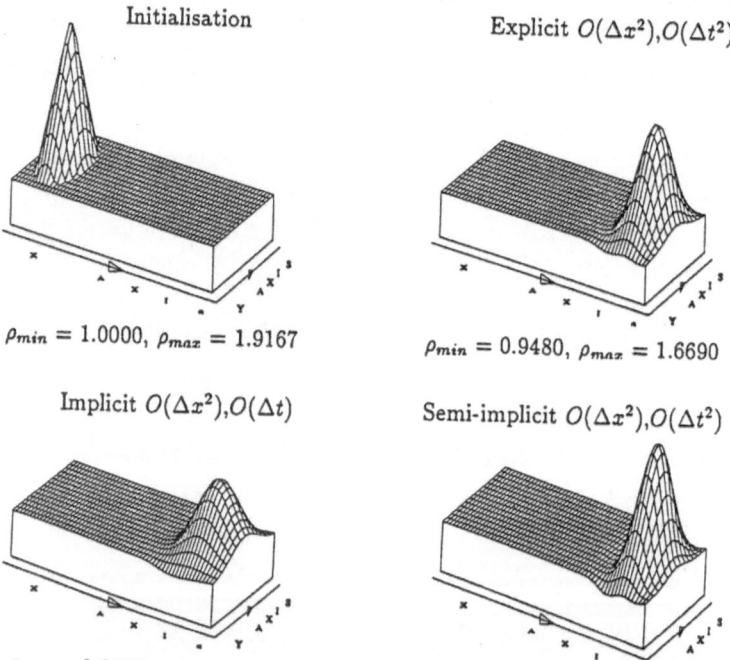

Initialisation

$\rho_{min} = 1.0000$, $\rho_{max} = 1.9167$

Explicit $O(\Delta x^2)$, $O(\Delta t^2)$

$\rho_{min} = 0.9480$, $\rho_{max} = 1.6690$

Implicit $O(\Delta x^2)$, $O(\Delta t)$

$\rho_{min} = 0.9555$, $\rho_{max} = 1.4482$

Semi-implicit $O(\Delta x^2)$, $O(\Delta t^2)$

$\rho_{min} = 0.9408$, $\rho_{max} = 1.7223$

Figure 3 : Propagation of a density hill , $Ma = 8.5\ 10^{-3}$.

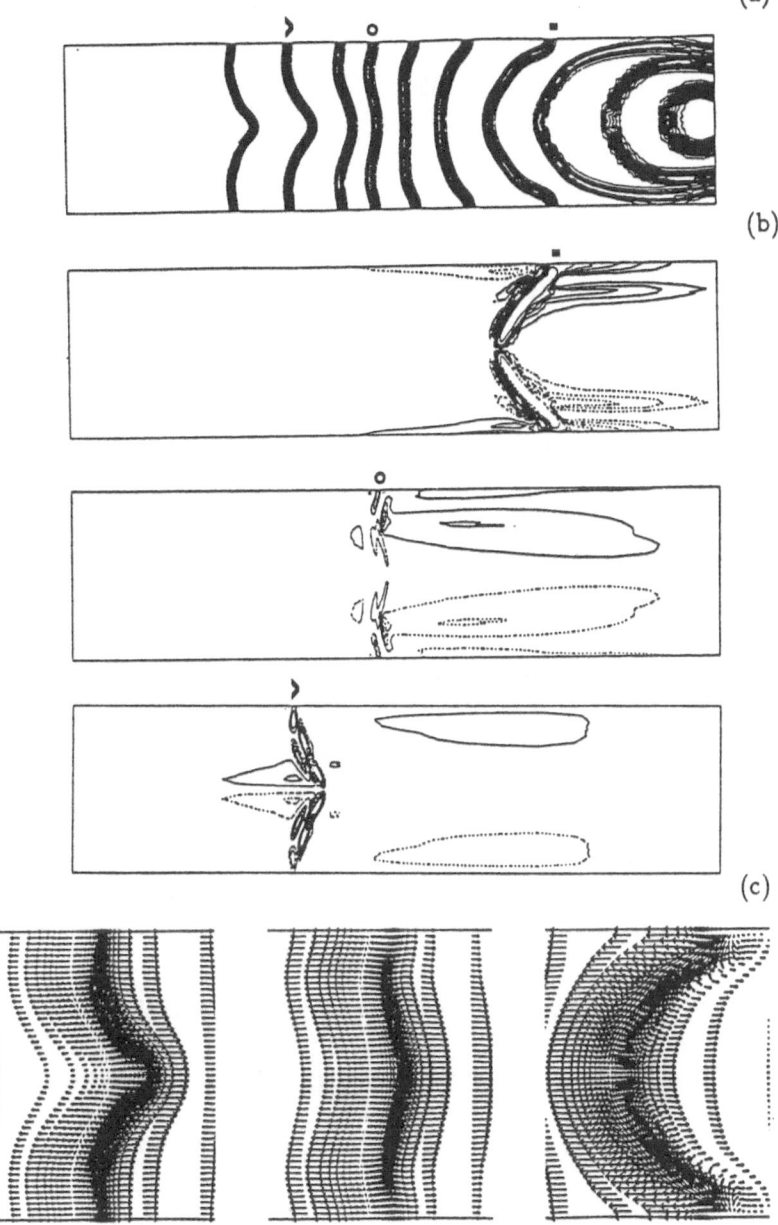

Figure 4: Tulip flame, (a) history of reaction rate

(b) vorticity

(c) velocity field

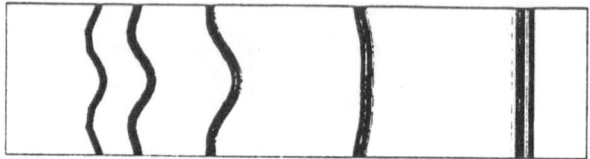

(a) Isocontours of the reaction rate

$t = 0$	$kt = 0$	$Max = 59$
$t = 0.079$	$kt = 240$	$Max = 119$
$t = 0.182$	$kt = 700$	$Max = 215$
$t = 0.262$	$kt = 1200$	$Max = 309$
$t \doteq 0.313$	$kt = 1600$	$Max = 428$

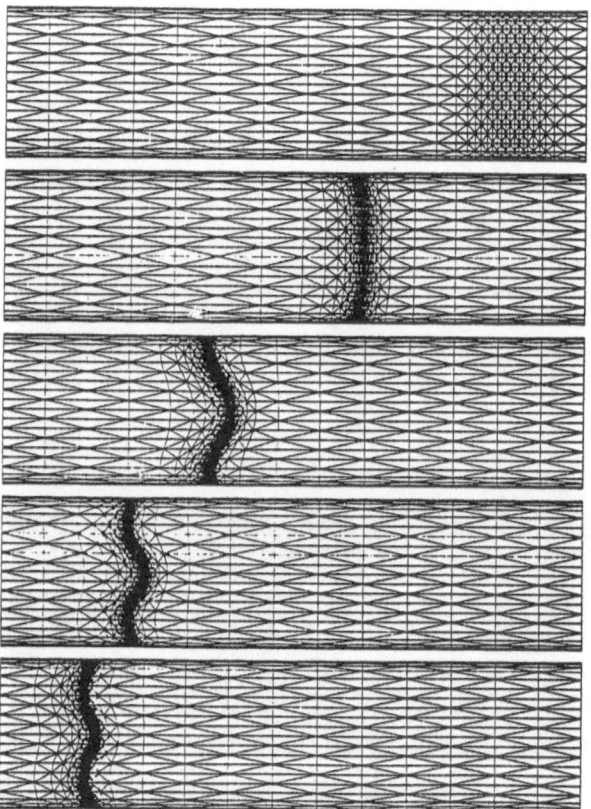

(b) Adaptive mesh at different times.

Figure 5: Tulip flame, plane flame front initialisation.

NUMERICAL STUDY OF CURVATURE EFFECTS
ON FLAME QUENCHING

G. FERNANDEZ, B. LARROUTUROU
INRIA Sophia-Antipolis
06560 Valbonne FRANCE

G.I. SIVASHINSKY
The Levich Institute
The City College of New-York
New-York NY 10031 USA

Abstract

The problem of inward propagating spherical premixed flames is studied numerically. It is shown that at Lewis number exceeding unity the burning rate and the flame speed gradually increase as the spherical flame approaches its center. At small Lewis numbers the picture is quite the opposite. The burning rate and the flame speed slow down and at some finite distance from the center the flame goes out. It is found that despite the quenching the flame manages to consume all the fuel available. There is no residual reactant after extinction. The cases of cylindrical and planar geometries are also discussed.

1 INTRODUCTION

The objective of this paper is to contribute to the investigation of the effects of flame curvature on flame extinction; such quenching effects appearing in relation with strong flame curvature can for instance be observed at the tip of a Bunsen flame (see [5]: for a Lewis number smaller than unity, the Bunsen flame opens up at the tip). To this end, we perform a detailed numerical study of inward propagating flames in spherical, cylindrical or planar geometries, in the framework of the constant density approximation, and carefully examine how the flame extinction occurs for small Lewis numbers.

Several formal studies have been proposed in the literature to describe this class of phenomena. The propagation of a converging spherical flame was investigated in [10]; somewhat different conclusions were obtained in [5]. We will briefly recall below the main hypotheses and results of these asymptotic studies, and see how the latter compare with our numerical results.

The inward propagation of a spherical flame has also been studied numerically by Flaherty, Frankel, Roytburd and Sivashinsky [4]. The results presented here extend their results, and allow us to give more detailed answers to the following essential questions: in a spherical converging flame with Lewis number less than unity, does the flame extinction occur at finite radius ? Does extinction occur with a positive flame speed ? Is there any finite residual of unburnt reactant ? How are the answers to these questions qualitatively or quantitavely affected by the change in the geometry (from spherical symmetry to cylindrical or planar symmetry) ?

We will first write the governing equations used to describe flame propagation and briefly recall the analytical results of [10] and [5] in Section 2. Then we describe in Section 3 our numerical approach, which uses an adaptive moving grid and a semi-implicit time integration; the numerical results are presented and discussed in Section 4.

2 FORMULATION

2.1 Constant density model

We consider a premixed flame with a single one step chemical reaction $\mathcal{R} \longrightarrow \mathcal{P}$. We will use the classical approximations of flame propagation theory: we neglect the viscous and external forces (such as gravity), we also assume that the spatial variations of the pressure are small (low Mach number approximation), and lastly we use the constant-density approximation, which essentially holds in the limit of a weak heat release. We refer to e.g. [2], [3], [9], [11] for the presentation and discussion of these assumptions.

With these hypotheses, the problem reduces to the simplified reaction-diffusion system:

$$
\begin{cases}
\rho C_p T_t = \lambda(\dfrac{\partial^2 T}{\partial r^2} + \dfrac{a}{r}\dfrac{\partial T}{\partial r}) + Q\omega(T, Z) \,, \\[4mm]
\rho Z_t = \rho D_Z(\dfrac{\partial^2 Z}{\partial r^2} + \dfrac{a}{r}\dfrac{\partial Z}{\partial r}) - m\omega(T, Z) \,,
\end{cases}
\tag{1}
$$

$$
\omega(T, Z) = A\frac{\rho Z}{m} \exp\left(\frac{E}{R^0 T}\right) \,,
\tag{2}
$$

associated with the boundary conditions:

$$
\frac{\partial T}{\partial r}\big|_{r=+\infty} = 0 \,, \quad \frac{\partial Z}{\partial r}\big|_{r=+\infty} = 0 \qquad ; \qquad \frac{\partial T}{\partial r}\big|_{r=0} = 0 \,, \quad \frac{\partial Z}{\partial r}\big|_{r=0} = 0
\tag{3}
$$

and with the initial conditions:

$$
T(r, 0) = T^0(r) \,, \quad Z(r, 0) = Z^0(r) \,.
\tag{4}
$$

In these equations, T is the temperature of the mixture, Z is the mass fraction of the reactant $\mathcal{R}$, ρ is the density of the mixture (assumed to be constant), and r is the spatial coordinate; the parameter a depends on the geometry ($a = 0; 1; 2$ for the planar, cylindrical and spherical geometry, respectively). Moreover, ω is the chemical reaction rate, A is the Arrhenius prefactor, E is the activation energy of the reaction and R^0 is the universal gas constant. Lastly, λ and D_Z denote the thermal conductivity and the molecular diffusion coefficient respectively, C_p is the specific heat at constant pressure of the mixture, Q is the heat release and m is the molecular weight of the reactant. The quantities C_p, λ, Q, D_Z, m, A and E are assumed to be constant.

Introduce the following dimensionless quantities:

$$\alpha = \frac{T_b - T_u}{T_b} , \quad \beta = \frac{E}{R^0 T_b} \frac{T_b - T_u}{T_b} \tag{5}$$

Here $T_b = T_u + \dfrac{Q Z_u}{m C_p}$ is the adiabatic temperature of the burnt gas, T_u and Z_u being the temperature and the mass fraction of the fresh mixture ; β is the so-called Zeldovich number. As units of length and velocity we use the flame thickness L_f and the normal flame speed V_f given by the high activation energy asymptotics; see e.g. [2], [3]):

$$L_f = \sqrt{\frac{\lambda}{2 A \mathcal{L} e \rho C_p}} \beta \exp\left(\frac{\beta}{2\alpha}\right) , \tag{6}$$

$$V_f = \sqrt{\frac{2 A \mathcal{L} e \lambda}{\rho C_p}} \frac{1}{\beta} \exp\left(-\frac{\beta}{2\alpha}\right) . \tag{7}$$

It is also convenient to introduce the normalized temperature Θ and mass fraction Y defined as:

$$\Theta = \frac{T - T_u}{T_b - T_u} , \quad Y = \frac{Z}{Z_u} , \tag{8}$$

The system (1)-(2) then can be rewritten in the following form:

$$\begin{cases} \Theta_t = \dfrac{\partial^2 \Theta}{\partial r^2} + \dfrac{a}{r} \dfrac{\partial \Theta}{\partial r} + \Omega(\Theta, Y) , \\[4mm] Y_t = \dfrac{1}{\mathcal{L} e}(\dfrac{\partial^2 Y}{\partial r^2} + \dfrac{a}{r} \dfrac{\partial Y}{\partial r}) - \Omega(\Theta, Y) , \end{cases} \tag{9}$$

$$\Omega(\Theta, Y) = \frac{\beta^2}{2 \mathcal{L} e} Y \exp\left(-\frac{\beta(1 - \Theta)}{1 - \alpha(1 - \Theta)}\right) . \tag{10}$$

This is the final model which will be the subject of our numerical study.

2.2 High activation energy asymptotics

The well-known starting point of any analytical work on flame propagation is that, in the limit $\beta \to \infty$, the reaction rate is negligible everywhere except in a very thin reaction zone, whose thickness is of the order of $\dfrac{L_f}{\beta}$. Thus, as $\beta \to \infty$ the reaction term (10) may be replaced by a Dirac delta function:

$$\Omega = \hat{\Omega} \delta(r - r_f) . \tag{11}$$

In this spirit, two studies dealing with inward propagating spherical flames have been proposed in the literature; we now briefly recall their results.

The slow-varying flame model [10]: The analysis in [10] uses instead of (1)-(2) an isobaric model [that is, a momentum equation and an equation of state of the form $\rho T = Constant$ are used in addition to (1)]; then, asymptotic expansions in powers of β^{-1} are sought for all variables, including the flame radius r_f appearing in (11). It is then shown that, in the leading order, the flame radius satisfies the second-order differential equation:

$$\mathcal{A}\ddot{R} + 2\dot{R}^3 \text{Log}(\dot{R}) = 2\frac{\mathcal{A}}{R}\dot{R} \,, \tag{12}$$

where $R = r_f$ and $\mathcal{A} = \beta\dfrac{1 - \mathcal{L}e}{\mathcal{L}e}$.

The quasi-steady flame model [5]: Starting from the constant-density model (1)-(2) and using high activation energy asymptotics and the additional assumptions that (i) the flame thickness is small compared to the flame radius, and that (ii) the flame structure is quasi-stationary in the frame of reference attached to the flame, Frankel and Sivashinsky [5] have obtained the following first-order differential equation for the flame radius:

$$\left(\dot{R} + \frac{2}{R}\right)\text{Log}\left[-\left(\dot{R} + \frac{2}{R}\right)\right] = \frac{1 - \mathcal{L}e}{R}\beta \,. \tag{13}$$

For $\mathcal{L}e < 1$ this equation yields the following expressions for the flame radius and velocity at the extinction point:

$$R_{ext} = \beta(1 - \mathcal{L}e)e \,, \quad \dot{R}_{ext} = e^{-1}\left(1 + \frac{2}{\beta(1 - \mathcal{L}e)}\right) \,. \tag{14}$$

3 NUMERICAL METHOD

We now present the adaptive numerical method used to solve problem (9)-(10).

We wish to simulate (with a non prohibitive computer time) an experiment where the flame front initially starts very far from the center of symmetry (i.e. $R \gg 1$) and eventually converges towards this center ($R \to 0$). For this purpose, we essentially use the adaptive gridding procedure proposed by B. Larrouturou [6], [7]. The main features of the method are (i) a dynamic adaption performed at each time step and allowing the nodes locations to vary smoothly with time, and (ii) a static adaption performed only at a few time levels during the computation in order to achieve a better redistribution of the nodes, i.e. a better adaptation of the grid to the solution. The method uses a finite-difference spatial approximation, and a semi-implicit time integration scheme which is employed in order to avoid a too severe stability restriction on the time step.

Some aspects of the numerical method are described below; the reader is referred to [6], [7] for more details.

3.1 Mesh strategy

The static adaption procedure used here is exactly similar to the one described in [7]. When this procedure is used, we evaluate a mesh function w which is essentially based on the leading order terms of the truncation error of the discrete scheme, and then construct a new grid (r_i^{new}) by equidistributing w, i.e. by requiring:

$$\int_{r_i^{new}}^{r_{i+1}^{new}} w = Constant \tag{15}$$

for all node index i. The solution is then projected from the old mesh onto the new mesh using the *conservative* interpolation described in [8], which preserves the positivity and monotonicity of the variables.

The present case also requires some modifications of the dynamic adaption procedure adopted in [6], [7]. We begin the calculation with a computational domain $[r_L^0, r_R^0]$ which does not contain the origin (i.e. $r_L^0 > 0$). As long as the flame front remains far enough from the origin, the flame structure appears to be quasi-stationary: as in [6], [7], we then move at each time level t all the nodes with the same velocity $V_0(t)$, which is chosen equal to the instantaneous average flame speed, given by:

$$V_0 = -\frac{1}{\Theta(r_R) - \Theta(r_L)} \int_{\mathcal{D}} \left(\Omega + \frac{a}{r} \Theta_r \right) dr \; ; \tag{16}$$

(in this expression, $\mathcal{D} = [r_L, r_R]$ is the moving computational domain). In practice, we apply this simple and efficient strategy (whose effect is to keep the flame at the same place inside the computational domain) as long as $r_L > 0$.

Once the domain has reached the origin, we separate it into three segments $[0, r_L']$, $[r_L', r_R']$, $[r_R', r_R]$ (r_L' and r_R' being chosen on each side of the flame front) and we define the velocity $V(r_i)$ of each node as a function of r which is linear in each of the three intervals $[0, r_L']$, $[r_L', r_R']$, $[r_R', r_R]$ and has the values:

$$V(0) = 0 \; , \quad V(r_L') = V_0 \; , \quad V(r_R') = V_0 \; , \quad V(r_R) = 0 \; , \tag{17}$$

V_0 being given by (16). This procedure has the advantage that all nodes in the neighbourhood of the flame still move with a velocity equal to the instantaneous average flame speed. Moreover, it has appeared (from numerical observation) to be important to keep the computational domain sufficiently long once it has reached the origin ($r_L = 0$), in particular because the flame becomes thicker and thicker when quenching occurs. This is why we keep the length of the domain $\mathcal{D}$ constant during the whole computation by setting $V(r_R) = V(r_L)$.

When r_L' itself becomes small (i.e. when the flame front is very close to the origin), we progressively decelerate the nodes [we substitute a fraction of V_0 instead of V_0 in (17)] in order to avoid large discrepancies between the velocities of neighboring nodes and to prevent all nodes coalescing towards the origin.

3.2 Approximation

Previous studies have shown that, when a fully explicit scheme is used for the present problem (which in particular does not involve complex chemistry), the most severe restriction on the time step comes from the diffusive terms. Since we intend to describe accurately the transient propagation of the flame, we do not attempt to use time steps which would be larger than the characteristic time of the chemical reaction (we refer to e.g. [1] for a discussion of this question). Thus, we use the following time integration scheme, where only the diffusive terms are treated implicitly:

$$
\begin{cases}
\dfrac{\Theta_i^{n+1} - \Theta_i^n}{\Delta t} = \delta_{rr}[(1-\theta)\Theta_i^{n+1} + \theta\Theta_i^n] + (\dfrac{a}{r_i^n} + V_i^n)\delta_r\Theta_i^n + \Omega_i^n \, , \\[4mm]
\dfrac{Y_i^{n+1} - Y_i^n}{\Delta t} = \dfrac{1}{\mathcal{L}e}\delta_{rr}[(1-\theta)Y_i^{n+1} + \theta Y_i^n] + (\dfrac{a}{r_i^n} + V_i^n)\delta_r Y_i^n - \Omega_i^n \, , \\[4mm]
r_i^{n+1} = r_i^n + V_i^n \Delta t \, ;
\end{cases}
\tag{18}
$$

here, V_i^n is the velocity of the ith node at the nth time step, and where the operators δ_r and δ_{rr} denote classical three-points expressions of the first and second spatial derivatives respectively.

A stability analysis of this scheme has been carried out, providing the restrictions on the time step insuring the positivity of the variables and the mesh monotonicity.

3.3 Flame radius evaluation

In the next section we will compare the results obtained via the scheme (18) with the predictions of the analytical studies [10], [5]. For this purpose, we need to deduce from the computed temperature, mass fraction and reaction rate profiles a reasonable estimate of the flame radius R. A natural definition would be to evaluate R as the point where the computed reaction rate Ω is maximal. Unfortunately, such a choice is not satisfactory in practice. Due to the discrete nature of the numerical approach, the evaluated radius does not vary smoothly with time; in particular, important difficulties arise when we try to evaluate $\dot{R}$, or to plot $\dot{R}$ as a function of R.

To remedy this, we use the following new definition of the flame radius; we set:

$$
R = \frac{\displaystyle\int_{\mathcal{D}} r\Omega(r)\, dr}{\displaystyle\int_{\mathcal{D}} \Omega(r)\, dr} \, .
\tag{19}
$$

This formula provides a smoother definition of R; moreover, this expression reduces to $R = r_f$ in the asymptotic limit (11) where Ω is proportional to $\delta(r - r_f)$. Nevertheless, some averaging procedure is still needed to have a smooth evaluation of $\dot{R}$ (we use several consecutive values of R and compute $\dot{R}$ as the slope of the straight line which realizes the best approximation of these values in the least square sense).

In the next section we will also consider the curves $(\dot{R},R)$ obtained from the differential equations (12) and (13); these curves are computed using the classical fourth-order Runge-Kutta scheme.

4 NUMERICAL EXPERIMENTS

4.1 Lewis number effects

The quenching phenomenon is very sensitive to the Lewis number. Figure 1 demonstrates this fact by showing the evolution of the temperature, mass fraction and reaction rate profiles in the spherical case for $\mathcal{L}e = 0.4$, $\mathcal{L}e = 1$, and $\mathcal{L}e = 1.5$.

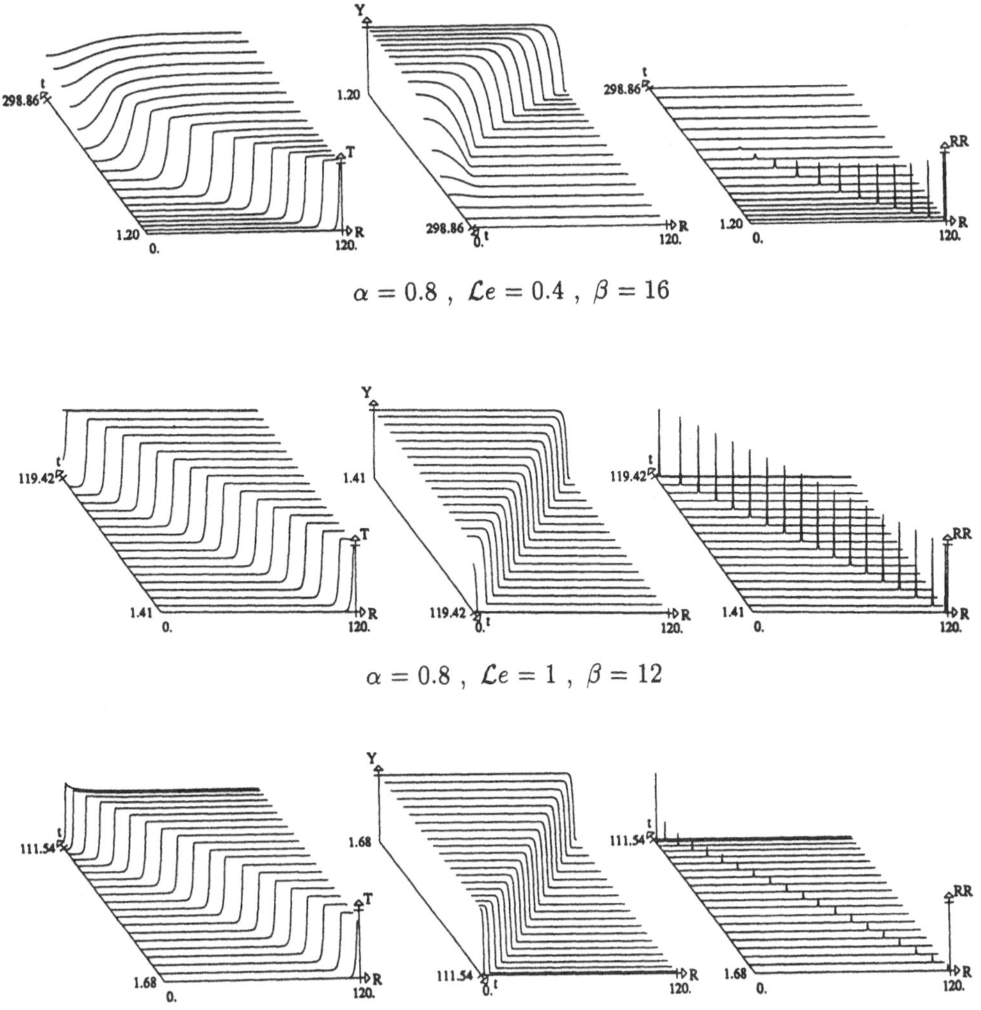

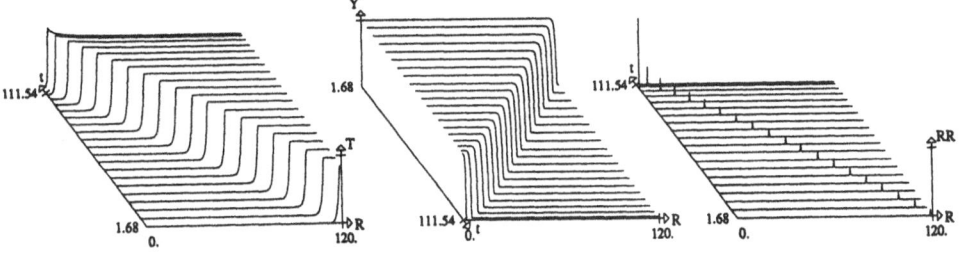

$$\alpha = 0.8 \ , \ \mathcal{L}e = 0.4 \ , \ \beta = 16$$

$$\alpha = 0.8 \ , \ \mathcal{L}e = 1 \ , \ \beta = 12$$

$$\alpha = 0.8 \ , \ \mathcal{L}e = 1.5 \ , \ \beta = 12$$

| Temperature profile | Mass fraction profile | Reaction rate profile |

Figure 1: Lewis number effects on quenching (spherical geometry).

The extinction of the flame at a positive flame radius is observed when $\mathcal{L}e = 0.4$. On the other hand, when $\mathcal{L}e = 1.5$ the peak value of the reaction rate continually grows ;

the temperature profile becomes sharper and sharper and the flame velocity increases. In the intermediate case $\mathcal{L}e = 1$, the flame retains the same structure propagating at constant velocity; no extinction is observed.

4.2 Geometrical effects

The planar, cylindrical or spherical geometry of the system also has an influence on flame quenching. Figure 2 shows the results corresponding to these three geometrical situations, the parameters α, β, and $\mathcal{L}e$ being kept fixed ($\alpha = 0.8$, $\beta = 12$, $\mathcal{L}e = 0.4$). It appears that quenching invariably occurs in all three cases, but is much more pronounced in the spherical flame where the curvature effect is stronger.

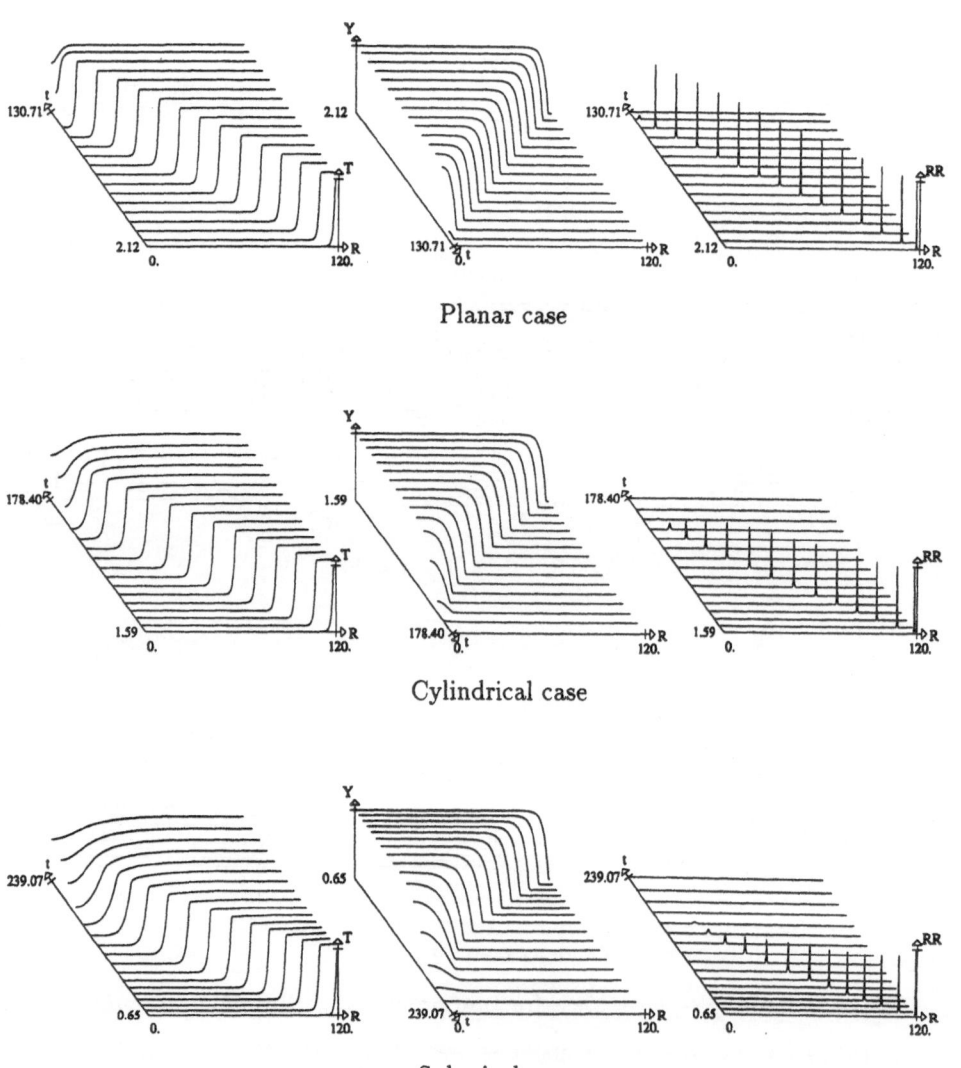

Planar case

Cylindrical case

Spherical case

Figure 2: Geometrical effects on quenching ($\alpha = 0.8$, $\beta = 12$, $\mathcal{L}e = 0.4$).

Different geometries also differ in their response to the parameter variations. For instance, the quenching radius is much more sensitive to the value of the Zeldovich number β in the spherical case than it is in the planar case.

Figure 3 shows that quenching actually exists in the plane case. In the spherical case with $\mathcal{L}e = 1$, the flame structure is conserved when the flame front reaches the origin, while quenching is clearly observed in the planar case with $\mathcal{L}e = 0.4$.

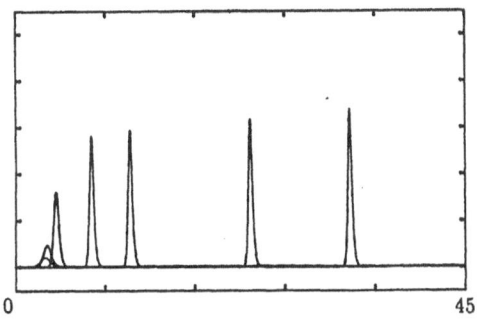
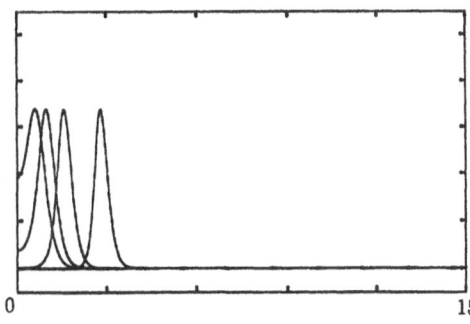

Planar case: $\mathcal{L}e = 0.4$, $\beta = 10$, $\alpha = 0.9$.   Spherical case: $\mathcal{L}e = 1$, $\beta = 10$, $\alpha = 0.8$.

Figure 3: Flame fronts near the origin.

4.3 Converging spherical flames

Restricting now our attention to the spherical case, we present in Table 1 the values of the flame radius R and flame speed $-\dot{R}$ at extinction, for different values of the reduced activation energy β. The computed values R_{ext} and $\dot{R}_{ext}$ can be compared to the values derived from the analytical studies [10], [5] (i.e. from equations (12), (13), (14)).

TABLE 1 : Flame radius and flame speed at extinction

β	$R_{ext}^{[10]}$	$\mid \dot{R}_{ext}^{[10]} \mid$	$R_{ext}^{[5]}$	$\mid \dot{R}_{ext}^{[5]} \mid$	R_{ext}	$\mid \dot{R}_{ext} \mid$
4	4.60	0.191	6.52	0.674	10	0.42
8	9.21	0.192	13.05	0.521	16	0.23
12	13.82	0.192	19.57	0.470	21	0.19
16	18.42	0.193	26.10	0.445	27	0.22

For the analysis proposed in [10], the extinction radius and velocity are taken at the inflexion point of the curve $(\dot{R},R)$ (see Figure 4). Similarily, in the numerical study, these values are calculated at the first inflexion point of the curve $(\dot{R},R)$. The oscillation appearing on this curve (a slight re-acceleration of the flame near extinction) has been observed in all numerical simulations, even defining the flame radius differently.

Table 1 shows that with regard to the extinction radius the analytical [5], [10] and numerical results are in quite a reasonable qualitative agreement (the correlation being better with the results of [5]). In all three cases the extinction takes place at a finite radius which becomes larger when the activation energy increases. With regard to the

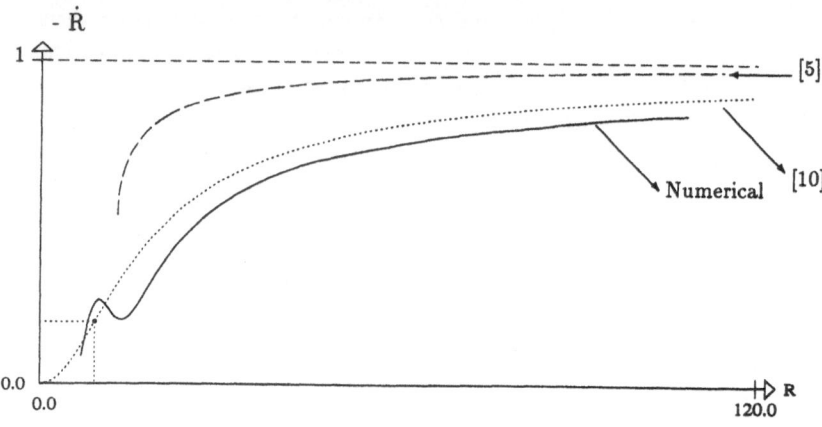

Figure 4: Phase diagram

extinction velocity, the numerical results are closer to [10], especially for high values of β. In all cases, the extinction occurs at a finite velocity.

In Figure 4 the phase diagram $(R,\dot{R})$ is shown for the analytical and numerical results corresponding to $\beta = 8$.

4.4 Concluding remarks

The numerical results show that the flame and the reaction zone thicknesses increase considerably as the flame approaches the origin. Thus, the basic premises of the asymptotic theories [5],[10] that the reaction zone is much smaller than the flame thickness and in turn, the flame thickness is much smaller than the flame radius are markedly violated near the extinction region. This may lead to a considerable loss of accuracy in the theoretical predictions. One of such effects (not covered by the asymptotic theories and apparently stemming from the widening of the reaction zone) is the full consumption of the deficient reactant. Inspite of the finite flame radius on extinction, there is no residual reactant after flame quenching (Figure 5).

It would be very interesting to understand whether this rather unexpected effect is merely a peculiar feature of a system with a limited amount of fuel involved or quite a generic phenomenon. For example, whether there exists any effective leakage of hydrogene through the open tip of the lean hydrogen-air Bunsen flame.

Acknowledgements: The third author was partially supported by the U.S. Department of Energy under grant No DE-FG02-88ER13822.

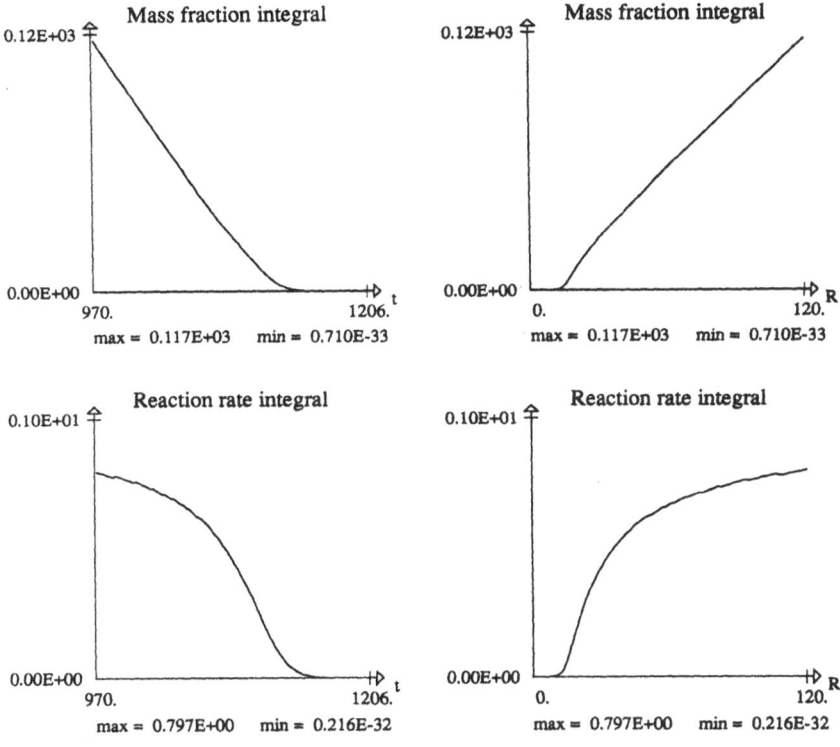

Figure 5

Mass fraction and reaction rate integrals
as functions of time and flame radius

References

[1] F. BENKHALDOUN, B. LARROUTUROU, "A finite-element adaptive investigation of curved stable and unstable flame fronts", to appear.

[2] J.D. BUCKMASTER & G.S.S. LUDFORD, "Theory of laminar flames", Cambridge Univ. Press, Cambridge, (1982).

[3] P. CLAVIN, "Dynamic behavior of premixed flame fronts in laminar and turbulent flows", Prog. Energ. Comb. Sci., 11, pp. 1-59, (1985).

[4] J.E. FLAHERTY, M.L. FRANKEL, V. ROYTBURD, G.I. SIVASHINSKY, "Numerical study of quenching of inward propagating spherical flames", Comb. Sci. Tech., 43, pp. 245-257, (1985).

[5] M.L. FRANKEL, G.I. SIVASHINSKY, "On quenching of curved flames", Comb. Sci. Tech., 40, pp. 257-268, (1984).

[6] B. LARROUTUROU, "Utilisation de maillages adaptatifs pour la simulation de flammes monodimensionnelles instationnaires", Numerical simulation of combustion phenomena, Glowinski, Larrouturou, Temam eds., pp. 300-312, Lecture Notes in Physics, 241, Springer Verlag, Heidelberg, (1985).

[7] B. LARROUTUROU, "Adaptive numerical methods for unsteady flame propagation", Combustion and chemical reactors, Ludford ed., pp. 415-435, Lecture in Appl. Math., 24, (2), AMS, Providence, (1986).

[8] B. LARROUTUROU, "A conservative adaptive method for unsteady flame propagation", SIAM J. Sci. Stat. Comp., (1989), to appear.

[9] B. LARROUTUROU, "Introduction to mathematical and numerical modelling in gaseous combustion", Applied Mathematics, Gordon and Breach, to appear.

[10] G.I. SIVASHINSKY, "On a converging spherical flame front", Int. J. Heat Mass Transfer, 17, pp. 1499-1506, (1974).

[11] G.I. SIVASHINSKY, "Instabilities, pattern formation and turbulence in flames", Ann. Rev. Fluid Mech., 15, pp. 179-199, (1983).

Computation of low Froude number turbulent diffusion flames

D. GARRETON, N. MECHITOUA, P.L. VIOLLET

EDF/DER/Laboratoire National d'Hydraulique, 6, quai Watier, 78400 Chatou, France.

ABSTRACT :

This study is related to the development of a model for the computation of pool fires of liquefied natural gas -sponsered by GAZ de FRANCE and SOCIETE NATIONALE d'EXPLORATION d'AQUITAINE-PRODUCTION. -As a first step, low Froude number flames which are similar to pool fires, are simulated.

A two-dimensional numerical model, HADES, using finite difference techniques, is presented : the combustion process involves a single chemical reaction, with a probability density function fitted using mean and RMS computed mixing rate values ; a modified discrete transfer method allows to determine radiative fluxes. The k-epsilon model is used for turbulence. Several tests about the soot production model show it remains the most sensitive point.

The model is compared to laboratory experiments on propane flames and a kerozene pool fire, performed in Poitiers (C.N.R.S.).

INTRODUCTION :

The purpose of the presented numerical study is the computation of axisymmetric low Froude number turbulent diffusion flames and the comparison with experimental results.

The model was previously applied to a number of incompressible flows (Viollet, 1987), (Simonin et al,1987), to high Froude number diffusion flames (Viollet et al, 1987) and to the computation of the flow and radiation transfer in natural gas heated furnaces (Mechitoua, Viollet, 1989).

1- DESCRIPTION OF THE MODEL

1.1- Basic equations

The equations for Favre averaged (symbols ~ and " for the fluctuations, symbols -- for time-averaging and ' for the fluctuations) velocity and enthalpy are written as :

State equation :
$$\bar{p} = \frac{\bar{\rho}\,R\,\tilde{T}}{M} \tag{1}$$

Continuity :
$$\frac{\partial \bar{\rho}}{\partial t} + \frac{\partial \bar{\rho}\,\tilde{u}_i}{\partial x_i} = 0 \tag{2}$$

Navier-Stokes :
$$\bar{\rho}\frac{\partial \tilde{u}_i}{\partial t} + \bar{\rho}\,\tilde{u}_j\frac{\partial \tilde{u}_i}{\partial x_j} = -\frac{\partial \bar{p}}{\partial x_i} + \bar{\rho}\,g_i + \frac{\partial}{\partial x_j}\left(\bar{\rho}\,\nu\,\frac{\partial \tilde{u}_i}{\partial x_j} - \overline{\rho\,u_i''\,u_j''}\right) \tag{3}$$

Energy balance :
$$\bar{\rho}\frac{\partial \tilde{H}}{\partial t} + \bar{\rho}\,\tilde{u}_i\frac{\partial \tilde{H}}{\partial x_i} = \frac{\partial}{\partial x_i}\left(\frac{\lambda}{C_p}\frac{\partial \tilde{H}}{\partial x_i} - \overline{\rho\,u_i''\,H''}\right) + \tilde{S} \tag{4}$$

H is the total enthalpy per unit of mass of the mixture ; if h_i and Y_i are mass enthalpy and mass fraction of the species i, and h_i^o its enthalpy of formation at 0K, H is also :

$$H = \Sigma\, Y_i\, h_i = \Sigma\, Y_i \left(\int_0^T C_{pi}\, dT + h_i^0 \right)$$

(5)

1.2- Turbulence model

An eddy viscosity k-ε model is used for the closure, assuming :

$$\overline{\rho\, u_i''\, u_j''} = -\overline{\rho}\, \nu_T \left(\frac{\partial\, \tilde{u}_i}{\partial\, x_j} + \frac{\partial\, \tilde{u}_j}{\partial\, x_i} \right) + 2/3\, (\overline{\rho}\, \tilde{k} + \overline{\rho}\, \nu_T\, \mathrm{div}\, (\tilde{u}))\, \delta_{ij} \quad \text{and} \quad \overline{\rho\, u_i''\, H''} = -K_T \frac{\partial\, \tilde{H}}{\partial\, x_i}$$

(6)

The eddy diffusivity is defined by : $\nu_T = \sigma_T K_T$, where σ_T is the turbulent Prandtl number ;

and the eddy viscosity is : $\nu_T = C_\mu \dfrac{\tilde{k}^2}{\varepsilon}$

(7)

The turbulent kinetic energy is defined as : $\tilde{k} = \dfrac{1}{2} \dfrac{\overline{\rho\, u_i''\, u_i''}}{\overline{\rho}}$ and the transport equation for $\tilde{k}$ is :

$$\overline{\rho} \frac{\partial\, \tilde{k}}{\partial\, t} + \overline{\rho}\, \tilde{u}_j \frac{\partial\, \tilde{k}}{\partial\, x_j} = -\frac{1}{2} \frac{\partial}{\partial\, x_j} (\overline{\rho\, u_j''\, u_i''\, u_i''}) - \overline{\rho\, u_j''\, u_i''} \frac{\partial\, \tilde{u}_i}{\partial\, x_j} + \frac{\overline{\rho u_i''}}{\overline{\rho}} \frac{\partial\, \overline{p}}{\partial\, x_i} - \overline{u_i'' \frac{\partial\, p'}{\partial\, x_i}} + \overline{u_i'' \frac{\partial}{\partial\, x_j} (\rho\, \nu \frac{\partial\, u_i}{\partial\, x_j})}$$

(8)

The terms are modeled as follows :

$$-\frac{1}{2} \frac{\partial}{\partial\, x_j} \overline{\rho\, u_j''\, u_i''\, u_i''} = \frac{\partial}{\partial\, x_j} (\overline{\rho} \frac{\nu_T}{\sigma_k} \frac{\partial\, \tilde{k}}{\partial\, x_j})$$

(9)

production terms : $\quad P = -\overline{\rho\, u_j''\, u_i''} \dfrac{\partial\, \tilde{u}_i}{\partial\, x_j} + \dfrac{\overline{\rho u_i''}}{\overline{\rho}} \dfrac{\partial\, \overline{P}}{\partial\, x_i}$

(10)

the first term in (10) is written as : $-\overline{\rho\, u_j''\, u_i''} \dfrac{\partial\, \tilde{u}_i}{\partial\, x_j} = \overline{\rho}\, \nu_T \left(\dfrac{\partial\, \tilde{u}_i}{\partial\, x_j} + \dfrac{\partial\, \tilde{u}_j}{\partial\, x_i} \right) \dfrac{\partial\, \tilde{u}_i}{\partial\, x_j} - \dfrac{2}{3} (\overline{\rho}\, \tilde{k} + \overline{\rho}\, \nu_T\, \mathrm{div}\, \vec{u}\,)\, \mathrm{div}\, \vec{u}$ (11)

The second term in (10) is related to the density fluctuations. It may be shown that this term includes inertia effects related to density fluctuations and gravity effects upon the turbulent kinetic energy balance. Only gravity effects are taken into account here, and this term is modeled as : $-K_T \dfrac{\partial\, \overline{\rho}}{\partial\, x_i}\, g_i$

(12)

The viscous dissipation is ε such as : $\overline{u_i'' \dfrac{\partial}{\partial\, x_j} (\rho\, \nu \dfrac{\partial\, u_i}{\partial\, x_j})} = -\overline{\rho}\, \varepsilon$

(13)

and its transport equation is modeled as : $\overline{\rho} \dfrac{\partial\, \varepsilon}{\partial\, t} + \overline{\rho}\, \tilde{u}_j \dfrac{\partial\, \varepsilon}{\partial\, x_j} = \dfrac{\partial}{\partial\, x_j} (\overline{\rho} \dfrac{\nu_T}{\sigma_\varepsilon} \dfrac{\partial\, \varepsilon}{\partial\, x_j}) + \dfrac{\varepsilon}{\tilde{k}} (C_{\varepsilon 1}\, P - C_{\varepsilon 2}\, \overline{\rho}\, \varepsilon)$

(14)

The values of the constants are taken from standard values used in incompressible flows, after Launder and Spalding (1974) : $\quad C_\mu = 0.09,\ C_{\varepsilon 1} = 1.44,\ C_{\varepsilon 2} = 1.92,\ \sigma_K = 1,\ \sigma_\varepsilon = 1.3.$

1.3- Combustion model

Following Jones and Whitelaw (1982), with the assumption of a very fast single chemical reaction -including only fuel, air and products-, a passive scalar, f, denoted here as the mixing rate, is defined as being equal to 1 in the pure fuel and to 0 in the pure air. Then, in the pure products, the mixing rate depends on the stoichiometry of the reaction and this value is noted f_s. It can be shown that the instantaneous mass fractions of fuel, air and products, $-Y_c$, Y_0 and Y_p respectively- depend only on the instantaneous value of the mixing rate :

$$\text{if } 0 \leq f \leq f_s \qquad\qquad \text{if } f_s \leq f \leq 1$$
$$Y_C = 0 \qquad\qquad\qquad Y_C = (f - f_s) / (1 - f_s)$$
$$Y_O = (f_s - f) / f_s \quad \text{and} \qquad Y_O = 0 \tag{15}$$
$$Y_p = f / f_s \qquad\qquad\qquad Y_p = (1 - f) / (1 - f_s)$$

In order to obtain averaged mass fractions, the probability density function, P(f), is introduced, so that :

$$\widetilde{Y}_i = \int_0^1 Y_i(f)\, P(f)\, df \tag{16}$$

P(f) is assumed to be a beta function and depends only on the mean and RMS values of the mixing rate, $\tilde{f}$ and $\widetilde{f''^2}$:

$$P(f) = f^{a-1}(1-f)^{b-1} \Big/ \int_0^1 f^{a-1}(1-f)^{b-1}\, df\,, \quad a = \tilde{f}\left[\tilde{f}(1-\tilde{f}) - \widetilde{f''^2}\right]/\widetilde{f''^2}\,, \quad b = (1-\tilde{f})\left[\tilde{f}(1-\tilde{f}) - \widetilde{f''^2}\right]/\widetilde{f''^2} \tag{17}$$

Transport equations for both quantities, $\tilde{f}$ and $\widetilde{f''^2}$, are written as :

$$\bar{\rho}\frac{\partial \tilde{f}}{\partial t} + \bar{\rho}\,\tilde{u}_i\frac{\partial \tilde{f}}{\partial x_i} = \frac{\partial}{\partial x_i}\left(\bar{\rho}\,K\frac{\partial \tilde{f}}{\partial x_i} - \overline{\rho\, u_i'' f''}\right) = \frac{\partial}{\partial x_i}\left(\bar{\rho}\,(K + K_T)\frac{\partial \tilde{f}}{\partial x_i}\right) \tag{18}$$

and

$$\bar{\rho}\frac{\partial \widetilde{f''^2}}{\partial t} + \bar{\rho}\,\tilde{u}_i\frac{\partial \widetilde{f''^2}}{\partial x_i} = -2\,\overline{\rho u_i'' f''}\frac{\partial \tilde{f}}{\partial x_i} - \frac{\partial}{\partial x_i}\overline{\rho u_i'' f''^2} + \frac{\partial}{\partial x_i}\left(\bar{\rho}K\frac{\partial \widetilde{f''^2}}{\partial x_i}\right) - \bar{\rho}\,\varepsilon_F \tag{19}$$

The closure assumptions, in (18) and (19), are similar to the ones previously used for the energy balance :

$$-\overline{\rho u_i'' f''}\frac{\partial \tilde{f}}{\partial x_i} = \bar{\rho}\,K_T\frac{\partial \tilde{f}}{\partial x_i}\frac{\partial \tilde{f}}{\partial x_i} \quad \text{and} \quad \frac{\partial}{\partial x_i}\overline{\rho u_i'' f''^2} = \frac{\partial}{\partial x_i}\left(\bar{\rho}\,K_T\frac{\partial \widetilde{f''^2}}{\partial x_i}\right) \tag{20}$$

The last term describes the turbulent diffusion effect - the turbulent Prandtl number is assumed to be equal to 1-.

The dissipation is modeled, assuming a constant ratio between the dynamics and scalar time scales, after Launder (1980),

as :
$$\varepsilon_F = \frac{2}{C_T}\frac{\widetilde{f''^2}}{\tilde{k}}\varepsilon \quad \text{and} \quad C_T = 1.6 \tag{21}$$

1.4- Radiation model

The source term S due to radiation in (4) is expressed by :

$$S = -\,\mathrm{div}\,\vec{q} = -\,\mathrm{div}\oint_{4\pi} L(\vec{x},\vec{S})\,\vec{S}\,d\Omega = -\oint_{4\pi}\frac{dL(\vec{x},\vec{S})}{dx_s}\,d\Omega \tag{22}$$

where the radiative flux $\vec{q}$ is obtained from the intensity $L(\vec{S})$ of radiation, function of the direction $\vec{S}$ of radiation propagation,
after integration for all directions : $\vec{q}(\vec{x}) = \oint_{4\pi} L(\vec{x},\vec{S})\,\vec{S}\,d\Omega \tag{23}$

Neglecting the diffusion of radiation, the radiation transfer equation for L is :

$$\frac{dL}{dx_s} = \mathrm{div}\,(L(\vec{x},\vec{S}).\vec{S}) = -\,K_{gaz}L(\vec{x},\vec{S}) + K_{gaz}\frac{\sigma T^4}{\pi} \tag{24}$$

The absorption coefficient K_{gaz} is obtained, following Modak (1978), as a function of the concentration in CO_2, H_2O and soot particles, with the grey gas assumption. The soot transport and production model, following Magnussen and Hjertager (1976), involves two equations for :

- precursor particles, P (resulting from fuel cracking) :

$$\frac{DP}{Dt} = \frac{1}{\bar{\rho}}\frac{\partial}{\partial x_j}\left(\bar{\rho}(K + K_T)\frac{\partial P}{\partial x_j}\right) + A_0\overline{Y_c}\exp(-E/RT) + (f - g)\,P\frac{Y_c}{Y_{co}} - g_0\,PN - A\,S(Y_c,Y_O)\frac{\varepsilon}{k}P \tag{25}$$

- soot particles, N :

$$\frac{DN}{Dt} = \frac{1}{\bar{\rho}}\frac{\partial}{\partial x_j}\left(\bar{\rho}(K + K_T)\frac{\partial N}{\partial x_j}\right) + (a - bN)\,P - A\,S(Y_c,Y_O)\frac{\varepsilon}{k}N \tag{26}$$

where $\dfrac{D}{Dt} = \dfrac{\partial}{\partial t} + \tilde{u}_i \dfrac{\partial}{\partial x_i}$

The combustion of the particles takes into account the competition with the combustion of fuel :

$$S(Y_c, Y_0) = \inf(1, \dfrac{\rho Y_0}{r_s m_s N + \rho Y_c r_f})$$

(27)

r_s and r_f are the mass of air required for the combustion of respectively 1kg of soot and of fuel ; m_s is the mass of a soot particle. (note that $r_s = 1 / f_s - 1 - f_s$ is introduced in (15) -)

The values of empirical constants, as used in the present study, are listed below :

$$A_0 = 1.16 \; 10^{31} \; ; \; \text{f-g} = 100 \; ; \; g_0 = 10^{-15} \; ; \; E/R = 90000 \; ; \; a = 10^5 \; ; \; b = 8 \; 10^{-14} \; ; \; A = 4 \; ;$$

2- NUMERICAL METHOD

2.1- Fluid dynamics and transport equations

The equations are solved in two dimensions and a finite difference rectilinear grid is used. Pressure is determined at the center of each mesh cell and all other quantities at each mesh point. All transport equations, (3-4-8-14-18-19-25-26), which are solved in the same way, can be written in a general form as :

$$\dfrac{G^{n+1} - G^n}{\delta t} + \tilde{u}_j \dfrac{\partial G^n}{\partial x_j} = \dfrac{S_G}{\rho} + \dfrac{1}{\rho} \dfrac{\partial}{\partial x_j} \left[\overline{\rho} \, (K + K_G) \dfrac{\partial G^{n+1}}{\partial x_j} \right]$$

(28)

where G^n and G^{n+1} are the values of G at the nth and $(n+1)^{th}$ time steps.

Between t^n and $t^n + \delta t$, using the fractional step method, the following steps are to be solved :

- advection step : $\qquad \dfrac{\widehat{G} - G^n}{\delta t} = - \tilde{u}_j^n \dfrac{\partial G^n}{\partial x_j}$

(29)

$\widehat{G}$ is obtained using a two dimensional characteristics method, with high order interpolation at the foot of the characteristics allowing to minimize numerical diffusion. -cf Viollet (1987)-

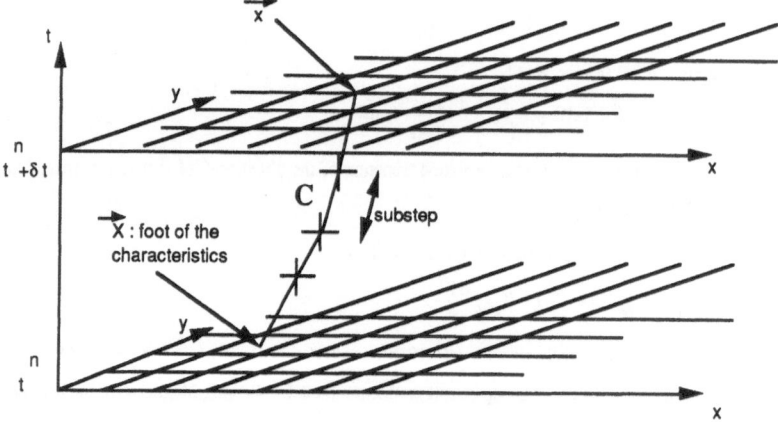

The characteristic line $\mathbf{C}$ is defined by : $\dfrac{d\vec{x}}{dt} = \vec{\tilde{u}}^n$ and $\widehat{G}(\vec{x}) = G^n(\vec{X})$

- diffusion step :

$$\overline{\rho}^n \dfrac{\delta G}{\delta t} = \left[\overline{\rho}^n \dfrac{\widehat{G} - G^n}{\delta t} + \dfrac{\partial}{\partial x_j} \left(\overline{\rho}^n \, (K + K_G)^n \dfrac{\partial G^n}{\partial x_j} \right) + S_G^n \right] + \dfrac{\partial}{\partial x_j} \left[\overline{\rho}^n \, (K + K_G)^n \dfrac{\partial (\delta G)}{\partial x_j} \right]$$

(30)

explicit balance

The implicit part of the diffusion term is written for the increment of G, $\delta G = (\widehat{\widehat{G}} - G^n)$ to minimize the splitting error.

For the enthalpy, the radiative source term is explicit while k and ε equations are solved simultaneously, coupled with implicit source terms.

$\widehat{\widehat{G}}$ is equal to G^{n+1} for all quantities, except for the velocity components : a third step is required in order to prescribe the continuity condition (2), leading to a Poisson equation (31) for the increment of pressure $\delta P = (P^{n+1} - P^n)$. The Laplacian operator is discretized partially (95%) on nine points and partially (5%) on five points in order to avoid the so-called checkerboard instabillity in pressure field. A S.O.R. method is used.

$$\text{div} (\delta t \, \text{grad} (\delta P)) = \text{div} (\overline{\rho}^n \, \widehat{\widehat{U}}) \text{ and } U^{n+1} = \widehat{\widehat{U}} - \delta t / \overline{\rho}^n \, \text{grad} (\delta P) \qquad (31)$$

In the presented results, the time-step is chosen in order to satisfy a Courant number (based on the local velocity) about 1 in high velocities regions of the mesh. Averaged CPU cost per time step is about 0.8s for 1000 points, on CRAY-1 computer.

2.2- Radiation model

A modified discrete transfer method is used to solve (23-24) : (23) is discretized along a finite number of directions (16 in our study) and (24) on the grid in the following way :

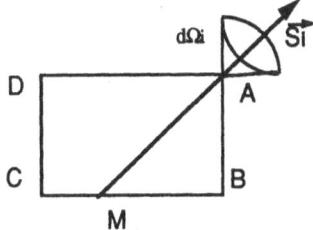

$L (A , \vec{S}_i)$ is computed with :

$$L (A , \vec{S}_i) - L (C , \vec{S}_i) \frac{MB}{BC} - L (B , \vec{S}_i) \frac{MC}{BC} = AM (-K_{gaz} L (A , \vec{S}_i) + K_{gaz} \frac{\sigma T^4 (A)}{\pi}) \qquad (32)$$

In the flame propane computations presented here, this model is used with 32 directions $\vec{S}_i$ only in the (r,z) plane.

3- BOUNDARY CONDITIONS

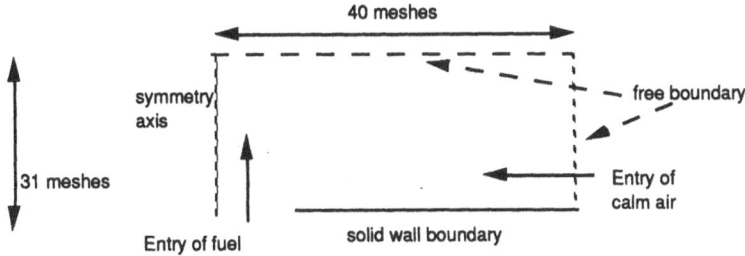

Definition sketch for the domain of computation in the (r,z) plane. (20 meshes for the entry of fuel)

Conditions of symmetry are written at the symmetry axis and the free boundaries. At the solid wall, the conditions for the velocity and the turbulent quantities are written using wall functions. -a logarithmic law is assumed for the tangential velocity component.- and adiabatic conditions for mixing rate and enthalpy. On the free boundaries (right and up), the pressure is prescribed ; at the mesh cells where the flow enters, the non turbulent pure air conditions are prescribed : f=0 ; k and ε are very weak and H is the enthalpy of air at 300°K.

The flow of fuel, from the pool or the burner, is very weak, and due to the existence of high diffusion processes very close to the fuel exit, a particular numerical treatment for the fuel entry condition is used. The first mesh is too coarse to represent the fine structure of the "boundary layer" above the pool or the burner. For this reason, the physical boundary condition $f = 1 = f_{fuel}$ cannot be prescribed at the entry of fuel. The total balance of mixing rate in the first mesh cell writes :

$$\bar{\rho}\, V_{lim}\, f_{lim}\, r\, \Delta r - \frac{\langle \rho(K_T + K)\rangle_{j=1} + \langle \rho(K_T + K)\rangle_{j=2}}{2\, \Delta y}\left(f_{int} - f_{lim}\right) r\, \Delta r = \bar{\rho}\, V_{lim}\, f_{fuel}\, r\, \Delta r \quad (33)$$

advection   diffusion   total physical flux from the pool or burner.

The flux is supposed to be uniform on all mesh cells at the entry. (33) is used as pool/burner boundary condition for the mixing rate computation.

In the case of a pool fire, fuel evaporates due to the radiation transfer ; the fuel flow rate Q from the pool is simply computed from a global energy balance :

$$\frac{1}{2\pi}\int_S \vec{q}\cdot\vec{n}\, dS = Q\left[L_v + \int_{T^o}^{T_v} C_{pl}\, dT\right] \quad \text{and if } Q = \int_0^R \bar{\rho}\, V_{lim}\, r\, dr \ , \text{ it reads :}$$

$$\bar{\rho}\, V_{lim}\, f_{lim}\, r\Delta r - \frac{\langle \rho(K_T + K)\rangle_{j=1} + \langle \rho(K_T + K)\rangle_{j=2}}{2\, \Delta y}\left(f_{int} - f_{lim}\right) r\Delta r = \bar{\rho}\, V_{lim}\, f_{fuel}\, r\Delta r$$

$$(34)$$

where L_v is the vaporization heat (J/kg) -the vaporization is assumed to occur at 420K- ;

C_{pl} is the liquid mass specific heat (J/kg/K) ;

and $\vec{q}$ is the radiative vector flux (W/m^2).$(\vec{q}\,\vec{x}) = \oint_{4\pi} L(\vec{x}, \vec{S})\, \vec{S}\, d\Omega$)

4- RESULTS

In the propane flame, the time-step is equal to 6.10^{-4} seconds. The time required for entrainment of air into the flame to get steady, is found to be pretty long, and the computational time is about one hour and a half on CRAY-1.

The computation is applied to the experimental studies from Vachon (1986), for a propane flame, and from Souil and al (1986), for a kerozene pool fire.

Flame	Diameter (m)	Froude number	pure fuel Flow rate (kg/s/rd)	pure fuel Velocity (m/s)	pure fuel ρ at 300 K (kg/m^3)
propane	0.15	2.10^{-6}	$0.86\ 10^{-5}$	$1.7.10^{-3}$	1.8
kerozene pool fire	0.30	$6.2\ 10^{-6}$	2.810^{-4}	$4.3\ 10^{-3}$	5.85

The RMS temperature fluctuations were measured ; in order to compare, an approximate RMS temperature fluctuation is deduced from the computation : as in the case of adiabatic flames, where the instantaneous temperature is a linear function of the mass fraction of products, such a law is written, but the coefficients are determined for each point of the computation domain, including a local pseudo adiabatic temperature of flame ; the RMS temperature is then known by integrating this relation, using the probability density function of the mixing rate. -the mass fraction of products is a function of the mixing rate.- see Viollet et al, (1987).

4.1- Propane flame

Measurements were available only on the flame axis.

Figure 1 presents the computed velocity field. Figure 2 shows the results obtained with the presented soot model : the maximum of temperature is too high in the computation -1500K instead of 1200K- and is located at 0.2m from the entry of fuel, instead of 0.16m.

To test the influence of the soot model, the volumic soot concentration, f_v, was imposed proportionally to the fuel mass fraction -$f_v = 3.10^{-5} Y_{c-}$. Results, figure 3, shows the improvement.

4.2- Kerozene pool fire

In order to compare with measured radiative fluxes out of the fire, a 3D radiative computation was used. A first computation was performed, the fuel flow rate, Q, being imposed from experimental data with the Magnussen soot model. In this computation, the axial temperature profile is rather well found out but the radiative fluxes are too important. -see figure 4- The pool receives a too high radiative energy which would generate a too high flow rate of fuel.

To introduce the relation between the fuel flow rate and the radiative energy received by the pool -eq. (34)-, the soot concentration was imposed to make the coupling possible -$f_v = 2.10^{-7} Y_{c-}$: in these conditions, the temperature in the fire is too high, -600K between computation and measurements-, and the computed radiative fluxes are too small.

There are still parameters in the radiative properties model which were not tested enough and some quite approximated data about the thermodynamic laws of the vaporized kerozene.

CONCLUSION

Low Froude number turbulent diffusion flames were computed and the results compared to measurements with an axisymmetric finite difference model. The results were found to be very sensitive to the presence of soot. The soot model, used in this study, does not seem to be valid for such flames, but the computation of a pool fire shows there are still tests to be performed about the radiative properties of gases.

However, fitting the soot concentration allows to get good comparisons for mean velocity and temperature and RMS temperature fluctuation fields. This suggests that the air flow and combustion phenomena are well taken into account, and that, for instance, the k-ε-f"^{2} turbulence model is sufficient for such flames (though more sophisticated turbulence models have now become available ; see Baron, Kanniche, Viollet, 1989).

Future additional work will be related to LNG pool fires in the presence of wind.

REFERENCES
P.L. VIOLLET (1987) : "The modelling of turbulent recirculating flows for the purpose of reactor thermal-hydraulic analysis". Nuclear Engineering and Design, 99.
O. SIMONIN, J. ROBINSON, M. BARCOUDA (1987) : "Measurements and computation of turbulent flows acoss a staggered tube bundle", presented at the 22nd IAHR Congress, Lausanne, Aug 1987.
P.L. VIOLLET, D. GARRETON, N. MECHITOUA (1987) : "Computation of trubulent diffusion flames with radiation", presented at the SIAM conference on Numerical Combustion, San Francisco, March 1987.
N. MECHITOUA P.L. VIOLLET (1989) : "Computations of turbulent diffusion flames with radiation in a gas fired cylindrical furnace", to be presented at the G.F.C. Congress, Rouen, Apr 1989.
B.E. LAUNDER, D.B. SPLADING (1974): "The numerical computation of turbulent flows". Computer Methods in Applied Mechanics and Engineering 1974, 3

B.E. LAUNDER (1980): "Turbulence transport models for numerical computation of complex turbulent flows". Von Karman Inst for fluid dynamics, lecture series 1980-3.

W.P. JONES, J.H. WHITELAW (1982) : "The numerical computation of turbulent flows : a review". Combustion and Flame, 48, 1-26.

A.T. MODAK (1978) : "Radiation from products of combustion". Fire Research,1 1978-1979 pp 339-361.

B.F. MAGNUSSEN B.H. HJERTAGER (1976) : "On mathematical modelling of turbulent combustion with special emphasis on soot formation and combustion". 16th International Symposium on Combustion 1976 ; The Combustion Institute, p719-729

M. VACHON (1976) : "Modélisation et étude expérimentale de flammes de diffusion turbulentes à bas nombre de Froude". thèse de docteur-ingénieur 1986, Université de Poitiers U.E.R.-E.N.S.M.A..

J.M. SOUIL, J.P. VANTELON, P. JOULAIN et W.L. GROSSHANDLERS (1986) : "Experimental and theorical study of thermal radiation from freely burning kerozene pool fires". AAA, 1986, pp 388-401.

F. BARON, M. KANNICHE, P.L. VIOLLET (1989) : "Computation of stably and unstably stratified turbulent shear flows using a second moment closure" ; to be presented at the 23nd IAHR Congress, Ottawa, Aug 1989.

on the whole computed domain

hades
→ 2 m/s

at the entry of fuel.

hades
→ 1 m/s

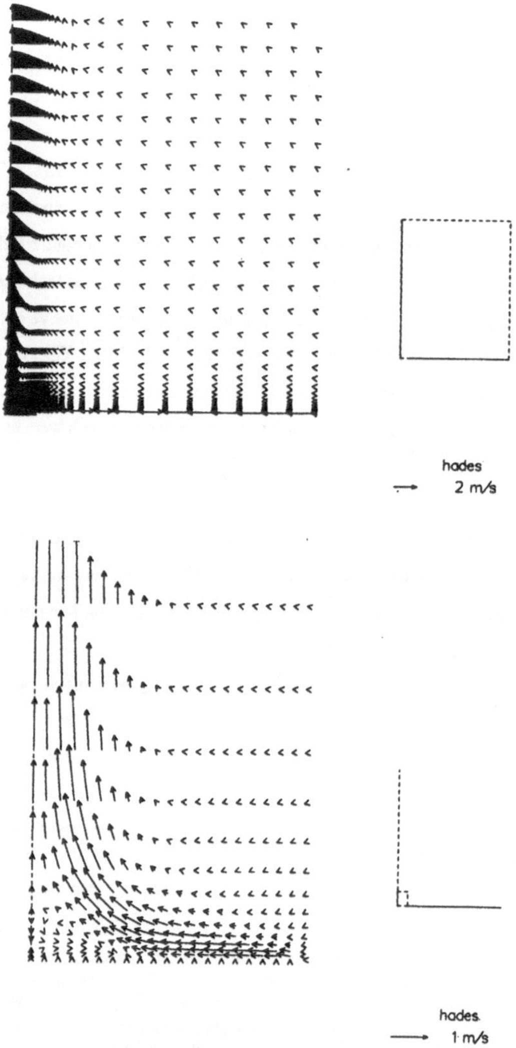

Figure 1 : Velocity field

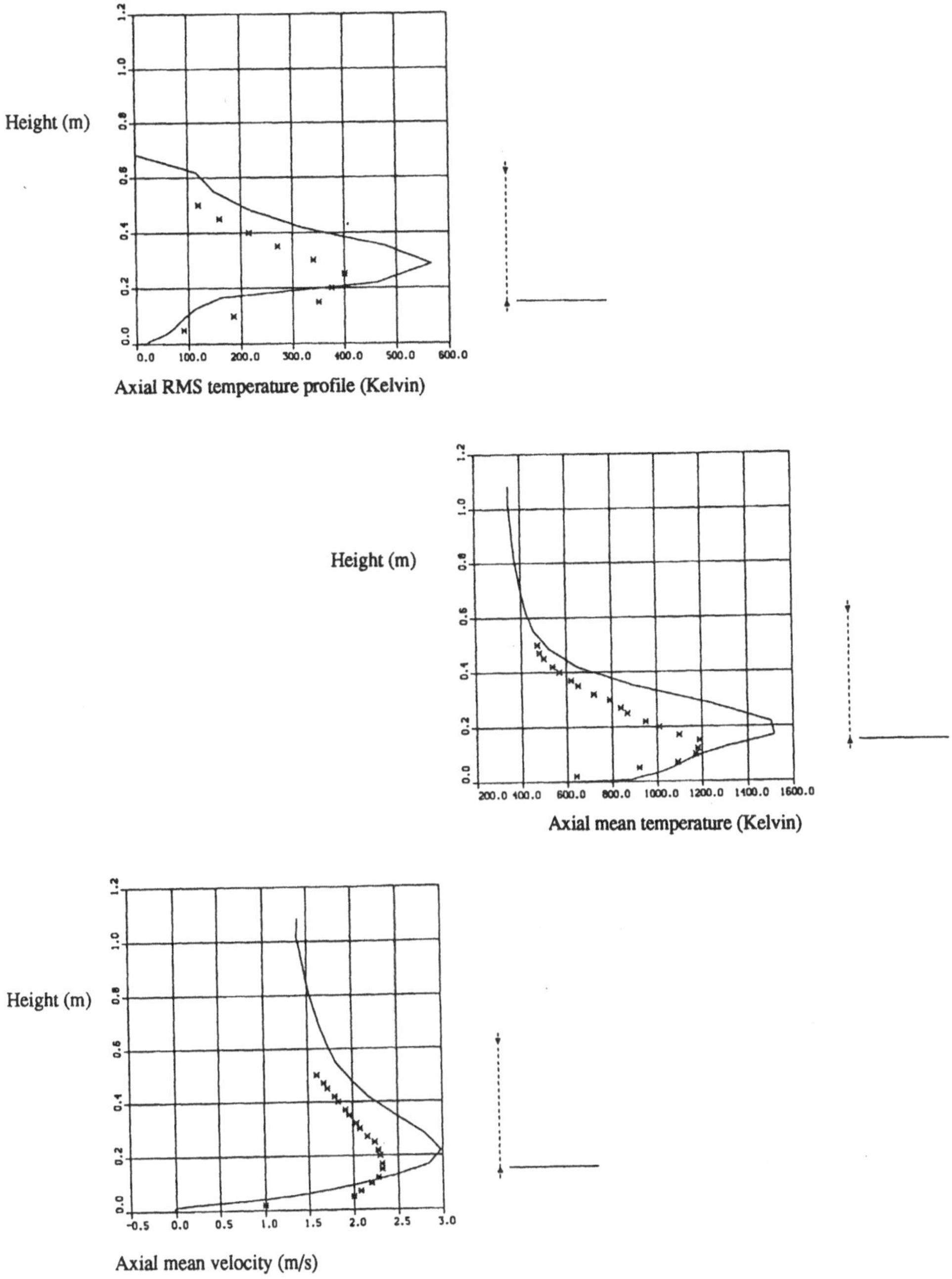

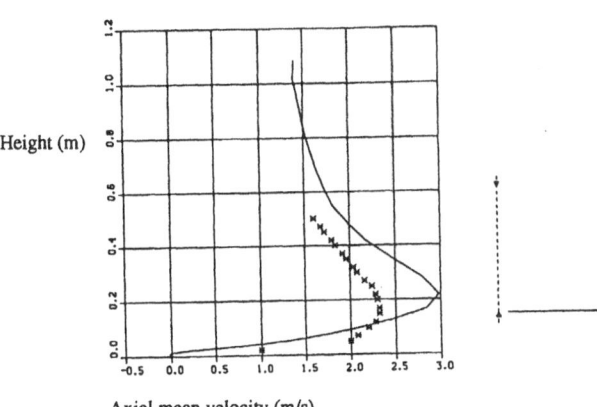

Axial RMS temperature profile (Kelvin)

Axial mean temperature (Kelvin)

Axial mean velocity (m/s)

solid lines : computation ; * : experimental data

Figure 2 : Propane flame ; results with the Magnussen soot model.

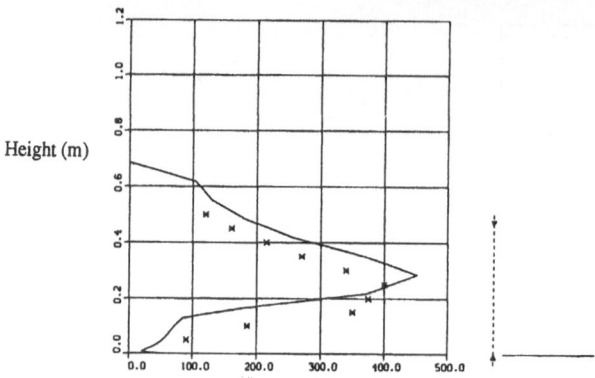

Axial RMS temperature profile (Kelvin)

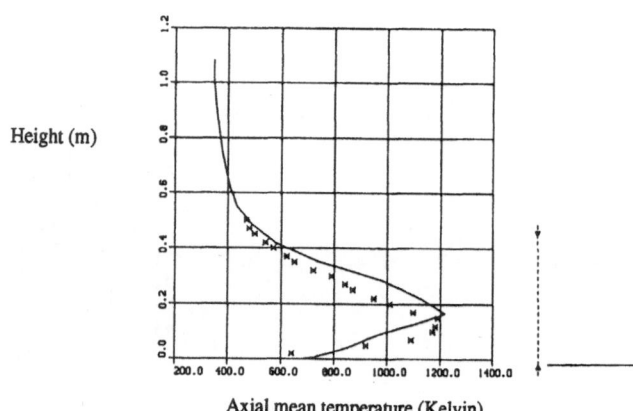

Axial mean temperature (Kelvin)

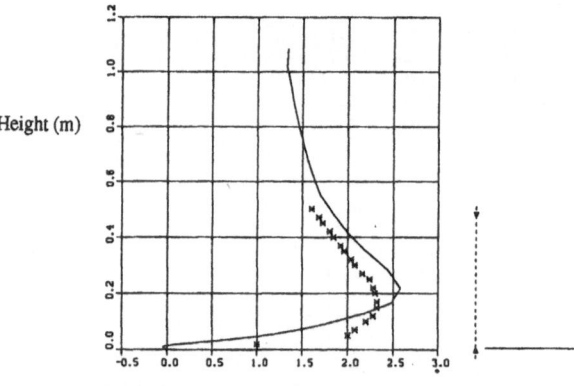

Axial mean velocity (m/s)

solid lines : computation ; * : experimental data

Figure 3 : Propane flame ; results when the soot concentration is a function of the mass fraction of fuel.

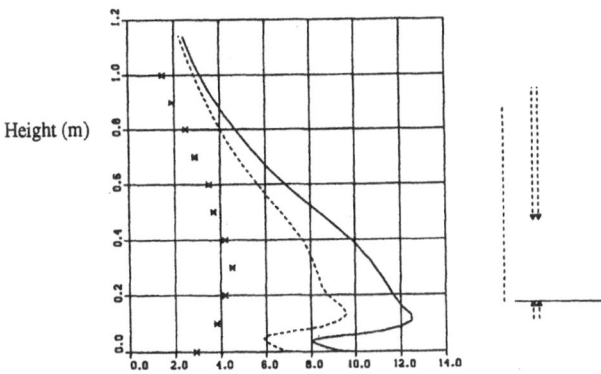

Radial radiative fluxes (W/m^2) at 0.5 m from the symmetry axis.

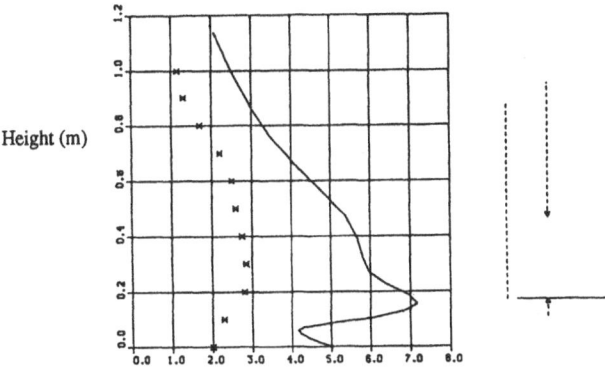

Radial radiative fluxes (W/m^2)at 0.6 m from the symmetry axis.

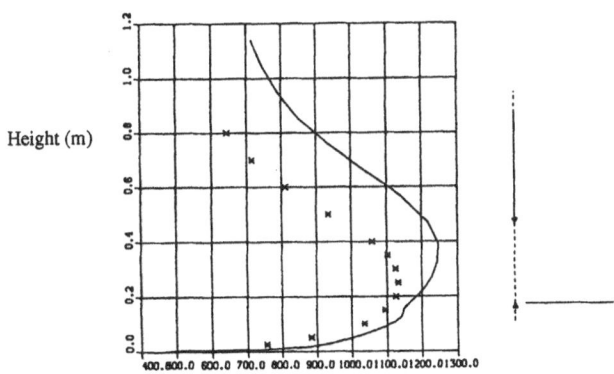

Axial mean temperature (Kelvin) on the symmetry axis.

solid lines : computation ; * : experimental data

Figure 4 : <u>Kerozene pool fire</u> ; results with the Magnussen soot model
The fuel flow rate is equal to the experimental data. The radiative fluxes are computed using a 3D discretization.

MASS CONSERVATION AND SINGULAR MULTICOMPONENT DIFFUSION ALGORITHMS

Vincent GIOVANGIGLI

Laboratoire d'Acoustique et Mécanique, UA 868 du CNRS,
Université Paris 6, T66-5e, 4 place Jussieu, 75232 Paris Cedex 05, and

Centre de Mathématiques Appliquées, UA 756 du CNRS
Ecole Polytechnique, 91128 Palaiseau Cedex, France

1   INTRODUCTION

The governing equations of multicomponent gaseous reacting flows are the hydrodynamic equations derived from the kinetic theory of gases [2] [6] [12]. Derivation of these equations shows that at any time and at any point of the physical space, various mass conservation constraints are satisfied. Example of such constraints are the relation $\sum_{k \in S} Y_k = 1$ between the species mass fractions Y_k, $k \in S = \{1, \ldots, N_S\}$ where S is the set of species indices and N_S is the number of species, and the relation $\sum_{k \in S} X_k = 1$ between the mole fractions X_k, or the constraints $\sum_{k \in S} Y_k V_k = 0$ between the species diffusion velocities V_k, $\sum_{k \in S} d_k = 0$ between the diffusion driving forces d_k and $\sum_{k \in S} W_k \omega_k = 0$ between the mass rate of production of the species $W_k \omega_k$ [2] [6] [12]. These relations imply that the N_S species conservation equations are linearly dependent and sum up to zero when written in nonconservative form.

An attractive approach in problems having one species which is always in excess is therefore to consider only $N_S - 1$ mass fractions as unknowns and to evaluate the excess species mass fraction by using the relation $\sum_{k \in S} Y_k = 1$ [22]. All of the mass conservation constraints are then automatically satisfied. However this approach is not always feasible. In a typical diffusion flame for instance, each species is deficient either on the fuel side or on the oxydant side and it is not accurate to evaluate one of the mass fractions by using $\sum_{k \in S} Y_k = 1$. An interesting approach however is to determine locally, at each computational cell, which species is in excess and to evaluate it by using $\sum_{k \in S} Y_k = 1$, thereby solving a solution-dependent set of equations.

An alternate approach — widely used in complex chemistry reacting flow solvers — is to consider all mass fractions as independent unknowns [7] [8] [9] [14] [15] [16] [17] [18] [20] [21]. In this situation, it is important to analyse which of the mass constraints are automatically satisfied and which are consequences of the governing equations. Indeed the various diffusion algorithms expressing the diffusion velocities V_k first show that the relation $\sum_{k \in S} Y_k V_k = 0$ is always satisfied. Similarly the relation $\sum_{k \in S} W_k \omega_k = 0$ between the chemical production rates ω_k is a consequence of their expression in terms of the reaction rates of progress. Usual relations expressing the mole fractions also yields the identity $\sum_{k \in S} X_k = 1$ so that the driving forces satisfy $\sum_{k \in S} d_k = 0$

under the approximation $d_k = \nabla X_k$. On the other hand, the relation $\sum_{k \in S} Y_k = 1$ between the species mass fractions Y_k must result from the conservation equations, the diffusion algorithm and the boundary conditions. The equations for $\sum_{k \in S} Y_k$ are indeed obtained by summing up the corresponding N_S species equations. However this latter set of equations may be artificially singular because of the constraints $\sum_{k \in S} Y_k V_k = 0$ and $\sum_{k \in S} X_k = 1$ which also imply that diffusion matrices are non invertible. These singularities may appear for instance with species flux boundary conditions or steady flows involving recirculation zones or stagnation points. In these situations, the Jacobian matrices of the discrete governing equations are singular, i.e., non invertible, and this may eventually lead to convergence difficulties or poor sensitivity informations.

Elimination of the singularities requires modifying the usual diffusion algorithlms. Modifications are proposed for three different diffusion algorithms, namely for the complex formalism of the kinetic theory of gases, for the Stefan-Maxwell equations and for the Hirschfelder-Curtiss expressions with mass correctors [2] [3] [4] [5] [6] [7] [12] [13] [14] [16] [19] [20] [21]. The modifications that we suggest are such that $\sum_{k \in S} Y_k V_k = -\mathcal{D} \sum_{k \in S} d_k$, where $\mathcal{D}$ is a positive diffusion coefficient, and $\sum_{k \in S} X_k = \sum_{k \in S} Y_k$. These modifications, of course, do not change the actual values of the diffusion velocities or the mole fractions. Only their mathematical expressions are changed.

The governing equations of gaseous laminar reacting flows which are relevant for our discussion are presented in section 2. In section 3, different singularities are exhibited in the governing equations. Finally the modified diffusion algorithms are introduced in section 4 and the properties of the corresponding modified governing equations are discussed.

2   SPECIES GOVERNING EQUATIONS

2.1   Conservation equations and boundary conditions

The species mass conservation equations in a gaseous laminar reacting flow may be written as

$$\rho \frac{\partial Y_k}{\partial t} + \rho(v.\nabla)Y_k = -\nabla.(\rho Y_k V_k) + W_k \omega_k, \qquad k \in S, \tag{1}$$

where ρ is the density, Y_k the mass fraction of the kth species, t the time, v the mass averaged flow velocity, V_k the diffusion velocity of the kth species, W_k the molecular weight of the kth species, ω_k the molar production rate of the kth species, $S = \{1, \ldots, N_S\}$ the set of species indices and N_S the number of species [2] [6] [12]. The total mass, momentum and energy conservation equations and the law of state, which are not needed in our analysis, are not presented. Equations (1) are usually simplified according to the problem under study, e.g., steady flow or boundary layer flow.

Typical boundary conditions for the species mass fractions can be of Dirichlet, Neumann or mixed type. Dirichlet boundary conditions are often involved with truncated infinite domains [8] [9] and may be written

$$Y_k = Y_k^0, \qquad k \in S, \tag{2}$$

where Y_k^0 denotes the specified mass fractions, which of course must be such that $\sum_{k \in S} Y_k^0 = 1$. Neumann boundary conditions often occur in symmetric problems [8] [9] or truncated infinite domains [18] and lead to the relations

$$\nabla Y_k . \boldsymbol{n} = 0, \qquad k \in \mathcal{S}, \tag{3}$$

where $\boldsymbol{n}$ is the unit normal vector at the boundary. On the other hand, typical flux boundary conditions arise in the general form [18] [21]

$$\rho Y_k (\boldsymbol{v} + \boldsymbol{V_k}).\boldsymbol{n} = \rho \boldsymbol{v} . \boldsymbol{n} Y_k^0 + W_k \widetilde{\omega}_k, \qquad k \in \mathcal{S}, \tag{4}$$

where Y_k^0, $k \in \mathcal{S}$, denotes the specified mass flux fractions, which of course are such that $\sum_{k \in \mathcal{S}} Y_k^0 = 1$, and $\widetilde{\omega}_k$ is the surface molar production rate of the kth species.

These equations have to be completed by the formulas expressing the diffusion velocities $\boldsymbol{V_k}$ and the chemical production rates ω_k and $\widetilde{\omega}_k$ in terms of the state variables T, p, Y_1, ... Y_{N_s}, and their gradients, where p is the pressure and T the absolute temperature. Only the relations involving $\boldsymbol{V_k}$, ω_k and $\widetilde{\omega}_k$ which are relevant for our discussion are written in the following.

2.3 Expressions for V_k

Different algorithms can be used to determine the diffusion velocities $\boldsymbol{V_k}$. A first possibility is to use the complex formalism of the kinetic theory of gases. In this situation, the diffusion velocities are written in the form [2] [4] [6] [19]

$$\boldsymbol{V_k} = - \sum_{l \in \mathcal{S}} D_{kl} (\boldsymbol{d_l} + \theta_l \nabla \log T) \tag{5}$$

with

$$\boldsymbol{d_l} = \nabla X_l + (X_l - Y_l) \frac{\nabla p}{p} + \frac{\rho}{p} \sum_{m \in \mathcal{S}} Y_l Y_m (\boldsymbol{F_m} - \boldsymbol{F_l}) \tag{6}$$

where $\boldsymbol{D} = (D_{kl})$ is the symmetric multicomponent diffusion coefficient matrix, $\boldsymbol{d_k}$ the diffusion driving force of the kth species, θ_k the thermal diffusion ratio of the kth species, T the absolute temperature, X_k the mole fraction of the kth species, p the pressure and $\boldsymbol{F_k}$ the external force per unit mass of the kth species. The mole fraction X_k of the kth species is given by

$$X_k = Y_k W / W_k, \tag{7}$$

where W is the molecular weight of the mixture [2] [6] [7] [12] [15] [22]

$$1/W = \sum_{k \in \mathcal{S}} Y_k / W_k. \tag{8}$$

Note that there is considerable variation among authors in the nomenclature and definition of the multicomponent diffusion and thermal diffusion coefficients. The diffusion coefficients D_{kl} of Equation (5) are defined with the constraints $\sum_{l \in \mathcal{S}} Y_l D_{kl} = 0$, see, e.g., Chapmann and Cowling [2], Curtiss [4] or Ferziger and Kaper [6]. These definitions are consistent with Onsager reciprocal relations of thermodynamics of irreversible process [6] [19] since $D_{kl} = D_{lk}$, i.e., $\boldsymbol{D} = \boldsymbol{D}^T$, and lead to the relations

$$\sum_{k \in \mathcal{S}} Y_k D_{kl} = 0, \qquad l \in \mathcal{S}, \tag{9}$$

i.e., $\boldsymbol{DY} = 0$ where $\boldsymbol{Y} = (Y_1, \dots, Y_{N_s})$, so that (D_{kl}) is not invertible, and to

$$\sum_{l \in \mathcal{S}} \theta_l = 0. \tag{10}$$

An alternate definition, due to Hischfelder, Curtiss and Bird [3] [5] [12] [14] [20] [21], imposes the constraints $D_{kk} = 0$ and breaks the symmetry of the diffusion process [4] [6] [19]. These later coefficients however are still such that (9) holds.

A direct consequence of Equations (9) is that the relation

$$\sum_{k \in S} Y_k V_k = 0, \tag{11}$$

is satisfied independently of the driving forces d_k and $\theta_k \nabla \log T$. An important property is also that the quadratic form $\xi \rightarrow \sum_{k,l \in S} D_{kl} \xi_l \xi_k$ in $\mathbb{R}^{Ns}$ is nonnegative and positive definite on the 'physical' hyperplane $\{\xi, \sum_{l \in S} \xi_l = 0\} = U^\perp$ where $U = (1, \ldots, 1)^T$ [6]. This is a direct consequence of the following expression for the entropy production due to particle collisions $(\partial s / \partial t)_c$ in a gas mixture [6]

$$\left(\frac{\partial s}{\partial t}\right)_c = (p/T) \sum_{k,l \in S} D_{kl} (d_k + \theta_k \nabla \log T) \cdot (d_l + \theta_l \nabla \log T) + \lambda (\nabla \log T) \cdot (\nabla \log T)$$

$$+ \quad (\eta/2T)(\nabla v + (\nabla v)^T - \tfrac{2}{3}(\nabla \cdot v)I) : (\nabla v + (\nabla v)^T - \tfrac{2}{3}(\nabla \cdot v)I), \tag{12}$$

where λ and η are the thermal conductivity and viscosity of the mixture. Strictly speaking, the quadratic form should be considered in $\mathbb{R}^{3*Ns}$, but using the canonical basis of $\mathbb{R}^3$ shows that it is equivalent to consider it in $\mathbb{R}^{Ns}$. Note also that $D_{kl} = O(1)$ for $k \neq l$ whereas $D_{kk} = O(1/X_k)$ so that $D_{kk} \rightarrow \infty$ when $X_k \rightarrow 0$ [2] [4] [6].

It is also interesting to introduce the dual relations

$$d_k + \theta_k \nabla \log T = - \sum_{l \in S} \Delta_{kl} V_l \tag{13}$$

where $\Delta = (\Delta_{kl})$ is the dual multicomponent diffusion coefficient matrix. These coefficients are also symmetric, $\Delta_{kl} = \Delta_{lk}$, i.e., $\Delta = \Delta^T$, and satisfy

$$\sum_{k \in S} \Delta_{kl} = 0, \qquad l \in S, \tag{14}$$

i.e., $\Delta U = 0$, so that the relation

$$\sum_{k \in S} d_k = 0, \tag{15}$$

is satisfied independently of the diffusion velocities. Note here that the matrices D and Δ are generalized inverses of each other. More specifically, D and Δ are generalized inverses with prescribed range and null space [1], i.e., Δ is the unique matrix with range $U^\perp$ and nullspace $\mathbb{R}U$ such that $D \Delta D = D$ and $\Delta D \Delta = \Delta$ and D is the unique matrix with range $Y^\perp$ and nullspace $\mathbb{R}Y$ such that $\Delta D \Delta = \Delta$ and $D \Delta D = D$ [1]. From these relations, one can show that ΔD and $D \Delta$ are projector matrices with range $U^\perp$ and $Y^\perp$ and nullspace $\mathbb{R}Y$ and $\mathbb{R}U$ respectively, which can be written $\Delta D = I - (Y U^T)/(U^T Y)$ and $D \Delta = I - (U Y^T)/(Y^T U)$ [1]. The former relation is well known but is usually written for first order approximations only and without the $Y^T U$ term, although it is generally valid [2] [4] [5] [6] [12]. Finally note that Δ is not the Moore-Penrose pseudo-inverse [1] [10] D^+ of D unless Y and U are proportional in which case $D \Delta$ and ΔD are orthogonal projectors.

The quadratic form $\zeta \rightarrow \sum_{k,l \in S} \Delta_{kl} \zeta_l \zeta_k$ in $\mathbb{R}^{Ns}$ is also nonnegative and positive definite on the 'physical' hyperplane $\{\zeta, \sum_{l \in S} Y_l \zeta_l = 0\} = Y^\perp$ where $Y =$

$(Y_1, \ldots, Y_{N_S})^T$. A motivation for introducing the dual relations is that evaluating the multicomponent diffusion coefficients D_{kl} requires solving large linear systems, namely of size jN_S*jN_S when j terms are retained in the Sonine polynomial expansion of the species perturbed density probability functions [2] [6]. This is not the case for the dual coefficients Δ_{kl} when $j = 1$, i.e., when only one term is retained in the latter expansion. Indeed, in this situation, the dual formulation reduces to the Stefan-Maxwell equations

$$d_k + \theta_k \nabla \log T = \sum_{\substack{l \in \mathcal{S} \\ l \neq k}} \frac{X_k X_l}{\mathcal{D}_{kl}} V_l - \Big(\sum_{\substack{l \in \mathcal{S} \\ l \neq k}} \frac{X_k X_l}{\mathcal{D}_{kl}}\Big) V_k, \qquad k \in \mathcal{S}, \qquad (16)$$

where $\mathcal{D}_{kl}$ denotes the usual binary diffusion coefficient for the species pair (k, l). These relations, which must be completed by the mass constraint $\sum_{k \in \mathcal{S}} Y_k V_k = 0$ in order to define uniquely the diffusion velocities, may now be inverted to yield the diffusion velocities V_k [2] [3] [5] [6] [12] [13] [17] [22].

Further note that some care must be taken when applying the Onsager reciprocal relations to the entropy quadratic form $\sum_{k \in \mathcal{S}} V_k d_k$ since the fluxes V_k, or the affinities d_k, are constrained. One must indeed use the reciprocal relations for N_S-1 fluxes or affinities and build the symmetric N_S*N_S diffusion matrices D and Δ from the corresponding $(N_S-1)*(N_S-1)$ symmetric positive definite matrices [12] [23]. This procedure, of course, again leads to matrices D and Δ which are nonnegative and positive definite on the proper hyperplanes.

Finally, comparisons between different mathematical approximations of the multicomponent transport properties have shown that the following expressions provide a good trade off between precision and computational costs [3] [16] [20] [21]. The diffusion velocity is written in the form

$$V_k = \mathcal{V}_k + V_{\mathrm{m}}, \qquad \mathcal{V}_k = -D_k^* \nabla \log X_k, \qquad D_k^* = (1 - Y_k) / \sum_{l \in \mathcal{S}, \, l \neq k} (X_l / \mathcal{D}_{kl}), \qquad (17)$$

where $\mathcal{V}_k$ is the Hirschfelder-Curtiss diffusion velocity due to species gradients [11], D_k^* the diffusion coefficient of the kth species in the mixture and V_{m} a correction velocity included to ensure that the mass is conserved so that [3] [7] [16] [18] [20] [21]

$$\sum_{k \in \mathcal{S}} Y_k (\mathcal{V}_k + V_{\mathrm{m}}) = 0. \qquad (18)$$

The resulting approximate diffusion matrix $D^a = (D_{kl}^a)$, with $V_k = \sum_{l \in \mathcal{S}} D_{kl}^a \nabla X_l$, is then given by $D_{kl}^a = (D_l^*/X_l)(\delta_{kl} - Y_l/\sum_{m \in \mathcal{S}} Y_m)$, where δ denotes the Kronecker symbol. The matrix D^a is therefore non symmetric so that the expressions (17)(18) are not clearly consistent with Onsager reciprocal relations [19]. However, although the quadratic form $\xi \to \sum_{k,l \in \mathcal{S}} D_{kl}^a \xi_l \xi_k$ associated to D^a, i.e., to its symmetric part, is nondefinite, i.e., has a negative eigenvalue, it still is positive definite on the 'physical' hyperplane $U^\perp$.

2.4 Expressions for ω_k and $\tilde{\omega}_k$

We consider $N_\mathcal{R}$ elementary reversible reactions involving N_S chemical species, which may be represented in the general form

$$\sum_{k \in \mathcal{S}} \nu'_{ki} \mathcal{X}_k \; \rightleftharpoons \; \sum_{k \in \mathcal{S}} \nu''_{ki} \mathcal{X}_k, \qquad i \in \mathcal{R}, \qquad (19)$$

where $\mathcal{R} = \{1, \ldots, N_{\mathcal{R}}\}$ denotes the set of reaction indices, $\mathcal{X}_k$ the symbol of the kth species and where the stoichiometric coefficients ν'_{ki} and ν''_{ki} are integers. Since no mass is created in chemical reactions, the stoichiometric coefficients are such that [22]

$$\sum_{k \in S} \nu'_{ki} W_k = \sum_{k \in S} \nu''_{ki} W_k, \qquad i \in \mathcal{R}. \tag{20}$$

The production rate of the kth species may now be written as [15] [22]

$$\omega_k = \sum_{i \in \mathcal{R}} (\nu''_{ki} - \nu'_{ki}) q_i, \tag{21}$$

and where q_i is the rate of progress variable of the ith reaction. We then easily get from (20)(21) that the constraint

$$\sum_{k \in S} W_k \omega_k = 0, \tag{22}$$

is automatically satisfied. Similarly, the surface production rates $\widetilde{\omega}_k$ are such that

$$\sum_{k \in S} W_k \widetilde{\omega}_k = 0, \tag{23}$$

provided there are no mass losses due to surface vapor deposition.

3 SINGULARITIES IN THE GOVERNING EQUATIONS

The singularities which may appear in the conservation equations or boundary conditions are related to the governing equations for $\sum_{k \in S} Y_k$. These governing equations are obtained by summing up the N_S species governing equations. The resulting equations may indeed degenerate, especially when written in discrete form.

3.1 The origin of singularities

To investigate the singular behavior due to the mass constraints, we assume that a set of discrete equations modeling a multicomponent reacting mixture has been derived and we consider its Jacobian matrix formally written

$$\Pi = \left(\frac{\partial(\mathcal{E}^H, \mathcal{E}^1, \ldots, \mathcal{E}^{N_S-1}, \mathcal{E}^{N_S})}{\partial(H, Y_1, \ldots, Y_{N_S-1}, Y_{N_S})} \right) \tag{24}$$

where H denote the dependent variables aside from the species, $\mathcal{E}^H$ the corresponding discrete equations and $\mathcal{E}^k$ the kth species discrete equations. The different dependent unknowns ρ, v, T, etc., are formally grouped into a single variable H and the dependence of discrete equations $\mathcal{E}$ on computational cells does not appear explicitly in (24) in order to avoid notational complexity. Furthermore, in this Jacobian matrix, the partial derivatives with respect to the mass fractions are assumed to be independent.

By adding now the lines corresponding to the equations $\mathcal{E}^1$, ..., $\mathcal{E}^{N_S-1}$ to the lines of $\mathcal{E}^{N_S}$, at each computational cell, and substracting the columns corresponding to the derivation with respect to Y_{N_S} to the columns of Y_1, ..., Y_{N_S-1}, at each computational cell, we get that

$$\mathcal{J} = \det(\Pi) = \det \left(\frac{\partial(\mathcal{E}^H, \mathcal{E}^1, \ldots, \mathcal{E}^{N_S-1}, \mathcal{E}^\sigma)}{\partial(H, Y_1, \ldots, Y_{N_S-1}, \sigma)} \right) \tag{25}$$

where $\sigma = \sum_{k \in S} Y_k$ and $\mathcal{E}^\sigma = \sum_{k \in S} \mathcal{E}^k$. Note that in this new expression, the partial derivatives with respect to $Y_1, \ldots, Y_{N_s-1}$ are taken with $\sigma = \sum_{k \in S} Y_k$ fixed. By regrouping then the lines corresponding to $\mathcal{E}^\sigma$ at the bottom of the later matrix in (25) and the columns corresponding to the partial derivatives with respect to σ on the right of this matrix, we obtain the following block decomposition

$$
\mathcal{J} = \det(\Pi) = \det \left(\begin{array}{c|c} \dfrac{\partial(\mathcal{E}^H, \mathcal{E}^1, \ldots, \mathcal{E}^{N_s-1})}{\partial(H, Y_1, \ldots, Y_{N_s-1})} & \dfrac{\partial(\mathcal{E}^H, \mathcal{E}^1, \ldots, \mathcal{E}^{N_s-1})}{\partial(\sigma)} \\ \hline \dfrac{\partial(\mathcal{E}^\sigma)}{\partial(H, Y_1, \ldots, Y_{N_s-1})} & \dfrac{\partial(\mathcal{E}^\sigma)}{\partial(\sigma)} \end{array} \right). \tag{26}
$$

On the other hand, for any physical solution, we have numerically $\sigma = \sum_{k \in S} Y_k = 1$, and in this situation one may check that the lower left block in (26) is zero. We thus deduce that for any solution for which $\sum_{k \in S} Y_k = 1$, we have the relation

$$
\mathcal{J} = \Gamma \Upsilon, \qquad \Gamma = \det \left(\frac{\partial(\mathcal{E}^H, \mathcal{E}^1, \ldots, \mathcal{E}^{N_s-1})}{\partial(H, Y_1, \ldots, Y_{N_s-1})} \right), \qquad \Upsilon = \det \left(\frac{\partial(\mathcal{E}^\sigma)}{\partial(\sigma)} \right). \tag{27}
$$

The determinant Γ in (27) is the determinant that would be obtained by using $Y_{N_s} = 1 - \sum_{k \neq N_s} Y_k$ in the governing equations. On the other hand, the determinant Υ in (27) is simply that of the Jacobian matrix of the discrete system whose continuous equations and boundary conditions are obtained by summing all the corresponding species governing equations, and in which the unknown function is $\sigma = \sum_{k \in S} Y_k$. The singularities in the equations due to the mass constraints arise now when the later determinant Υ is zero. These singularities may lead to convergence difficulties, or decrease the domain of convergence of various numerical methods or poor sensitivity information. They may also lead to artificial mass creation in such a way that $\sigma = \sum_{k \in S} Y_k$ is not uniformly unity so that the lower left block in (26) is no more zero. Different types of singularities may in fact lead to $\Upsilon = 0$ as detailed in the following where we investigate the governing equations for $\sigma = \sum_{k \in S} Y_k$.

3.2 Singularities for steady flows

By summing up the N_S equations (1) we get that

$$
\rho \frac{\partial}{\partial t} \left(\sum_{k \in S} Y_k \right) + \rho v . \nabla \left(\sum_{k \in S} Y_k \right) = 0, \tag{28}
$$

which specializes to

$$
\rho v . \nabla \left(\sum_{k \in S} Y_k \right) = 0, \tag{29}
$$

for steady flows. A direct consequence is that at a stagnation point where $v = 0$, the equation (29) degenerates. In this situation, the N_S species equations, and thus the governing equations, are numerically linearly dependent. Of course, this is still true when the N_S species equations are written in conservative form since then the sum of the species equations is proportional to the total mass equation. For similar reasons, it still holds if mole fractions or molar concentrations or number densities are used to describe the species. From a discrete point of view, the Jacobian matrix Π is then singular at grid points which exactly coincide with stagnation points. However if v is small but non zero, the Jacobian matrix Π will be ill conditioned since we deduce from (27) (29) that $\kappa(\Pi)$ will behave like $O(1/|v|)$, where $\kappa(\Pi)$ denotes the condition number of Π. A typical example of steady flow involving a stagnation point is provided for instance by the flow obtained with two counterflowing jets [8] [9]. Any flow with recirculation zones also involve stagnation points and thus leads to singular behavior.

Note also that when the simplified expressions (17) are used, the correction velocity V_m must be evaluated with (18) which leads to $V_m = -(\sum_{k\in S} Y_k V_k)/(\sum_{k\in S} Y_k)$. If V_m is evaluated with the simplified expression $V_m = -\sum_{k\in S} Y_k V_k$ [3] [7] [13] [15] [18] then $\sum_{k\in S} Y_k V_k$ does not sum up algebraically to 0 but to $(\sum_{k\in S} Y_k - 1)V_m$ so that (29) is modified such that

$$\rho v . \nabla(\sum_{k\in S} Y_k) = -\nabla.(\rho(\sum_{k\in S} Y_k - 1)V_m), \tag{30}$$

which fortuitously may suppress the singularity at stagnation points. Nevertheless, if $\nabla.(\rho V_m) = 0$ and $v + V_m = 0$ we still have singular behavior. Similarly, when $\nabla.V_m$ and $v + V_m$ are small, the discrete equations are still ill conditioned. Our numerical experience indeed confirms that the term $(\sum_{k\in S} Y_k - 1)V_m$ does not satisfactorily stabilize the governing equations.

3.3 Singularities in the boundary conditions

First, Dirichlet boundary conditions do not introduce any difficulty and provide that $\sum_{k\in S} Y_k = 1$ at that boundary. By summing up the N_S species Neumann boundary conditions (3) we then get that

$$\nabla(\sum_{k\in S} Y_k - 1).n = 0, \tag{31}$$

which usually does not lead to singular discrete equations. However, a classical finite difference technique used to obtain centered second order accurate, discrete boundary conditions consists in introducing ghost points, writing the centered discrete boundary conditions and governing equations at the boundary point and eliminating the ghost point values from the resulting set of equations. In one dimension for instance and for steady flows, the discrete Neumann boundary conditions lead to $(Y_k)_h = (Y_k)_{-h}$, where the subscripts h and $-h$ refer to the first and ghost grid points respectively. We then deduce that $(\rho Y_k V_k)_{h/2} = -(\rho Y_k V_k)_{-h/2}$ where the mass diffusion fluxes are estimated with centered finite differences and where $h/2$ and $-h/2$ refer to the corresponding midpoints. Using then the discrete equation $((\rho Y_k V_k)_{h/2} - (\rho Y_k V_k)_{-h/2})/h + (W_k \omega_k)_0 = 0$, where we have used $v = 0$ and where the subscript 0 refer to the boundary point, we obtain the following species centered second order accurate, Neumann discrete boundary condition for one dimensional steady flows

$$-\frac{2}{h}(\rho Y_k V_k)_{h/2} + (W_k \omega_k)_0 = 0, \qquad k \in \mathcal{S}. \tag{32}$$

These equations again lead to a singularity since generally the discrete flux velocities $(\rho Y_k V_k)_{h/2}$ at $h/2$ algebraically sum up to zero. Here again, if the simplified expressions (17) are used with the incorrect formulation $V_m = -\sum_{k\in S} Y_k V_k$ for V_m, as in Equation (30), then summing up the N_S species equations (32) leads to the boundary condition $((\sum_{k\in S} Y_k - 1)V_m)_{h/2} = 0$. In this situation, the non singular behavior relies on the poorly known term V_m, and our numerical experience confirms that this term does not satisfactorily stabilize the centered boundary conditions (32).

Finally, by summing up the N_S species mixed boundary conditions (4) we get that

$$\rho v.n(\sum_{k \in \mathcal{S}} Y_k - 1) = 0, \tag{33}$$

so that when $v.n = 0$ the equation degenerates for both steady and unsteady flows. Note that a typical situation where $v.n = 0$ and $\rho Y_k V_k.n = W_k \tilde{\omega}_k$ is for instance that of a solid-gas interface with no surface vapor deposition. On the other hand, for a plane laminar flame, we have $m = \rho u = v.n \neq 0$ where $m = \rho u$ is the laminar flame eigenvalue which is positive [18] and there is no singularity.

4   MODIFIED DIFFUSION ALGORITHMS

In this section we modify the various diffusion algorithms in order to eliminate the singularities exhibited in the previous section. These singularities are indeed due to the absence of suitable diffusion terms in the various governing equations for $\sum_{k \in \mathcal{S}} Y_k$. Diffusion terms are indeed missing because diffusion matrices are not invertible due to the mass conservation constraints. In order to locate the origin of the problems, it is very instructive to assume for a while that $d_k + \theta_k \nabla \log T = \nabla X_k$. In this situation the numerical difficulties are due to the singular diffusion matrix M such that $\rho Y_k V_k = -\sum_{l \in \mathcal{S}} M_{kl} \nabla Y_l$, which may be written $M = \rho \, \mathbf{Diag}(Y_1, \ldots, Y_{N_S}) \, D \, E$ where $\mathbf{Diag}(Y_1, \ldots, Y_{N_S})$ is the matrix whose nonzero entries are the diagonal elements $(Y_1, \ldots, Y_{N_S})$, D the diffusion matrix of Equation (5) and $E = (E_{lm}) = ((W/W_m)(\delta_{lm} - Y_l W/W_l))$ is such that

$$\nabla X_l = \sum_{m \in \mathcal{S}} E_{lm} \nabla Y_m. \tag{34}$$

But the matrix M is not invertible since, first, D is not invertible and, second, E is not invertible, because $\sum_{k \in \mathcal{S}} X_k = 1$ implies that $\sum_{k \in \mathcal{S}} \nabla X_k = 0$ and thus $\sum_{k \in \mathcal{S}} E_{kl} = 0$. Therefore we deduce from this simple analysis that part of the difficulties are due to the matrix E, i.e., to the relations between mole and mass fractions, and part are due to the matrix D, i.e., to the diffusion algorithms.

4.1   Mole and mass fractions

The matrix E, which relates the gradients of the mole fractions to those of the mass fractions, is singular because Equation (8) imposes the relation $\sum_{k \in \mathcal{S}} X_k = 1$ independently of the mass fractions. Similarly, the dual relations between X_k and Y_k [14], usually expressed with (7) and

$$W = \sum_{k \in \mathcal{S}} X_k W_k, \tag{35}$$

lead to the relation $\sum_{k \in \mathcal{S}} Y_k = 1$ independently of the mole fractions so that the matrix F, which relates the gradients of the mass fractions to these of the mole fractions,

$$\nabla Y_l = \sum_{m \in \mathcal{S}} F_{lm} \nabla X_m \tag{36}$$

is also singular. In both cases, the singularity of E and F is due to the fact that the corresponding constraints $\sum_{k \in \mathcal{S}} X_k = 1$ and $\sum_{k \in \mathcal{S}} Y_k = 1$ are imposed *a priori*.

The correct formulation for W is indeed

$$\left(\sum_{k \in \mathcal{S}} Y_k\right)/W = \sum_{k \in \mathcal{S}} Y_k/W_k, \tag{37}$$

since it leads to invertible relations between X_k and Y_k and provides the identity

$$\sum_{k \in \mathcal{S}} X_k = \sum_{k \in \mathcal{S}} Y_k. \tag{38}$$

The dual formulation, which now can be deduced from (7) and (37), is then

$$\left(\sum_{k \in \mathcal{S}} X_k\right) W = \sum_{k \in \mathcal{S}} X_k W_k. \tag{39}$$

With these new relations the matrices E and F become invertible and are inverse of each other and one may easily check that $\det(E) = \prod_{k \in \mathcal{S}}(W/W_k)$. Finally, from the relations (6) (10) and (38) we get the important relation

$$\sum_{k \in \mathcal{S}}(d_k + \theta_k \nabla \log T) = \sum_{k \in \mathcal{S}} \nabla X_k = \nabla\left(\sum_{k \in \mathcal{S}} X_k\right) = \nabla\left(\sum_{k \in \mathcal{S}} Y_k\right), \tag{40}$$

since the thermal diffusion, pressure and external force terms sum up to zero.

4.2 Diffusion velocities

As for the matrices E and F, the singularity of the matrices D and Δ is due to the fact that the corresponding mass constraints (11) and (15) are imposed *a priori*. Now any modification of the diffusion matrix D should take into account the symmetry of D, leave unchanged the physical hyperplane $\{ \xi, \sum_{l \in \mathcal{S}} \xi_l = 0 \} = U^\perp$ and promote the positivity of D. We must thus modify D in the form

$$\tilde{D} = D + \alpha U U^T, \tag{41}$$

where $\tilde{D}$ denotes the modified matrix, U the vector $U = (1, \dots, 1)$ and α a positive function. Note that the resulting matrix $\tilde{D}$ is positive definite since D is positive on $U^\perp$ and $\alpha U U^T$ is positive on $\mathbb{R}U$. The corresponding diffusion velocities satisfy now

$$\sum_{k \in \mathcal{S}} Y_k V_k = - \mathcal{D} \sum_{k \in \mathcal{S}}(d_k + \theta_k \nabla \log T), \tag{42}$$

where $\mathcal{D} = \alpha(\sum_{k \in \mathcal{S}} Y_k)$ is positive, which combined to (40) yields the important relation

$$\sum_{k \in \mathcal{S}} Y_k V_k = - \mathcal{D} \nabla\left(\sum_{k \in \mathcal{S}} Y_k\right). \tag{43}$$

Further note that the matrix $\rho \, \mathbf{Diag}(Y_1, \dots, Y_{N_S}) \tilde{D}$ relating the mass fluxes $\rho Y_k V_k$ to the mole fraction gradients ∇X_k, which is not symmetric, has bounded coefficients and has positive eigenvalues as product of two positive definite symmetric matrices [10].

The dual relations must also be modified into

$$\tilde{\Delta} = \Delta + \beta Y Y^T, \tag{44}$$

where $\tilde{\Delta}$ denotes the modified matrix, Y the vector $Y = (Y_1, \dots, Y_{N_S})$ and β a positive function. The resulting matrix $\tilde{\Delta}$ is positive definite since Δ is positive on $Y^\perp$ and $\beta Y Y^T$ is positive on $\mathbb{R}Y$. One may easily check that when α and β are related through the relation

$$\alpha\beta(\sum_{k\in\mathcal{S}} Y_k)^2 = 1, \tag{45}$$

where $\boldsymbol{U}^T\boldsymbol{Y} = \boldsymbol{Y}^T\boldsymbol{U} = \sum_{k\in\mathcal{S}} Y_k$, we then have $\widetilde{\boldsymbol{\Delta}}\widetilde{\boldsymbol{D}} = \boldsymbol{I}$ where $\boldsymbol{I}$ is the identity matrix. A consequence of (44) is that the following modified Stefan-Maxwell equations

$$\boldsymbol{d}_k + \theta_k \boldsymbol{\nabla}\log T = \sum_{\substack{l\in\mathcal{S}\\l\neq k}} \left(\frac{X_k X_l}{\mathcal{D}_{kl}} - \beta Y_k Y_l\right)\boldsymbol{V}_l - \left(\sum_{\substack{l\in\mathcal{S}\\l\neq k}} \frac{X_k X_l}{\mathcal{D}_{kl}} + \beta Y_k Y_k\right)\boldsymbol{V}_k, \qquad k\in\mathcal{S}, \tag{46}$$

define uniquely the diffusion velocities $\boldsymbol{V}_k$ and automatically handle mass conservation constraints in the sense that $-\beta(\sum_{k\in\mathcal{S}} Y_k)\sum_{k\in\mathcal{S}} Y_k \boldsymbol{V}_k = \sum_{k\in\mathcal{S}}(\boldsymbol{d}_k + \theta_k\boldsymbol{\nabla}\log T)$.

To our knowledge, the modified expressions (41) (43) (44) and (46) have not previously been written although related ideas may be found in the literature. For instance $(N_S+1)*(N_S+1)$ regular diffusion matrices are considered in Chapman and Cowling [2]. These matrices, however, unnecessarily increase the size of the linear systems. Linearly independent diffusion driving forces $\boldsymbol{d}_k$ are also considered in Ferziger and Kaper for different purposes [6].

Similarly, the simplified expressions (17) must now be completed with

$$\sum_{k\in\mathcal{S}} Y_k(\boldsymbol{V}_k + \boldsymbol{V}_m) = -\mathcal{D}\boldsymbol{\nabla}(\sum_{k\in\mathcal{S}} Y_k), \tag{47}$$

which corresponds to the modified approximated diffusion matrix $\widetilde{\boldsymbol{D}}^a = \boldsymbol{D}^a + \alpha\boldsymbol{U}\boldsymbol{U}^T$. The associated quadratic form can be shown to be positive definite for $\mathcal{D} = \alpha(\sum_{k\in\mathcal{S}} Y_k)$ large enough. Moreover, since $\boldsymbol{D}^a$ is not symmetric, one may also introduce non symmetric modifications such that $\widetilde{\boldsymbol{D}}^a = \boldsymbol{D}^a + \boldsymbol{C}\boldsymbol{U}^T$ where $\boldsymbol{C}$ is a somewhat arbitrary vector. Jones and Boris have indeed introduced an $O(N_S^2)$ iterative algorithm [13] [17] to invert the Stefan-Maxwell equations (16) based on the simplified expressions (17)(18) and these authors pointed out that using modified matrices like $\widetilde{\boldsymbol{D}}^a = \boldsymbol{D}^a + \boldsymbol{C}\boldsymbol{U}^T$ where $\boldsymbol{C}$ is arbitrary was feasible. However they have found convenient to use $\boldsymbol{C} = 0$ [13] [17].

In summary, the mutlticomponent diffusion matrix from the kinetic theory of gases (4) should be modified as in (41), the stefan-Maxwell equations (16) should be modified as in (46), the modified expressions (47) should replace the relations (18) and the expressions (37)(39) should be used to relate the mole and mass fractions.

4.3   Nonsingular behavior

We must now investigate the properties of the governing equations when the modified relations (38) and (43) are used. Summing up the N_S species equations (1) for a steady flow, we first get that

$$\rho\boldsymbol{v}.\boldsymbol{\nabla}(\sum_{k\in\mathcal{S}} Y_k) = -\boldsymbol{\nabla}.(\mathcal{D}\boldsymbol{\nabla}(\sum_{k\in\mathcal{S}} Y_k)), \tag{48}$$

and the new artificial diffusion term suppress the singular behavior at stagnation points. Heuristically, $\sum_{k\in\mathcal{S}} Y_k$ behaves now like a new species which tends to diffuse until the equilibrium state $\sum_{k\in\mathcal{S}} Y_k = 1$, normally imposed by the boundary conditions, is reached. But once the equilibrium state is reached, we then have $\sum_{k\in\mathcal{S}} X_k = 1$ from (38) and $\sum_{k\in\mathcal{S}} Y_k \boldsymbol{V}_k = 0$ from (43) so that the mass constraints are ultimately satisfied. Note that the origins of small deviations of $\sum_{k\in\mathcal{S}} Y_k$ from unity are the iteratives processes devoted to solve the discrete governing equations like for instance Newton's method [18].

Note also that the extra diffusion term $\nabla.\left(\mathcal{D}\nabla(\sum_{k\in\mathcal{S}}Y_k)\right)$ stabilizes the equations more satisfactorily than the term involved in Equation (30).

By summing up the $N_\mathcal{S}$ species centered second order accurate, Neumann discrete boundary condition (32) for one dimensional steady flows, we also get

$$\left(\rho\nabla(\sum_{k\in\mathcal{S}}Y_k)\right)_{h/2}=0, \tag{49}$$

which suppress the singular behavior observed with (32).

Similarly, for the boundary conditions (4) when $v.n=0$, we now get, by summing up the $N_\mathcal{S}$ species equations, that

$$\nabla(\sum_{k\in\mathcal{S}}Y_k).n=0, \tag{50}$$

which again suppress the artificial singular behavior. The species $\sum_{k\in\mathcal{S}}Y_k$ of course appears as non reactive in (50).

More generally, the boundary value problem in $\sigma=\sum_{k\in\mathcal{S}}Y_k$ obtained by summing up all the species equations and boundary conditions — whose Jacobian matrix has determinant Υ — is a linear convection-diffusion problem which is generally well posed. Of course, by modifying the mole/mass fractions relations and the diffusion matrices, we have only suppressed the artificial singularities due to the mass constraints which arise through the Υ term in (27). Other types of singularities, like those which appear at extinction limits, i.e., simple turning points, may still occur [8] [9] and arise through the Γ term in (27). Furthermore, suppressing the artificial singularities due to the mass conservation constraints does not suppress other numerical problems which may arise in (5) (16) or (17) like for instance those of vanishing small concentrations [14] [16]. Finally, we have tested the modified expressions (47) by computing various twin premixed strained flames [8] [9] and we have found that the modified expressions improve both the accuracy and the robustness of our numerical algorithms.

5 CONCLUSION

We have investigated mass conservation in multicomponent diffusion algorithms. Various singularities in the governing equations, due to mass conservation constraints, have been exhibited when all mass fractions are considered as independent unknowns. Modifications of the usual diffusion algorithms have been introduced to eliminate these artificial singularities. These modifications, of course, do not change the actual values of the diffusion velocities. Only their mathematical expressions are changed.

ACKNOLEDGEMENT

I would like to thank Dr. B. Laboudigue for interesting discussions concerning this material.

REFERENCES

[1] A. Ben-Israel and T. N. E. Greville, *Generalized Inverses, Theory and Applications*, John Wiley & Sons, Inc., New York, (1974).

[2] S. Chapman and T. G. Cowling, *The Mathematical Theory of Non-Uniform Gases*, Cambridge University Press, Cambridge, (1970).

[3] T. P. Coffee and J. M. Heimerl, *Transport Algorithms for Premixed, Laminar Steady-state Flames*, Comb. and Flame **43** (1981) 273–289.

[4] C. F. Curtiss, *Symmetric Gaseous Diffusion Coefficients*, J. Chem. Phys. **49**, No 7, (1968) 2917–2919.

[5] G. Dixon-Lewis, *Flame Structure and Flame Reaction Kinetics, II. Transport Phenomena in Multicomponent Systems*, Proc. Roy. Soc. **A 307** (1968) 111–135.

[6] J. H. Ferziger and H. G. Kaper, *Mathematical Theory of Transport Processes in Gases*, North Holland Pub. Co., (1972).

[7] Giovangigli V. and Darabiha N., *Vector Computers and Complex Chemistry Combustion*, Mathematical Modeling in Combustion and Related Topics, C. Brauner and C. Schmidt-Laine Eds., M. Nijhoff Pub., NATO ASI Series **140**,1988, p. 491.

[8] V. Giovangigli and M. D. Smooke, *Extinction Limits of Strained Premixed Laminar Flames with Complex Chemistry*, Comb. Science and Tech. **53** (1987) 23–49.

[9] V. Giovangigli and M. D. Smooke, *Adaptive Continuation Algorithms with Application to Combustion Problems*, submitted to Applied Numerical Mathematics.

[10] G. H. Golub and C. F. Van Loan, *Matrix Computations*, The Johns Hopkins University Press, Baltimore, (1983).

[11] J. O. Hirschfelder and C. F. Curtiss, *Flame Propagation in Explosive Gas Mixtures*, in : Third Symposium (International) on Combustion, Reinhold, New York, (1949) 121–127.

[12] J. O. Hirschfelder, C. F. Curtiss and R. B. Bird, *Molecular Theory of Gases and Liquids*, John Wiley & Sons, Inc., New York, (1954).

[13] W. W. Jones and J. P. Boris, *An Algorithm for Multispecies Diffusion Fluxes*, Comp. Chem. **5** (1981) 139–146.

[14] R. J. Kee, G. Dixon-Lewis, J. Warnatz, M. E. Coltrin and J. A. Miller, *A Fortran Computer Code Package for the Evaluation of Gas-Phase Multicomponent Transport Properties*, SANDIA National Laboratories Report, SAND86-8246, (1986).

[15] R. J. Kee, J. A. Miller and T. H. Jefferson, *CHEMKIN: A General-Purpose, Problem-Independent, Transportable, Fortran Chemical Kinetics Code Package*, SANDIA National Laboratories Report, SAND80-8003, (1980).

[16] R. J. Kee, J. Warnatz and J. A. Miller, *A Fortran Computer Code Package for the Evaluation of Gas-Phase Viscosities, Conductivities, and Diffusion Coefficients*, SANDIA National Laboratories Report, SAND83-8209, (1983).

[17] E. S. Oran and J. P. Boris, *Detailed Modeling of Combustion Systems*, Prog. Energy Combust. Sci. **7** (1981) 1–72.

[18] M. D. Smooke, *Solution of Burner-Stabilized Premixed Laminar Flames by Boundary Value Methods*, J. Comp. Phys. **48** (1982) 72–105.

[19] J. Van de Ree, *On the Definition of the Diffusion Coefficients in Reacting gases*, Physica, **36**, (1967), 118–126.

[20] J. Warnatz, *Calculation of the Structure of Laminar Flat Flames I : Flame Velocity of Freely Propagating Ozone Decomposition Flames*, Ber. Bunsenges. Phys. Chem. **82** (1978) 193–200.

[21] J. Warnatz, *Influence of Transport Models and Boundary Conditions on Flame Structure* in Numerical Methods in Laminar Flame Propagation, N. Peters and J. Warnatz, eds., Vieweg Verlag, Braunschweig, (1982).

[22] F. A. Williams, *Combustion Theory*, Second ed., The Benjamin/Cummings Pub. Co. Inc., Melo park, (1985).

[23] L. C. Woods, *The Thermodynamics of Fluid Systems*, Oxford Engineering Science Series **2**, Clarendon press, Oxford, (1986).

UPWIND METHODS FOR FLOWS WITH NON–EQUILIBRIUM CHEMISTRY AND THERMODYNAMICS

B. Grossman and P. Cinnella

Department of Aerospace and Ocean Engineering
Virginia Polytechnic Institute and State University
Blacksburg, Virginia 24061 USA

ABSTRACT

The numerical computation of gas flows with non-equilibrium thermodynamics and chemistry is considered. Several thermodynamic models are discussed, including an equilibrium model, a general non-equilibrium model and a simplified model based upon vibrational relaxation. The effects of the various models on the state equation and the homogeneity property of the Euler equations is described. Flux-splitting procedures are developed for the fully-coupled inviscid equations involving fluid dynamics, chemical production and internal energy relaxation processes. New forms of flux-vector split and flux-difference split algorithms valid for non-equilibrium flow, are embodied in a fully coupled, implicit, large-block structure. Several numerical examples in one space dimension are presented, including high-temperature nozzle flows with hydrogen-air chemistry.

INTRODUCTION

The numerical prediction of combustion in a high-speed flow environment remains a challenging computational problem. Flow fields such as those to be encountered in the SCRAMJET engine of future hypersonic vehicles may produce significant departures from equilibrium chemistry and thermodynamics. The design of these vehicles will require very accurate solutions to the entire three-dimensional non-equilibrium flow field.

Accurate solutions of shock-wave dominated flows have been obtained using the class of algorithms referred to as *upwind* or *flux split*, (e.g., see the survey papers of Harten, Lax and Van Leer [1] and Roe [2]). These methods which include flux-vector splitting and flux-difference splitting, utilize difference procedures which are biased in the direction determined by the signs of the characteristic speeds. These approaches were originally developed for perfect gases and rely on the simplicity of the equation of state to develop algebraic relationships for the split fluxes and their associated Jacobians.

In this paper, we seek to develop upwind methods for the computation of non-equilibrium flows. We will concentrate only on the convective algorithms and will not consider the other important features of this flow regime, such as viscous effects and radiative transfer. The algorithms will be developed for the one-dimensional Euler equations. Extensions to multidimensional flows may be accomplished through finite-volume methods using procedures such as those developed by Walters and Thomas [3]. We will also limit our discussion to the frequently-used flux-vector split algorithms of Steger and Warming [4] and Van Leer [5] and the flux-difference split algorithm of Roe [6].

The computation of flows with non-equilibrium chemistry and thermodynamics is considerably more complex than the perfect gas situation. The number of dependent variables and partial differential equations increases, with production equations necessary for each species mass density (or mass fraction), and, for the case of non-equilibrium internal energy, a production equation for the portion of energy not in equilibrium, e.g., vibrational relaxation. Difficulties appear, due to the often disparate time scales associated with the fluid motion and the non-equilibrium chemistry and thermodynamics. One method to avoid this *stiffness* problem and the approach taken here, is to utilize a fully-coupled procedure, where the equations governing the fluid dynamics and non-equilibrium chemistry and thermodynamics are solved simultaneously with implicit numerical methods. This results in a very complex, large block structure for the solution algorithm which, however, fully accounts for all the non-equilibrium effects.

In previous efforts, [7], [8], the present authors developed flux-split algorithms for a specific non-equilibrium energy model, which included a relaxing vibrational energy and negligible electronic excitation. This model, which we will call the *simplified vibrational energy model*, led to a simplified equation of state which allowed a non-iterative evaluation of the pressure and temperature from the conserved variables. We now extend this approach to more general energy models, which will include the previous simplified vibrational energy model, along with an equilibrium energy model as special cases. An important feature of this approach is that by performing the calculations with different energy models, we can assess the importance of non-equilibrium internal energy on high-temperature flows.

In the next section, we discuss our chemical and thermodynamic models along with the corresponding equation of state. We derive the flux Jacobian for the Euler equations, and develop the corresponding eigenvalues and eigenvectors. We then develop non-equilibrium flow versions of Steger-Warming and Van Leer flux-vector split schemes and derive an approximate Riemann solver, which results in a flux-difference split scheme of the Roe type. We will present several elementary numerical examples of non-equilibrium flow computations in order to illustrate the accuracy and wave-capturing properties of our methodology. The test cases involve the steady combustion of a hydrogen-air mixture in a supersonic diffuser, including an embedded shock wave. The effects of non-equilibrium thermodynamics will be discussed for these cases. Also the beneficial effect of utilizing implicit computational procedures on the *stiff* set of governing equations will be discussed. Future research will be required to fully evaluate the accuracy of the algorithms and the efficiency of the fully-coupled approach in multi-dimensional flows.

FORMULATION

Chemical and Thermodynamic Models

At high temperatures, chemical reactions will occur in gas flows resulting in changes in the amount of mass of each chemical species. We consider a system containing N species and assume that our chemical kinetics is limited to homogeneous reactions, neglecting any solid or liquid surface reactions and excluding all photochemical reactions. We assume that all the chemical reactions are known along with their corresponding forward and backward reaction rates. For non-equilibrium chemistry, the rate of production of species i due to chemical reaction may be written as

$$\dot{w}_i \equiv \frac{d\rho_i}{dt} = f(\rho_1, \rho_2, \ldots, \rho_N, T), \tag{1}$$

where ρ_i is the species mass density.

At high temperatures, imperfect gas effects are due to chemical changes in the amount of mass of each species and to the activation of internal energy modes which behave non-linearly with temperature. As long as the pressure is sufficiently low, away from the gas triple point, then it has been found that each species of the gas mixture will behave as a thermally perfect gas. That is, $e_i = e_i(T_i)$ and $p_i = \rho_i R_i T_i$, where T_i is the translational temperature, e_i is the internal energy per unit mass, p_i, the partial pressure, ρ_i the density and R_i the gas constant for species i. The gas constant for species i may be expressed as $R_i = \mathcal{R}/M_i$ where $\mathcal{R}$ is the universal gas constant (per mole) and M_i is the mass per mole of species i.

The thermodynamic model that will be considered here is to assume that a portion of the internal energy of each species is in thermodynamic equilibrium, and that the remaining portion is in a non-equilibrium state. The non-equilibribrium part of the energy is assumed to be modeled by appropriate production rates, c.f., Herzberg [9], and Jaffe [10]. We will also make the simplifying assumption that the mass of each species is approximately the same, whereby the translational temperature of all species will assumed to be the same, or $T_i = T$. Thus, the analysis will not be applicable to flows with free electrons.

We begin by defining the energy per unit mass of species i as

$$e_i = \tilde{e}_i(T) + e_{n_i}, \tag{2}$$

where $\tilde{e}_i$ is the equilibrium portion of the energy. It is convenient to express $\tilde{e}_i$ in terms of a specific heat as

$$\tilde{e}_i = \int_{T_{ref}}^{T} \tilde{c}_{v_i}(T)\, dT + h_{f_i}, \tag{3}$$

where the specific heat at constant volume, $\tilde{c}_{v_i} = d\tilde{e}_i/dT$, and h_{f_i} is the heat of formation of species i.

We consider a gas mixture composed of N species, with the first M species assumed to contain a non-equilibrium portion of their internal energy. The internal energy per unit mass of the mixture may then be written as

$$e = \sum_{i=1}^{N} \frac{\rho_i}{\rho} e_i = \tilde{e} + \sum_{i=1}^{M} \frac{\rho_i}{\rho} e_{n_i}, \tag{4}$$

where we have introduced the definition of reduced specific internal energy

$$\tilde{e} \equiv \sum_{i=1}^{N} \frac{\rho_i}{\rho} \tilde{e}_i. \tag{5}$$

It is convenient to define the reduced specific heats of the mixture as

$$\tilde{c}_v \equiv \sum_{i=1}^{N} \frac{\rho_i}{\rho}\tilde{c}_{v_i}, \qquad \tilde{c}_p \equiv \sum_{i=1}^{N} \frac{\rho_i}{\rho}\tilde{c}_{p_i}. \qquad (6a, b)$$

We may also write the mixture gas constant as

$$\tilde{R} \equiv \sum_{i=1}^{N} \frac{\rho_i}{\rho}R_i = \tilde{c}_p - \tilde{c}_v, \qquad (7)$$

and it is convenient to define

$$\tilde{\gamma} \equiv \tilde{c}_p/\tilde{c}_v. \qquad (8)$$

Equation of State

For the conditions specified here, each individual species behaves as a thermally perfect gas. Then, the pressure, according to Dalton's law, is

$$p = \sum_{i=1}^{N} \rho_i R_i T = \rho\tilde{R}T, \qquad (9)$$

where the mass density of the mixture is

$$\rho = \sum_{i=1}^{N} \rho_i. \qquad (10)$$

The state relationship of the pressure to the specific internal energy occurs implicitly through the temperature. For a given chemical composition, internal energy and non-equilibrium energy, the temperature must be evaluated from

$$e = \sum_{i=1}^{N} \frac{\rho_i}{\rho}\left[\int_{T_{ref}}^{T} \tilde{c}_{v_i}(T)\, dT + h_{f_i} \right] + \sum_{i=1}^{M} \frac{\rho_i}{\rho}e_{n_i}. \qquad (11)$$

Iterative procedures may be used to solve for T. Once T is found, the pressure is determined from (9).

The appropriate speed of sound for the models prescribed here is the frozen speed of sound, defined as

$$a^2 \equiv \left(\frac{\partial p}{\partial \rho} \right)_{s,\,\rho_i/\rho,\,e_{n_i}} = \left(\frac{\partial p}{\partial \rho} \right)_{e,\,\rho_i/\rho,\,e_{n_i}} + \left(\frac{\partial p}{\partial e} \right)_{\rho,\,\rho_i/\rho,\,e_{n_i}} \left(\frac{\partial e}{\partial \rho} \right)_{s,\,\rho_i/\rho,\,e_{n_i}}, \qquad (12)$$

where s is the entropy per unit mass. Utilizing the First Law of Thermodynamics, along with (9) and (11), we find the interesting result that

$$a^2 = \tilde{\gamma}\tilde{R}T = \tilde{\gamma}\left(\frac{p}{\rho}\right). \qquad (13)$$

The above result is not approximate, but corresponds to the frozen speed of sound for this chemically reacting, non-equilibrium flow.

Governing Equations

The governing equations for an inviscid, non-heat-conducting one-dimensional flow with non-equilibrium chemistry and non-equilibrium internal energy may be written in vector conservation form as

$$\frac{\partial U}{\partial t} + \frac{\partial F}{\partial x} = W, \tag{14}$$

where U is the vector of conserved variables, F is the flux vector and W is the vector of production rates given by

$$
U = \begin{pmatrix} \rho_1 \\ \rho_2 \\ \vdots \\ \vdots \\ \rho_N \\ \rho u \\ \rho_1 e_{n_1} \\ \vdots \\ \rho_M e_{n_M} \\ \rho e_0 \end{pmatrix}, \quad
F = \begin{pmatrix} \rho_1 u \\ \rho_2 u \\ \vdots \\ \vdots \\ \rho_N u \\ \rho u^2 + p \\ \rho_1 e_{n_1} u \\ \vdots \\ \rho_M e_{n_M} u \\ \rho u h_0 \end{pmatrix}, \quad
W = \begin{pmatrix} \dot{w}_1 \\ \dot{w}_2 \\ \vdots \\ \vdots \\ \dot{w}_N \\ 0 \\ \rho_1 \dot{e}_{n_1} + e_{n_1} \dot{w}_1 \\ \vdots \\ \rho_M \dot{e}_{n_M} + e_{n_M} \dot{w}_M \\ 0 \end{pmatrix}. \tag{15a,b,c}
$$

Equation (14) represents $N + M + 2$ conservation equations, with the first N corresponding to species continuity, followed by momentum conservation, M non-equilibrium energy conservation equations and the total energy conservation equation. In the above equations, u is the velocity in the x direction, ρ is the global mass density, p is the pressure, e_0 is the total energy per unit mass, h_0 is the stagnation enthalpy per unit mass, ρ_i is the mass density of species i and e_{n_i} is the non-equilibrium portion of the internal energy per unit mass of species i. The chemical production terms are $\dot{w}_i$ are assumed to be a given function of temperature and species densities, and the non-equilibrium energy production terms are denoted $\dot{e}_{n_i}$ and are assumed to be a given function of pressure, temperature and non-equilibrium energy.

The system is completed by the density definition, (10) and the equation of state, defined implicitly through (9) and (11), giving the pressure in terms of conserved variables. The stagnation enthalpy, is defined by $h_0 = e_0 + p/\rho$.

Flux Jacobian Matrix

From the above description, the flux vector F may be considered a function of the vector of conserved variables U, or $F = F(U)$. A quantity which plays a major role in flux-vector split algorithms and in any implicit formulation is the Jacobian matrix

$\mathbf{A} \equiv \partial F / \partial U$. We have evaluated $\mathbf{A}$, in general, as

$$- \rho \mathbf{A} =$$

$$\begin{pmatrix}
(\rho_1-\rho)u & \rho_1 u & \cdots & \cdots & \rho_1 u & -\rho_1 & 0 & \cdots & 0 & 0 \\
\rho_2 u & (\rho_2-\rho)u & \cdots & \cdots & \rho_2 u & -\rho_2 & 0 & \cdots & 0 & 0 \\
\vdots & \vdots & \vdots & \vdots & \vdots & \vdots & \vdots & \vdots & \vdots & \vdots \\
\vdots & \vdots & \vdots & \vdots & \vdots & \vdots & \vdots & \vdots & \vdots & \vdots \\
\rho_N u & \rho_N u & \cdots & \cdots & (\rho_N-\rho)u & -\rho_N & 0 & \cdots & 0 & 0 \\
a_1 & \cdots & a_M & \cdots & a_N & a_{N+1} & \rho\ell & \cdots & \rho\ell & -\rho\ell \\
\rho_1 e_{n_1} u & \cdots & \rho_1 e_{n_1} u & \cdots & \rho_1 e_{n_1} u & -\rho_1 e_{n_1} & -\rho u & \cdots & 0 & 0 \\
\vdots & \vdots & \vdots & \vdots & \vdots & \vdots & \vdots & \vdots & \vdots & \vdots \\
\rho_M e_{n_M} u & \cdots & \rho_M e_{n_M} u & \cdots & \rho_M e_{n_M} u & -\rho_M e_{n_M} & 0 & \cdots & -\rho u & 0 \\
b_1 & \cdots & b_M & \cdots & b_N & b_{N+1} & \rho u \ell & \cdots & \rho u \ell & -\tilde{\gamma}\rho u
\end{pmatrix}$$

$$(16)$$

where $a_i = \rho(u^2 - \partial p/\partial\rho_i)$, $b_i = \rho u(h_0 - \partial p/\partial\rho_i)$, for $i = 1,\dots,N$ along with $a_{N+1} = -\rho u(3 - \tilde{\gamma})$, $b_{N+1} = \rho(u^2\ell - h_0)$ and $\ell \equiv (\tilde{\gamma}-1)$.

It can be shown that for this set of equations, the Jacobian matrix $\mathbf{A}$ has the homogeneity property

$$F = \mathbf{A} U. \tag{17}$$

Of course, this property was noted for thermally perfect gases in [4].

The $N + M + 2$ eigenvalues of the Jacobian matrix $\mathbf{A}$ are found to be

$$\lambda_i = \begin{cases} u & i = 1,\dots,N+M \\ u + a & i = N+M+1 \\ u - a & i = N+M+2 \end{cases} \tag{18}$$

The corresponding set of $N + M + 2$ linearly independent right eigenvectors have been evaluated as:

$$E_i = \begin{pmatrix} 0 \\ \vdots \\ \rho_i/\rho \\ \vdots \\ 0 \\ \rho_i u/\rho \\ 0 \\ \vdots \\ 0 \\ (u^2 - \xi_i)\rho_i/\rho \end{pmatrix}, \quad E_{j+N} = \begin{pmatrix} 0 \\ \vdots \\ 0 \\ \vdots \\ 0 \\ 0 \\ \vdots \\ \rho_j e_{n_j}/\rho \\ 0 \\ \rho_j e_{n_j}/\rho \end{pmatrix}, \quad E_{k+N+M} = \begin{pmatrix} \rho_1/\rho \\ \rho_2/\rho \\ \vdots \\ \vdots \\ \rho_N/\rho \\ u \pm a \\ \rho_1 e_{n_1}/\rho \\ \vdots \\ \rho_M e_{n_M}/\rho \\ h_0 \pm ua \end{pmatrix}$$

$$i = 1,\dots,N \qquad\qquad j = 1,\dots,M \qquad\qquad k = 1,2.$$

$$(19a,b,c)$$

In (19a), we have introduced the definition from (9)–(11) of

$$\xi_i \equiv \frac{1}{\tilde{\gamma}-1}\frac{\partial p}{\partial \rho_i} = \frac{R_i T}{\tilde{\gamma}-1} - \tilde{e}_i(T) + \frac{u^2}{2}. \tag{20}$$

Flux-Vector Splitting

Utilizing the homogeneity property of the governing equations as determined in (17) we may develop a flux-vector splitting along the lines of Steger and Warming, [4], for perfect gases.

The essential feature of this approach is to diagonalize the Jacobian matrix $\mathbf{A}$ given in (16). We can write

$$\Lambda = \mathbf{S}^{-1} \mathbf{A} \mathbf{S}, \tag{21}$$

where Λ is a diagonal matrix whose diagonal elements correspond to the eigenvalues of $\mathbf{A}$ and the rows of matrix $\mathbf{S}$ are composed of the right eigenvectors of $\mathbf{A}$. A flux-splitting may be developed by splitting Λ into a non-negative matrix Λ^+ and a non-positive matrix Λ^-, where the diagonal elements of $\Lambda^\pm$ are $\lambda^\pm = (\lambda \pm |\lambda|)/2$. Utilizing the homogeneity property, (17),

$$\mathbf{F} = \mathbf{A} \mathbf{U} = \mathbf{S} \left(\Lambda^+ + \Lambda^- \right) \mathbf{S}^{-1} \mathbf{U} = \mathbf{F}^+ + \mathbf{F}^-. \tag{22}$$

We obtain after some algebraic rearrangement

$$\mathbf{F}^\pm = \left(\frac{\tilde{\gamma}-1}{\tilde{\gamma}}\right)\rho\lambda_A^\pm \mathbf{F}_A + \frac{1}{2\tilde{\gamma}}\rho\lambda_B^\pm \mathbf{F}_B + \frac{1}{2\tilde{\gamma}}\rho\lambda_C^\pm \mathbf{F}_C, \tag{23}$$

where

$$\mathbf{F}_A = \begin{pmatrix} \rho_1/\rho \\ \rho_2/\rho \\ \vdots \\ \vdots \\ \rho_N/\rho \\ u \\ \rho_1 e_{n_1}/\rho \\ \vdots \\ \rho_M e_{n_M}/\rho \\ h_0 - a^2/(\tilde{\gamma}-1) \end{pmatrix}, \quad \mathbf{F}_{B,C} = \begin{pmatrix} \rho_1/\rho \\ \rho_2/\rho \\ \vdots \\ \vdots \\ \rho_N/\rho \\ u \pm a \\ \rho_1 e_{n_1}/\rho \\ \vdots \\ \rho_M e_{n_M}/\rho \\ h_0 \pm ua \end{pmatrix}, \tag{24a,b,c}$$

and $\lambda_A = u$, $\lambda_B = u + a$ and $\lambda_C = u - a$. This represents a Steger-Warming-type flux-vector splitting for a vibrationally-relaxing, chemically-reacting flow.

An alternate flux-vector splitting has been developed for perfect gases by Van Leer, [5]. His formulation has continuously differentiable flux contributions and has been shown to result in smoother solutions near sonic points. For the non-equilibrium flow model considered here, the procedures of [5] may be directly utilized by considering the flux vector as $F = F(\rho, a, M, \rho_i/\rho, e_{n_i})$, where M is the Mach number u/a. Then the mass flux, $\rho u = \rho a M$ may be split as $\rho u = f_m^+ + f_m^-$, where

$$f_m^\pm = \pm\rho a\left(\frac{\pm M + 1}{2}\right)^2. \tag{25}$$

The remaining fluxes when written in the above functional form may be split to yield a Van Leer-type flux-vector splitting for a non-equilibrium flow:

$$F^\pm = f_m^\pm \begin{pmatrix} \rho_1/\rho \\ \rho_2/\rho \\ \vdots \\ \vdots \\ \rho_N/\rho \\ [(\tilde{\gamma}-1)u\pm 2a]/\tilde{\gamma} \\ \rho_1 e_{v_1}/\rho \\ \vdots \\ \rho_M e_{v_M}/\rho \\ f_e^\pm \end{pmatrix}, \tag{26}$$

where

$$f_e^\pm = h_0 - m(-u \pm a)^2. \tag{27}$$

and m is an arbitrary parameter. For consistency with previous developments [8], m has been assigned the value $1/(\tilde{\gamma}+1)$. Numerical experiments have shown no sensitivity to the value of this parameter.

Flux-Difference Splitting

The essential features of flux-difference split algorithms involve the solution of local Riemann problems arising from the consideration of discontinuous states at cell interfaces on an initial data line. The scheme developed for perfect gases, by Roe [6], falls into this category and has produced excellent results for both inviscid and viscous flow simulations (c.f., [11]). We now extend this approach for the present non-equilibrium thermodynamic and chemistry model.

We follow the procedure outlined by Glaister [12] for real gases, which followed from the approach of Roe and Pike [13] for perfect gases. In their approach, the first step corresponds to obtaining a solution for the linearized, approximate Riemann problem, valid for small jumps in the interface states. The second step involves the determination of the appropriate averaging procedures, (often referred to as Roe-averages), in order to render the solution valid for arbitrary jumps in the cell interface values. We may extend this procedure directly to our non-equilibrium flow model, following the details outlined in [8] for the simplified vibrational energy model. The resulting algorithm is now summarized.

We initially seek a solution to the approximate Riemann problem for two states, U_r, U_ℓ, which represent the conserved variables evaluated at the cell interface using data from the right and left, respectively. This involves the determination of the relationship between the *jump* in U, denoted as $[\![U]\!] \equiv U_r - U_\ell$, to average eigenvectors $\hat{E}_i$ of the Jacobian matrix, and average wave strengths $\hat{\alpha}_i$, $i = 1, \ldots, N + M + 2$, such that

$$[\![U]\!] = \sum_{i=1}^{N+M+2} \hat{\alpha}_i \, \hat{E}_i. \tag{28}$$

The *jump* in the flux vector F is related similarly as

$$[\![F]\!] = \sum_{i=1}^{N+M+2} \hat{\alpha}_i \, \hat{\lambda}_i \, \hat{E}_i. \tag{29}$$

The eigenvalues $\hat{\lambda}_i$ and the eigenvectors $\hat{E}_i$ are found to be those given in (18) and (19), with averages denoted by a *caret* over each dependent variable. We also adopt the notation

$$\hat{\rho}_i \equiv \left(\widehat{\frac{\rho_i}{\rho}} \right), \qquad \hat{e}_{n_i} \equiv \left(\widehat{\frac{\rho_i e_{n_i}}{\rho}} \right). \tag{30a, b}$$

The corresponding wave strengths $\hat{\alpha}_i$ are found to be:

$$\hat{\alpha}_i = \frac{[\![\rho_i]\!]}{\hat{\rho}_i} - \frac{[\![p]\!]}{\hat{a}^2}, \qquad i = 1, \ldots, N, \tag{31a}$$

$$\hat{\alpha}_{j+N} = \frac{[\![\rho_j e_{n_j}]\!]}{\hat{e}_{n_j}} - \frac{[\![p]\!]}{\hat{a}^2}, \qquad j = 1, \ldots, M, \tag{31b}$$

$$\hat{\alpha}_{k+N+M} = \frac{1}{2\hat{a}^2} \left([\![p]\!] \pm \hat{\rho}\hat{a}[\![u]\!] \right), \qquad k = 1, 2. \tag{31c}$$

The solution of the approximate Riemann problem involves the algebraic determination of the averages $\hat{\rho}$, $\hat{u}$, $\hat{a}$, $\hat{p}$, $\hat{h}_0$, $\hat{\rho}_i$, $\hat{e}_{n_i}$, $\hat{\xi}_i$, such that (28) and (29) are satisfied.

We define the Roe-average of a quantity f to be $\Re(f) \equiv (f_r\sqrt{\rho_r} + f_\ell\sqrt{\rho_\ell})/(\sqrt{\rho_r} + \sqrt{\rho_\ell})$. Most of the averages turn out to be Roe-averages. Namely, $\hat{u} = \Re(u)$, $\hat{h}_0 = \Re(h_0)$, $\hat{\rho}_i = \Re(\rho_i/\rho)$, $\hat{e}_{n_j} = \Re(\rho_j e_{n_j}/\rho)$, $\hat{R} = \Re(\tilde{R})$, $\hat{T} = \Re(T)$ and $\hat{e}_i = \Re(\tilde{e}_i)$. The remaining quantities needed are the usual density average, $\hat{\rho} = \sqrt{\rho_r \rho_\ell}$ and

$$\hat{c}_v^* = \sum_{i=1}^{N} \hat{\rho}_i c_{v_i}^*, \tag{32a}$$

where

$$c_{v_i}^* \equiv \frac{1}{[\![T]\!]} \int_{T_\ell}^{T_r} \tilde{c}_{v_i} \, dT. \tag{32b}$$

We also find

$$\hat{\xi}_i = R_i \hat{T} - \hat{e}_i + \frac{\hat{u}^2}{2}, \tag{33}$$

and

$$\hat{\gamma} - 1 = \frac{\hat{R}}{\hat{c}_v^*}, \tag{34}$$

and

$$\hat{a}^2 = (\hat{\gamma} - 1)\left[\hat{h}_0 - \frac{\hat{u}^2}{2} + \hat{c}_v^*\hat{T} - \sum_{i=1}^{N}\hat{\rho}_i\hat{e}_i - \sum_{j=1}^{M}\hat{e}_{n_j}\right]. \tag{35}$$

We have not proven that the average sound speed, defined above, will be positive for arbitrary left and right states. However, numerical experiments with very strong discontinuities confirm its utility.

We have now defined the averages so that (29) constitutes an approximate Riemann solver. We may rewrite (29) by grouping the repeated eigenvalues and upon algebraic simplification we obtain

$$[\![F]\!] = [\![F]\!]_A + [\![F]\!]_B + [\![F]\!]_C \tag{36}$$

where the $[\![F]\!]_A$ term arises from the first $N+M$ terms of the sum in (29), corresponding to the repeated eigenvalue $\lambda_i = \hat{u}$ and may be written as

$$[\![F]\!]_A = \left([\![\rho]\!] - \frac{[\![p]\!]}{\hat{a}^2}\right)\hat{u}\begin{pmatrix}\hat{\rho}_1 \\ \hat{\rho}_2 \\ \vdots \\ \vdots \\ \hat{\rho}_N \\ \hat{u} \\ \hat{e}_{n_1} \\ \vdots \\ \hat{e}_{n_M} \\ \hat{h}_0 - \hat{a}^2/(\hat{\gamma}-1)\end{pmatrix} + \hat{\rho}\hat{u}\begin{pmatrix}[\![\rho_1/\rho]\!] \\ [\![\rho_2/\rho]\!] \\ \vdots \\ \vdots \\ [\![\rho_N/\rho]\!] \\ 0 \\ [\![\rho_1 e_{n_1}/\rho]\!] \\ \vdots \\ [\![\rho_M e_{n_M}/\rho]\!] \\ \sum^M[\![\rho_j e_{n_j}/\rho]\!] - \sum^N \hat{\xi}_i[\![\rho_i/\rho]\!]\end{pmatrix} \tag{37a}$$

Similarly the $[\![F]\!]_B$ and $[\![F]\!]_C$ arise from the last two terms in the summation of (29), corresponding to the eigenvalues $\hat{u} + \hat{a}$ and $\hat{u} - \hat{a}$, and are found to be

$$[\![F]\!]_{B,C} = \frac{1}{2\hat{a}^2}\left([\![p]\!] \pm \hat{\rho}\hat{a}[\![u]\!]\right)(\hat{u} \pm \hat{a})\begin{pmatrix}\hat{\rho}_1 \\ \hat{\rho}_2 \\ \vdots \\ \vdots \\ \hat{\rho}_N \\ \hat{u} \pm \hat{a} \\ \hat{e}_{n_1} \\ \vdots \\ \hat{e}_{n_M} \\ \hat{h}_0 \pm \hat{u}\hat{a}\end{pmatrix}. \tag{37b, c}$$

The approximate Riemann solver is implemented in a numerical integration of the governing equations (14), by computing the cell interface flux as a summation over wave speeds, [2], as

$$F_{i+1/2} = \frac{1}{2}\left(F_r + F_\ell\right) - \frac{1}{2}\sum_{i=1}^{N+M+2} \hat{\alpha}_i \, |\hat{\lambda}_i| \, \hat{E}_i, \tag{38}$$

Numerical Formulation

The implementation of the non-equilibrium flux-split algorithms follows procedures similar to those for perfect gases, e.g., [3]. Since we anticipate *stiffness* problems in the numerical solution of the governing equations, we seek to compute solutions using an Euler-implicit time integration.

The implementation of the implicit algorithm in a flux-vector split scheme involves the splitting of the flux, $F = F^+ + F^-$, and the splitting of the flux Jacobian, $\mathbf{A} = \mathbf{A}^+ + \mathbf{A}^-$. The spatial derivatives of these terms are evaluated using the usual upwind differencing, (in MUSCL form), appropriate for flux split codes. The details may be found, for example, in [3]. The split flux Jacobians, $\mathbf{A}^\pm = \partial F^\pm/\partial U$, have been analytically determined from the flux-vector splittings given in (23), (24) for the Steger-Warming-type scheme and given in (25)–(27) for the Van Leer-type scheme.

The Jacobian of the chemical and thermodynamic source terms, $\partial W/\partial U$, are needed for the computation and depend upon the specific chemical and thermodynamic models. Some examples are given in [8]. In addition, for a quasi-one-dimensional flow computation, the effect of area variation on the governing equations (14) may be included by adding to the source vector W, (15c), a vector W_a, where

$$W_a = -\frac{u}{A}\frac{dA}{dx}\begin{pmatrix} \rho_1 \\ \vdots \\ \rho_N \\ \rho u \\ \rho_1 e_{n_1} \\ \vdots \\ \rho_M e_{n_M} \\ \rho h_0 \end{pmatrix}, \tag{39}$$

and $A(x)$ is the cross-sectional area distribution. The Jacobian $\partial W_a/\partial U$ must also be included in an implicit formulation.

NUMERICAL SOLUTIONS

We include several numerical examples of flow computations with non-equilibrium chemistry and thermodynamics, in order to illustrate the accuracy and wave-capturing properties of our methodology. The test cases involve the steady combustion in a supersonic diffuser, including an embedded shock wave. In all the test cases presented here, we will utilize the simplified vibrational model. Thus the internal energy of each (non-monatomic) species will contain an equilibrium portion which is linear in T and a non-equilibrium vibrational energy whose production rates come from a Landau-Teller model, [14].

The flow entering the diffuser is composed of hydrogen mixed with air. For the hydrogen–air chemistry, we utilized the same model used in [15], the simple, two-reaction hydrogen-oxygen model of Rogers and Chinitz [16]. It consists of the two basic reactions:

$$H_2 + O_2 \rightleftharpoons 2OH$$
$$2OH + H_2 \rightleftharpoons 2H_2O$$

(41)

and five species $(N = 5)$ are utilized, N_2, O_2, H_2, OH and H_2O, all of which are considered to be vibrationally relaxing $(M = 5)$. The reactions of nitrogen are neglected in this model. The particular geometry, an axisymmetric, rapid-expansion supersonic diffuser, has a radius defined as $r/L = 0.125[1 + \sin(\pi x/2L)]$ for $0 \leq x \leq L$, with a length $L = 2.0$ meters. The flow at the inlet to the diffuser $(x = 0)$, has a velocity of 1245 m/sec., a temperature of 1884.3° K and a pressure of $0.8026 \times 10^5 \, N/m^2$. The initial chemical composition of the flow consists of an equivalence ratio Φ, the ratio of the mass fraction of H_2 to the mass fraction of O_2, normalized by the stoichiometric mass fraction ratio, of $\Phi = 0.29841$. This composition causes very steep concentration gradients near the inlet of the duct.

Computations were performed using three different energy models. The first one utilized the simple vibrational model, which did not require any iterations for the determination of pressure and temperature from the conserved variables. In the second case, the vibrational energy was considered to be in equilibrium, distributed according to an equilibrium harmonic oscillator. The resulting e_i are the sum of linear and exponential contributions, and iterations are required to recover the value of the temperature. For a typical solution using Newton's method, two to three iterations were needed to produce values accurate to seven significant digits. For the third case, a curve-fit e_i distribution using a quadratic temperature profile for each species was incorporated. The resulting model is the same one used by Drummond, Hussaini and Zang [15]. In this model, one can directly solve a quadratic relationship for the temperature from the conserved variables.

All our computations were performed using 161 uniformly-spaced, axial grid points and the equations were solved iteratively, using our fully-coupled implicit method, with the Steger-Warming-type flux-vector splitting, given by (23) and (24). The computation was performed with a *CFL* number of 10 and convergence to a steady state, with a residual reduction of 10^{-10} in $110 \sim 130$ cycles. It may be noted that an explicit solution of the same problem would have required *CFL* numbers of the order of 10^{-3} and at least one order of magnitude more computer time, due to the *stiffness* of the combustion model.

In Fig. 1 we present the comparison of our computations with a linear c_{v_i} model along with the spectral and finite-difference computations of Drummond, Hussaini and Zang [15]. It should be noted that their spectral calculation used 17 nodes, and their finite-difference calculation used 101 grid points (although only every second value is indicated on our plots). The three results for pressure (given in bars, $1 \, bar = 10^5 \, N/m^2$), and temperature are in excellent agreement.

In Fig. 2 we present the comparison of our computations using the three energy models for this case. The three cases do not show any significant differences for the pressure distribution. The temperature distribution however, indicates that substantial

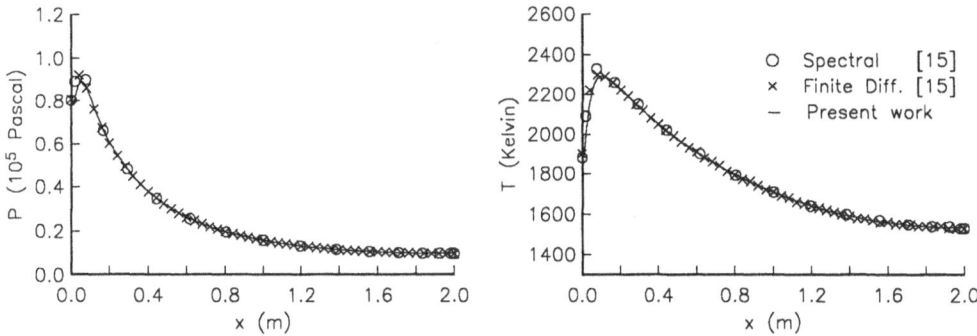

Figure 1. Non-equilibrium flow of an H_2-air mixture through a supersonic diffuser. Comparison with other numerical methods.

non-equilibrium energy effects are present under these conditions. The two equilibrium energy models are, as expected, in agreement. Plots of the vibrational temperatures for each species, (not presented here), show that the flow has not reached vibrational equilibrium. It appears that a non-equilibrium internal energy model is needed to obtain accurate temperature distributions for this case. Whether the simple vibrational energy model is adequate remains to be determined.

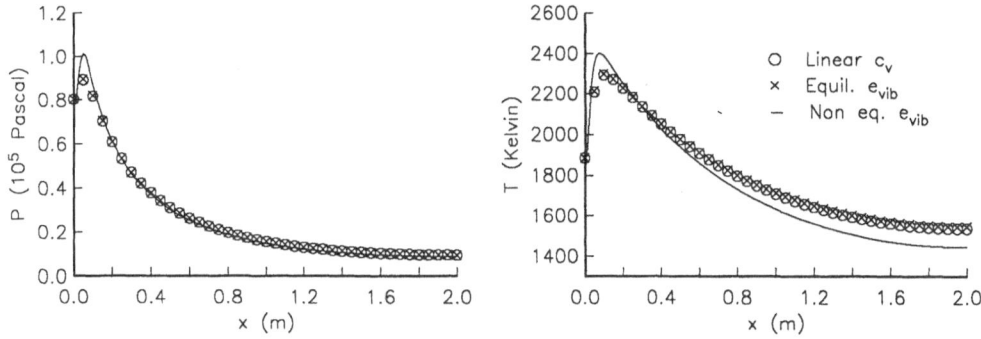

Figure 2. Non-equilibrium flow of an H_2-air mixture through a supersonic diffuser. Comparison between three thermodynamic models.

Another computation was performed with this same supersonic diffuser, for conditions where a normal shock wave stands inside the duct. The exit pressure was raised to 10.02 times the perfectly expanded value. Using the identical inlet conditions and chemistry model, discussed previously, the computations were performed with an exit pressure specified to be 10^5 N/m^2. The conserved variables, $\rho_i, \rho_i e_{n_i}$ and ρu were extrapolated to the exit boundary. The Van Leer-type flux-vector splitting, (25)–(27), with second-order upwind, spacial differencing and a *MIN-MOD* limiter, was used for this computation. We utilized a *CFL* number of 5 for the Euler-implicit time integration.

The resulting pressure and temperature distributions, as plotted in Fig. 3, show the computation performed with the same three thermodynamic models and the same grid as in the previous example. The results for the two equilibrium energy models are again in good agreement, and they both depart from the simplified vibrational model results in the lower temperature region. After the shock, the flow reaches vibrational equilibrium after a short relaxation zone. The slight discrepancy in the temperature profile between the exponential and the quadratic e_i cases are likely due to the differences in the thermodynamic models.

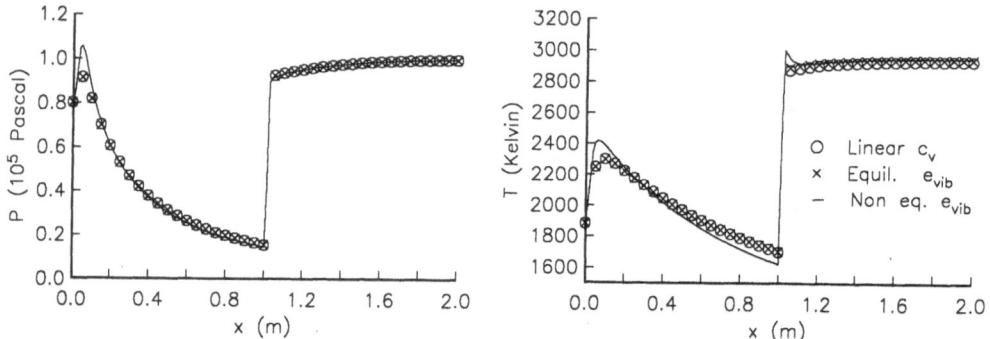

Figure 3.   Non-equilibrium flow of an H_2-air mixture through a supersonic diffuser with an embedded shock wave. Comparison between three thermodynamic models.

The computed mass fractions for both flows considered here, with solutions for the three thermodynamic models are shown in Fig. 4. The left hand figure corresponds to the shockless diffuser flow and the right hand figure is for the flow with an imbedded shock wave. The results for the linear c_{v_i} model are in very good agreement with those of [15].

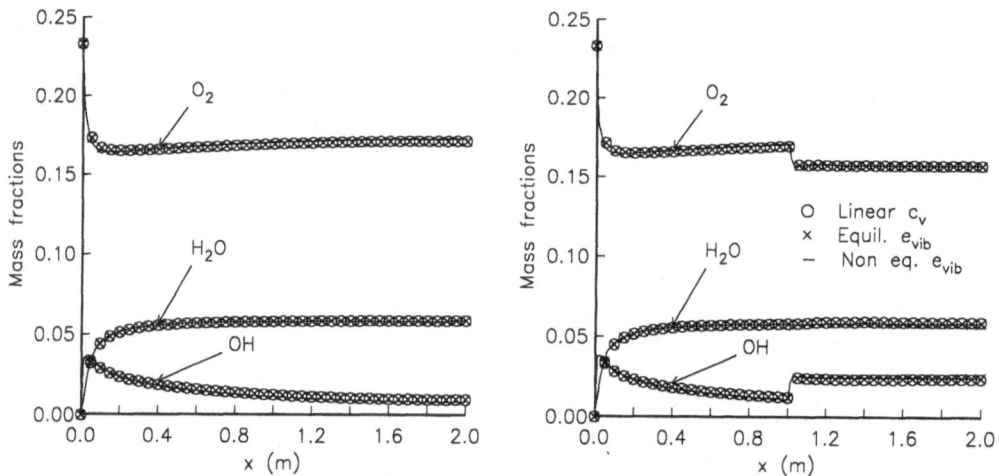

Figure 4.   Computed mass fractions for the non-equilibrium flow of an H_2-air mixture through a supersonic diffuser.

SECOND-MOMENT COMPUTATION OF STRONGLY-SWIRLING REACTING FLOW IN A MODEL COMBUSTOR

by

S. Hogg and M. A. Leschziner

University of Manchester

Institute of Science and Technology, UK

SUMMARY

The paper reports a computational effort directed towards the modelling of a turbulent, swirling and reacting propane-air flow in a laboratory combustor. The study focuses, in particular, on the performance of a Favre-averaged second-moment closure, applied in conjunction with an equilibrium-combustion model and the so-called β-probability density function. The above model is applied within the framework of a finite-volume scheme incorporating a quadratic approximation for convection. Some results are presented for non-reacting conditions, but the majority of comparisons relate to the reacting case. These comparisons demonstrate that, in contrast to previous experience with non-reacting flows, no decisive advantages are gained from second-moment modelling in the particular case examined, and this is attributed to the dominance of combustion-model defects.

NOMENCLATURE

A_P	Finite-volume coefficient related node at cell centre	$\tilde{U}_i$	Density-weighted mean velocity in tensor direction i
a, b	Constants in β-p.d.f.	$\widetilde{u_i''c''}$	Density-weighted scalar-flux
$\tilde{C}$	Density-weighted mean scalar	$\widetilde{u_i''u_j''}$	Density-weighted Reynolds-stress
$C_c, C_D, C_k,$	Constants in second-moment	V	Radial velocity component
$C_\epsilon, C_{\epsilon_1},$	closure	Vol	Volume of finite-volume cell
$C_{\epsilon_2}, C_{\epsilon_3}, C_\mu$		W	Swirl velocity component
D	Combustor diameter	x	Axial coordinate
f	Fuel mixture fraction	x_i	Coordinate in tensor direction i
$G_{ij}/G_{i,c}$	Additional stress/flux generation terms in density-weighted closure		
k	Turbulence kinetic energy	GREEK	
m	Iteration level		
n	Time level	α_c, α_t	Relaxation parameters
P	Pressure	δ_{ij}	Kronecker delta
P(f)	Probability density function	Δr	Radial height of control-volume
$P_{ij}/P_{i,c}$	Stress/flux generation terms	Δt	Time step
$R_{ij}/R_{i,c}$	Additive stress/flux contributions arising from axial symmetry and swirl	Δx	Axial length of control-volume
		ϵ	Turbulence energy dissipation rate
r	Radial coordinate	ζ	Constant in equations (7)
S_{ij}	Source term in finite-volume equation	μ_{app}	Pseudo-viscosity
		ρ	Density
U	Axial velocity component	Φ	General flow variable

Although we have been successful in obtaining efficient computations of steady-state solutions in one space dimension with an implicit algorithm, these results may not necessarily carry over to multi-dimensional flows. The penalty associated with solving very large block-structure tri-diagonal (or penta-diagonal) systems may overrun the benefits in *CFL* number for an implicit method.

ACKNOWLEDGMENT

This research was funded by the NASA Langley Research Center under grant NAG–1–776.

REFERENCES

1. A. Harten, P. D. Lax and B. van Leer, *SIAM Rev.*, **25**, 35, 1983.
2. P. L. Roe, *Ann. Rev. Fluid Mech.* **18**, 337, 1986.
3. R. W. Walters and J. L. Thomas, "Advances in Upwind Relaxation Methods", in *State-of-the-Art Surveys on Computational Mechanics*, ed. A. K. Noor, (A.S.M.E. Publication, New York, 1988).
4. J. L. Steger and R. F. Warming, *J. Comp. Phys.*, **40**, 263, 1981.
5. B. van Leer, in *Lecture Notes in Physics*, **170**, (Springer-Verlag, Berlin, 1982), p. 507.
6. P. L. Roe, *J. Comp. Phys.*, **43**, 357, 1981.
7. B. Grossman and P. Cinnella, "The Development of Flux-Split Algorithms for Flows with Non-Equilibrium Thermodynamics and Chemical Reactions", AIAA Paper No. 88–3596, 1st National Fluid Dynamics Congress, July 1988.
8. B. Grossman and P. Cinnella, "Flux-Split Algorithms for Flows with Non-equilibrium Chemistry and Vibrational Relaxation", ICAM Report 88–08–03, Virginia Polytechnic Institute and State University, Blacksburg VA, August 1988.
9. G. Herzberg, *Molecular Spectra and Molecular Structure, I. Spectra of Diatomic Molecules*, (D. van Nostrand, Inc., Princeton, N. J., 1950).
10. R. L. Jaffe, "The Calculations of High Temperature Equilibrium and Nonequilibrium Specific Heat Data for N_2, O_2 and NO", AIAA Paper No. 87-1633, 1987.
11. B. van Leer, J. L. Thomas, P. L. Roe and R. W. Newsome, "A Comparison of Numerical Flux Formulas for the Euler and Navier-Stokes Equations, AIAA Paper No. 87-1104-CP, 1987.
12. P. Glaister, *J. Comp. Phys.*, **74**, 382, 1988.
13. P. L. Roe and J. Pike, "Efficient Construction and Utilisation of Approximate Riemann Solutions", in *Computing Methods in Applied Sciences and Engineering VI*, edited by R. Glowinski and J.-L. Lions (North-Holland, Amsterdam, 1984), p. 499.
14. W. G. Vincenti and C. J. Kruger Jr., *Introduction to Physical Gas Dynamics*, (John Wiley and Sons, Inc., New York, 1965), p. 135.
15. J. P. Drummond, M. Y. Hussaini and T. A. Zang, *AIAA J.* **24**, 1461 (1986).
16. R. C. Rogers and W. Chinitz, *AIAA J.* **21**, 586 (1983).

1. INTRODUCTION

Combustor flows form the most important group of fluids-engineering applications in which swirl is introduced deliberately in order to achieve enhanced operational characteristics - in this case, improved turbulent mixing and flame stability. While swirl might be beneficial from a practical point of view, it leads to a considerable complication of the fundamental flow processes at play, and makes the task of describing these processes - be it by experimental or theoretical methods - much more demanding than it otherwise would be.

On the basis of the considerable evidence available for *non-reacting* flows, swirl - when imparted to *reacting* combustor flow - must be expected to strongly modify the structure of turbulence and hence the level of mixing, through a complex interaction between swirl-related strains and the turbulent stresses. Because this interaction proceeds at the level of Reynolds-stress generation and preferential turbulence-energy sharing between the anisotropic normal-stress components, the influence of swirl can only be captured properly if the Reynolds-stress components are treated individually – via a second- (or higher-) moment closure – without the intervention of an isotropic eddy-viscosity formulation.

The proven favourable sensitivity of second-moment closures to streamline curvature in a range of attached curved and recirculating flows [1-3] has naturally encouraged the expectation that highly swirled flows might benefit most strongly from being computed with such models. Some of the few recent studies adopting second-moment closures for such conditions have, indeed, shown this expectation to be justified to some extent, although predictive performance has by no means been found to be uniformly good.

Confined combustor-*type* flows in which the rotational motion contains a significant forced-vortex component, and in which swirl-induced recirculation is not a dominant feature, seem to derive much of the expected benefits due to the tendency of stress closures to properly attenuate turbulence activity in response to swirl-related strains. Recent studies by Boysan et al [4] and Hogg and Leschziner [5] serve to exemplify this response for constant-density conditions, while Hogg and Leschziner [6] demonstrate very encouraging model capabilities in the simultaneous presence of strong swirl and density gradients, the later arising from helium-air mixing.

In the last study, a Favre-averaged version of Gibson & Launder's Reynolds-stress/flux closure [7] was applied, with the modelling of pressure-strain and diffusion, originally proposed within an unweighted framework, retained. While differences between computed and experimental helium concentrations were not insignificant, very satisfactory agreement was obtained for the aerodynamic field. In particular - and in common with isothermal test calculations - the peculiarly subcritical character of the flow was well represented by the second-moment closure.

The above observations have encouraged the present authors to explore the performance of second-moment closure, in conjunction with an equilibrium-chemistry model, in a reacting propane/air model combustor examined experimentally by Wilhelmi [8]. This paper focuses on the essential modelling elements employed and presents the outcome of this exploration. Some preliminary isothermal calculations are also presented for the same geometry to constrast stress- and eddy-viscosity model performance in isolation from combustion-model uncertainties.

2. ESSENTIAL MODELLING ELEMENTS

2.1 MATHEMATICAL MODEL

Turbulence is modelled by a mass-weighted (Favre-averaged) version of Gibson & Launder's second-moment closure, with appropriate wall-correction terms accounting for the influence of wall-reflected pressure fluctuation on the redistribution of stresses and fluxes. This approach - rather than one involving explicit density decomposition within an unweighted formulation - has been taken so as to arrive at a tractable model in the context of a complex, practically relevant combustor flow. It must be acknowledged, however, that the above rests on the tacit (and bold) assumption that modelling proposals devised for diffusion, dissipation and pressure/mean-gradient interaction within the *unweighted* framework carry over without change to the *density-weighted* formulation. A limited a-posteriori justification for this assumption is the encouraging behaviour returned by the model in variable-density, but isothermal, conditions observed in ref [6].

In all, the model involves 11 differential equations: six for the stresses, three for fluxes, one for dissipation and one for the scalar variance. Space constrains do not permit a presentation of the model in its entirety, with all details included, and therefore rather abbreviated forms of the Reynolds-stress, flux, dissipation and scalar-variance equations are given here, written in Cartesian tensor notation with additive corrections (R_{ij} and $R_{i,c}$ below) accounting for contributions arising from axial symmetry and swirl:

$$\frac{1}{r}\frac{\partial \bar{\rho} r \, \tilde{U}_k \widetilde{u_i'' u_j''}}{\partial x_k} = \frac{1}{r}\frac{\partial}{\partial x_k}\left[r \, C_k \bar{\rho} \, \widetilde{u_k'' u_\ell''} \frac{k}{\epsilon} \frac{\partial \widetilde{u_i'' u_j''}}{\partial x_\ell} \right] + P_{ij} + G_{ij} - \frac{2}{3}\bar{\rho}\delta_{ij}\epsilon + \Phi_{ij} + R_{ij} \quad (1)$$

$$\frac{1}{r}\frac{\partial \bar{\rho} r \, \tilde{U}_k \widetilde{u_i'' c''}}{\partial x_k} = \frac{1}{r}\frac{\partial}{\partial x_k}\left[r \, C_c \bar{\rho} \, \widetilde{u_k'' u_\ell''} \frac{k}{\epsilon} \frac{\partial \widetilde{u_i'' c''}}{\partial x_\ell} \right] + P_{i,c} + G_{i,c} + \Phi_{i,c} + R_{i,c} \quad (2)$$

$$\frac{1}{r}\frac{\partial \bar{\rho} r \, \tilde{U}_k \epsilon}{\partial x_k} = \frac{1}{r}\frac{\partial}{\partial x_k}\left[C_\epsilon r \, \bar{\rho} \, \widetilde{u_k'' u_\ell''} \frac{k}{\epsilon} \frac{\partial \epsilon}{\partial x_\ell} \right] + C_{\epsilon_1} \frac{\epsilon}{k} P_{kk} - C_{\epsilon_2} \bar{\rho} \frac{\epsilon^2}{k} + C_{\epsilon_3} \frac{\epsilon}{k} G_{kk} \quad (3)$$

$$\frac{1}{r} \frac{\partial \bar{\rho} r \ U_j \widetilde{c''^2}}{\partial x_j} = \frac{1}{r} \frac{\partial}{\partial x_k}\left[C_c \ r \ \bar{\rho} \ \frac{k}{\epsilon} \ \overline{u_k'' u_\ell''} \ \frac{\partial \widetilde{c''^2}}{\partial x_\ell}\right] - 2 \ \bar{\rho} \ \widetilde{u_\ell'' c''} \ \frac{\partial \widetilde{C}}{\partial x_\ell} - 2 \ \bar{\rho} \ C_D \ \widetilde{c''^2} \ \frac{\epsilon}{k} \qquad (4)$$

In the above, P_{ij}, $P_{i,c}$ and P_{kk} represent generation of density-weighted stresses, fluxes and turbulence energy, respectively; R_{ij} and $R_{i,c}$ account for contributions arising from the cylindrical nature of the flow and from swirl, while Φ_{ij} and $\Phi_{i,c}$ identify pressure-strain and pressure/scalar-gradient interaction processes, respectively. A detailed exposition of all terms so abbreviated may be found in refs [6,10].

While Favre-averaging leads to a very considerable simplification of the modelling task (albeit at the expense of uncertainties on the validity of constant-density modelling approximations in variable-density situations), it does not entirely avoid the appearance of explicit density-related terms, which are identified by G_{ij}, $G_{i,c}$ and G_{kk} in equations (1), (2) and (3). In the case of stresses, these terms involve products of density-weighted velocity fluctuations and pressure gradients, viz.

$$G_{ij} = - \ \overline{u_i''} \ \frac{\partial P}{\partial x_j} - \ \overline{u_j''} \ \frac{\partial P}{\partial x_i} \qquad (5)$$

while, in the case of fluxes, corresponding products of scalar fluctuations and pressure gradients arise as follows,

$$G_{i,c} = - \ \overline{c''} \ \frac{\partial P}{\partial x_i} \qquad (6)$$

In variable-density but non-reacting conditions, the velocity and scalar fluctuations can be uniquely related to the fluxes and scalar variance by:

$$\overline{u_i''} = \zeta \ \bar{\rho} \ \widetilde{u_i'' c''}$$
$$\overline{c''} = \zeta \ \bar{\rho} \ \widetilde{c''^2} \qquad (7)$$

and these may thus be determined from the flux equations within the set of eleven mentioned above. However, the above does not apply in the presence of reaction, and this raises the need to solve additional transport equations for the fluctuations. In the above-mentioned recent study focusing on variable-density, non-reacting conditions [6], the present authors have demonstrated that the additional density-related terms given above do not contribute significantly to the predictive properties of the Favre-averaged second-moment closure. This observation, coupled with the high level of uncertainties associated with modelling the transport equations for the mass-weighted velocity fluctuations, have led the authors to omit the terms in the presence of reaction.

The flow under consideration involves the combustion of propane and air which are introduced as separate streams into the combustion chamber and hence react under the control of diffusion processes. To account for heat release, density variations and turbulence/combustion interaction, the present turbulence closure has been applied in conjunction with a modified version of Gordon & McBrides [9] fast-chemistry (equilibrium) model and the so-called "beta" probability density function

$$P(f) = \frac{f^{a-1}(1-f)^{b-1}}{\int_0^1 f^{a-1}(1-f)^{b-1} df} \tag{8}$$

In essence, the chemistry model provides instantaneous values for density and enthalpy as functions of the instantaneous fuel mixture fraction, f, a conserved scalar which varies between the limits of of zero and one at the air and fuel inlets respectively. The modification referred to above amounts to the introduction of a flamability limit due to a saturation of the reactants mixture by unburnt fuel. Details of the modification may be found in ref [10]. Use of the functional relations provided by the chemistry model together with the β-pdf allows the determination of the mean density and the value of any mass-weighted or unweighted flow property, according to

$$\bar{\rho} = \left[\int_0^1 \frac{P(f)}{\rho(f)} df \right]^{-1} ; \quad \bar{\Phi} = \int_0^1 \Phi(f)P(f)df ; \quad \tilde{\Phi} = \rho \int_0^1 \frac{\Phi(f)}{\rho(f)} P(f)df \tag{9}$$

respectively

2.2 NUMERICAL SOLUTION

The numerical framework is based on the conservative finite-volume procedure TEAM [11], in which discretized versions of the equations of motion and scalar transport are solved over a staggered grid arrangement, with the principle of mass-flux continuity imposed directly via the solution of pressure correction equations according to the SIMPLE [12] algorithm. A staggered arrangement is adopted not only for velocity components but also for the stresses, as shown in Fig. 1. This arrangement, although complicated, is of decisive assistance to stability by virtue of the strong coupling it establishes between the stresses/fluxes and their associated primary 'driving' strains/mean-scalar-gradients. The underlying rationale of the arrangement shown in Fig. 1 is to locate the stress or flux at a position which is straddled centrally by the nodal property values (velocities or scalar) used to evaluate approximations to the property gradients associated with the particular stress or flux in question. For example, the stress $\overline{u''v''}$ is positioned between those velocity nodes used to approximate $(\partial \tilde{U}/\partial r + \partial \tilde{V}/\partial x)$. A useful by-product of this practice is that no interpolation is involved in evaluating stress differences required for the finite-volume equations.

The convective fluxes appearing in the finite-volume equations are approximated by the power-law differencing scheme (PLDS) of Patankar [13], or, alternatively, by the quadratic upstream-weighted

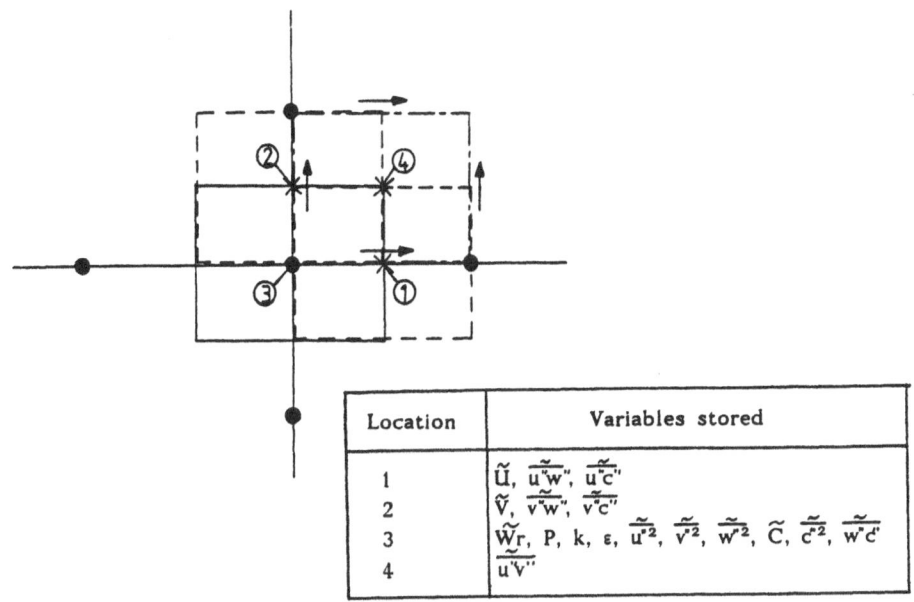

Location	Variables stored
1	$\tilde{U}$, $\overline{u''\tilde{w}''}$, $\overline{u''\tilde{c}''}$
2	$\tilde{V}$, $\overline{v''\tilde{w}''}$, $\overline{v''\tilde{c}''}$
3	$\tilde{W}_r$, P, k, ε, $\overline{\tilde{u'^2}}$, $\overline{\tilde{v'^2}}$, $\overline{\tilde{w'^2}}$, $\tilde{C}$, $\overline{\tilde{c'^2}}$, $\overline{w'\tilde{c'}}$
4	$\overline{u'\tilde{v}''}$

Fig. 1: Staggered finite-volume arrangement

scheme (QUICK) of Leonard [14], while central differences are used for diffusion. Because of the high swirl intensity, the complexity of the closure and the finess of the meshes employed, the level of computational stability is low, especially when numerically non-diffusive QUICK is used. Therefore, in order to prevent divergence, a range of special-stability promoting measures had to be introduced into the solution algorithm, in addition to source term linearization and under-relaxation which are normally used to stabilize numerical schemes of this type.

One important measure, proposed by Huang and Leschziner [3], involved the representation of a particular portion of each of the stresses/fluxes by means of an associated pseudo-viscosity/diffusivity. The Reynolds-stresses enter the momentum equations as spatial gradients such as $\partial \overline{u''\tilde{v}''}/\partial y$, $\partial \overline{u''^2}/\partial x$ etc. Taking the normal stress $\overline{\tilde{u''^2}}$ as an example, the corresponding finite-volume equation can be written in short-hand notation as;

$$A_p \overline{\tilde{u''^2}} = \sum_i A_i \overline{\tilde{u''_i^2}} + \left[\ldots -2\rho \overline{\tilde{u''^2}} \frac{\partial \tilde{U}}{\partial x} + \ldots \right] Vol \tag{10}$$

where i represents the positions of the neighbouring control-volume nodes, ie. N,S,E and W, and the bracketed portion contains all source-like contributions, of which only one part of the production term has been written explicitly. If equation (10) is divided by the finite-volume coefficient A_P, one part of the balance for $\widetilde{\overline{u''^2}}$ becomes;

$$\widetilde{\overline{u''^2}} = \ldots \; -2\rho\frac{\widetilde{\overline{u''^2}}}{A_P}\text{Vol}\frac{\partial \tilde{U}}{\partial x} \tag{11}$$

The term shown in equation (11) allows part of the normal stress $\widetilde{\overline{u''^2}}$ to be expressed as the product of a velocity gradient and a pseudo exchange coefficient. Substitution of this term into the $\tilde{U}$-momentum equation, accompanied by a suitable subtraction from the source term in order to restore the original balance, gives rise to the following pseudo-viscosity;

$$\mu_{app} = 2\rho^2\frac{\widetilde{\overline{u''^2}}}{A_P}\text{Vol}\Bigg|_{11} \tag{12}$$

where the subscript $_{11}$ at the end of the expression indicates that in this case, the coefficient A_P and cell volume are evaluated in accordance with the finite-volume equation for $\widetilde{\overline{u''^2}}$.

Terms analogous to equation (12) can also be extracted from the remaining stress/flux equations, yielding x- and y-directed pseudo-viscosities/diffusivities for each of the momentum equations and the scalar equation. Obviously, these terms will only enhance stability provided the pseudo-exchange coefficients remain positive at all times. In practice, this condition is satisfied by ensuring that physically unrealistic negative values for the normal stresses and scalar variance never occur during the iteration sequence. This is achieved by making a local discrimination between positive and negative source terms in the related transport equations, with contributions arranged according to;

$$S_{ij} = S_{U,ij} + S_{P,ij}\widetilde{\overline{u''_i u''_j}} \qquad \text{(no summation)} \tag{13}$$

where S_U combines all positive and S_P all negative contributions.

Two further unrelated measures were used to assist stability and rate of convergence. The first of these involved a relaxation of the coupling between the swirl and radial momentum equations during the iteration sequence. This was introduced by adopting Gosman et al's [15] representation of the centrifugal acceleration term in the $\tilde{V}$-momentum equation;

$$\frac{\overline{\rho \tilde{W}^2}}{r}\Bigg|^{(m)} \rightarrow \frac{\overline{\rho \tilde{W}^2}}{r}\Bigg|^{(m)}\left[1 + \frac{\alpha_c}{\tilde{W}(m-1)}\left[\tilde{V}(m-1)-\tilde{V}(m)\right]\right] \tag{14}$$

where m denotes the iteration level and α_c is an ad-hoc numerical 'relaxation parameter' which was assigned a value of unity in the present calculations. The above is seen to become an identity when the iteration process has converged.

The second measure uses the well-known analogy between under-relaxation and time marching. Using a simple two-time level approximation, the discretized time-dependent term (TDT) appearing in the axisymmetric form of the finite-volume equation for the general variable Φ, can be written;

$$TDT = \rho r \Delta r \Delta x \frac{(\Phi(n) - \Phi(n-1))}{\Delta t} \tag{15}$$

where n and n-1 are two time levels Δt apart.

For stationary flows such as those examined here, if n is interpreted as an iteration level rather than a time level, then the term shown in equation (15) is clearly finite prior to convergence. This term can be included in the flow solver as a cell-dependent variable under-relaxation, which, provided a suitable (pseudo-) time step is chosen, can in some cases maintain stability where conventional under-relaxation cannot. In the present calculations, rather than setting the time step directly, a relaxation parameter α_t was chosen such that $\alpha_t = \Delta x / \Delta t$, which was set equal to the approximate mean axial velocity of the combustor flow.

The introduction of the measures described above allowed stable computations to be performed, albeit at high CPU expense (the combined effect typically increasing the CPU time per iteration by 50%). Additionally, despite the introduction of these measures, convergence rates were still very slow. Calculations were performed using meshes of 40*32 and 78*62 grid-lines. Execution times for Reynolds-stress calculations using these grids were approximately 2.5s per iteration on a Cyber 205 for the former, rising to over 12.0s for the finer grid. This resulted in total CPU times of up to 30 hours for reaching fully converged solutions in the most complicated calculations.

3. APPLICATION AND RESULTS

The computed geometry is shown in Fig. 2, together with the finest 78x62 mesh used to cover the solution domain. A thin non-swirling sheet of propane is injected into the circular model chamber together with an outer annular air stream swirled by 45° vanes. Inlet conditions for velocity were prescribed in accordance with experimental measurements, where possible, with minor adjustments introduced to satisfy the prescribed mass-flow rates given in Table 1. The level of turbulence energy in the air stream at inlet was set to $k = 0.03 \tilde{U}^2$, while the dissipation rate was prescribed on the basis that the length scale $\ell = C_\mu^{0.75} k^{1.5} / \epsilon$ was 6% of the distance between the inner and outer swirler radii. Inlet shear-stress values were set to zero and the normal stresses were assumed to be isotropic.

Tests in which both k and ϵ were varied over a wide range of values showed the solutions to be largely insensitive to the inlet turbulence levels

	Mass flow (kg/s)	Temperature (^{O}K)
Propane	1.284×10^{-3}	294
Air	30.4×10^{-3}	313

Table 1: Inlet flow conditions for reacting flow

Computations were performed with the k-ϵ eddy-viscosity model and the above second-moment closure. In order to identify the influence of numerical errors, sensitivity to mesh density was investigated, and calculations for a given set of flow conditions were almost invariably carried out with both the quadratic and the power-law approximations for convection.

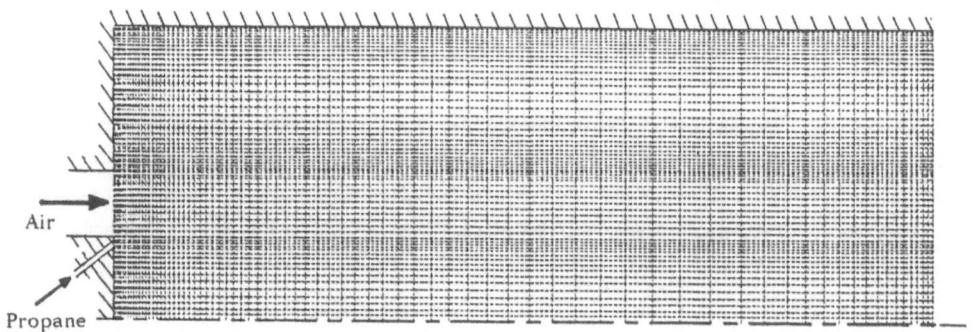

Fig. 2: Computed reacting-flow geometry and numerical mesh

In the isothermal conditions, there was no flow through the fuel injector, and a baffle was placed at the chamber exit so as to produce flow acceleration and with it a local transition from a 'subcritical' to a 'supercritical' state. Experiments have shown that reverse flow along the entire chamber in the vicinity of the centre line occured with the baffle absent.

Currently available results for *isothermal* conditions, computed with a 40x32 grid, suggest that a significant level of sensitivity to the turbulence representation occurs only in respect of the swirl-velocity component, and this sensitivity is illustrated in Fig. 3. Agreement with the rather scant experimental data is not satisfactory with either model, but the second-moment closure is clearly seen to return profiles which contain a much weaker forced-vortex component, implying a much lower level of diffusive transport. This behaviour is fully consistent with results obtained by the authors for other confined flows.

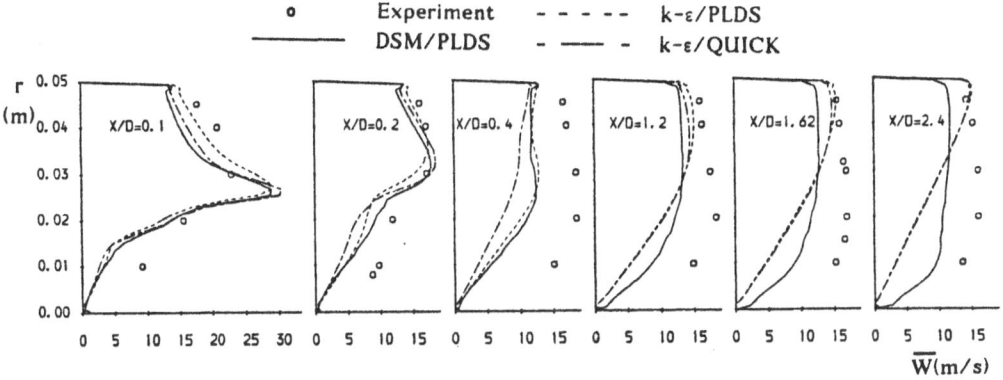

Fig. 3: Swirl-velocity profiles for isothermal flow

A much more complete set of computational results has now been assembled for the reacting case. Figs. 4 and 5 provide overall views of the predicted streamfunction and temperature contours, obtained with the 78x62 mesh of Fig. 2 and the quadratic approximation, while Figs. 6 to 8 show comparisons of predicted and experimental profiles for velocity, temperature and turbulence-intensity components.

The significant level of experimental uncertainties must be borne in mind when making comparisons between the measured and predicted flow data. It is clear, however, that important flow features are not well represented whatever turbulence closure is employed. The most influential defects, from which other aerodynamic defects appear to spring, are the seriously under-estimated temperature level *within* the flame envelope and the inappropriate sharp peak in temperature close to the fuel-injection point, Fig. 7. Both are suspected to be rooted in serious departures from chemical-equilibrium and other combustion-model defects, rather than due to turbulence-model inadequacies.

The predicted temperature peak at x/D=0.1 obviously results in a corresponding density trough, and this can be assumed to cause a strong local forward acceleration counteracting the tendency towards the reverse motion which is clearly observed in the experimental variation shown in Fig. 6 at x/D=0.1. Reverse-flow features can, of course, be expected to exert a decisive influence on the temperature within the region bounded by the flame, and the under-prediction of these features is, in part, responsible for the much too low temperatures predicted in this region. Reference to Fig. 7 shows that the measured temperatures at any position underneath the (predicted) flame front are as high as the maximum temperature within the flame. This is a curious observation which can, in the presence of a rather weak reverse motion, only be explained by a sustained combustion over a much larger flame region than that predicted. Here again, departures from chemical equilibrium must be suspected as contributory causes. One might also also suspect that the very high temperature within the above region is due to large-scale coherent motions in the reaction region producing a far greater level of 'mixing' than that predicted with the statistical model. Curiously, however, the level of turbulence intensities shown in Fig. 8 are broadly in accord with the measurements close to the inlet region, although somewhat further downstream evidence exists which suggests that the real turbulence activity is significantly more vigorous than the level predicted.

As regards swirl, it is important to note first that the computational scheme must satisfy swirl momentum by the very nature of the conservative formulation employed. The reduction of swirl momentum by the action of wall shear is known to be minor in this case, and it follows that the serious disagreements observed in Fig. 6 at x/D=1 cannot be due to a related computational mis-representation of swirl conservation. Rather, the differences can be attributed to distortions in the predicted $\tilde{U}$-velocity and density fields relative to the experimental distributions. On the positive side, it may be observed that the swirl component in downstream flow regions is, here again, better represented by the second-moment closure due to a swirl-induced reduction in the shear-stress level. This reduction is clearly observed from comparisons of k-ϵ-model and stress-closure predicted variations of the shear stresses, which are not included here, but which may be found in ref [10].

4. CONCLUSIONS

The major conclusion emerging from the study is that second-moment closure, at least the form used here which is often observed to be clearly superior to eddy-viscosity models in isothermal conditions, does not appear to offer decisive advantages in highly complex conditions involving combustion. In such circumstances, combustion modelling appears to be the primary factor dictating predictive realism. Specifically, the assumption of chemical equilibrium appears to be suspect. It is also likely that the omission of wall heat transfer and radiation have contributed to errors, although the former is unlikely to be a serious error source in view of the fact that predicted near-wall temperatures are relatively low, at least close to the inlet plane.

ACKNOWLEDGEMENT

The authors gratefully acknowledge the financial support provided through a CASE award made jointly by SERC and Rolls Royce plc.

REFERENCES

1. Rodi W and Scheuerer C (1983), Calculation of curved shear layers with two-equation turbulence models", Phys. Fluids, 26, p. 1422.

2. Leschziner M (1986), "Finite-volume computation of recirculating flow with Reynolds-stress closures, Proc. 3rd Symp. on Numerical Methods for Non-Linear Problems, Dubrovnik, p. 847.

3. Huang P G and Leschziner M A (1985), "Stabilization of recirculating flow computations using second-moment closures and third-order discretization, Proc. 5th Turbulent Shear Flow Symposium, Cornell, pp. 20.7-20.12.

4. Boysan F, Zhou T, Vasquez-Malebran S and Swithenbank J (1983), "Calculation of turbulent swirling flows with a second order Reynolds-stress closure", Int. Report, Dept. of Chemical Engineering and Fuel Technology, University of Sheffield.

5. Hogg S and Leschziner M A (1989), "Computation of highly swirling confined flow with a Reynolds stress turbulence model", J AIAA, 27, pp. 57-63.

6. Hogg S and Leschziner M A (1988), "Second-moment closure calculation of strongly-swirling confined flow with large density gradients", Int J Heat & Fluid Flow (in press).

7. Gibson M M and Launder B E (1978), "Ground effects on pressure fluctuations in the atmospheric boundary layer", JFM, 86, pp 491-511.

8. Wilhelmi J (1984), "Axisymmetric swirl stabilized combustion", Ph.D. Thesis, University of London.

9. Gordon S and McBride B J (1971), "Computer program for calculation of complex equilibrium composition, rocket performance, incident and reflected shocks and Chapman-Jouguet Detonations", NASA SP-273.

10. Hogg S (1988) "Second-moment-closure calculations of strongly-swirling confined flows with and without density variations", Ph.D. Thesis, University of Manchester.

11. Huang P G and Leschziner M A (1983), "An introduction and guide to the computer code TEAM", Report TFD/83/9(R), Thermofluids Div., Dept. of Mech. Eng., UMIST.

12. Patankar S V and Spalding D B (1972), "A calculation procedure for heat, mass and momentum transfer in three-dimensional parabolic flows", Int. J. Heat and Mass Transfer, 15, pp. 1787-1806.

13. Patankar S V (1980), "Numerical heat transfer and fluid flow", Hemisphere Publishing Corp., McGraw-Hill.

14. Leonard B P (1979), "A stable and accurate convective modelling procedure based on quadratic upstream interpolation", Comp. Meths. Appl. Mech. Eng., 19, pp. 59-98.

15. Gosman A D, Koosinlin M L, Lockwood F C and Splading D B (1976), "Transfer of heat in rotating systems", ASME Paper No. 76-GT-25.

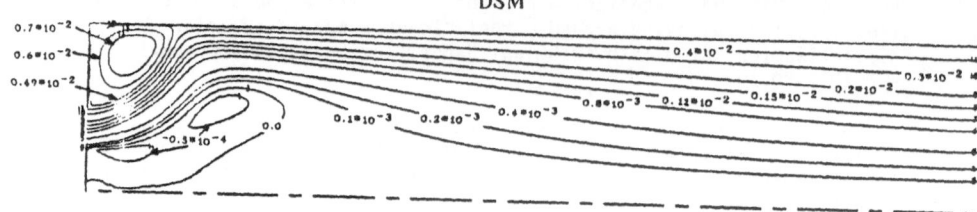

Fig. 4: Streamfunction contours for reacting flow: response to
turbulence model

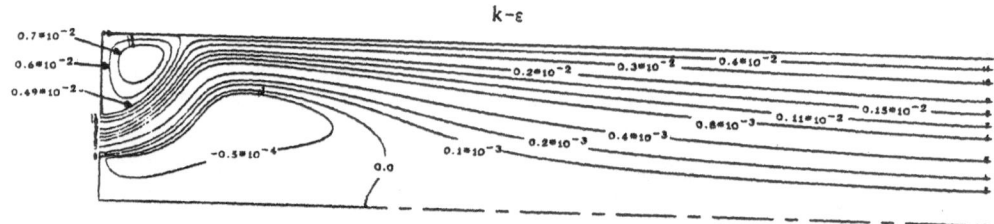

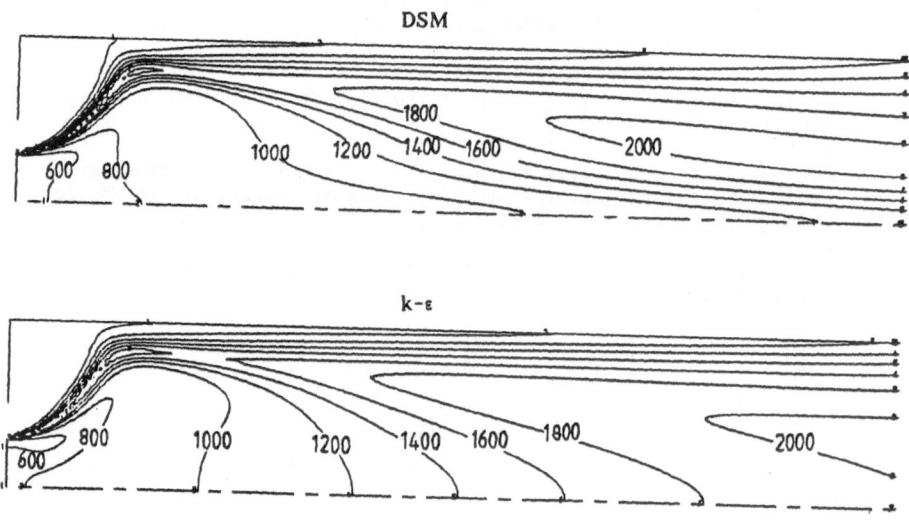

Fig. 5: Temperature contours for reacting flow: response to
turbulence model

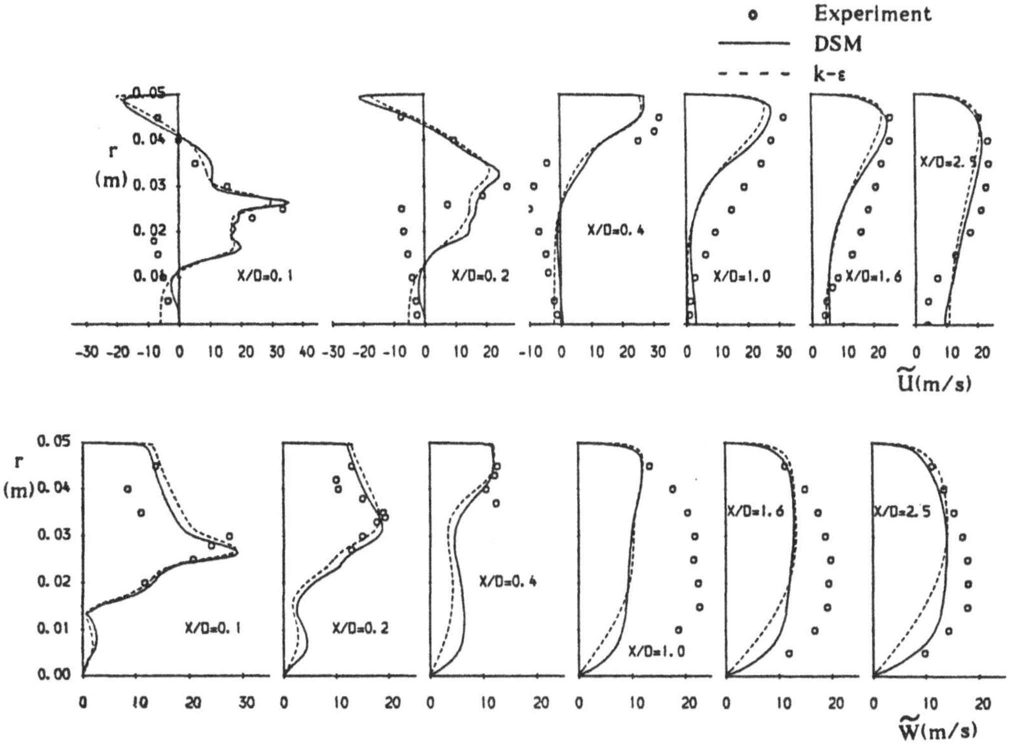

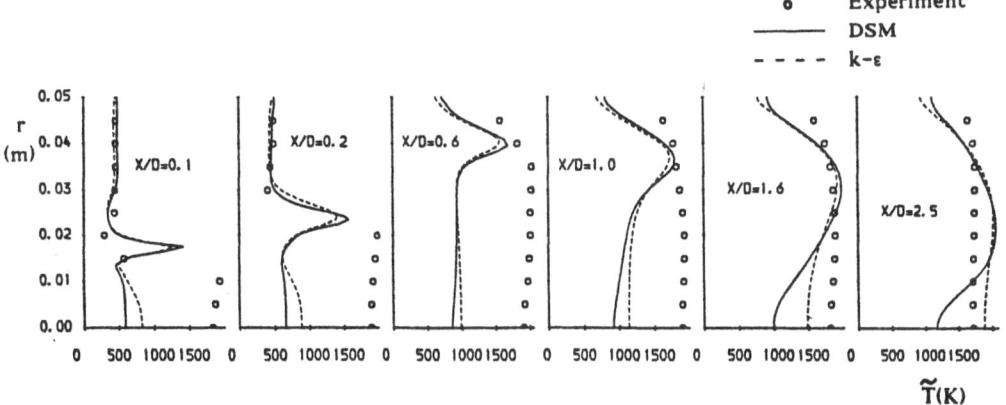

Fig. 6: Axial and swirl-velocity profiles for reacting flow

Fig. 7: Temperature profiles for reacting flow

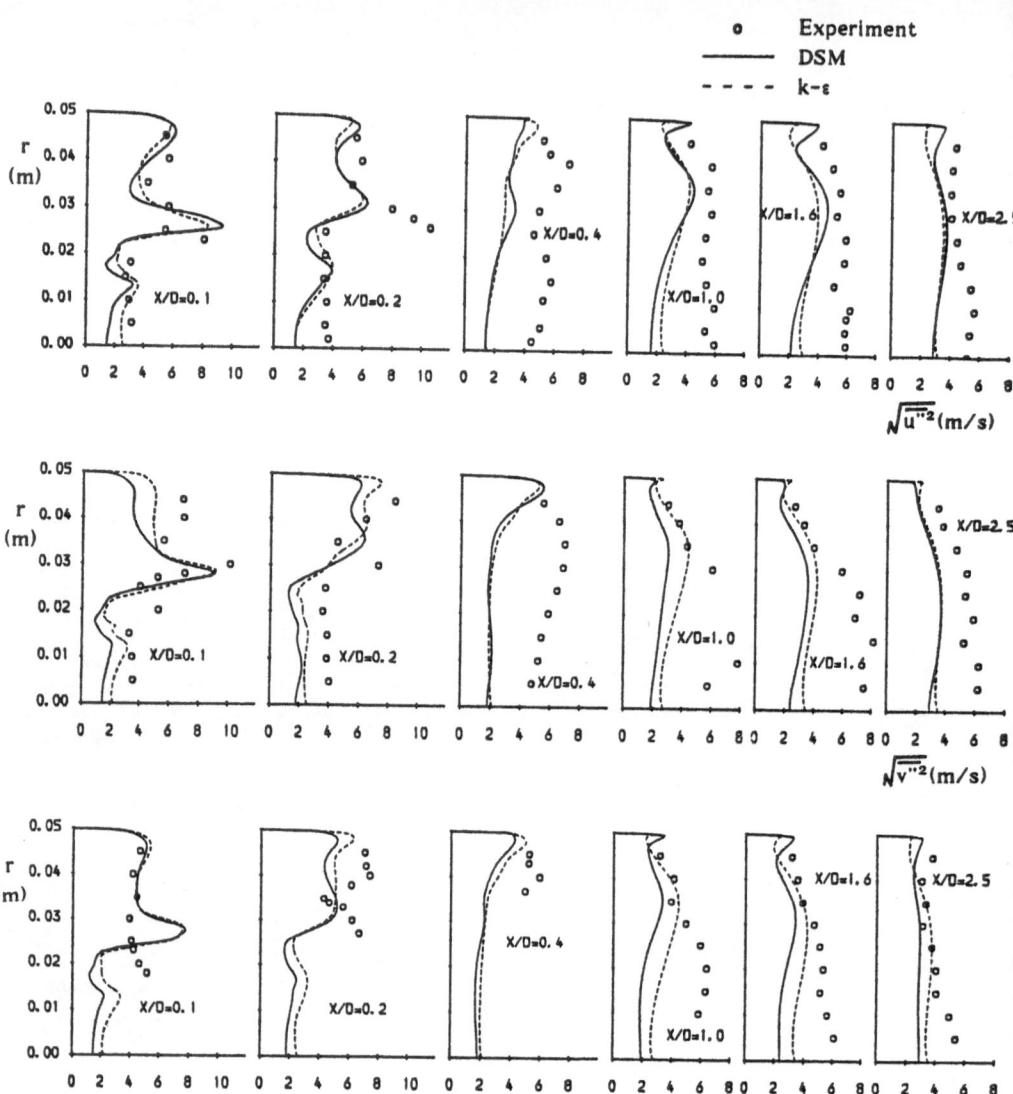

Fig. 8: Turbulence-intensity profiles for reacting flow

A NUMERICAL SIMULATION OF PULSED REACTING JETS

Ashvin Hosangadi, Charles L. Merkle, and Stephen R. Turns
The Pennsylvania State University
Department of Mechanical Engineering
University Park, PA 16802

ABSTRACT

Numerical simulations of unsteady combusting jets are performed using a dual-time, fully coupled, implicit procedure that enables time step sizes to be based on the particle velocity. Results are presented for a range of Froude numbers that span the momentum-dominated to buoyancy-dominated regimes. At high Froude number conditions, heat release is shown to be stabilizing while at low Froude numbers it is destabilizing to the point that the flames become spontaneously unstable. Dimensional evaluation of the buoyancy induced unsteady effects agrees well with the 10-15 Hz. flickering observed in laminar flame experiments.

INTRODUCTION

The numerical solution of unsteady diffusion flames involves the solution of the compressible equations in regions where the eigenvalues are generally quite stiff. For steady calculations, this stiffness can be circumvented by appropriate modification of the temporal derivatives[1-5], but in unsteady problems these arbitary changes cannot be made. As a means around this, we describe here a dual-time stepping approach that allows the physical time step to be scaled by the particle velocity even when the eigenvalues are highly stiff. Implicit procedures are used to provide control in highly stretched grids and in the presence of buoyancy.

We demonstrate the technique by simulating the oscillations in an axisymetric jet diffusion flame which is either spontaneously unstable or is forced sinusoidally. The aim of the paper is to determine the parametric effects of Froude number on vortex inception and growth in these unsteady flames. Previous calculations of unsteady mixing with hear release have been reported by McMurtry et al[6], Mahalingam et al[7], and Laskey et al[8], although none included the effect of buoyancy at very low Froude numbers. Our procedure differs from theirs in that the equations are solved in a fully coupled form rather than a sequential manner, a quality essential for low Froude number flames.

PROBLEM FORMULATION

To simplify the problem, we restrict our attention to a single one step reaction with infinite chemistry rates. consequently, a thin flame front is constrained to exist at a location where the fuel and oxidizer must be in stoichiometric ratio. We designate the fuel by F, the oxidizer by O, and the product by P, and the stoichiometric mass ratio of oxidant to fuel by r. The reaction between fuel and air is then represented as follows:

$$F + rO = (r + 1)P \tag{1}$$

We also simplify the gas dynamics by dropping the higher order terms in Mach number so that the usual conserved scalar formulation[9] is obtained. The axisymetric formulation then reduces to a set of four equations: continuity, two momentum, and the conserved scalar equation. Of particular importance from the numerical viewpoint is the fact that the conserved scalar removes the discontinuity at the flame sheet and provides continuous variables across the flame.

Upon making these approximations, the equations governing the dynamics of unsteady diffusion flames can be expressed in vector form as,

$$\partial_t Q + \partial_\xi E + \partial_\eta F = H + \partial_\xi R_{\xi\xi}\partial_\xi(JQ_v) + \partial_\xi R_{\xi\eta}\partial_\eta(JQ_v)$$
$$+ \partial_\eta R_{\eta\xi}\partial_\xi(JQ_v) + \partial_\eta R_{\eta\eta}\partial_\eta(JQ_v) \tag{2}$$

Here Q is the primary dependent vector, E and F are the inviscid flux vectors, Q_v is the viscous vector, H is a source term and the second derivative terms represent diffusion effects. The equations expressed here are written in an arbitary nonorthogonal coordinate system (ξ, η) which is related to the original axisymmetric coordinate system (x, y) by,

$$\xi = \xi(x, y) \qquad \eta = \eta(x, y) \tag{3}$$

The Jacobian of the transformation is $J = \xi_x\eta_y - \eta_x\xi_y$.

In Eqn.(2), the primary dependent variable takes the form,

$$Q = J^{-1}y(\rho, \rho u, \rho v, \rho\varsigma)^T \tag{4}$$

where ς is the conserved scalar, ρ is the density, and u and v are velocity components in the x and y drection respectively. The flux vector, E, is given by,

$$E = J^{-1}y(\rho U, \rho uU + \xi_x P_1, \rho vU + \xi_y P_1, \rho U\varsigma)^T \tag{5}$$

where U is the contravariant velocity, and P_1 is the gauge pressure. The corresponding vector F is obtained from Eq.(5) by replacing U with V, and ξ by η. The viscous vector, Q_v, is defined as,

$$Q = J^{-1}(0, u, v, \varsigma)^T \tag{6}$$

The source term, H, contains effects of the axisymmetric geometry and bouyant force, while the viscous matrices R_1 through R_4 depend on the metrics of transformation, the viscosity μ, and the diffusivity D (see Ref.(10) for details).

NUMERICAL SOLUTION

The numerical solution of the above system of equations becomes more and more difficult as Mach number is reduced because of the growing stiffness of the eigenvalues. To circumvent this difficulty, we have chosen to use an implicit iterative procedure at each time step which allows very large values of CFL to be taken for the characteristics (while retaining CFL's of order unity for the more significant convective speeds). In formulating this iterative procedure, we express the iteration as a dual time-stepping procedure analogous to that used for both compressible[11] and incompressible[12] flows. The modified form for the dual time-stepping procedure is,

$$\Gamma \partial_\tau \tilde{Q} + \partial_t Q + \partial_\xi E + \partial_\eta F = \text{RHS}_2 \tag{7}$$

where τ is the psuedo-time and RHS_2 signifies the right-hand side of Eqn.(2). For low speeds where eigenvalue stiffness prevents economical solutions, the vector $\tilde{Q}$ is defined as,

$$\tilde{Q} = J^{-1}y(P_1, u, v, \varsigma)^T \tag{8}$$

and Γ is defined as,

$$\Gamma = \begin{pmatrix} 1 & 0 & 0 & 0 \\ u & \rho & 0 & 0 \\ v & 0 & \rho & 0 \\ \varsigma & 0 & 0 & \rho \end{pmatrix}$$

in concert with low Mach number procedures for steady flows[13].

A general time marching procedure (in psuedo-time) can now be applied to Eqn.(7) wherein the unsteady solution at each physical time step is obtained by converging to a 'steady-state' in the psuedo-time, τ.

The numerical solution of Eqn.(7) is obtained by using a three point backward differencing in physical time, central differencing in space, and Euler implicit differencing in psuedo-time. This type of discretization provides a scheme that is second order accurate in space and time. We have employed an ADI procedure[14,15] to solve the resulting matrix operator efficiently. Performing the indicated discretization, Eqn(7) now becomes,

$$[S + \Delta\tau(\partial_\xi A. - \partial_\xi R_1 \partial_\xi J/y.)]S^{-1}$$
$$[S + \Delta\tau(\partial_\eta B. - \partial_\eta R_4 \partial_\eta J/y.)]^P \Delta\tilde{Q}y \tag{9}$$
$$= -\Delta\tau(R)^P$$

where the matrix R is the residual at each time step and is given by,

$$R = \frac{(3.Q^{n+1} - 4.Q^n + Q^{n-1})}{(2.\Delta t)} + \partial_\xi E + \partial_\eta F - \text{RHS}_2 \tag{10}$$

Here, n represents the current physical time step, and p represents the iteration number (psuedo-time step). The matrices A,B, and D are Jacobians of the vectors E,F, and H, while S is the matrix

$$S = \Gamma - D\Delta\tau + (3/2)(\Delta\tau/\Delta t)T \tag{11}$$

where T is the Jacobian $\frac{\partial \tilde{Q}}{\partial Q}$. When Eq.(9) is iterated to convergence at a given physical time step ($\Delta\tilde{Q} = 0$), we obtain $Q^{p+1} = Q^{n+1}$ and the right-hand side of Eq.(9) provides the time accurate solution.

The computational cost of obtaining the unsteady solution over a complete time period is a function of both the number of iterations needed to converge in psuedo-time, as well as the number of time steps comprising an unsteady cycle. Employing larger number of physical time steps per cycle reduces the required number of iterations in psuedo-time τ. However, this does not necessarily imply a more economical solution procedure since it is the product of the two quantities which determines the total cost. In the unsteady results presented in the following sections, the number of time steps within one time period was taken as 40, and the number of iterations needed to converge to a level of 5×10^{-4} varied between 25 and 50 depending on the flow conditions. The above combination for the two parameters was found to yield the most efficient solution procedure.

BOUNDARY CONDITIONS

The boundary conditions for both the steady solution that is used as initial condition and the ensuing unsteady perturbation are as follows. At the upstream end, the velocity profiles in both the fuel jet and the co-flowing oxidizer stream are specified. The axial velocity profile is taken as uniform in both streams except for a thin 'boundary layer' both inside and outside the nozzle lip. To preclude numerical difficulties, the minimum velocity in the wake of the nozzle lip is taken as 20% of the freestream value. The upstream boundary conditions are completed by specifying the radial distribution of mixture fraction and by setting the radial component of the velocity to zero. A Gaussian function with width equal to the aforementioned boundary layer thickness is used to smooth the transition from fuel to oxidizer on the inlet line. For the steady flow calculations, the fuel jet velocity is constant in time. For the pulsed jet cases, a sinusoidal variation is superimposed on the mean axial velocity. For both unsteady and steady calculations, these upstream boundary conditons are augmented by information from an upstream-running characteristic[16].

At the downstream boundary a constant back pressure is specified as the single boundary condition. This condition is then augmented by information from three outrunning characteristics. This inviscid technique has proven adequate for calculations, to date, although the choice of a constant back pressure is not as meaningful for unsteady cases as it is for steady computations. We do note, however, that outflow boundary conditions remain a difficulty in most unsteady flow calculations. In some cases (for example, Ref.8) this has necessitated extending the computational domain much further downstream of the region of interest, with only the results in the smaller physical domain being presented. In addition to being uneconomical the more distant placement of a difficult boundary condition does not remove the problem but only delays the time at which it will contaminate the solution. The present procedure, however, allows us to make the downstream boundary coincident with the physical region of interest and accordingly all plots described in the following sections extend all the way to the end of the numerical boundary.

At the outer boundary provisions are made for mass influx and standard inflow boundary conditions are applied. The radial velocity was therefore set to zero, while the axial velocity component and mixture fraction were set to their respective freestream values.

COMPARISON WITH EXPERIMENT

As most experiments on forced flames, at present, are restricted to qualitative flow visualization studies with quantitative data being unavailable, we use the steady state velocity and temperature profiles of Santoro et al[17] to validate the accuracy of the numerical code developed. Santoro et al[17] considered a coflowing ethene-air laminar diffusion flame. For the conditions chosen they considered a fuel velocity of 4.0 cm/s and a coflowing velocity of 8.9 cm/s for a fuel passage which was 11.1 mm in inside diameter. Their data consisted of radial profile measurements of two components of velocity and temperature at a series of axial distances downstream of the nozzle exit plane. Comparison with these quantitative measurements provides an unique oppurtunity to verify the code in a realistic environment. Figure(2) compares the numerical and experimental measurements for the axial component of velocity at various axial locations, while Fig.(3) compares the corresponding temperature profiles. The plots on the left half of these figures represent line segments joining the individual data points while those on the right show the numerical results.

The numerical results computed compare well with experimental measurements up until flame radiation loss becomes substantial. In the first 20 mm ($x/R \leq 3.6$) of the flame, the axial velocity profiles are in good quantitative agreement with the measurements and have the proper slope as well as the right magnitude. Good agreement is also obtained between numerical and experimental temperature values in the same region. Further downstream ($x/R \geq 3.6$) heat loss due to soot radiation starts to become significant and causes the experimental temperature to drop substantially below the adiabatic value. In contrast, the numerical results continue to predict the peak at the adiabatic flame temperature in keeping with the conserved scalar formulation. This discrepancy in the temperature field then leads to stronger buoyant acceleration in the numerical calculation and consequently the values of the axial velocity component are increasingly overpredicted as we go downstream.

EFFECTS OF FROUDE NUMBER ON UNSTEADY FLAMES

Calculations have been performed for unsteady jet diffusion flames at three Froude numbers: 5000, 50 and 5. Corresponding Reynolds numbers ranged from 1000 to 100. The higher Froude number case is momentum dominated, while buoyancy begins to get increasingly dominant as the Froude number drops. For all these calculations the co-flow velocity was one third the jet velocity. The two higher Froude number cases (5000, and 50) were pulsed to induce unsteady effects, while the last one proved to be unstable without pulsing. For brevity, only the results for the Fr=50, and 5 cases will be presented.

The results for the momentum dominated case (Fr=5000), which are in agreement with other workers[6,7], indicate that heat release has a stabilizing influence when buoyancy is negligible. The increased stability of the jet has also been demonstrated by parallel flow stability calculations[7,10], and is attributed to the density profile.

At the intermediate Froude number (Fr=50) buoyancy becomes significant and causes acceleration in the hot regions of the flow. This acceleration of the axial velocity component gives rise to a dual shear layer which upon pulsing exhibits counterrotating vortices on either side of the flame, which are accompanied by undulations in the flame surface as seen in Fig.(4) for a Strouhal number of 0.65. Here, the presence of buoyancy is destabilizing and leads to increased vortex growth.

At the lowest Froude number (Fr=5) the effects of buoyancy become so sever that the steady state solution degenerates spontaneously into an oscillating flame with a nearly periodic nature. Once again this unsteady flame is characterized by counterrotating vortices and the temperature contours begin to exhibit strong roll-up as shown in Fig.(5).

SUMMARY

In summary, the flowfield solutions computed encompass Froude numbers ranging from momentum dominated to the buoyancy dominated regime. The results show that heating is stabilizing at high Froude numbers but becomes destabilizing at low Froude numbers where buoyancy effects dominate. The unstable frequencies observed agree well with the commonly observed 10-15 Hz flickering in experimental flames.

ACKNOWLEDGEMENT

This work was supported by the Gas Research Institute under Contract No. 5086-260-1308 with J.A. Kezerle as Program Manager.

REFERENCES

1. Choi, Y.H., Computation of Low Mach Number Compressible Flow, Ph.D. Dissertation, Department of Mechanical Engineering, Pennsylvania State University, 1989.
2. Briley, W.R., McDonald, H., and Shamroth, S.J., "A Low Mach Number Euler Formulation and Application to Time-Iterative LBI schemes", AIAA Journal, Vol.21, No.4, 1983, p.1467.
3. Turkel, E., "Preconditioned Methods for solving the Incompressible and Low Speed Compressible Equations", J. Comput. Phys., Vol.72, 1987, p.277.
4. Guerra, J., and Gustafsson, B., "Numerical Method for Incompressible and Compressible Flow Problems with Smooth Solutions", J. Comp. Phys., Vol.63, 1986, p.377.
5. Majda, A., and Sethian, J., "The Derivation and Numerical Solution of the Equations for Zero Mach Number Combustion", Comb. Sci. and Tech., Vol.42, 1985, p.185.
6. McMurtry, P.A., Jou, W.H., Riley, J.J., and Metcalfe, R.W., "Direct Numerical Simulations of a Reacting Mixing Layer with Heat Release", AIAA Journal, Vol.24, No.6, 1986, p.962.
7. Mahalingam, S., Cantwell, B., and Ferziger, J., "Effects of Heat Release on the Structure and Stability of a Coflowing, Chemically Reacting Jet", AIAA Paper No. 89-0661, 1989.
8. Laskey, K.J., Ellzey, J.L., and Oran, E.S., "A Numerical Study of an Unsteady Diffusion Flame", AIAA Paper No. 89-0572, 1989.
9. Bilger, R.W., "The Structure of Diffusion Flames," Combustion Sci. and Technology, Vol.(13), 1976, p.155.
10. Hosangadi, A., Merkle, C.L., and Turns, S.R., "Analysis of Forced Combusting Jets", AIAA Paper No. 89-0662, 1989.

11. Rai, M.M., "Navier Stokes Simulations of Blade Vortex Interactions using Higher Order Accurate Upwind Schemes", AIAA Paper No. 87-6543, 1987.

12. Merkle, C.L., and Athavale, M., "A Time Accurate Unsteady Incompressible Algorithm Based on Artificial Compressibility", AIAA Paper No. 87-1137, AIAA 8th Computational Fluid Dynamics Conference, JUne 9-11, 1987, Honolulu, HA.

13. Merkle, C.L., and Choi, Y.H., "Computation of Low-Speed Flow with Heat Addition", AIAA Journal, Vol.25, No.6, 1987, p.831.

14. Douglas, J. and Gunn, J.E., "A General Formulation of Alternating Direction Method-Part I. Parabolic and Hyperbolic Problems",
Numerische Mathematik, Vol.(82), 1964, p.428.

15. Beam, R.M., and Warming, R.F., "An Implicit Finite-Difference Algorithm for Hyperbolic Systems in Conservation-Law Form", J. Comput. Phys.,
Vol.(22), 1976, p.87.

16. Merkle, C.L., and Choi, D., "Application of Time-Iterative Schemes to Incompressible Flow", AIAA Journal, Vol.23, 1985, p.1518.

17. Santoro, R.J., Yeh, T.T., Horvath, J.J., and Semerjian, H.G., "The Transport and Growth of Soot Particles in Laminar Diffusion Flames",
Comb. Sci. and Tech., Vol.53, 1987, p.89.

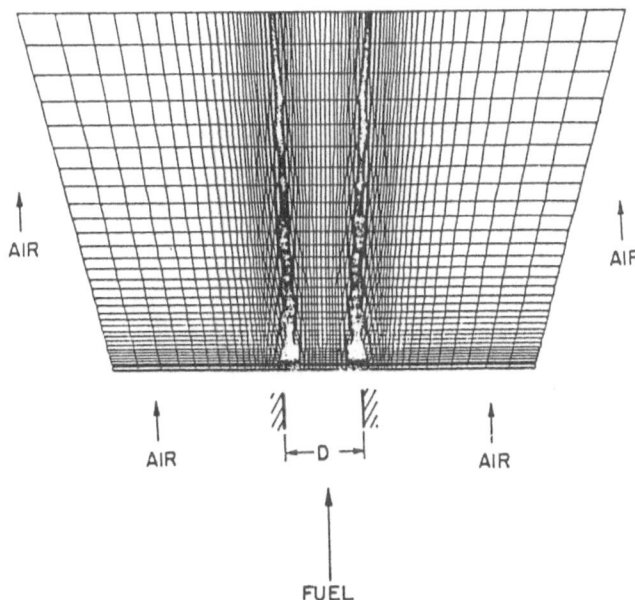

Fig. 1   Flow geometry for a combusting axisymmetric fuel jet diffusing into a co-flowing stream of air. The dotted line shows the typical location of the flame front.

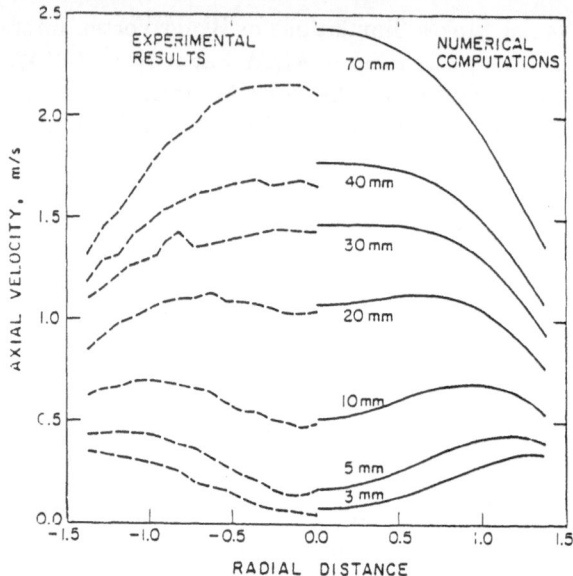

Fig. 2 Comparison of predicted axial velocity profiles with experimental results of Santoro et al.[17]. Left side shows experimental results while right side contains numerical predictions.

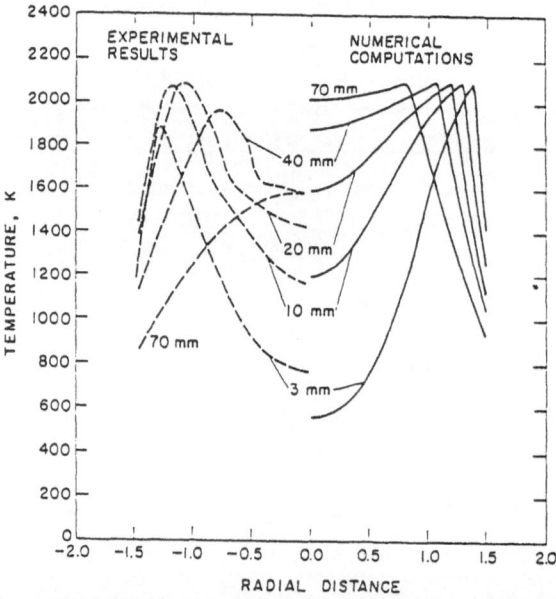

Fig. 3 Comparison of predicted temperature profiles with measurements of Santoro et al.[17]. Left side shows experimental measurements; right side contains numerical predictions.

Phase Angle=0   Phase Angle=π/2   Phase Angle=π

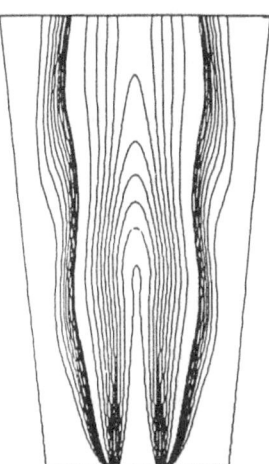
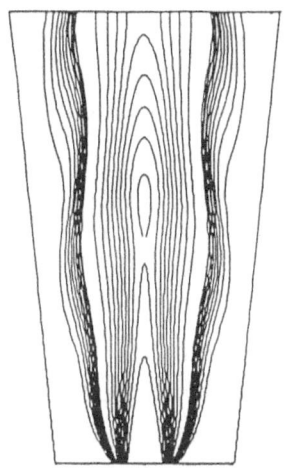
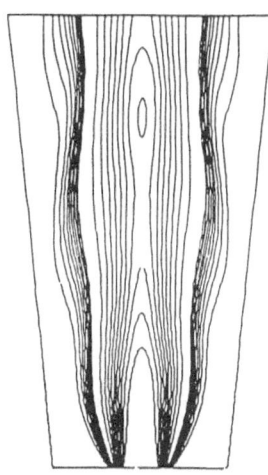

Fig. 4   Pulsed combusting jet with buoyancy Fr = 50.
          Isothermal contours at various phase angles for
          pulsing at St$_R$ = 0.65. The solution domain extends up
          to 20 radii in the axial direction and from -4.0 to
          +4.0 at the inlet in the radial direction.

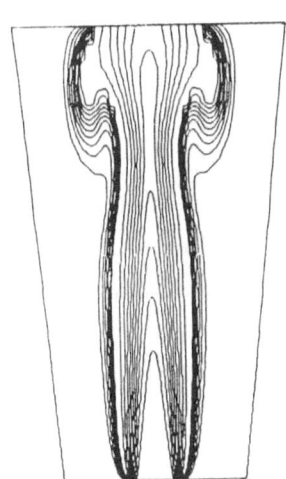
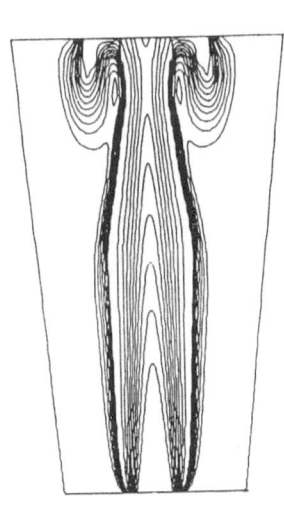
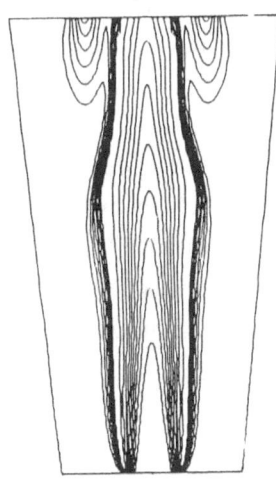

Fig. 5   Buoyancy induced fluctuations in unforced combusting
          jet, Fr = 5. Isothermal contours at various time
          steps. Time increases from top to bottom. Time
          increment between each plot is the same as that used
          in Figs. 4 and 5. The solution domain extends up to
          20 radii in the axial direction and -4.0 to +4.0 radii
          at the inlet in the radial direction.

EFFECT OF HEAT RELEASE AND EQUIVALENCE RATIO ON THE INVISCID SPATIAL STABILITY OF A SUPERSONIC REACTING MIXING LAYER

T.L. Jackson and C.E. Grosch
Old Dominion University
Norfolk, Virginia 23529   USA

Abstract. We present some preliminary results of a numerical study of the stability of compressible mixing layers in which a diffusion flame is embedded. In this study we have approximated the mean velocity profile by a hyperbolic tangent profile and taken the limit of infinite activation energy which reduces the diffusion flame to a flame sheet. The addition of combustion in the form of a flame sheet was found to have important, and complex, effects on the flow stability.

1. Introduction. Quite recently it has been realized that an understanding of the stability characteristics of reacting compressible mixing layers is extremely important in view of the projected use of the scramjet engine for the propulsion of hypersonic aircraft. For example, Drummond and Mukunda (1988) suggest that "... a single supersonic, spatially developing and reacting mixing layer serves as an excellent physical model for the overall flow field." Thus knowledge of the stability characteristics may allow one, in principle, to control the downstream evolution of such flows in the combustor. This is particularly important because of the observed increase in the flow stability at high Mach numbers (Brown and Roshko, 1974; Chinzei, Masuya, Komuro, Murakami, and Kudou, 1986; and Papamoschou and Roshko, 1986). Because of the gain in stability, natural transition may occur at downstream distances which are larger than practical combustor lengths. A number of techniques which may enhance mixing are discussed by Kumar, Bushnell and Hussaini (1987).

Despite the fact that understanding of the flow field in a reacting compressible mixing layer in a scramjet engine is extremely important, there appears to be very few studies on the stability of such flows. Menon, Anderson and Pai (1984) studied the inviscid spatial stability of a compressible wake in which there was a H_2-O_2 reaction. When the reaction was turned on the flow became completely unstable. The phase speed was found to be a monotonically increasing function of frequency. It seems that their results show a complete absence of neutral or stable disturbances. This result appears to be in conflict with that of Drummond and Mukunda (1988). They carried out a numerical simulation using the two dimensional, compressible, time dependent Navier-Stokes equations with combustion in a mixing layer. The reaction was the burning of a 10% H_2, 90% N_2 fuel in air. They found that the nonreacting flow was very stable and that turning on the combustion had little effect. But it should be noted that the authors were not carring out a stability calculation, per se, and did not excite the flow with disturbances with a fixed frequency. They relied on the "natural" disturbances to perturb the flow.

We have begun a numerical study of the inviscid spatial stability of compressible mixing layers in which a diffusion flame is embedded. The basic steady flow with which we began is that calculated by Jackson and Hussaini (1988). In their study the limit of infinite activation energy was used and the diffusion flame reduced to a flame sheet. The flame sheet model is a standard approximation and has been used in the study of the burning of a fuel particle and of the flame at the mouth of a tube, for example (Buckmaster and Ludford, 1982; and Williams,1985). We began our study by taking the mean velocity profile in the mixing layer to be that of the Lock profile. In the course of carrying out the stability calculations for this profile with different sets of values of the basic parameters we noted some quite interesting features of the solutions, particularly at higher Mach numbers. In order to examine these features in more detail, we replaced the Lock profile with a hyperbolic tangent profile. This is a reasonable approximation to the Lock profile and has been used in studying the stability of compressible mixing layers, for example, Tam and Hu (1988), Ragab and Wu (1988), and Zhuang, Kubota and Dimotakis (1988).

2. **Formulation.** The nondimensional equations governing the steady two dimensional flow of a compressible, reacting mixing layer which lies between streams of reactants with different speeds and temperatures are given by Jackson and Hussaini (1988). In the limit of infinite activation energy, the thin diffusion flame reduces to a flame sheet. The nondimensional temperature T and mass fractions C_i profiles are given by

$$T = 1 - (1 - \beta_T - \beta \phi^{-1})(1-U) + \frac{\gamma-1}{2} M^2 U (1-U), \tag{2.1a}$$

$$C_1 = 1 - (1 + \phi^{-1})(1-U), \quad C_2 = 0, \tag{2.1b,c}$$

for $\eta > \eta_f$ in the moving stream, and

$$T = \beta_T + (1 - \beta_T + \beta)U + \frac{\gamma-1}{2} M^2 U (1-U), \tag{2.2a}$$

$$C_1 = 0, \quad C_2 = r[\phi^{-1} - (1 + \phi^{-1})U], \tag{2.2b,c}$$

for $\eta < \eta_f$ in the stationary stream. Here, η_f gives the location of the flame sheet where both reactants vanish, β_T is the ratio of the temperature of the stationary stream to that of the moving stream, γ is the ratio of specific heats, β is the nondimensional heat release parameter, r is a ratio involving stoichiometry, and ϕ is the equivalence ratio, defined by

$$\phi = (C_{1,\infty} / C_{2,-\infty}) / r, \tag{2.3}$$

which is the ratio of the mass fraction C_1 in the moving stream to the mass fraction C_2 in the stationary stream divided by the ratio of their molecular weights times their stoichiometric coefficients. Note that if $\phi = 1$, then the mixture is said to be stoichiometric, and if $\phi > 1$ it is C_1 rich, while if $\phi < 1$ it is C_1 lean. As discussed in the introduction, we assume here that

$$U = (1 + tanh(\eta)) / 2. \tag{2.4}$$

The flow field is perturbed by introducing wave disturbances in the velocity, pressure, temperature, density and mass fractions on either side of the flame sheet with amplitudes which are functions of η. In addition, the flame sheet location must also be perturbed with a wave disturbance. For example, the pressure perturbation is

$$p = \Pi(\eta) \, exp\,[i\,(\alpha x - \omega t)], \tag{2.5}$$

with Π the amplitude. Here, for spatial stability α is complex. The real part of α is the wave number in the x direction, while the imaginary part of α indicates whether the disturbance is amplified, neutral, or damped depending on whether α_i is negative, zero, or positive. The frequency ω is taken to be real. Substituting the expression (2.5) for the pressure perturbation and similar expressions for the other flow quantities, it is straightforward to derive a single equation governing Π, given by

$$\Pi'' - \frac{2U'}{U-c}\Pi' - \alpha^2 T\,[T - M^2(U-c)^2]\,\Pi = 0, \qquad c = \frac{\omega}{\alpha}, \tag{2.6}$$

which is valid on either side of the flame sheet. Here c is the complex wave velocity and primes indicate differentiation with respect to the similarity variable η. The phase speed is given by ω / α_r, and for a neutral wave will be denoted by c_N.

The boundary conditions for Π are obtained by considering the limiting form of equation (2.6) as $\eta \rightarrow \pm\infty$. The solutions to (2.6) are of the form

$$\Pi \rightarrow exp\,(\pm\Omega_\pm \eta), \tag{2.7}$$

$$\Omega^2_+ = \alpha^2[1 - M^2(1-c)^2], \qquad \Omega^2_- = \alpha^2 \beta_T\,[\beta_T - M^2 c^2]. \tag{2.8}$$

Let us define $c_\pm$ to be the values of the phase speed for which $\Omega^2_\pm$ vanishes. Thus,

$$c_+ = 1 - 1/M, \qquad c_- = \sqrt{\beta_T}/M. \qquad (2.9)$$

Note that c_+ is the phase speed of a sonic disturbance in the moving stream and c_- is the phase speed of a sonic disturbance in the stationary stream. Across the flame sheet, Π and Π' are continuous.

The nature of the disturbances can be illustrated by reference to Figure 1, where we plot $c_\pm$ versus M. In what follows we assume that $\alpha^2_r > \alpha^2_i$. These curves divide the $c_r - M$ plane into four regions. If a disturbance exists with a M and c_r in region 1, then Ω^2_+ and Ω^2_- are both positive, and the disturbance is subsonic at both boundaries. In region 3, both Ω^2_+ and Ω^2_- are negative and hence the disturbance is supersonic at both boundaries. In region 2, Ω^2_+ is positive and Ω^2_- is negative, and the disturbance is subsonic at $+\infty$ and supersonic at $-\infty$, and we classify it as a fast mode. Finally, in region 4, Ω^2_+ is negative and Ω^2_- is positive so the disturbance is supersonic at $+\infty$ and subsonic at $-\infty$, and we classify it as a slow mode. One can now see that the boundary conditions in the moving and stationary streams are, respectively,

$$\Pi \to e^{-\Omega_+ \eta}, \quad \text{if } c_r > c_+, \qquad \Pi \to e^{-i\eta\sqrt{-\Omega^2_+}}, \quad \text{if } c_r < c_+, \qquad (2.10a)$$

$$\Pi \to e^{\Omega_- \eta}, \quad \text{if } c_r < c_-, \qquad \Pi \to e^{-i\eta\sqrt{-\Omega^2_-}}, \quad \text{if } c_r > c_-. \qquad (2.10b)$$

To solve the disturbance equation (2.6), we first transform it to a Riccati equation by setting

$$G = \Pi' / (\alpha T \Pi). \qquad (2.11)$$

Thus, (2.6) becomes

$$G' + \alpha T G^2 - [\frac{2U'}{U-c} - \frac{T'}{T}]G = \alpha [T - M^2 (U-c)^2]. \qquad (2.12)$$

The boundary conditions can be found from (2.10) and (2.11), with G continuous across the flame sheet. The stability problem is thus to solve equation (2.12) for a given real frequency ω and Mach number M, with U and T defined by (2.1)-(2.4). The eigenvalue is the wavenumber α. Because this equation has a singularity at $U = c_N$, we shall integrate it along the complex contour (-6,-1) to $(\eta_f,0)$ and (6,-1) to $(\eta_f,0)$. Using a Runge-Kutta scheme with variable step size, we choose an initial α and compute the boundary conditions from (2.10). We then iterate on α, using Muller's method, until the boundary conditions are satisfied and the jump in G at $(\eta_f,0)$ is less than 10^{-6}. All calculations were done in 64 bit precision.

3. Results. In all of our calculations reported here we have taken $\gamma = 1.4$, $\beta_T = 2$, $0 \le \beta \le 5$, and $0 \le M \le 10$. In Figures 2 and 3 we show the phase speeds of the neutral waves and the corresponding maximum growth rates, respectively, for the nonreactive ($\beta = 0$) mixing layer (Jackson and Grosch, 1988). From Figure 2 one can see that there is only a single subsonic neutral mode in region 1. This modes crosses over the sonic curve at the Mach number (M_s) at which its phase speed equals that of a sonic wave, and is transformed into a supersonic neutral mode in region 2. In addition, a slow supersonic neutral mode appears in region 4. In regions 2 and 4 there are unstable modes with phase speeds between that of the supersonic neutral mode and that of the sonic neutral mode. The band of unstable modes in region 2 is a group of fast unstable modes and that in region 4 is a group of slow unstable modes. The phase speeds of both the fast and slow modes have a small range about the average, so that little dispersion of wave packets is expected. From Figure 3 it is apparent that an increase in the Mach number results in a decrease of the growth rate of the subsonic mode by a factor of three or four up to about M_s, and for higher Mach numbers the growth rate of the fast modes approaches a limiting value. The growth rate of the slow mode is comparable to that of the fast mode for Mach numbers greater than M_s, and for higher Mach numbers it increases slightly.

The addition of heat release ($\beta > 0$) was found to have important, and complex, effects on the flow stability. In order to show these effects clearly, we first consider zero Mach number and plot the phase speeds and maximum growth rates in Figures 4 and 5. Figure 4 shows the existence of multiple subsonic neutral modes. There are two subsonic modes, fast and slow, and a singular neutral mode adjacent to each with phase speeds independent of β. The fast neutral modes exists for all values of β, but the slow modes only exist for $\beta \ge \beta_c = 1$. We note here that as ϕ increases the phase speed of the fast neutral mode approaches that of the fast singular mode, which is independent of ϕ. The phase speeds of the slow neutral modes is independent of ϕ except β_c increases with increasing ϕ. One can see from Figure 5 that an increase in β results in a reduction in the maximum growth rate of the fast mode and an increase in the growth rate of the slow mode. For sufficiently large β these curves cross so that the slow mode becomes the most unstable. Finally, an increase in ϕ causes an increase of the maximum growth rate of the fast modes and a decrease for the slow modes.

The phase speed of the neutral modes as a function of Mach number for $\beta = 1, 2, 5$ and $\phi = 1$ is shown in Figure 6. An increase in the value of β causes an increase in the phase speed of the fast subsonic mode in region 1. In all cases, the fast subsonic neutral modes are transformed into fast supersonic neutral modes in region 2 at M_s. In region 1 there are unstable modes between the subsonic neutral mode and its corresponding singular neutral mode. For higher Mach numbers, there are unstable waves with phase speeds between that of the supersonic neutral mode and that of the sonic neutral mode. It can be seen that the addition of heat release has a major effect on the slow modes. For $\beta \geq 1$, there are now both regular and singular subsonic neutral modes in region 1. Again, there are unstable waves with phase speeds between that of the slow neutral modes in region 1 and between that of the supersonic neutral mode and the sonic mode in region 4. The effect of ϕ on the neutral phase speeds is the same as in the zero Mach number case.

Even with combustion heating, the maximum growth rates of the unstable waves, shown in Figures 7-8, decrease by a factor of three to four as the Mach number approaches M_s. As in the nonreacting case, for Mach numbers greater than M_s, the growth rates level off and those of the slow modes eventually begin to increase with increasing Mach number while those of the fast modes approach a limiting value. Also, increasing the heat release increases the growth rate of the slow modes and decreases that of the fast modes. Finally, Figure 8 shows that the growth rate of the slow modes decreases with increasing ϕ while that of the fast modes increases.

Figures 9 and 10 are plots of the fast supersonic neutral eigenfunctions for $\phi = 1$, Mach 5, and $\beta = 0$ and 5, respectively. These have been normalized to a maximum amplitude of unity. They have exponential decay in the subsonic region and oscillations with constant amplitude and linear phase in the supersonic region. An increase in β causes the rate of decay of the amplitude in the subsonic region to increase, the variation of the amplitude near the center of the shear layer to increase, and the rate of change of phase in the supersonic region to increase. This effect is similar to that of increasing the Mach number with $\beta = 0$ (Jackson and Grosch, 1988).

In this study we have assumed unit Prandtl and Lewis numbers, used Chapman's linear relation between viscosity and temperature, and approximated the mean velocity profile by a hyperbolic tangent. In addition, we have taken the limit of infinite activation energy which reduces the diffusion flame to a flame sheet. We do not know how sensitive our results are to these assumptions. We believe that this study is an important first step in classifying and understanding the complex effects that chemistry has on the stability of compressible free shear layers. The next step is to consider a more realistic model of the chemistry and the thermodynamics. For this case the calculation of the mean field as well as the perturbation solution will be a more difficult numerical problem since now the velocity, temperature, and mass fraction equations are coupled. We have begun this study with large, but finite θ, and hope to report the results at a later date and will compare those results to the benchmark results reported here.

Acknowledgements. We wish to acknowledge helpful conversations with and comments from J. P. Drummond. This work was supported by the National Aeronautics and Space Administration under NASA Contract Nos. NAS1-18107 and NAS1-18605 while the authors were in residence at the Institute for Computer Applications in Science and Engineering, NASA Langley Research Center, Hampton, VA 23665 USA.

References

[1] Brown, G. L. & Roshko, A. 1974 On Density Effects and Large Structure in Turbulent Mixing Layers. J. Fluid Mech., 64, pp. 775-816.

[2] Buckmaster, J. D. & Ludford, G. S. S. 1982 *Theory of Laminar Flames*. Cambridge University Press, Cambridge.

[3] Chinzei, N., Masuya, G., Komuro, T., Murakami, A. & Kudou, D. 1986 Spreading of Two-Stream Supersonic Turbulent Mixing Layers. Phys. Fluids, 29, pp. 1345-1347.

[4] Drummond, J. P. & Mukunda, H. S. 1988 A Numerical Study of Mixing Enhancement in Supersonic Reacting Flow Fields. AIAA Paper 88-3260.

[5] Jackson, T. L. & Grosch, C. E. 1988 Spatial Stability of a Compressible Mixing Layer. ICASE Rep. No. 88-33. Also, submitted to J. Fluid Mech.

[6] Jackson, T. L. & Hussaini, M. Y. 1988 An Asymptotic Analysis of Supersonic Reacting Mixing Layers. Comb. Sci. Tech., 57, pp. 129-140.

[7] Kumar, A., Bushnell, D. M. & Hussaini, M. Y. 1987 A mixing augmentation technique for hypervelocity scramjets. AIAA Paper No. 87-1882.

[8] Menon, S., Anderson, J. D. and Pai, S. I. 1984 Stability of a Laminar Premixed Supersonic Free Shear Layer With Chemical Reactions. Int. J. Engng. Sci., 22(4), pp. 361-374.

[9] Papamoschou, D. & Roshko, A. 1986 Observations of supersonic free-shear layers. AIAA Paper No. 86-0162.

[10] Ragab, S. A. & Wu, J. L. 1988 Instabilities in the Free Shear Layer Formed by Two Supersonic Streams. AIAA Paper 88-0038.

[11] Tam, C. K. W. & Hu, F. Q. 1988 Instabilities of Supersonic Mixing Layers Inside a Rectangular Channel. AIAA Paper 88-3675.

[12] Williams, F. A. 1985 *Combustion Theory*, 2nd Ed., The Benjamin/Cummings Pub. Co., Menlo Park, Ca.

[13] Zhuang, M., Kubota, T. & Dimotakis, P. E. 1988 On the Instability of Inviscid, Compressible Free Shear Layers. AIAA Paper 88-3538.

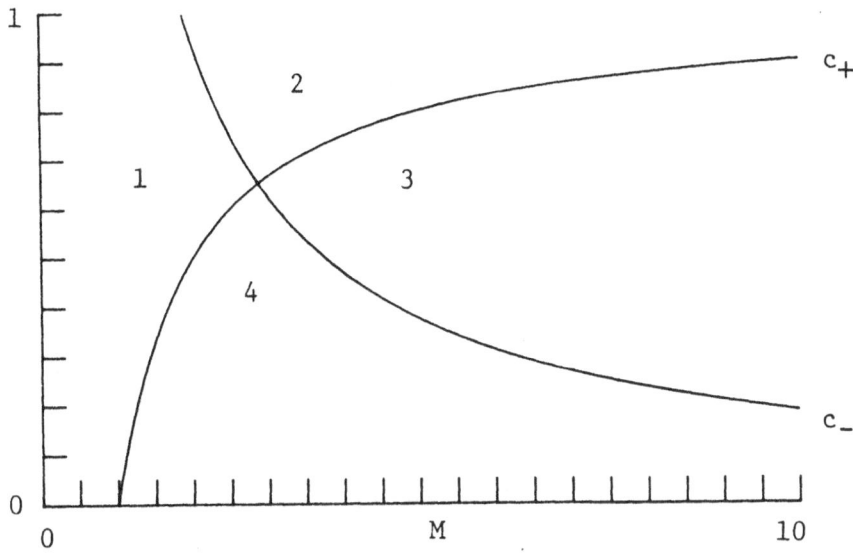

Figure 1. Plot of the sonic speeds $c_\pm$ versus Mach number.

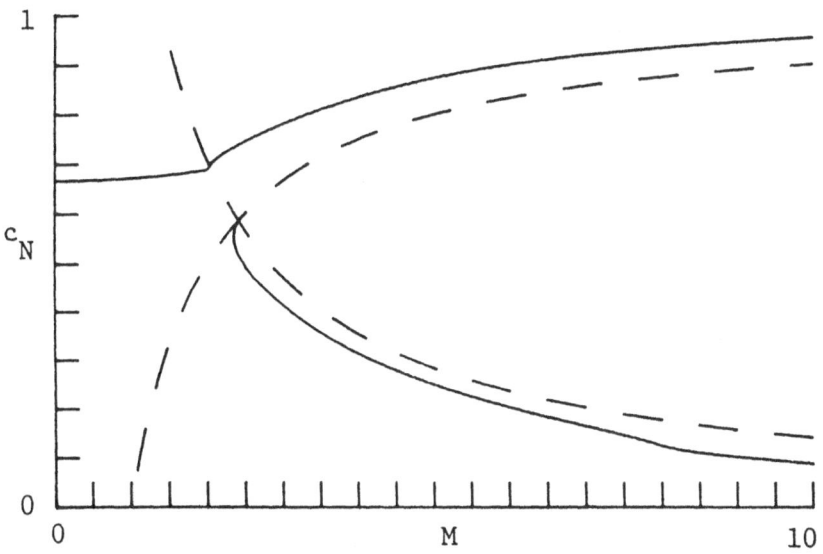

Figure 2. Plot of neutral phase speeds (solid) and sonic speeds (dashed) versus Mach number for $\beta_T = 2$, $\beta = 0$, and $\phi = 1$.

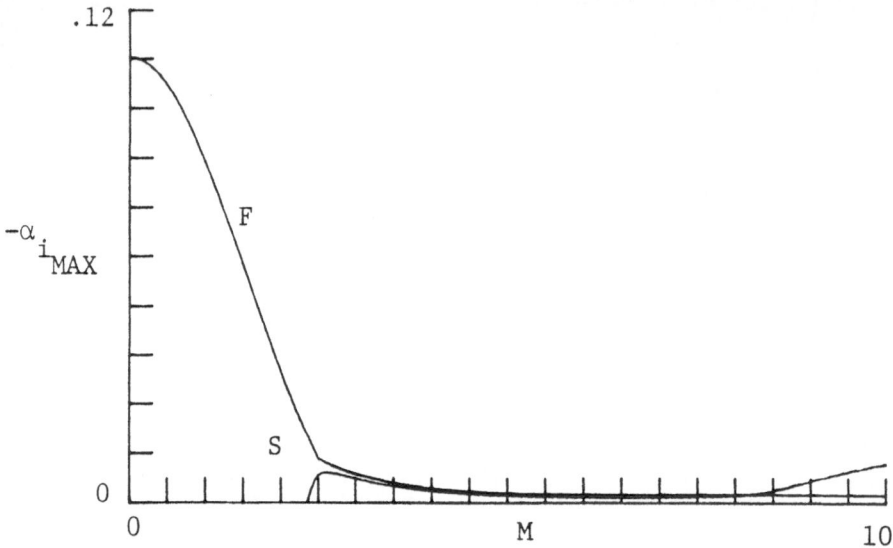

Figure 3. Plot of maximum growth rates of the fast and slow modes versus Mach number for $\beta_T = 2$, $\beta = 0$, and $\phi = 1$.

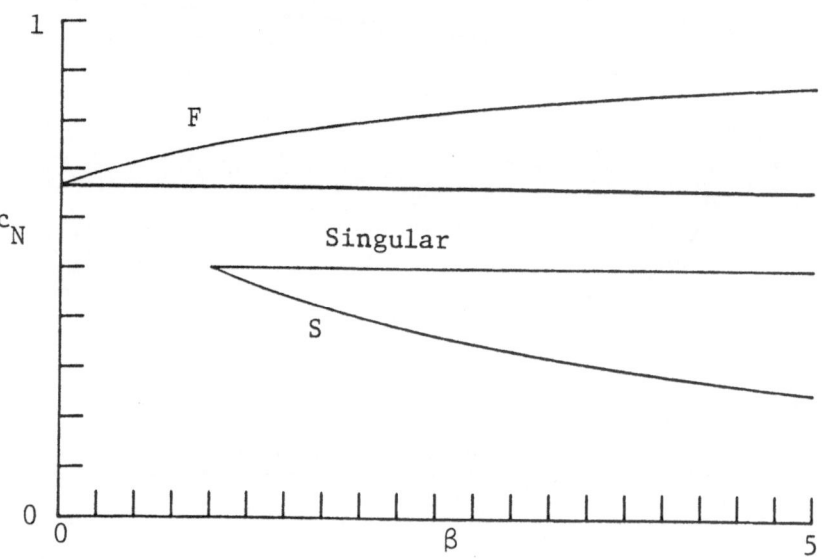

Figure 4. Plot of neutral phase speeds versus β for $\beta_T = 2$, $\phi = 1$, and $M = 0$.

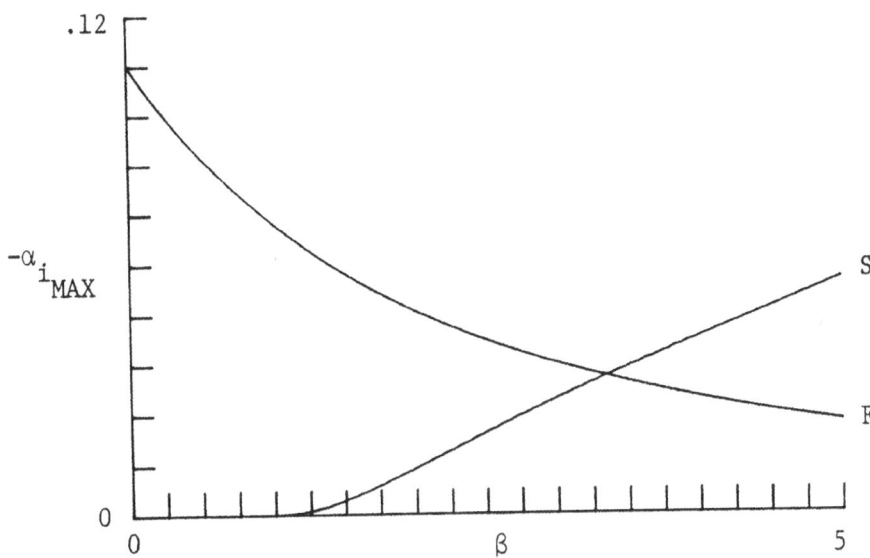

Figure 5. Plot of maximum growth rates of the fast and slow modes versus β for $\beta_T = 2$, $\phi = 1$, and $M = 0$.

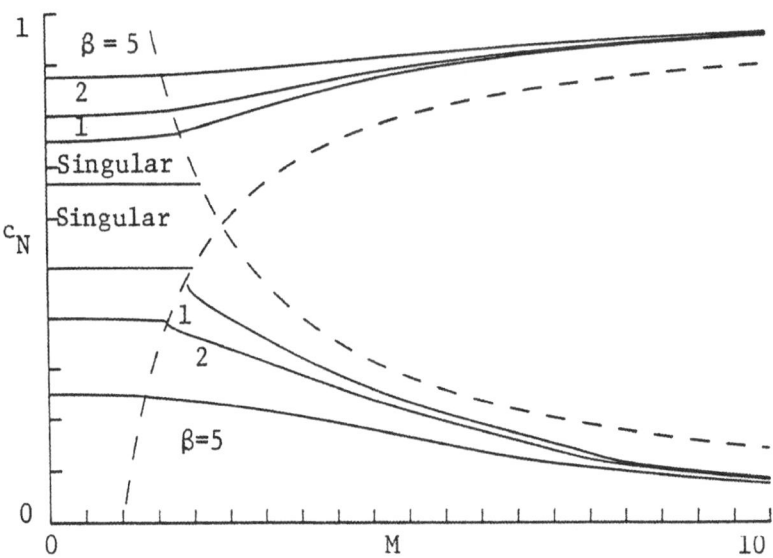

Figure 6. Plot of neutral phase speeds (solid) and sonic speeds (dashed) versus Mach number for $\beta_T = 2$, $\beta = 1, 2, 5$, and $\phi = 1$.

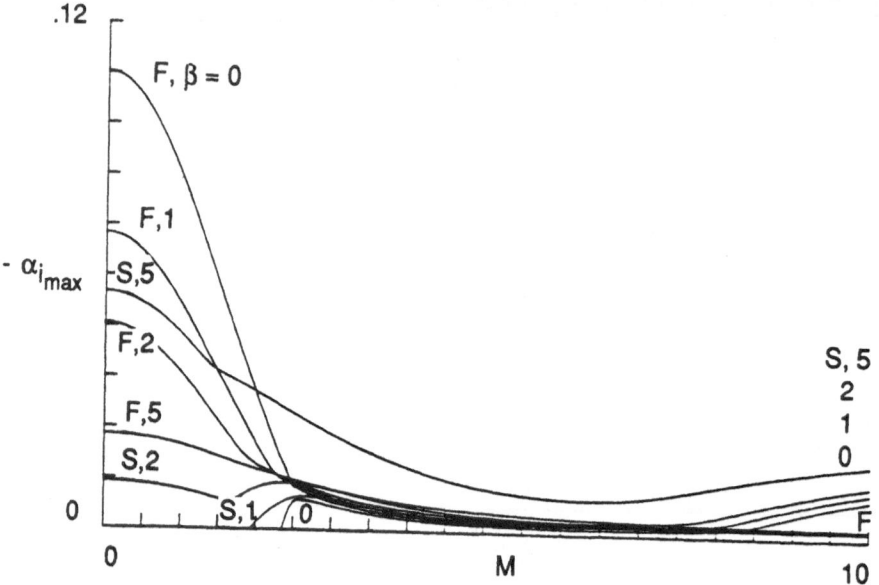

Figure 7. Plot of maximum growth rates of the fast and slow modes versus Mach number for $\beta_T = 2$, $\beta = 0, 1, 2, 5$, and $\phi = 1$.

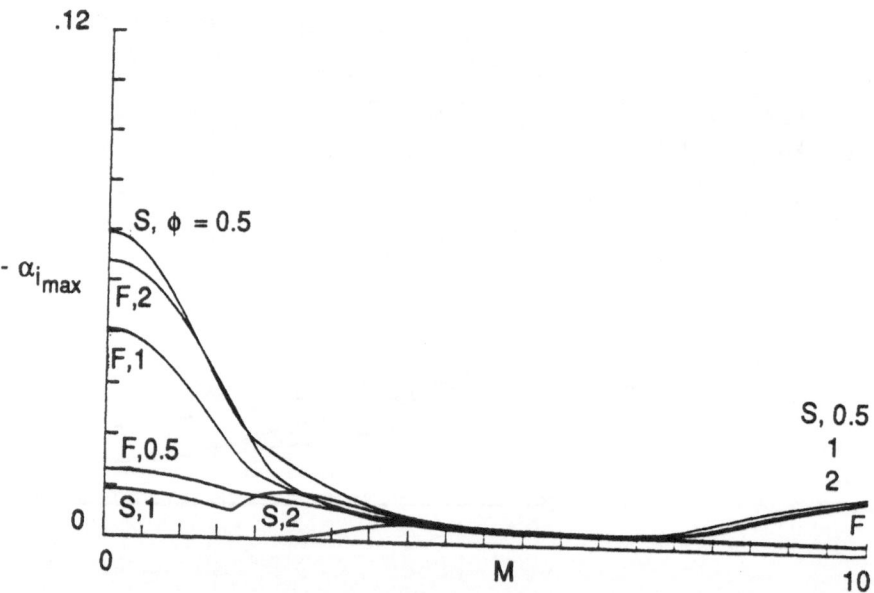

Figure 8. Plot of maximum growth rates of the fast and slow modes versus Mach number for $\beta_T = 2$, $\beta = 2$, and $\phi = 0.5, 1, 2$.

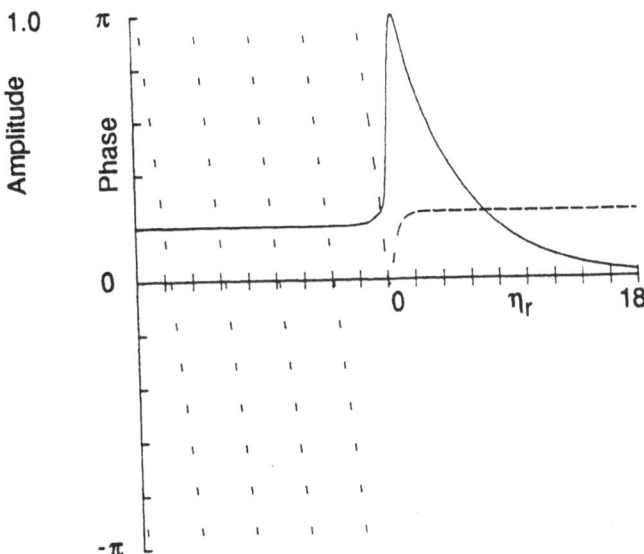

Figure 9. Plot of the two dimensional fast supersonic neutral eigenfunction $\Pi(\eta)$ along the contour $\eta = \eta_r - i$. The solid curve corresponds to the amplitude and the dashed curve to the phase. $M = 5$, $\beta_T = 2$, $\phi = 1$, $\beta = 0$, with $\omega_N = 0.240932$, $\alpha_N = 0.276046$, $c_N = 0.872797$.

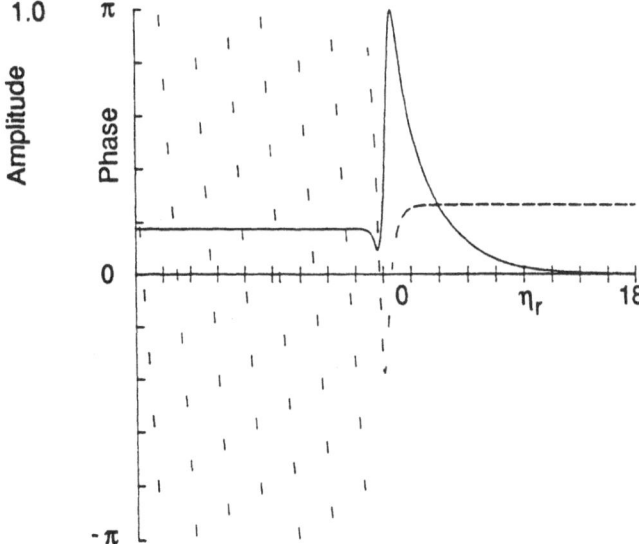

Figure 10. Plot of the two dimensional fast supersonic neutral eigenfunction $\Pi(\eta)$ along the contour $\eta = \eta_r - i$. The solid curve corresponds to the amplitude and the dashed curve to the phase. $M = 5$, $\beta_T = 2$, $\phi = 1$, $\beta = 5$, with $\omega_N = 0.374146$, $\alpha_N = 0.407399$, $c_N = 0.918378$.

TRANSITION TO DETONATION: A NUMERICAL STUDY

A. Kapila and V. Roytburd
Department of Mathematical Sciences
Rensselaer Polytechnic Institute, Troy, New York 12180-3590, USA

ABSTRACT

Details of transition to detonation in a planar, one-dimensional configuration are examined via a numerical simulation of the reactive Euler equations. The initial state of the explosive is taken to involve a small temperature nonuniformity. It is observed that the eventual emergence of a ZND (Zeldovich, von Neumann, Döring) detonation is preceded by a rich sequence of events, including the creation of a localized thermal explosion, the birth of a supersonic reaction wave, and the appearance of a quasisteady weak detonation.

INTRODUCTION

This paper is concerned with a fundamental problem in the theory of reactive flow, namely, the emergence of a detonation wave in an initially quiescent, homogeneous reactive fluid, subsequent to the application of a sufficiently strong stimulus such as a piston-driven shock. Of primary interest are the details of the transient processes that occur prior to the establishment of a full-fledged detonation wave.

Our study begins by considering a model problem for a polytropic fluid, capable of undergoing an exothermic decomposition reaction governed by one-step, irreversible Arrhenius kinetics. We consider a planar configuration in which the fluid occupies the half space $x > 0$. If the initial state of the fluid is spatially uniform and motionless, then it will remain so as the reaction evolves in time. We have in mind a situation in which the initial temperature is sufficiently high (due to the passage of a preconditioning shock, for example) and the activation energy sufficiently large so that the temporal evolution is a well-defined thermal explosion, with a moderate induction time.

In order to promote the development of a wave, let us now assume that a spatial nonhomogeneity is present in the system at the initial time, say in the form of a small temperature perturbation, with a maximum at $x = 0$ and decaying monotonically with x. Nonhomogeneities of this kind appear naturally when a precursor shock passes through the spatially uniform material and switches on a chemical reaction behind it. Temperature will be the highest at $x = 0$ because the material there has had the most time to react.

With the initial state of the fluid so defined, we expect a localized thermal explosion to occur at $x = 0$. If the parameter range is appropriate the thermal explosion will involve, in addition to a temperature rise, a substantial pressure rise as well. A reaction wave, generated at the boundary, will propagate into the interior and transform into a detonation in due course. We propose to investigate the intermediate transient events.

THE MATHEMATICAL PROBLEM

In the conservative form the equations of motion are

$$\rho_t + (u\rho)_x = 0, \tag{1}$$

$$(\rho u)_t + (\rho u^2 + (1/\gamma)p)_x = 0, \tag{2}$$

$$(\rho E)_t + (\rho u E + (1/\gamma)up)_x = 0, \tag{3}$$

$$(\rho Z)_t + (\rho u Z)_x = -w, \tag{4}$$

$$p = \rho T, \tag{5}$$

where

$$E = \frac{1}{\gamma - 1}\frac{p}{\rho} + \frac{u^2}{2} + \beta Z, \tag{6}$$

and

$$w = \frac{1}{\beta\theta}\rho Z exp(\theta - \theta/T). \tag{7}$$

In these equations, p, ρ, T, u, E and Z are, respectively, the gas pressure, density, temperature, velocity, specific internal energy and reactant mass fraction. The variables have been rendered dimensionless by referring them to a constant state. Velocity is referred to the frozen acoustic speed at the reference state and time to the spatially homogeneous induction time at that state (derived from classical thermal explosion theory for large activation energy). The dimensionless parameters appearing above are the polytropic exponent γ, the heat release parameter β and the activation energy θ.

The equations are examined in the region $x > 0$, $t > 0$. Since only one forward characteristic enters the region from the boundary $x = 0$, only one condition needs to be imposed there, viz,

$$u(0,t) = 0. \tag{8}$$

The initial state is taken to be a small, $O(\theta^{-1})$ perturbation of the spatially homogeneous and stationary reference state, i.e.,

$$u(x,0) = 0, \quad p(x,0) = Z(x,0) = 1, \tag{9}$$

$$T(x,0) = 1 + \theta^{-1}T_1(x,0), \tag{10}$$

where $T_1(x,0)$ is a monotonically decreasing function of x. Note that the corresponding profile of ρ is fixed by the equation of state (5).

THE NUMERICAL SCHEME

For numerical integration of the system (1)-(7), ρ, u, E and Z are selected as the primitive variables, while p, T and w are determined from the constitutive relations (5)-(7). A rather standard fractional-step operator splitting procedure is employed. The algorithm involves two ingredients per time step. In the first half step the hydrodynamic portion of the problem is solved, i.e., equations (1)-(3) are advanced. In the second half step the species equation (4) is advanced by means of an explicit ODE solver, with the temperature field known from the previous half step. Thus the mass fraction Z

is advected as a passive scalar and does not affect directly the distribution of other hydrodynamic variables.

Full details of the numerical method have been recorded elsewhere [1]. Suffice it to say that a uniform sampling procedure is used for the hydrodynamic step. The initial data are approximated by piecewise constant functions, and the Euler equations are then solved exactly by a Riemann solver, the time step being restricted by the CFL condition. The resulting solution is, of course, not piecewise constant. In order to obtain piecewise constant data for the next time interval, the solution is sampled at some point within the spatial mesh interval. At a given time level, the relative position of the sampling point is taken to be the same for all the mesh intervals. It changes with time so that the sequence of sampling points is uniformly distributed.

The use of the Riemann solver requires that the full set of Riemann data be available for the problem at the boundary. This is achieved by extrapolating the solution symmetrically to the left of $x = 0$. The right boundary is chosen to lie at a sufficiently large value of x to ensure that it does not affect the solution upto the time level of interest.

It is important to observe that the events being simulated occur on widely disparate time scales. For example, all the "action" in Fig. 1 occurs for $0.5310 < t < 0.5318$. Thus, the "explosion" time of 8×10^{-4} is a very small fraction of the homogeneous induction time, which is the time unit adopted as the reference. In order to deal with this situation the early-time calculation is carried out on a course grid, which is refined when large gradients appear. For the results displayed here, the coarsest grid had a mesh size of $1/320$ and the finest grid a mesh size of $(1/6000)$. Our results show that even this fine a mesh fails to resolve satisfactorily the reaction zone behind the lead shock once the shock strength approaches that of the steady, CJ (Chapman Jouguet) detonation. The use of adaptive, nonuniform gridding to overcome this difficulty is currently under investigation.

THE RESULTS

Computations were performed for the following data:

$$\beta = 3.0, \quad \gamma = 1.4, \quad \theta = 18.5,$$

$$T_1(x, 0) = 0.5(1 - 4\pi^2 x^2).$$

The results of the study are presented in several different sets of figures, to identify the various stages of evolution. Fig. 1 corresponds to the time interval $0 < t < 0.5318$, and displays the profiles of temperature, pressure, density and reactant mass fraction at successive time levels. During most of this interval, the state of the combustible is found to depart only slightly from the initial state. At the end of the interval, however, rapid variations occur in a thin layer near the boundary while the state is essentially stationary elsewhere. In the layer the reactant concentration falls sharply from near-unity to zero, accompanied by substantial increases in both temperature and pressure. Changes in density within the layer are negligible.

Fig. 1 presents two distinct stages of evolution. In the first stage chemical reaction is relatively weak, as are gasdynamic effects, and the basic balance is chemico-acoustic. Although changes are small, they occur over a broad domain, whose size is such that

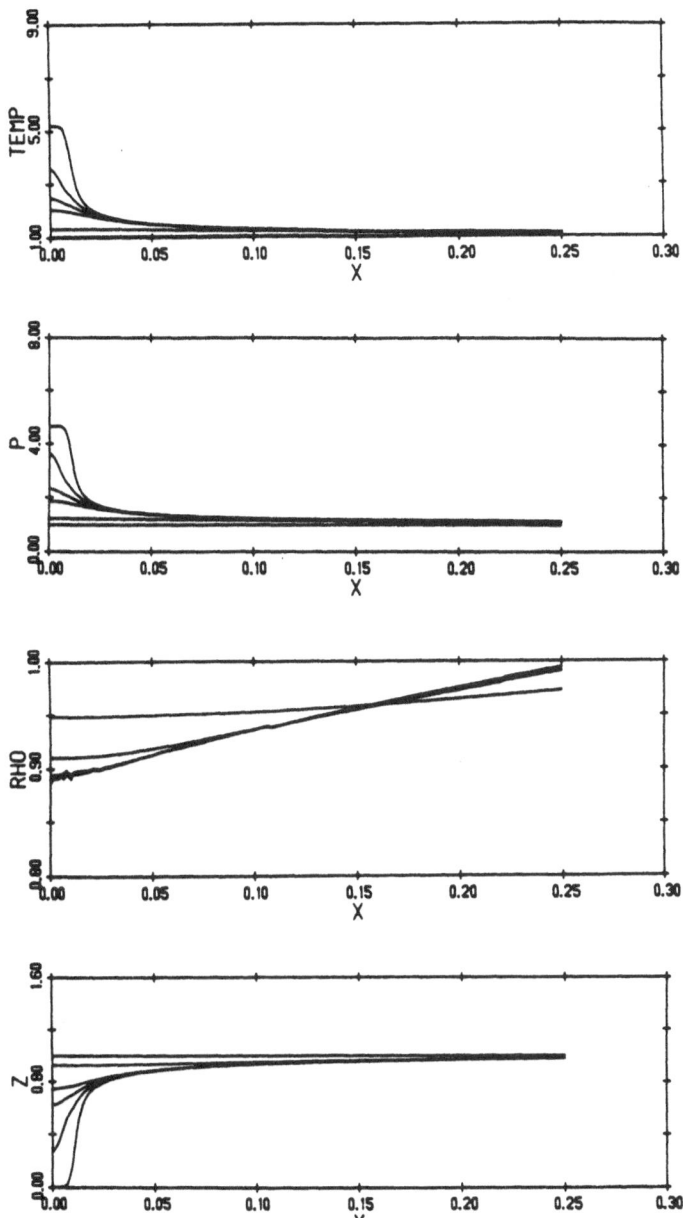

Fig. 1 Profiles of temperature, pressure, density and reactant mass fraction in the interval $0 < t < 0.5318$. The five curves refer to the following time levels: (1) 0, (2) 0.5310, (3) 0.5314, (4) 0.5316, and (5) 0.5318. Note that density is practically constant at the last three time levels.

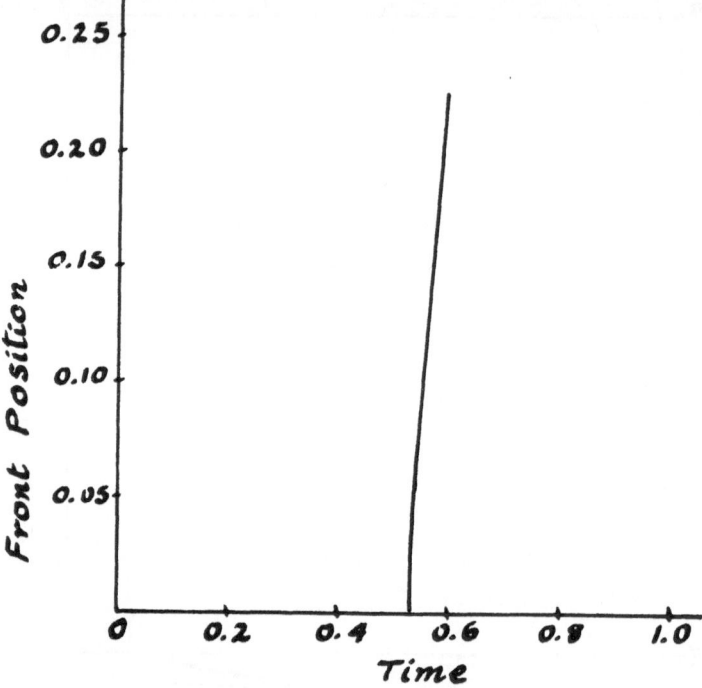

Fig. 2 The wave path (locus of $Z = 0.5$) in the xt-plane. Note the deceleration at earlier times.

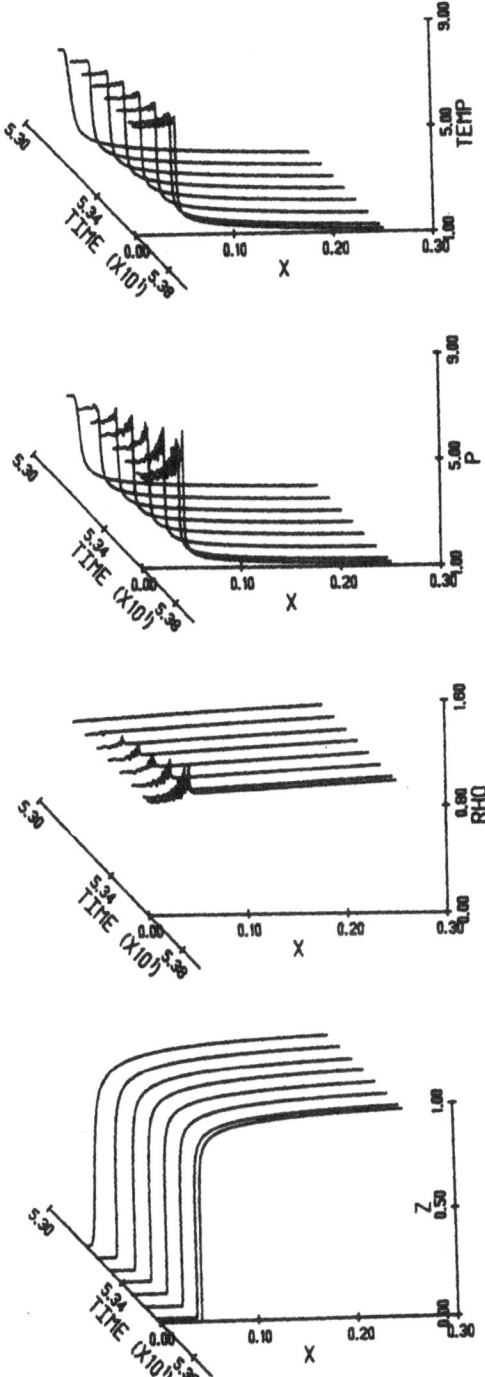

Fig. 3 Profiles of temperature, pressure, density and reactant mass-fraction in the time interval $0.5318 < t < 0.5359$.

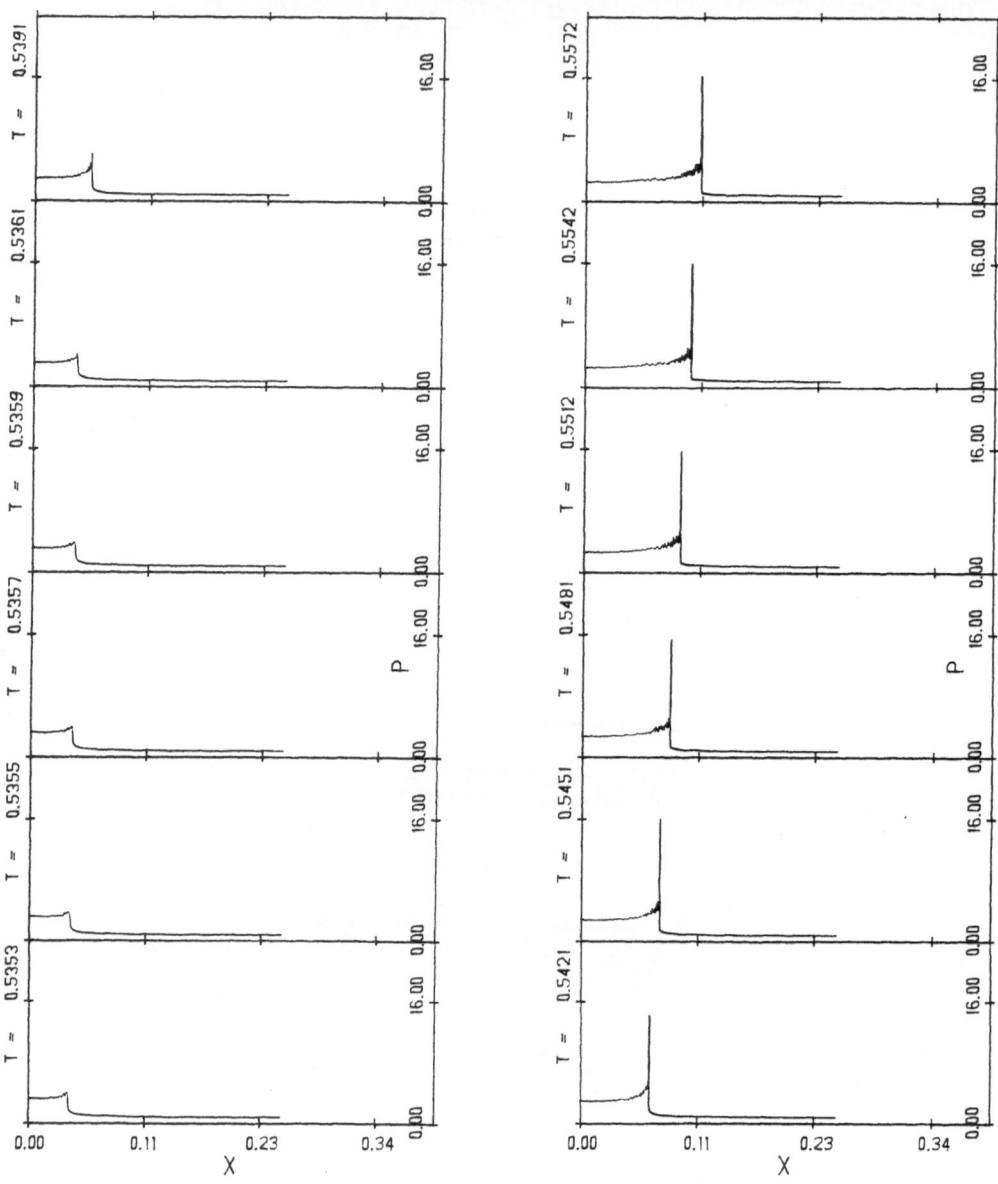

Fig. 4 Profiles of pressure for $t > 0.5353$. Note the birth of a shock at $t = 0.5361$ and its rapid stregthening.

acoustic time and initial chemical time are comparable within it. This stage can be called the induction stage. At the end of this stage, chemical reaction at the boundary has strengthened sufficiently to provide a localized thermal explosion. Local reaction times now become exceedingly small compared to the acoustic times (inertial confinement), so that thermal expansion is negligible and explosion occurs without appreciable change in density. Increase in pressure is therefore directly proportional to increase in temperature. Each particle, oblivious of its neighbors, undergoes a thermal explosion slightly later than the one to its left, and the result is the birth of a purely reactive wave propagating to the right, its speed depending upon the thermal gradient established at the end of the induction stage. The wave path (the locus of $Z = 1/2$) is shown in Fig. 2.

Fig. 3 depicts events occuring in the interval $0.5318 < t < 0.5359$. The wave is now well-defined and the material no longer inertially confined; gradients of density have appeared. As the wave travels to the right the pressure, temperature and density behind it continue to rise. The wave is followed by a rarefaction and its speed continues to decline (Fig. 2). A detailed analysis of the data indicates, however, that the speed remains higher than the CJ value and the flow through the wave supersonic, until the end of this stage.

The final stage is shown by the sequence of profiles in Figure 4, drawn for $t > 0.5360$. For the sake of brevity only the pressure profiles have been shown; the other gasdynamic properties behave in a similar way. A weak shock appears just barely at $t = 0.5361$. As time evolves the shock quickly strengthens and then the wave profile appears to propagate essentially undistorted at the CJ speed. We did not carry the calculations far enough to observe any galloping instability. As already stated, the resolution of the chemical zone in the final stage is unsatisfactory and needs to be improved.

DISCUSSION

The numerical study has shown that the eventual emergence of a conventional, ZND detonation is preceeded by the appearance of a localized thermal explosion and a supersonic reaction wave. These findings are consistent with recent asymptotic studies [2,3] and provide missing details in the existing transition scenario [4].

Because the fast waves can be taken to be quasisteady (except in the thermal explosion and shock-formation stages), the entire picture can be understood by referring to the Hugoniot diagram (Fig. 5). The point C on the unreacted Hugoniot represents the initial state. The constant-density explosion at the boundary occurs along the vertical line CW_1. The subsequent progress of the supersonic reaction wave (which is nothing but a weak detonation!) towards the CJ point is illustrated, at intermediate times, by the sequence of Rayleigh lines CW_2, CW_3 and CJ. The swift transformation near the CJ point from the weak detonation structure (CJ) to the ZND structure (CSJ) occurs through a succession of intermediate stages (such as CS_1S_2J) as the shock strengthens.

Acknowledgements

This work was supported by the National Science Foundation under Grant No. DMS-8603506, by the United States Army Research Office, and by the Los Alamos National laboratory under Contract No. DOE- LANL-9XG9-4906Y-1. The paper was written while the second author was on leave at Princeton University. He gratefully acknowledges the hospitality and financial support of Professors A. Majda and S. Orszag.

References

[1] P. Colella, A. Majda and V. Roytburd, "Fractional step methods for reacting shock waves," *Reacting Flows: Combustion and Chemical Reactors, Part 2, Lectures in Applied Mathematics*, Vol. 24, G. S. S. Ludford, Editor, pp. 459-477 (1986). American Mathematical Society, Providence, Rhode Island.

[2] J. B. Bdzil and A. K. Kapila, "Shock-to-detonation transition: a model problem," to appear (1989).

[3] A. K. Kapila and J. W. Dold, "A theoretical picture of shock-to-detonation transition in a homogeneous explosive," to appear (1989).

[4] A. W. Campbell, W. C. Davis and J. R. Travis, "Shock- initiation of detonation in liquid explosives," *Phys. Fluids*, 4 (1961), pp. 498-510.

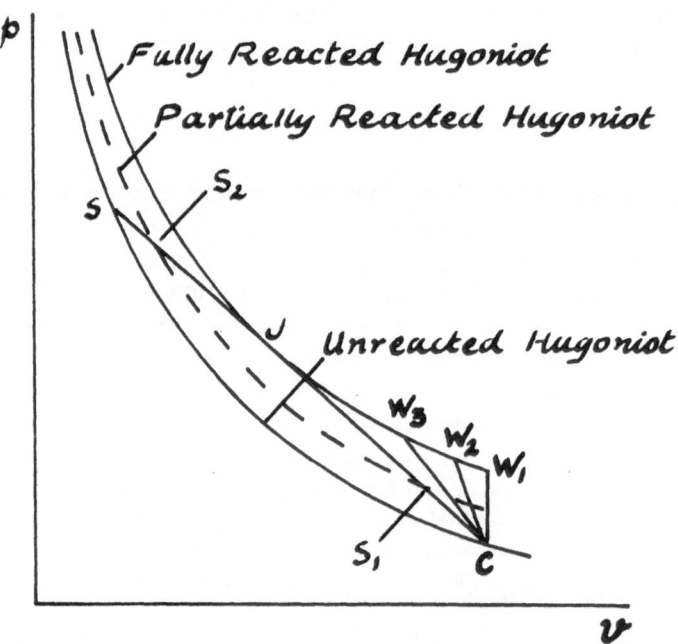

Fig. 5 Evolution towards ZND detonation on the pressure- specific volume Hugoniot diagram.

SHOCK INDUCED SELF-IGNITION

OF A REACTIVE GAS MIXTURE IN AN L–SHAPED DUCT

Justus J. Klöker

Aerodynamisches Institut, RWTH Aachen

Templergraben 55, D-5100 Aachen

Introduction

For the design of engines the interaction of gasdynamic waves due to chemical reactions with a complex geometry is an important aspect. Calculations of complex geometries with complete reaction mechanisms results in program codes, which are to expensive for practical problems. Therefore, recent studies of combustion processes concentrate either on the calculation of really complex reaction mechanism in the one-dimensional case (see Oran *et al.* [1], Kailasanath *et al.* [2]) or of a complex geometry with a simple reaction model (see Toro *et al.* [3], Klein [4]). Here a two-dimensional model, a L–shaped duct, is investigated to study the wave-boundary interactions and the self-ignition of a homogenous, reactive gas-mixture. A plane shock-wave enters perpendicular to the open side of the duct (Fig. 1). For the solution of this unsteady, two-dimensional problem a piecewise-linear-method (PLM) is employed to the Euler equations. It is a second-order Godunov-type method, described by Colella *et al.* [5], which is an refinement of the MUSCL-scheme by van Leer [6]. In this paper an extension of this method is presented which includes source terms for a reactive gas-mixture. The data needed for the reaction model, developed by Korobeinikov *et al.* [7], are taken from experiments.

Mathematical Model of Reactive Flow

The flow is described by the Euler-equations for mass, momentum and energy densities extended by a source term for the chemical heat release and the reaction rates of the components α and β.

$$U_t + F(U)_x + G(U)_y = Q \tag{1}$$

$$
U = \begin{pmatrix} \varrho \\ \varrho u \\ \varrho v \\ e \\ \varrho\beta \\ \varrho\alpha \end{pmatrix} \quad
F = \begin{pmatrix} \varrho u \\ \varrho u^2 + p \\ \varrho u v \\ u(e+p) \\ \varrho u \beta \\ \varrho u \alpha \end{pmatrix} \quad
G = \begin{pmatrix} \varrho v \\ \varrho u v \\ \varrho v^2 + p \\ v(e+p) \\ \varrho v \beta \\ \varrho v \alpha \end{pmatrix} \quad
Q = \begin{pmatrix} 0 \\ 0 \\ 0 \\ \varrho w_\beta Q \\ -\varrho w_\beta \\ -\varrho w_\alpha \end{pmatrix} \tag{2}
$$

where ϱ is the density, u, v the velocity components in x and y-direction, respectively, e the energy density, p the pressure and w_k the reaction rates. Instead of using a complete set of real elementary reactions, a global two-step model reaction, for simulating a stoichiometric hydrogen-air mixture, is utilized [7]. So, α and β are reaction progress parameters, where α expresses the fuel concentration and β is similar to the radical concentration after the induction period. The following range of parameters is chosen for the induction period and the exothermic reaction period, respectively:

$$0 < \alpha < 1 \quad and \quad \beta = 1 \tag{3}$$

$$\alpha = 0 \qquad and \quad 0 < \beta < 1 \tag{4}$$

Here, $\alpha = 0$ stands for the end of the induction period and the start of the exothermic reaction. The reaction rates are represented by:

$$w_\alpha = -\frac{d\alpha}{dt} = k_1 \, p^{n_1} \, \varrho^{l_1} \, exp\left(-\frac{E_1}{R\,T}\right) \tag{5}$$

$$w_\beta = -\frac{d\beta}{dt} = k_2 \, p^{n_2} \, \varrho^{l_2} \, \beta^{m_1} \, exp\left(-\frac{E_2}{R\,T}\right) - k_3 \, p^{n_3} \, \varrho^{l_3} \, (1-\beta)^{m_2} \, exp\left(-\frac{E_2+Q}{R\,T}\right) \tag{6}$$

The constants k express the frequency factors and E is the activation energy. The rate constants k, E, l, m, n correspond to [8]. Due to the global reaction:

$$2H_2 + O_2 \longrightarrow 2H_2O \tag{7}$$

the heat release Q is calculated by the enthalpy-difference of the reactants i and the products j:

$$\Delta H_{0m} = \sum_j \nu_j H_{0mj} - \sum_i \nu_i H_{0mi} \tag{8}$$

with the corresponding stochiometric coefficients ν and the molar enthalpy of formation H_{0m}. The equations of state for an ideal gas complete the mathematical model:

$$p = \varrho \, T \tag{9}$$

$$e \equiv \varrho \left(\varepsilon + \frac{u^2 + v^2}{2} \right) \qquad , \qquad \varepsilon = \frac{p}{(\gamma - 1)} \varrho \tag{10}$$

Solution Method

Godunov-type methods are based on the solution of the Riemann-problem, which is an initial value problem of a one-dimensional jump discontinuity. The *operator splitting* divides the fundamental equation-system (1) in two unsteady, quasi-one-dimensional equation-systems:

$$U_t + F(U)_x = \frac{1}{2} Q \tag{11}$$

$$U_t + G(U)_y = \frac{1}{2} Q \tag{12}$$

The source term is also splitted and is calculated in parts in each space direction. Now the explicit conservative difference scheme can be written as:

$$U_i^{n+1} = U_i^n + \frac{\Delta t}{\Delta x_i} \left(F_{i-\frac{1}{2}}^{n+\frac{1}{2}} - F_{i+\frac{1}{2}}^{n+\frac{1}{2}} \right) + \frac{\Delta t}{2} Q_i^{n+\frac{1}{2}} \tag{13}$$

where Δt is the time increment and Δx is the spatial increment. The non-conservative variables to determine the source term are given by:

$$V_i^{n+\frac{1}{2}} = \frac{V_{i+\frac{1}{2}}^{n+\frac{1}{2}} + V_{i-\frac{1}{2}}^{n+\frac{1}{2}}}{2} \tag{14}$$

with $V = (\varrho, u, v, p, \alpha, \beta)^T$. At the time t^n a piecewise linear distribution over each grid zone for the non-conservative variables is assumed, which is accomplished by the TVD-slope limiter:

$$\Delta q_i^n = \begin{cases} 0 & \text{for } (q_{i+1}^n - q_i^n)(q_i^n - q_{i-1}^n) \leq 0 \\ q_i^n - q_{i-1}^n & \text{for } |q_{i+1}^n - q_i^n| > |q_i^n - q_{i-1}^n| \\ q_{i+1}^n - q_i^n & \text{otherwise} \end{cases} \tag{15}$$

where q and Δq express the non-conservative variables and their slopes, respectively. The slopes lead to $\Delta V = (\Delta \varrho, \Delta u, \Delta v, \Delta p, \Delta \alpha, \Delta \beta)^T$. Further slope limiters, which lead to TVD-schemes, are described by Munz [9].

For the determination of the fluxes the characteristic method is employed to construct the variables of state at the zone edges for the time $t^{n+\frac{1}{2}}$. Here the description is only made for the x-direction. The calculation in y-direction is analogous. First Eq.(11) is transformed into the non-conservative form:

$$V_t + C(V)V_x = \frac{1}{2} S \tag{16}$$

with the transformation matrix P:

$$P = \frac{dU}{dV} \tag{17}$$

Then the left and the right states at the zone edge are constructed as follows:

$$V_{i+\frac{1}{2},L}^{n+\frac{1}{2}} = \tilde{V}_L + P_> \left(\frac{\Delta t}{2\Delta x_i} \left(\lambda_i^+ E - C_i^n \right) \Delta V_i^n \right) + P_> \left(\frac{\Delta t}{4} S_i^n \right) \tag{18}$$

$$V_{i+\frac{1}{2},R}^{n+\frac{1}{2}} = \tilde{V}_R + P_< \left(\frac{\Delta t}{2\Delta x_{i+1}} \left(\lambda_{i+1}^- E - C_{i+1}^n \right) \Delta V_{i+1}^n \right) + P_< \left(\frac{\Delta t}{4} S_{i+1}^n \right) \tag{19}$$

with the operators defined by:

$$P_>(w) = \sum_{\substack{\nu \\ \lambda_i^{(\nu)} > 0}} \left(L_i^{(\nu)} \cdot w \right) R_i^{(\nu)} \tag{20}$$

$$P_<(w) = \sum_{\substack{\nu \\ \lambda_{i+1}^{(\nu)} < 0}} \left(L_{i+1}^{(\nu)} \cdot w \right) R_{i+1}^{(\nu)} \tag{21}$$

where λ are the eigenvalues and L and R are the corresponding left and right eigenvectors, respectively. The reference states read:

$$\tilde{V}_L = V_i^n + \frac{1}{2} \left(1 - max(\lambda_i^+, 0) \frac{\Delta t}{\Delta x_i} \right) \Delta V_i^n \tag{22}$$

$$\tilde{V}_R = V_{i+1}^n - \frac{1}{2} \left(1 + min(\lambda_{i+1}^-, 0) \frac{\Delta t}{\Delta x_{i+1}} \right) \Delta V_{i+1}^n \tag{23}$$

The left and right states are only developed by waves, whose characteristics reach the point $P(x_{i+\frac{1}{2}}, t^{n+\frac{1}{2}})$. The reference state guarantees a state at supersonic flow, if from one side there is no characteristic reaching the zone edge (see Fig. 2). This construction is developed from the Taylor-approximation of Eq.(16) added by some modifications, which only effect on discontinuities (see Colella *et al.* [5]). With this states the Riemann-problem is defined:

$$V(t = 0, x < 0) = V_L = const.$$

$$V(t = 0, x > 0) = V_R = const. \tag{24}$$

and will be solved for $t > 0$ at $x = x_{i+\frac{1}{2}}$. Therefore an exact Riemann-solver, described by Chorin [10], with a Newton-iteration:

$$p^{\star^{n+1}} = \frac{u_R - u_L - \frac{p_R}{M_R^{n+1}} + \frac{p_L}{M_L^{n+1}}}{\frac{1}{M_L^{n+1}} - \frac{1}{M_R^{n+1}}} \tag{25}$$

and with:

$$M_k^{n+1} = \sqrt{\varrho_k p_k} \cdot \Phi\left(\frac{p^{\star^n}}{p_k}\right) \qquad k = L, R \tag{26}$$

is employed, where $\Phi(x)$ is a function describing the Rankine-Hugoniot- or the compatibility-condition, respectively:

$$\Phi(x) = \begin{cases} \sqrt{\frac{\gamma+1}{2}x + \frac{\gamma-1}{2}} & x > 1 \\ \sqrt{\gamma} & x = 1 \\ \frac{\gamma-1}{2\sqrt{\gamma}} \frac{x-1}{1-x^{\frac{\gamma-1}{2\gamma}}} & x < 1 \end{cases} \tag{27}$$

The Riemann problem is sketched in Fig. 3. After the pressure $p^\star$ between the shock wave and the rarefaction fan (Fig. 3) is determined, the velocity $u^\star$ yields as:

$$u^\star = \frac{p_L - p_R + u_R M_R - u_L M_L}{M_R - M_L} \tag{28}$$

To accelerate the convergence rate of the iteration after each 20 iteration steps the following procedure is utilized:

$$p^{\star^{n+1}} = a_j\, p^{\star^{n+1}} + (1 - a_j) p^{\star^n} \tag{29}$$

with:

$$a_0 = 1 \qquad a_j = \frac{a_{j-1}}{2} \qquad j = 1, 2, ... \tag{30}$$

With a selection algorithm (see Fig. 4) the state (ϱ, p, u) at the zone edge is determined applying again either the shock jump equations, the compatibility equations or an interpolation inbetween the rarefaction fan, respectively. The state values v, α, β, which are not involved in the Riemann-solver, are found by using:

$$q_G = \begin{cases} q_{i+\frac{1}{2}, L}^{n+\frac{1}{2}} & \text{for } u_G \geq 0 \\ q_{i+\frac{1}{2}, R}^{n+\frac{1}{2}} & \text{for } u_G < 0 \end{cases} \tag{31}$$

where q expresses the missing state variables. With the non-conservative states $V_{i-\frac{1}{2}}^{n+\frac{1}{2}}$ and $V_{i+\frac{1}{2}}^{n+\frac{1}{2}}$ the fluxes can be calculated for the conservative difference step.

At an extreme local state the reaction rate may become so large that the whole fuel burns within one time-step. Then the reaction front propagates with grid velocity. To avoid this, the local signal velocity of the cell is used to limit the reaction rate. The fundamental idea is that every event in a cell moves forward into the next cell with the signal velocity [11]. If the characteristic time of the event is smaller then the time of the local signal velocity, the event (i.e. the formation of a reaction front) continues in the neighbour cell, since due to the flow it has already left the cell behind. This yields the following reaction rate calculation:

$$w_z = min\left(w_z(\varrho, p, T, z), \frac{c + \sqrt{u^2 + v^2}}{\Delta x}\right) \qquad z = \alpha, \beta \qquad (32)$$

The time step for the calculation of the problem is determined according to the CFL-condition.

Numerical Results

The method described above has been implemented and tested for various shock Mach numbers. The calculations were performed with a Cartesian mesh of 150×47 or 100×47 grid points for the channel inlet and 45×102 for the branch, respectively. The division into two connected grids was necessary for reasons of computer storage. Due to the small, but long branch a cell width ratio:

$$\frac{\Delta x}{\Delta y} = \frac{1}{3} \qquad (33)$$

is used. Only the end part of the L–shaped duct has been calculated and in the following pictures it is rotated by $90°$ degrees, so the incoming shock entering through the open side starts at the bottom side. The calculation is performed for a stoichiometric hydrogen-air mixture with the following initial data:

$$p_1 = 0.1 \ bar$$

$$T_1 = 298 \ K$$

$$\varrho_1 = 0.08472 \ kg/m^3.$$

With the width of the channel inlet (the open side of the L–shaped duct) of $16 \ mm$ the unity of the dimensionless time is equivalent to:

$$t_1 = \sqrt{\frac{\varrho_1}{p_1}}L = 46.571 \ \mu s.$$

The calculation with $M_S = 2.5$ until $t = 3.0$ needs a CPU-time of about 11.8 hours with a time step of $\Delta t = 0.0008 - 0.0016$. The algorithm requires about 1.5 ms CPU time per time step and per grid cell. With further optimizations of the program code a CPU-time reduction of about 20% could be obtained.

The first case with the shock Mach number $M_S = 2.5$ is shown in Fig. 5. The time count starts when the incoming shock wave reaches the convex corner of the channel. Now the shock expands into the branch (Fig. 5a, $t = 0.1$) and reflects at the opposite wall, while a pressure wave moves into the branch (Fig. 5b, $t = 0.2$). After the reflection the entropy increases as shown in Fig. 5c ($t = 0.2$). At the convex corner an entropy peak is obtained in the vortex due to numerical diffusion, which, however, does not result in an temperature increase. For $t = 0.4$ the gas mixture ignites behind the reflected shock wave (Fig. 5d, $t = 0.5$), also seen in a 3-D-surface plot in Fig. 5e. The width of the reaction front can be seen in Fig. 5f, which is a contour plot of the exothermic parameter β. The heat release causes an entropy increase (Fig. 5g), which means a temperature increase. The temperature level, just as the entropy level, remains constant behind the reaction front. For $t = 0.75$ the detonation wave catches the shock wave, which is accelerated (Fig. 5h). For $t = 0.95$, when the outgoing shock wave reaches the mesh boundary, the pressure wave in the branch reflects at the wall and a shock is generated (Fig. 5i). At the same time the mixture ignites at a second location behind the forming Mach-stem near the convex corner (Fig. 5j), which produces a new detonation wave (Fig. 5k, $t = 1.05$). At this time the mixture also ignites at the end wall of the branch (Fig. 5l). From the shock a triple configuration (Fig. 5m, $t = 1.4$) with a complex reaction front (Fig. 5n) arises in the branch. During the propagation of this triple a vortex peels off (Fig. 5o, $t = 1.6$). At this time the gas mixture has almost reacted completely, except for the stretched vortex at the convex corner, where the temperature remains too cold for a reaction (Fig. 5p).

For $M_S = 2.8$ the reaction occurs immediately behind the reflected shock wave (Fig. 6a, $t = 0.2$) due to the high pressure raise (Fig. 6b in a 3-D-surface plot), which induces also a high temperature level. The reaction front follows the outgoing shock after the very short induction length (Fig. 6c). When the second ignition occurs at the end wall of the branch after the reflection of the incoming pressure wave, only a small rest of unburnt gas mixture is left there (Fig. 6d, $t = 0.9$).

On the other hand for $M_S = 2.25$ the first ignition occurs in the branch at $t = 1.45$ after a long induction period, when the shock already has generated a triple configuration (Fig. 7a). Again the mixture ignites at different points at the same time (Fig. 7b). At $t = 1.6$ (Fig. 7c) a complex reaction front can be observed (Fig. 7d), where some unburnt gas pockets are formed. When a second ignition is examined at the right side of the main channel a strong vortex follows the triple configuration in the branch (Fig. 7e, $t = 2.05$). In this vortex exists a big pocket of unburnt gas (Fig. 7f). Due to the chemical–gasdynamical interactions a complex wave system has been formed. For $t = 2.3$ a shock wave leaves the branch into the main channel (Fig. 7g), which rises the temperature and accordingly initiates the reaction in this part of the L–shaped duct followed by a complex reaction front system due to two vortices (Fig. 7h). This, finally, leads to a very strong pressure peak in the upper left corner of the channel

(Fig. 7i, $t = 2.65$) due to an ignition induced by the reflected shock front from the branch (Fig. 7j).

Calculations for $M_S = 1.6$ did not reveal any ignition with the employed initial data.

The algorithm has been tested and was compared with experimental results by Ritzerfeld [12] for the case of nitrogen as an ideal gas without a source term. The investigations were performed for a circular and a quadratic cylinder in a shock tube. The agreement of the calculated density contours with the experimental interferograms was good.

Conclusions

The presented calculations give an impression of the complexity of shockwave interactions due to the geometry and the chemical reactions. Similar processes may occur in piston engines leading to the knock phenomena, which still can't be explained satisfactory. Although in y-direction only a coarse spatial resolution could be realized due to the limited central memory, no stability problems arised. Due to the exact Riemann solution the code is robust, but its efficiency can be improved by an approximate Riemann solver (i.e. see Roe [13]). Further improvements may probably be obtained with adaptive gridding, which is already developed for the one-dimensional case, and with reduced reaction schemes, which simulate the complete reaction by a very small number of quasi-global reaction steps. Recent developments in this fields can be found in Hyman et al. [14] and Paczko et al. [15], respectively.

Acknowledgements

This work was accomplished at the Institut für Allgemeine Mechanik at the RWTH Aachen. I would like thank Rupert Klein and Günter Paczko for their helpful discussions.

References

[1] Oran,E.S., Young,T.R., Boris,J.P., Cohen,A. : *Weak and Strong Ignition. I.Numerical Simulations of Shock Tube Experiments*, Comb. Flame 48, 1982,pp135

[2] Kailasanath,K., Oran,E.S. : *Time-Dependent Simulations of Laminar Flames in Hydrogen-Air Mixtures*, Complex Chem. Reaction Systems, Springer Series Chem. Phys. 47, 1986, pp 243

[3] Toro,E.F., Clarke,J.F. : *Application of the Random Choice Method to Computing Problems of Solid-Propellant Combustion in a Closed Vessel*, Cranfield, CoA Report NFP 85/16,1985

[4] Klein,R. : *Stoßinduzierte Zündung und der Übergang zur Detonation in engen Spalten*, Dissertation, RWTH Aachen, 1988

[5] Colella,P., Glaz,H.M. : *Efficient Solution Algorithms for the Riemann Problem for Real Gases*, J.Comp.Phys. 59, 1985, pp 264

[6] van Leer,B. : *Towards the Ultimate Conservative Difference Scheme, V.A Second-Order-Sequel to Godunov's Method*, J.Comp.Phys. 32, 1979, pp 101

[7] Korobeinikov,V.P., Levin,V.A., Markov,V.V., Chernyi,G.G. : *Propagation of Blast Waves in a Combustible Gas*, Astron. Acta 17, 1972, pp 529

[8] Taki,S., Fujiwara,T. : *Numerical Analysis of Two-Dimensional Nonsteady Detonations*, AIAA Journal 16, 1978, pp 73

[9] Munz,C.-D. : *On the Comparison and Construction of Two-Step Schemes for the Euler Equations*, Notes on Numerical Fluid Mech. 14, 1986, pp 195

[10] Chorin,A.J. : *Random Choice Solution of Hyperbolic Systems*, J.Comp.Phys. 22, 1976, pp 517

[11] Oran,E.S., Boris,J.P. : *Detailled Modelling of Combustion Systems*, Prog. Energy Combust. Sci., 2, 1981, pp 1

[12] Ritzerfeld,E. : *Numerische Simulation der Wechselwirkung einer Stoßwelle mit einfachen Hinderniskörpern unter Verwendung eines Godunov-Verfahrens höherer Ordnung*, Diplomarbeit, RWTH Aachen, 1986, unpublished

[13] Roe,P.L. : *Characteristic-Based Schemes for the Euler Equations*, Ann. Rev. Fluid Mech. 18, 1986, pp 337

[14] Hyman,J.M., Larrouturou,B. : *On the Use of Adaptive Moving Grid Methods in Combustion Problems*, Complex Chem. Reaction Systems, Springer Series Chem. Phys. 47, 1986, pp 222

[15] Paczko,G., Lefdal,P.M., Peters,N. : *Reduced Reaction Schemes for Methane, Methanol and Propane Flames*, 21st Symp. (Int.) on Combustion, 1986, pp739

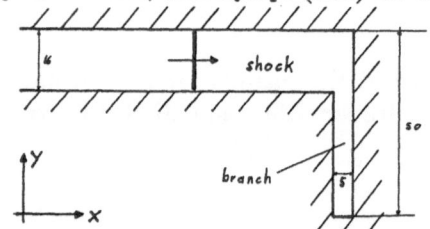

Figure 1: configuration of the considered problem

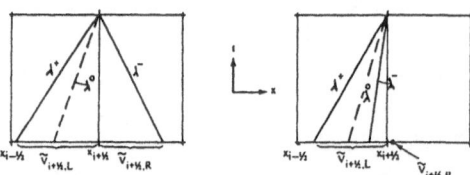

Figure 2: characteristic waves reaching the zone edge at $t^{n+\frac{1}{2}}$

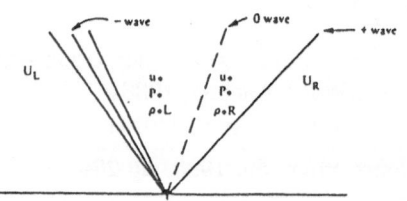

Figure 3: solution of the Riemann problem

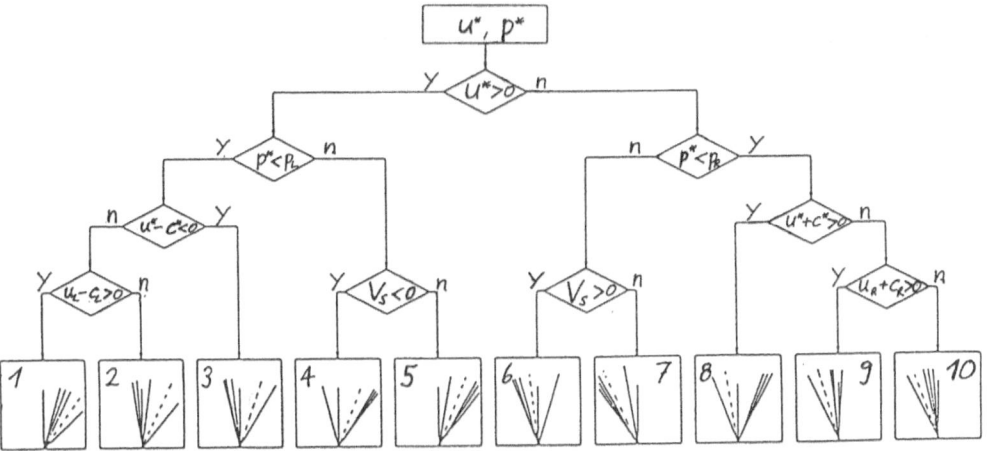

Figure 4: selection algorithm of the exact Riemann solution at the zone edge

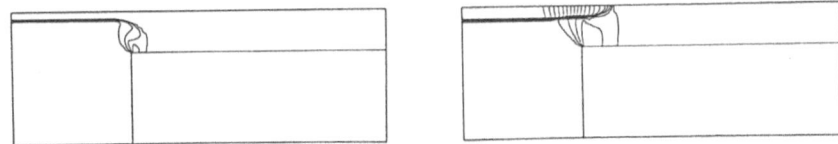

Figure 5: $M_S = 2.5$: pressure-contours at a) $t = 0.1$, b) $t = 0.2$

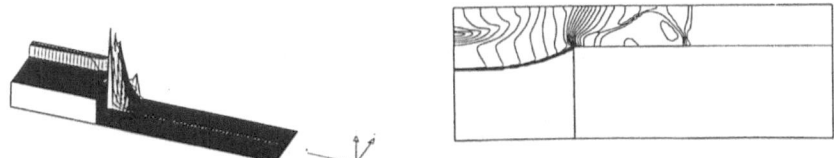

Figure 5: c) entropy-surface at $t = 0.2$, d) pressure-contours at $t = 0.5$

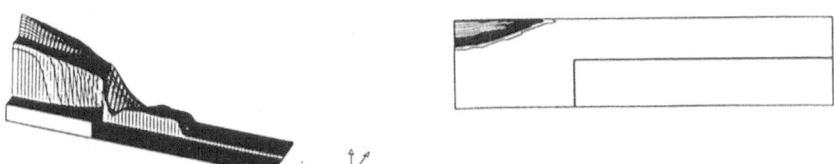

Figure 5: e) pressure-surface at $t = 0.5$, f) β-contours at $t = 0.5$

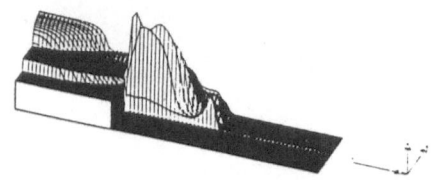
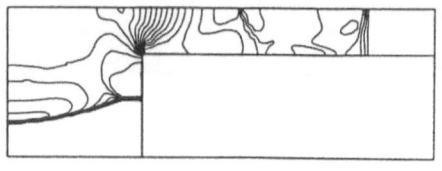

Figure 5: g) entropy-surface at $t = 0.5$, h) pressure-contours at $t = 0.75$

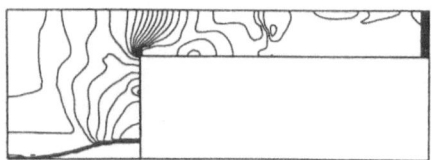
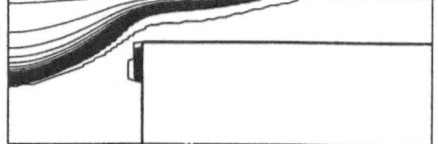

Figure 5: i) pressure-contours at $t = 0.95$, j) β-contours at $t = 0.95$

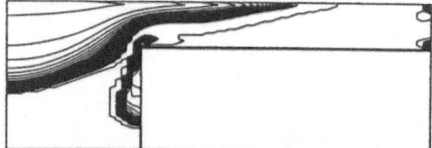

Figure 5: k) pressure-contours at $t = 1.05$, l) β-contours at $t = 1.05$

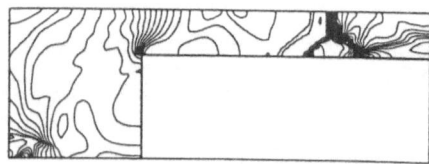
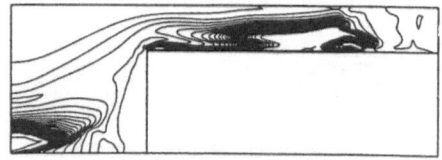

Figure 5: m) pressure-contours at $t = 1.4$, n) β-contours at $t = 1.4$

Figure 5: o) pressure-contours at $t = 1.6$, p) β-contours at $t = 1.6$

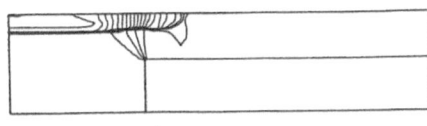
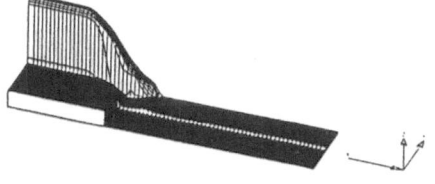

Figure 6: $M_S = 2.8$: a) pressure-contours at $t = 0.2$, b) pressure-surface at $t = 0.2$

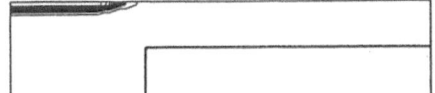
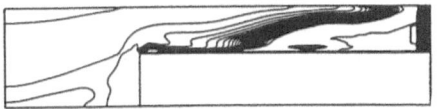

Figure 6: c) β-contours at $t = 0.2$, d) β-contours at $t = 0.9$

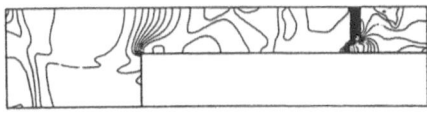

Figure 7: $M_S = 2.25$: a) pressure-contours at $t = 1.45$, b) β-contours at $t = 1.45$

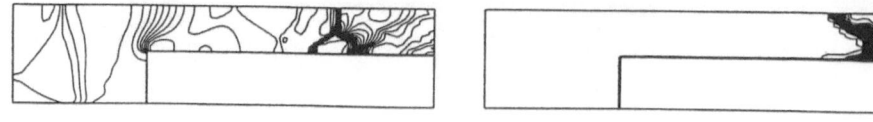

Figure 7: c) pressure-contours at $t = 1.6$, d) β-contours at $t = 1.6$

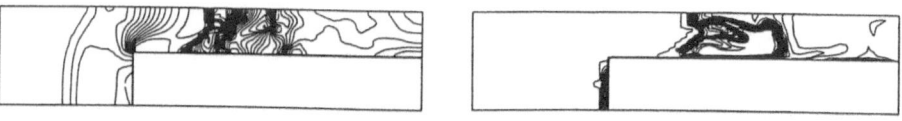

Figure 7: e) pressure-contours at $t = 2.05$, f) β-contours at $t = 2.05$

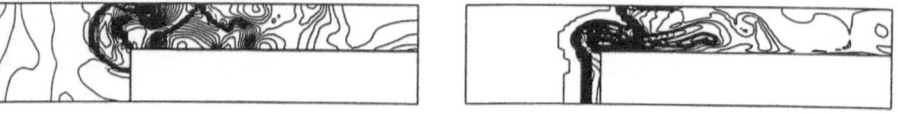

Figure 7: g) pressure-contours at $t = 2.3$, h) β-contours at $t = 2.3$

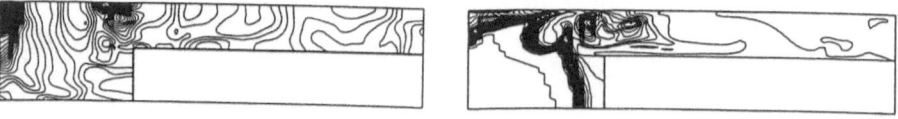

Figure 7: i) pressure-contours at $t = 2.65$, j) β-contours at $t = 2.65$

A NUMERICAL STUDY OF PROPAGATING PREMIXED TURBULENT FLAMES

F. Lacas *, D. Veynante and S.M. Candel

Laboratoire E.M2.C., C.N.R.S.

F92295 Chatenay Malabry, France

(* Also with S.E.P. Vernon - France)

I- Introduction

Numerical modelling of turbulent premixed flames is an important problem in many technological applications, such as turbojets, rocket motors or other industrial burners. Among the possible descriptions of turbulent combustion, those based on the flamelet concept are particularly attractive (see **Williams**[1], **Peters**[2-3], **Clavin et al.**[4], **Bray**[5-6]). In flamelet models the reaction zone is viewed as a collection of laminar flame elements embedded in the turbulent flow. The Coherent Flame Model due to **Marble and Broadwell**[7], is derived from the flamelet hypothesis. Good results have been obtained with this model in several steady flow geometries (see **Darabiha et al.**[8] for premixed flames and **Veynante**[9] or **Lacas et al.**[10] for nonpremixed combustion). Our aim is to study the capability of this model in prediction of nonsteady reactive flows using a finite difference algorithm. The characteristics of the numerical code used and the elements of the basic Coherent Flame Model are first reviewed. Results in one and two-dimensional geometries are presented, and a new model describing the transition between premixed and non premixed combustion is explored.

2-Numerical treatment

The formulation of the problem is based on statistical mass averages of conservation equations, for mass, momentum, enthalpy, chemical species turbulent kinetic energy and its dissipation. These equations may be written in the general form, for any of the standard variables, $\widetilde{\Phi}$:

$$\frac{\partial \widetilde{\Phi}}{\partial t} = \dot{\omega}_\Phi - \nabla \left(\rho V \widetilde{\Phi} - \frac{\mu_t}{\sigma} \nabla \widetilde{\Phi} \right) = F\left(\widetilde{\Phi}\right) \tag{1}$$

These equations are discretized on a cartesian two-dimensional mesh and solved using a finite volume technique. The computation of the fluxes components is split following the two space directions: x1 and x2 (see **Beam and Warming**[11], or **Yanenko**[12]) After discretization using a centered scheme we have:

$$F\left(\widetilde{\Phi}\right) = \left[A^{x1} + A^{x2} \right] \left[\widetilde{\Phi} \right] \tag{2}$$

where A^{x1} and A^{x2} are block tridiagonal matrices . As our purpose is to simulate unsteady phenomena, the time integration must be treated with special care. A semi implicit fractional step method was chosen (see **Briley and Mac Donald**[13] or **Dupoirieux**[14]). Equation (1) may be transformed into:

$$\Phi_{n+1} - \Phi_n = \Delta t \left(\beta F^{n+1} + (1-\beta) F^n \right) \tag{3}$$

where β is a parameter varying between 0 (corresponding to a fully explicit technique) and 1 (fully implicit). Thus:

$$\frac{\Phi_{n+1} - \Phi_n}{\Delta t} = \left[A^{x1} + A^{x2} \right] \left(\beta \Phi_{n+1} + (1-\beta) \Phi_n \right) \tag{4}$$

To the first order, it may be written:

$$\left([I] - \left[A^{x1} \right] \beta \Delta t \right) \left([I] - \left[A^{x2} \right] \beta \Delta t \right) \frac{\Phi_{n+1} - \Phi_n}{\Delta t} = \left[A^{x1} + A^{x2} \right] \Phi_n \tag{5}$$

where $[I] = [\delta_{ij}]$ is the unity matrix. The solution is then obtained by computing successively:

$$\delta \Phi^n = \left[A^{x1} + A^{x2} \right] \Phi_n \tag{6a}$$

$$\delta \Phi^{n+1/2} = \left[I - A^{x1} \beta \Delta t \right]^{-1} \delta \Phi^n \tag{6b}$$

$$\delta \Phi^{n+1} = \left[I - A^{x2} \beta \Delta t \right]^{-1} \delta \Phi^{n+1/2} \tag{6c}$$

$$\Phi_{n+1} = \Phi_n + \Delta t \, \delta \Phi^{n+1} \tag{6d}$$

It is important to choose an optimum value of β and of the time step. This choice is based on numerical tests like the propagation of an acoustic wave in a closed tube, with reflexion on the extremity. The exact solution of this problem is well known. The time step is defined by its ratio to the Δt° corresponding to a CFL number equal to unity. The CFL number is defined as the ratio of the smallest cell size over the sound velocity and time step:

$$CFL = \frac{\Delta x_{min}}{c \, \Delta t} \tag{7}$$

Figure 1 presents the results of this calculation as the variation of the pressure at one extremity of the domain over time. As it could be easily foreseen, it may be noticed that the increasing of the time step corresponds to an increasing of the numerical viscosity. The same behaviour can be seen (Fig. 1) for the influence of the coefficient β on the quality of the unsteady result: the time integration scheme becomes all the more dissipative than it is implicit. As a consequence, an optimum has to be chosen in order to have a good accuracy for a reasonable computation time (i.e. a time step as large as possible). The preceding numerical experiments led us to choose the following values:

$$\beta = 0.6 \qquad \text{and} \qquad \Delta t = 1.5 \times CFL$$

These values are the one used in all the following calculations. On a Cray XMP, this method requires 3.10^{-5} s for each time step, each node and each equation.

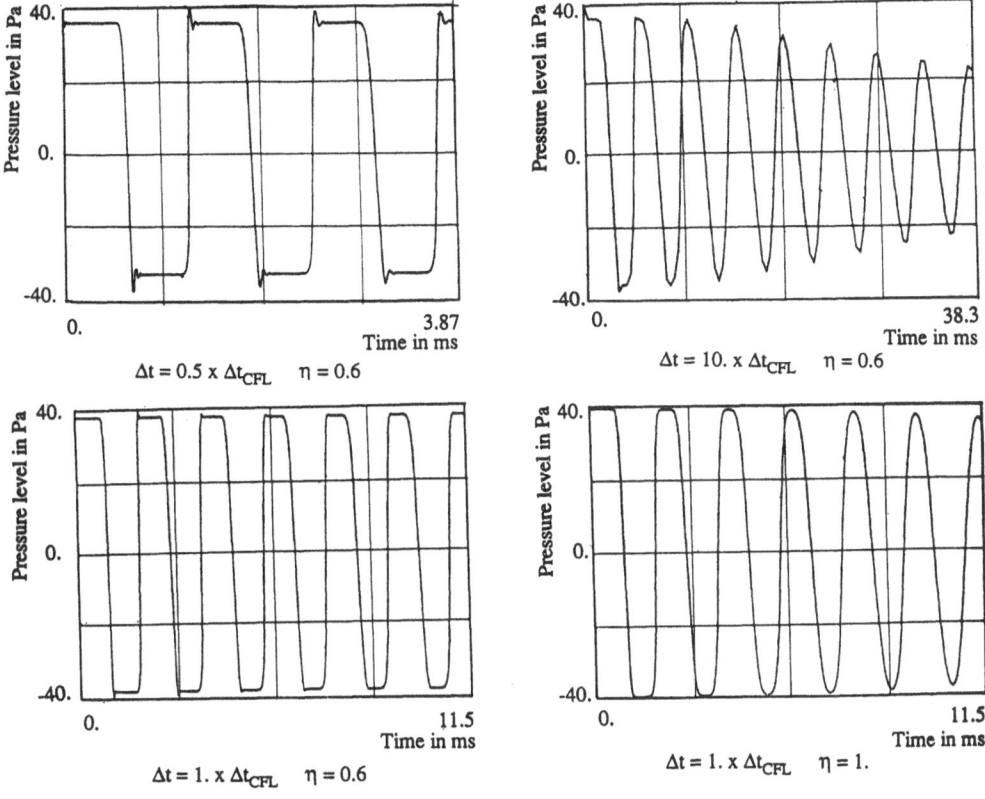

Figure 1: Time evolution of the mean pressure level at one end of a closed tube during a pressure wave propagation and reflexion , for several values of the numeriacl parameters.

3- turbulent combustion modelling

In the Coherent Flame Model, the reaction field is described as a collection of laminar flame elements. These elements are convected and distorted by the turbulence motion but retain an identifiable structure (flame elements remain "coherent" i.e. organized). This assumption is valid only if the reaction zone thickness is small compared to the typical turbulence scale. A rough sketch of this situation is given in Fig. 2. A flame element is mainly affected by the local strain rate which acts in the plane of the flame, modifies its structure and changes the local reaction rate. This local consumption rate per unit of flame area may be obtained from an asymptotic or numerical analysis of strained laminar flames including multi species transport and complex chemistry features. The available flame area is increased by stretching (due to the turbulent fluctuations) and shortened by various destruction mechanisms such as mutual interaction of adjacent elements by consumption of the intervening reactants or flame quenching (due, for example, to an excessively large strain rate). As a consequence, the mean reaction rate per unit volume ω_i of any chemical species i may be expressed as the product of the consumption rate per unit area V_i by the mean density of flame area Σ (i. e. the mean flame area per unit volume) thus :

$$\dot{\omega}_i = \rho\, V_i\, \tilde{\Sigma} \tag{8}$$

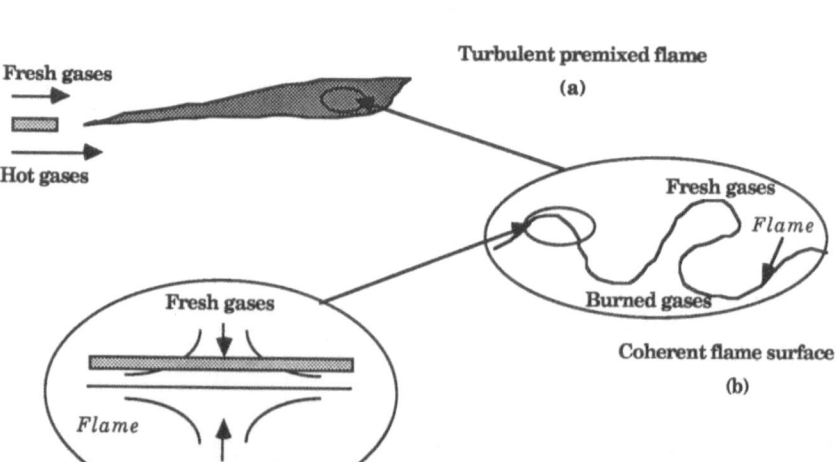

Figure 2 : schematic representation of physical processes described in the Coherent Flame Model :

    (a)   Turbulent premixed flame

    (b)   Flame elements

    (c)   Stained laminar flamelet model

In summary, the Coherent Flame Model requires :

a - <u>a model of the turbulent flow</u> comprising a standard set of Reynolds or mass average dynamic equations and a closure model, as described in section 2-.

b - <u>a local model for laminar strained flame elements</u> providing a local flame laminar velocity V_i and taking into account the influence of stretching and mixture ratio. This model is decoupled from the turbulent flow balance equations, and it may be treated with asymptotic methods. In practical applications, this local model is used to construct a flamelet library.

c- <u>a balance equation for the mean flame area per unit volume</u>, written in the standard transport form in Eulerian coordinates (i.e. t and x_k). For a one, single step and irreversible reaction such as :

$$O + sF \;\text{------>}\; P$$

where s is the stoechiometric ratio, Marble and Broadwell proposes :

$$\frac{\partial \rho \, \tilde{\Sigma}}{\partial t} + \frac{\partial \rho \, \widetilde{u_k \Sigma}}{\partial x_k} = \frac{\partial}{\partial x_k}\left(\frac{\mu_t}{\sigma_\Sigma}\frac{\partial \Sigma}{\partial x_k}\right) + \alpha \rho E_s \tilde{\Sigma} - \beta \rho \left(\frac{V_f}{Y_f} + \frac{V_o}{Y_o}\right)\tilde{\Sigma}^2 \qquad (9)$$

where V_f and V_o are the consumption rates per unit flame area, Y_i are the chemical mass fractions, indices o and f respectively denote oxydizer and fuel, and E_s represents the strain rate. Further details about the model and this equation may be found in **Marble and Broadwell**[7], **Darabiha et al.**[8], **Veynante et al.**[9], **Lacas et al.**[10], **Lacas**[15].

4- Turbulent premixed flame propagation

The first test case concerns the behaviour of a flame propagating along the x space coordinate. This axis is perpendicular to the flame plane. The flame is ignited near a wall which is also perpendicular to the propagation direction. The geometrical mesh used for the computations is one-dimensional and comprises 500 nodes. The boundary conditions correspond to a wall on the left of the computational domain, while the other side is a constant pressure outflow. Initial conditions are defined by a zero mean velocity field and imposed values for turbulent quantities (i.e. k and ε), while the domain is filled with a Propane-Air mixture with a stoichiometric mixture ratio (Φ=1.). At time t=0., the flame is ignited by imposing a value of the mean density of flame area in the first cell following the closed end of the mesh.

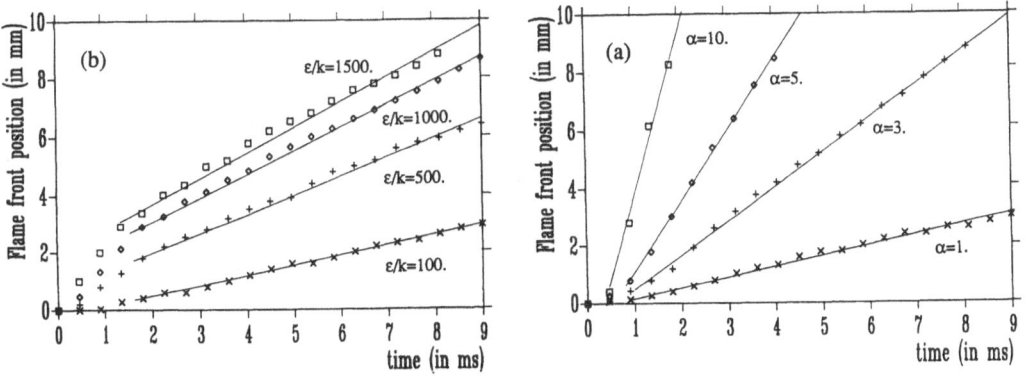

Figure 3: Time history of the flame front position in an open geometry for several values of the a constant [a] and of the initial strain rate [b].

Calculations of steady premixed or non-premixed flows (see references quoted above) have shown a good agreement with experiments for fixed values of the ratio α/β where α and β are the two constants of the flame area density balance equation (9). The value obtained from these steady cases is $\alpha/\beta = 0.5$. Figure 3(a) shows a plot of the time history of the flame front position for several values of α, while the other quantities, including the ratio α/β, remain fixed. The position of the flame area density maximum is displayed in the x-t plane, the spatial origin being the ignition point. The simplicity of this problem enables systematic studies of the influence of several parameters on the flame front velocity. It can be noted that after an initiation delay where the flame motion is non linear, the flame front displacement exhibits a nearly linear behaviour with a constant flame propagation velocity V_p. Further calculations show that V_p is linearly linked to α [fig. 4(a)]. As a consequence, the experimental measurements of Vp may be used to find the precise values of α and β, while α/β is obtained from a comparison between theory and experiments in steady configurations.

Figure 3(b) shows the same trends for variations of the initial ε/k ratio. Once more, the same limiting linear behaviour is observed after a nonlinear initiation phase. But Figure 4(b) shows that the link between the limit flame velocity and the initial value of ε/k is not linear. In fact a kind of saturation process is observed and V_p reaches a limit value. The mean flame front velocity increases with the initial value assigned to ε/k, but then tends to a limit for large values of the initial strain rate.

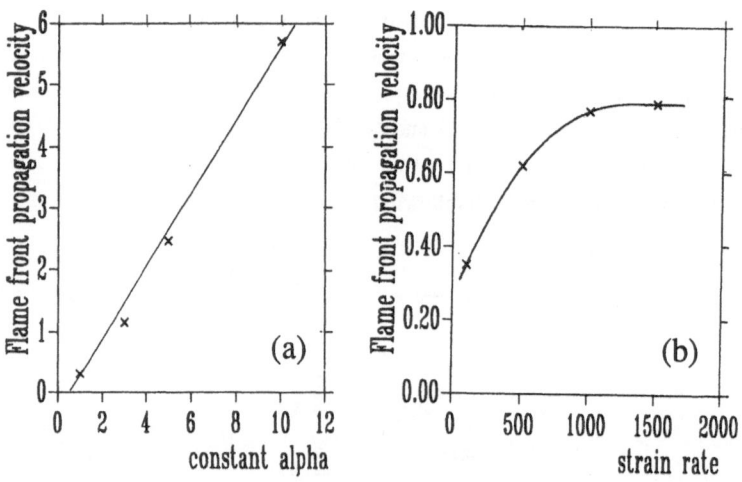

Figure 4: Variation of the flame front velocity Vp over the value of a constant [a] and of the initial strain rate [b].

The model predictions, regarding propagating premixed flames, have been compared with experimental results in another test-geometry. Consider a closed rectangular chamber of 90mm long, 30mm width and 28mm depth, initially filled with turbulent mixture of propane and air. These conditions correspond to experiments of **Shimizu et al.**[16] where Laser Doppler Velocity measurements and Schlieren type visualizations were reported. Ignition occurs in the 90x28mm symmetry plane, near one of the back walls. Figure 5 shows the results of the calculation. The evolution of the temperature field is plotted at a few instants, allowing a comparison with Schlieren visualizations performed on the experimental device. The initial temperature is 373K and the burned gas temperature is 1600K, with one contour curve every 200K. Only one half of the total domain is shown, the other half being symetric. Results are plotted every 4.5 ms. As in the one-dimensionnal case, two different phases may be observed. At first the flame is

t=4.5 m s

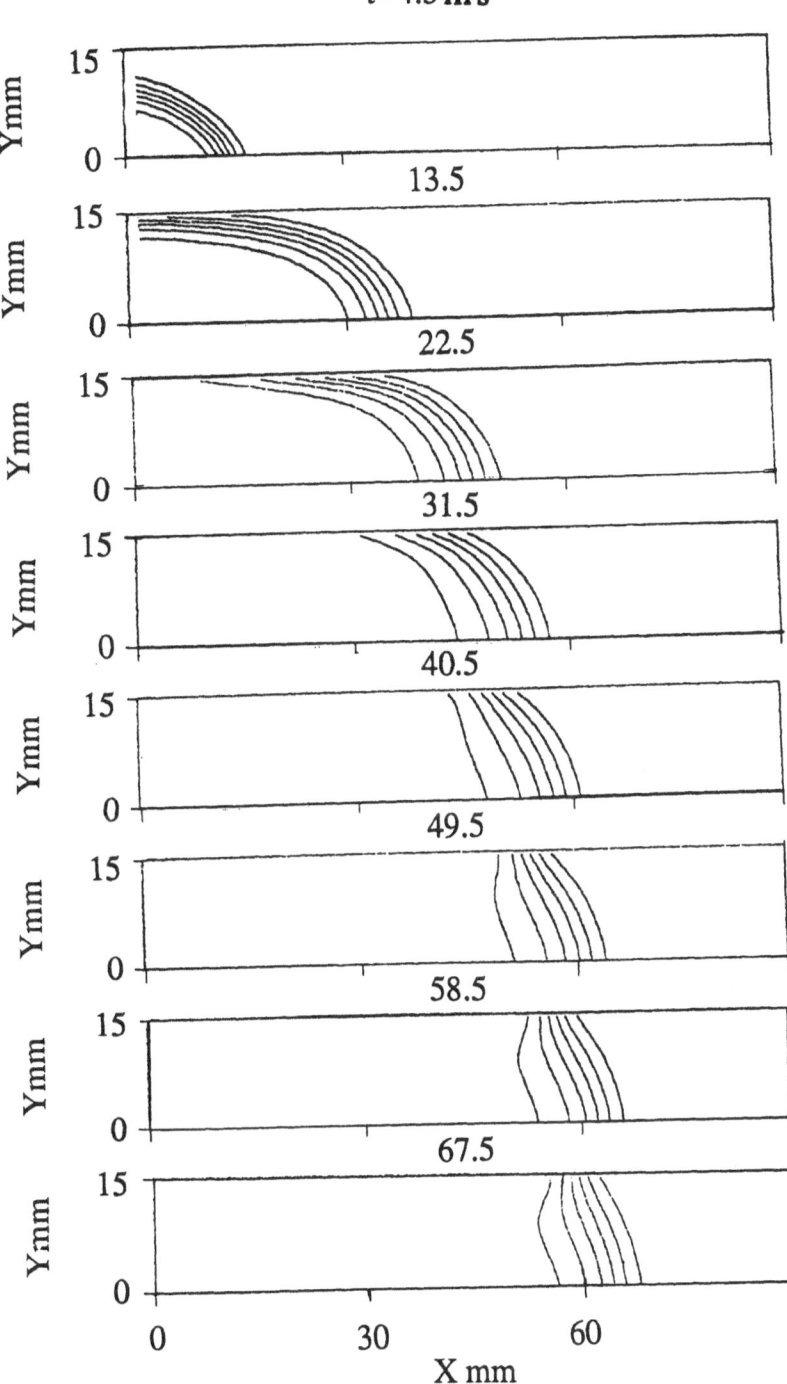

Figure 5 : Computed evolution of the temperature field in a rectangular combustion chamber

"self propelled" by the hot burned gases with a large flame front velocity . When the flame comes closer to the back wall, where the influence of velocity field and recirculations is less important, the flame velocity is reduced. A good agreement regarding the propagation of the mean flame front is achieved with the model.

5- Presentation of the combined model.

To compute the transient ignition of a diffusion flame, such as the ignition of a rocket engine, the simple Coherent Flame Model is unsufficient because in a first step, before ignition, a premixing of the two reactants occur. Then, these premixed reactants are ignited by spark, auto ignition or mixing with hot gases. In the last step of this process, the premixed flame disappears and is replaced the diffusion flame between pure reactants. Instead of one single flame surface density, we now introduce two different flame areas: Σ^d for diffusion flamelets and Σ^p for premixed flamelets. A third area density, Σ^c , is needed : it represents a contact area between the two reactants, in the absence of reaction. Such a situation may happen when the combustion is not yet initiated, or through holes in the flame front, due to quenching. Through this inert surface, the two reactants may form some premixed gases. Two different mass fractions have to be considered : Y_i^d for the mass fraction of reactant i in nonpremixed form, Y_i^p for the premixed form. New balance equations are written for the five quantities introduced, in the same form as equation (9), by changing only the expression of the source terms. For the balance equation for unpremixed reactants, this source term is written :

$$\dot{\omega}_{Y_i^d} = -\rho V_i^d \widetilde{\Sigma}^d - \rho V_i^c \widetilde{\Sigma}^c \qquad (10)$$

where the first term represents the consumption of nonpremixed reactant i in the diffusion flame and the second one the mixing of reactants through the inert surface. The reaction rate per unit flame surface V_i^d and the mixing rate through the contact surface V_i^c may be determined from local models. In the same way, for premixed reactants, we have:

$$\dot{\omega}_{Y_i^p} = -\rho V_i^p \widetilde{\Sigma}^p + \rho V_i^c \widetilde{\Sigma}^c \qquad (11)$$

expressing the consumption of premixed reactants in the premixed flame and creation of mixing through the inert surface.

Balance equations are also needed for the three densities, Σ^d, Σ^p and Σ^c. The source term of the balance equation for diffusion flame surface density is :

$$\dot{\omega}_{\Sigma^d} = \alpha^d \rho E_s \Sigma^d - \beta^d \rho \left(\frac{V_o^d}{Y_o^d} + \frac{V_f^d}{Y_f^d}\right)\left(\widetilde{\Sigma}^d\right)^2 + \gamma^d \rho \left(\frac{V_o^p}{Y_o^p} + \frac{V_f^p}{Y_f^p}\right)\left(\widetilde{\Sigma}^p\right)^2 \qquad (12)$$

where the first two terms have the same meaning as in the general model, and the last one represents the creation of diffusion flame by consumption of premixed reactants. The source term of premixed flame surface looks like the source term of the simple Coherent Flame Model :

$$\dot{\omega}_{\Sigma}^{p} = \alpha^{p} \rho E_{s} \Sigma^{p} - \beta^{p} \rho \left(\frac{V_{o}^{p}}{Y_{o}^{p}} + \frac{V_{f}^{p}}{Y_{f}^{p}} \right) \left(\widetilde{\Sigma^{p}} \right)^{2} \tag{13}$$

For the inert contact surface, the source terms becomes:

$$\dot{\omega}_{\Sigma}^{c} = \alpha^{c} \rho E_{s} \Sigma^{c} - \beta^{c} \rho \left(\frac{V_{o}^{c}}{Y_{o}^{d}} + \frac{V_{f}^{c}}{Y_{f}^{d}} \right) \left(\widetilde{\Sigma^{c}} \right)^{2} - \gamma^{d} \rho \left(\frac{V_{o}^{p}}{Y_{o}^{p}} + \frac{V_{f}^{p}}{Y_{f}^{p}} \right) \left(\widetilde{\Sigma^{p}} \right)^{2} \tag{14}$$

In this expression the last term describes the destruction of mixing surface when the disparition of premixed reactants lead to the ignition of a diffusion flame. The reaction rates or diffusion rates may be obtained by local analysis or by calculations involving complex chemistery and thermophysics but, in our simple test case, these reaction rates are assumed to be constant. Further details on this extended model may be found in **Lacas**[15] or **Veynante et al.**[17]

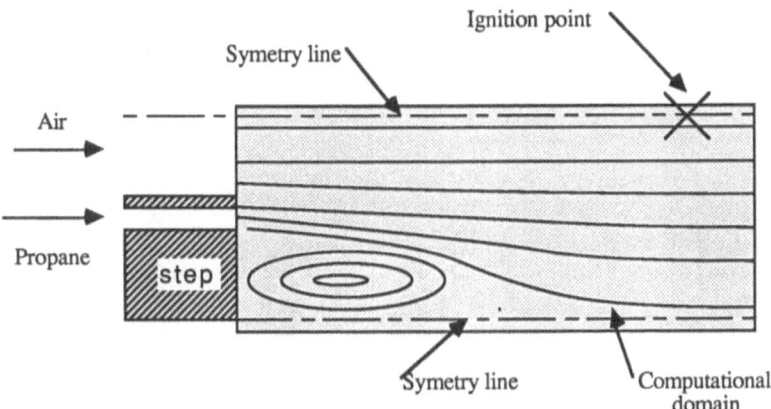

Figure 6 : Geometry of the computed test case for premixed to unpremixed unsteady combustion.

This new formulation was tested in the case of the non-premixed configuration described on Figure 6. Propane is injected through a slot in an air flow beside a step. The ignition occurs in a steady non burning flow where the most part of the reactants are mixed throught the contact surface. The combustion begins with an unsteady premixed flame where the initial mixture burns. In a second phase, the non-premixed flame is ignited and comes to a stationary configuration while the premixed flame disappears. These coupled behaviours may be seen on Figure 6 where density of premixed and non premixed flame areas are plotted. The temperature field is also shown: only one front, corresponding to both premixed and nonpremixed flames may be seen during the unsteady phase. This front crosses the flow and propagates

404

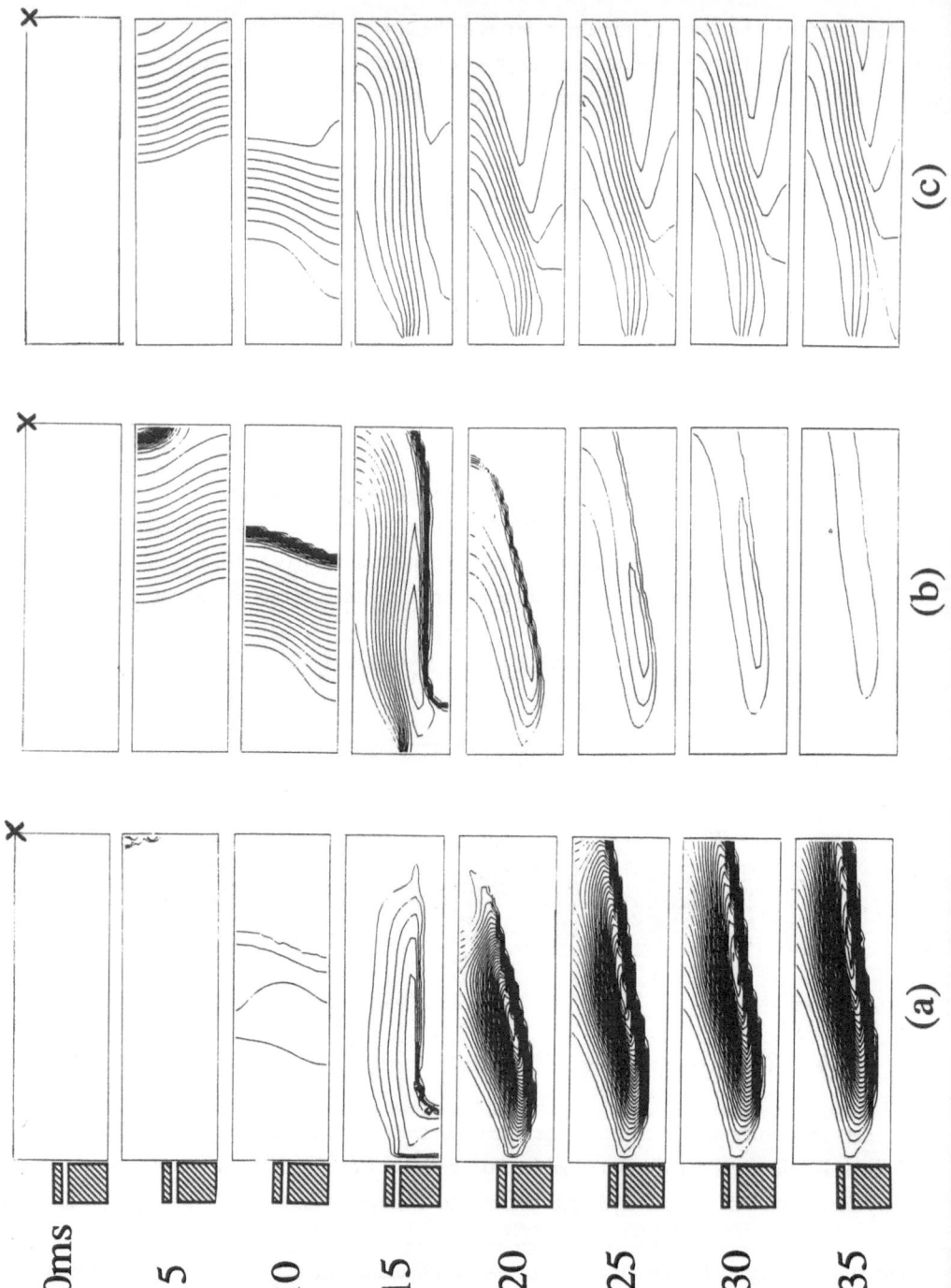

t=0ms

5

10

15

20

25

30

35

(a)   (b)   (c)

Figure 7 :   Isovalues of unpremixed flame area density (a), premixed flame area density (b) and temperature (c) during the ignition of a non premixed combustion.

upstream through the recirculation zone. The dynamical behaviour computed in this case is in agreement with experimental visualizations performed at our laboratory (**Zikikout et al.**[18]). The well established diffusion flame obtained at the end of our computation is the same than the one computed with the simple Coherent Flame Model for diffusion flame with the same geometry under steady flow conditions (**Lacas et al.**[10], **Lacas**[15])

6- Conclusion

Further studies have to be carried out to improve the description of the dominant processes which govern the propagation of turbulent flames. Of special interest is the influence of laminar flame speed in the balance equations for the flame areas. But the present calculations prove the ability of the extended flamelet description to compute nonsteady phenomena, and to describe the transition between a premixed and nonpremixed flame configuration.

References

1- **Williams F.A.** "A review of some theoretical combustion of turbulent flame structure," AGARD conf. Proc. 164, p. II 1,1 ,1975

2- **Peters N.** "Laminar diffusion flamelets model in non-premixed turbulent combustion," Prog in Energy and Comb. Sci. 10, p319,1984

3- **Peters N.** "Laminar flamelet concepts in turbulent combustion," 21st Symposium (International) on combustion, The Combustion Institute,1986

4- **Clavin P. and Williams F.A.** "effects of molecular diffusion and thermal expansion on the structure and dynamics of premixed flames in turbulent flows of large scale and low intensity," J. Fluid. Mech.116, 215, 1982

5- **Bray K.N.C., Libby P.A.and Moss J.B.** "Flamelet crossing frequencies and mean reaction rates in premixed turbulent combustion models," Comb. Sci. Tech., 41, 143, 1984

6- **Bray K.N.C.** "Method of including realistic chemical reaction mechanisms in turbulent combustion models," 2nd Workshop on modelling of chemical reaction systems, Heidelberg, 1986

7- **Marble F.E. and Broadwell J.E.** "The Coherent Flame Model for turbulent chemical reactions," Project Squid Rep. TRW-9-PU, 1977

8- **Darabiha N.,Giovangigli V.,Trouvé A.,Candel S.M. and Esposito E.** "Coherent Flame description of turbulent premixed ducted flames," Workshop on turbulent combustion, Rouen 1987

9- **Veynante D., Candel S.M. and Martin J.P.** "Coherent Flame Modelling of Chemical reaction in a turbulent mixing layer," 2nd Workshop on modelling of chemical reaction systems, Heidelberg, 1986

10-   **Lacas F., Zikikout S. and Candel S.M.** "A comparison between calculated and experimental mean source terms in non premixed turbulent combustion", AIAA 23rd joint Prop. Conf.,87-1782, 1987

11-   **Beam R.M.and WarningR.F.,** "Alternating direction implicit methods for parabolic equations with a mixed erivative", SIAM J. Sci. Stat. Comp., Vol 1, 1979

12-   **Yanenko,** "Méthodes à pas fractionnaires", Armand Colin, Interscience, 1968

13-   **Brilley W.R. and McDonald,H.** "On the structure and use of linearized block implicit schemes", J. of Comp. Phs., J. of Comp. Phys., vol 24, p372, 1975

14-   **Dupoirieux F.and Scherrer D.,** " Methodes numeriques à convergence rapide utilisées pour le calcul des écoulements réactifs", Proc. of the Symposium on Numerical Simulation of Combustion phenomena, Sophia-Antipolis, May 1985

15-   **Lacas F.,** "modelisation et simulation numérique de la combustion turbulente dans les moteurs fusée cryotechniques", PhD Thesis, ECP, Juin 1989

16-   **Shimizu S., Wakai K. and Hibino Y.** "The velocity distribution of gases and the wedge shaped flame front formation in a square combustion chamber," 3rd International Symposium on Laser Doppler Velocimetry, Lisboa, 1986

17-   **Veynante D., Lacas F. and Candel S.,** "A new flamelet combustion model combining premixed and non-premixed turbulent flames", AIAA-89-0487, AIAA 27th Aerospace Sciences meeting, Reno, 1989

18-   **Zikikout S., Poinsot T., Trouvé A., Veynante D., Candel S., Esposito E.,** "Mécanismes d'allumage dans un foyer à flammes multiples non prémélangées", Joint Meeting French and British Sections of the Combustion Institute, Rouen, April 1989.

Modélisation de la combustion dans un moteur diesel d'automobile

Philippe PINCHON
Institut Français du Pétrole
1 & 4 av. de Bois Préau — BP 311 — 92506 RUEIL-MALMAISON CEDEX

Résumé

La plupart des moteurs diesels utilisés pour la traction des véhicules automobiles sont des moteurs à préchambre. Le combustible y est introduit sous forme liquide, à haute pression, se vaporise, auto-inflamme et brûle dans un espace confiné dans lequel règnent des vitesses d'écoulement très élevées. La vitesse de combustion, et donc le rendement du moteur, sont très sensibles à divers paramètres liés à l'injection ou à la géométrie de la chambre. Cette dernière joue un rôle prédominant sur l'aérodynamique interne et son interaction avec la propagation de la flamme.

Un modèle de calcul tridimensionnel a été utilisé pour simuler le déroulement de la combustion dans un moteur diesel à préchambre dite "RICARDO". La turbulence a été décrite à l'aide d'un modèle classique de type k-ε et les conditions limites traitées par une loi de paroi. Les transferts thermiques aux parois ont été pris en compte et calculés à l'aide d'un modèle basé sur la formulation k-ε. Un mécanisme de cinétique chimique simplifié a été développé et intégré dans le modèle général, afin de décrire le processus d'auto-inflammation du mélange air-combustible. Cependant, le schéma cinétique utilisé pour décrire l'oxydation du combustible à relativement basse température n'est plus adapté pour traduire la chimie de la combustion à haute température. Par ailleurs, on a supposé que le régime de combustion turbulente succédant à l'auto-inflammation était contrôlé principalement par la diffusion des espèces chimiques et de la chaleur. Cette hypothèse a ainsi autorisé l'utilisation du modèle de Magnussen et Hjertager.

Le modèle de calcul a été appliqué à la simulation de l'écoulement gazeux, de l'injection du combustible et de la combustion dans la chambre de combustion du moteur. Les rôles respectifs de la préchambre et de la chambre principale ont été analysés et les résultats du calcul comparés à des résultats expérimentaux de visualisation de la combustion.

Les moteurs diesels peuvent être classés en deux grandes catégories : les moteurs à injection directe et les moteurs à injection indirecte ou moteurs à préchambre. Dans un moteur à préchambre, la chambre de combustion est divisée en une préchambre et une chambre principale connectées l'une à l'autre par un canal de transfert (figure 1). Le combustible est injecté dans la préchambre, au sein d'un écoulement de gaz très rapide, généré par le déplacement du piston. Ce processus donne lieu à des interactions très fortes entre le combustible et l'air, qui contrôlent le déroulement de la combustion et particulièrement la vaporisation du combustible, son mélange avec l'air et la propagation de la flamme.

Les performances du moteur ainsi que les émissions de polluants sont fortement influencées par la combustion. On comprend donc l'intérêt que peut présenter la modélisation mathématique pour l'optimisation de systèmes aussi complexes.

L'objectif de cette communication est de présenter un exemple d'application de la modélisation tridimensionnelle de la combustion au cas du moteur diesel à préchambre d'automobile.

Tous les calculs qui sont présentés ici ont été effectués à l'aide du code KIVA qui a été réalisé par le laboratoire national de Los Alamos aux USA [1, 2, 3]. Il résout les équations tridimensionnelles de l'aérothermochimie instationnaire et est particulièrement adapté à l'application aux moteurs à piston.

En particulier, les sous-modèles d'injection liquide et de vaporisation, intégrés dans la version de base, n'ont nécessité que des modifications mineures en vue de l'adaptation au cas considéré ici. En revanche, des développements plus importants ont été nécessaires pour améliorer la flexibilité du code vis-à-vis de géométries complexes et pour traiter plus convenablement la turbulence et la combustion dans le cas du moteur diesel à préchambre.

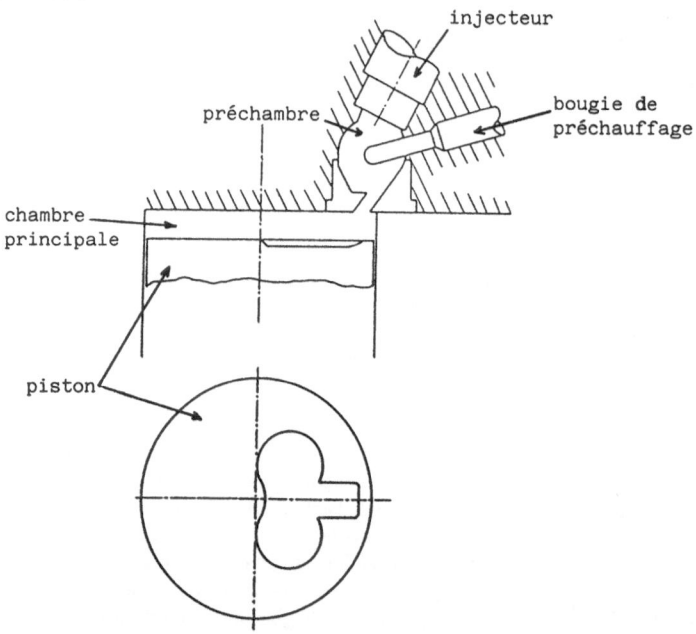

Fig. 1 : Chambre de combustion d'un moteur diesel
à préchambre RICARDO

Les algorithmes numériques qui sont utilisés dans KIVA sont adaptés par nature au traitement de géométries arbitraires mais la version originale du code et du mailleur ne le permettaient pas. Les maillages possibles étaient soit quasi-axisymétriques et pseudo-polaires, soit quasi-cubiques et cartésiens. Il a donc été nécessaire de modifier le code de calcul de manière à ce qu'il accepte des géométries beaucoup plus complexes. On l'a ensuite couplé à un nouveau mailleur beaucoup plus performant.

Plusieurs sous-modèles physiques ont dû être modifiés ou remplacés.

Tous les phénomènes mis en jeu dans le processus de combustion sont très fortement influencés par la turbulence. Dans ces conditions, le modèle de turbulence joue un rôle-clef, non seulement pour la détermination des coefficients de diffusion, mais aussi pour fournir des données aux autres sous-modèles. En particulier, les sous-modèles d'injection, de vaporisation, de combustion et de transferts thermiques ont besoin d'informations sur les caractéristiques de la turbulence comme l'intensité, les échelles temporelles, etc...

Dans la présente application, on a utilisé un modèle standard k-ϵ, dont on trouvera une description des équations à la référence [4]. Les conditions-limite de vitesses sont décrites par une loi de paroi turbulente [5]. Il en est de même pour le modèle de transfert thermique aux parois. Ce dernier est basé sur une corrélation directe entre le flux thermique et l'intensité de turbulence. Dans une publication précédente, P. Gilaber et P. Pinchon [4] ont obtenu un bon accord entre les flux thermiques instantanés calculés à l'aide de ce modèle et ceux mesurés dans un moteur à allumage commandé. Cependant, comme le modèle néglige les transferts thermiques par rayonnement, ce qui pourrait être important dans le cas du moteur diesel, on peut craindre que les pertes thermiques aux parois ne soient ici sous-évaluées.

Dans le moteur diesel, la combustion démarre par l'auto-inflammation d'une fraction du mélange air-combustible qui est provoquée par la compression due au déplacement du piston.

L'influence de la chimie a été prise en compte à l'aide d'un modèle semi-empirique de cinétique chimique comprenant quatre étapes pseudo-élémentaires et un radical généralisé R chargé de représenter l'ensemble des radicaux effectivement mis en oeuvre dans le processus d'auto-inflammation. Ce modèle a été mis au point et testé par M. Zellat et H. Zeller [6]. Les quatres étapes sont les suivantes :

(1) Etape d'initiation :   Fuel $\longrightarrow$ 2R
(2) Etape de ramification :   Fuel + R $\longrightarrow$ 3R
(3) Etape de rupture :   2R + M $\longrightarrow$ produits + M
(4) Etape de propagation :   Fuel + O_2 + R $\longrightarrow$ produits + R + Δ Q

où   M   est une molécule quelconque
    ΔQ est l'énergie dégagée par la réaction de propagation.

On a supposé ici que le temps chimique, caractéristique des pré-réactions conduisant à l'auto-inflammation était grand en comparaison avec le temps de mélange turbulent. Il devenait donc légitime de remplacer les variables instantanées (températures et concentrations) dans l'expression des taux de réactions par leur valeur moyenne dans la maille considérée pendant un pas de temps. Les taux de réactions de chacune des étapes élémentaires (i) précédentes est donc :

$$\dot{\omega}_i = A_i T^{n_i} e^{\left(-\frac{T_i}{T}\right)} \left(\frac{\rho_A}{W_A}\right)^{a_{A_i}} \cdot \left(\frac{\rho_B}{W_B}\right)^{a_{B_i}} \cdots \left(\frac{\rho_Z}{W_Z}\right)^{a_{Z_i}}$$

où :

$\dot{\omega}_i$	est le taux de réaction de l'étape n^0(i)
T	est la température des gaz
$\rho_A, \rho_B, ..., \rho_Z$	sont les densités des espèces A, B, ..., Z
$W_A, W_B, ..., W_Z$	sont les masses molaires des espèces A, B, ..., Z
$A_i, T_i, n_i, a_{A_i}, ..., a_{Z_i}$	sont des coefficients caractéristiques de la réaction considérée.

Certaines de ces constantes ont été déterminées sur la base d'une analyse phénoménologique et les autres ont été ajustées en fonction des résultats expérimentaux concernant l'injection de combustible diesel dans une bombe.

Cependant, ce modèle n'est plus adapté pour décrire la phase de combustion succédant à l'auto-inflammation. Le modèle de cinétique chimique qui est utilisé pour décrire l'oxydation du combustible à relativement basse température n'est plus valide pour traduire la chimie de la combustion à haute température. De plus, l'hypothèse d'un temps caractéristique de la chimie beaucoup plus grand que le temps de mélange turbulent n'est plus vérifiée et une nouvelle expression pour le taux de réaction moyen turbulent est donc requise. On a utilisé, pour simuler cette phase de combustion, un autre modèle initialement conçu pour les flammes de diffusion. Il s'agit du modèle de B.F. Magnussen et B.H. Hjertager [7] qui est aussi simple que le modèle Eddy Break Up, déjà appliqué à la simulation de la combustion dans les moteurs à allumage commandé [8]. Le modèle de Magnussen a été utilisé avec succès das le cas des flammes atmosphériques [7] et également pour des applications au moteur diesel à injection directe [9, 10, 11]. Selon ce modèle, le taux de combustion du combustible est supposé contrôlé par la plus basse des concentrations moyennes en réactants ou en produits de combustion et par un temps de mélange turbulent qui est pris ici comme proportionnel à k/ϵ.

Six expèces chimiques ont été prises en compte : le fuel, O_2, N_2, CO_2 H_2O, CO.

La réaction chimique mettant en oeuvre la consommation du fuel dans la phase vapeur est la suivante :

$$\text{fuel} + \alpha(O_2 + 3.76 N_2) \longrightarrow \beta CO_2 + \gamma H_2O$$

et le taux de réaction moyen :

$$R_f = A \min\left(\bar{C}_F, \frac{\bar{C}_{O_2}}{rf}, \frac{B\bar{C}_p}{1+rf}\right) \frac{\epsilon}{k}$$

où $\bar{C}_F$ est la concentration moyenne locale en fuel

$\bar{C}_{O_2}$ est la concentration moyenne locale en oxygène

$\bar{C}_p$ est la concentration moyenne locale en produits de combustion

rf   est le rapport stoechiométrique de la masse d'oxygène sur la masse de fuel

A et B   sont des constantes du modèle.

Ce modèle de combustion se complète par la prise en compte de l'équilibre suivant :

$$CO_2 \longleftrightarrow CO + 1/2O_2$$

Une représentation schématique du moteur diesel qui a été modélisé est donnée à la figure 1. Il s'agit d'un moteur diesel à préchambre, typique de l'application automobile, de 8.3 cm d'alésage et 23:1 de rapport volumétrique. Il est équipé d'une préchambre dite Ricardo Comet. La figure 2 montre comment la chambre de combustion, c'est-à-dire la préchambre, la chambre principale et le canal de transfert ont été maillés. Du fait de la symétrie par rapport à un plan, le domaine de calcul a été réduit à une demi-chambre. Le maillage, de forme structurée, comporte 22 x 13 x 28 mailles mais beaucoup d'entre-elles sont des mailles inactives situées hors du domaine de calcul proprement dit. Les résultats présentés ici correspondent au point de fonctionnement suivant :

régime du moteur :	1500 tours/minute
quantité de fuel injectée :	0,024 g/cycle
début d'injection :	5 degrés de rotation du vilebrequin avant le Point Mort Haut
durée d'injection :	20 degrés de rotation du vilebrequin.

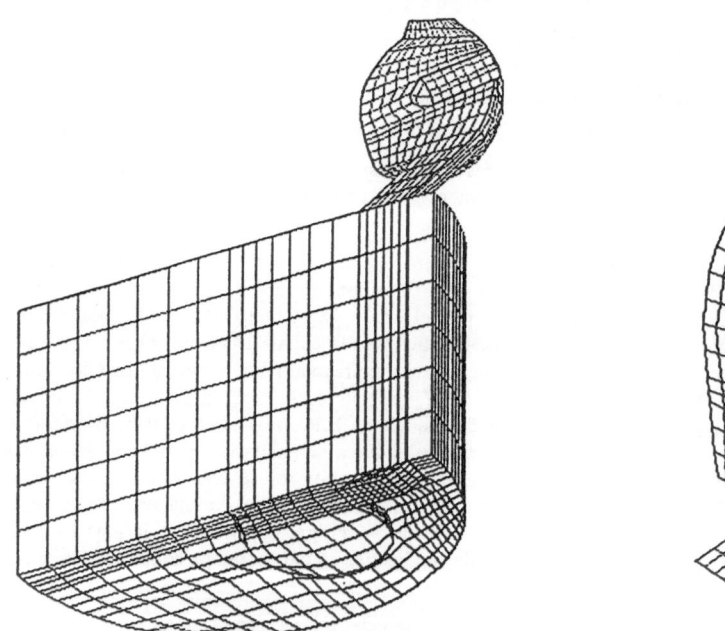
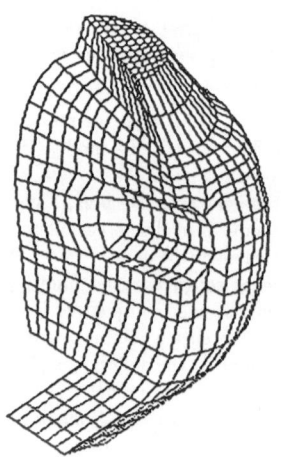

Fig. 2 : Maillage de la chambre de combustion

Les premiers calculs ont été effectués sans injection de combustible, de manière à caractériser les propriétés de l'aérodynamique interne de ce type de chambre. Les figures 3 et 4 présentent les champs de vitesse calculés, dans des plans de coupe horizontaux (A) et verticaux (B) respectivement aux angles de 20 degrés avant le Point Mort Haut (PMH) et 20 degrés après le PMH. La compression par le piston chasse le gaz de la chambre principale vers l'orifice de transfert (fig. 3.A). Ceci entraîne la création d'un écoulement à grande vitesse dans le canal de liaison puis la génération d'un mouvement tourbillonnaire dans

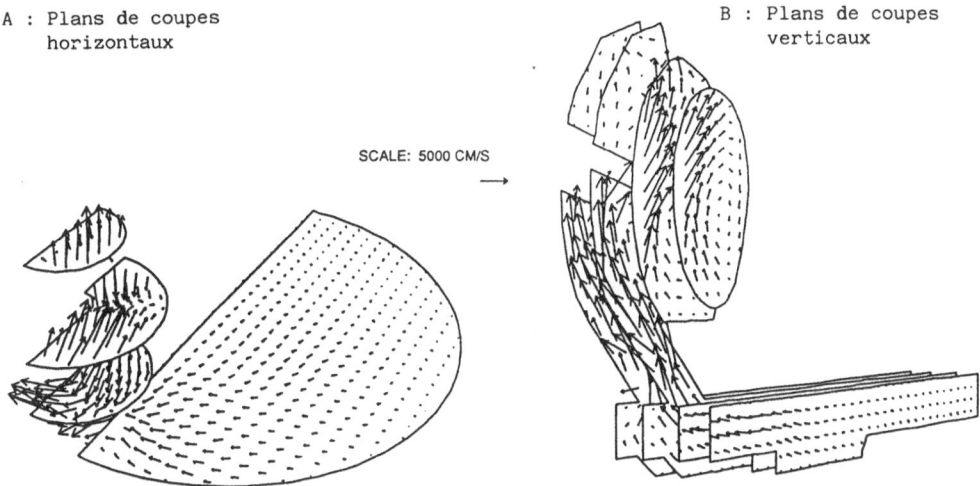

A : Plans de coupes
    horizontaux

B : Plans de coupes
    verticaux

SCALE: 5000 CM/S

Fig. 3 : Champs de vitesses - 20 deg. avant le PMH

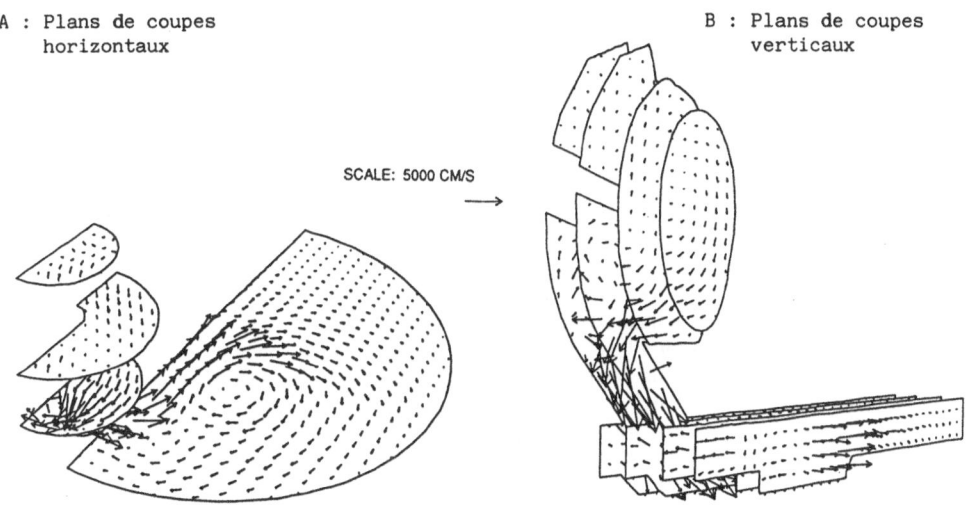

A : Plans de coupes
    horizontaux

B : Plans de coupes
    verticaux

SCALE: 5000 CM/S

Fig. 4 : Champs de vitesses - 20 deg. après le PMH

la préchambre, ce qui est le but recherché. On notera cependant que la présence de la bougie de préchauffage gène considérablement la génération du tourbillon, par la déviation du jet d'entrée qu'elle provoque et par le freinage de l'écoulement interne (fig. 3.B). A 20 degrés après le PMH, le mouvement tourbillonnaire est quasiment amorti dans la préchambre. La redescente du piston entraîne une aspiration inverse au travers du canal de transfert. Un jet de gaz violent est éjecté hors de la préchambre et percute la surface du piston où l'interaction avec l'empreinte ménagée à cet effet dans la tête du piston, favorise la génération de deux vortex symétriques (fig. 4.A). Ces résultats mettent en évidence le rôle joué par l'aérodynamique sur la combustion dans ce type de moteur : le tourbillon dans la préchambre favorise le mélange air-combustible et la combustion, puis le combustible imbrûlé passe dans la chambre principale où les deux vortex assurent la diffusion des gaz chauds et accélèrent la propagation de la flamme. Ce processus est mis en évidence par les résultats suivants, qui concernent la modélisation de la combustion. Le combustible est introduit sous forme liquide (figure 5) ; le jet n'est pas notablement dévié par l'écoulement, ce qui s'explique par l'interaction entre l'écoulement tourbillonnaire et la bougie de préchauffage dans cette zone. On note qu'une fraction relativement faible du combustible se dépose sur la paroi.

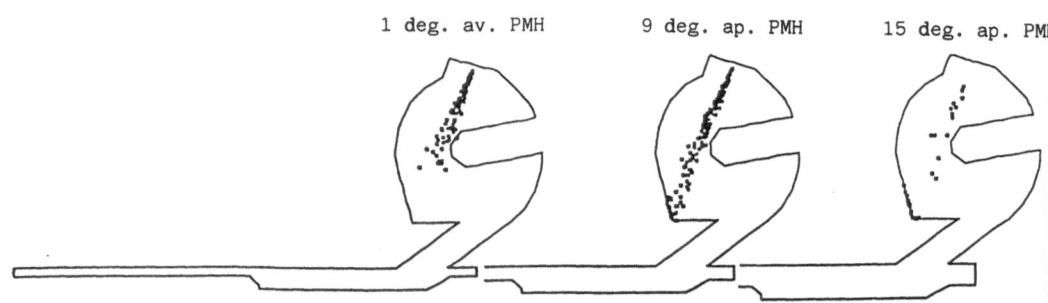

Fig. 5 : Evolution du jet liquide pendant l'injection

La combustion dans la préchambre s'effectue suivant deux régimes distincts ; le premier est contrôlé par le jet et les caractéristiques d'injection et l'autre par le transport des gaz chauds par l'écoulement. On peut observer ces deux régimes sur la figure 6 qui représente les isothermes à différents angles de rotation du vilebrequin et dans des plans de coupe verticaux. Après l'autoinflammation (1,2 deg. avant le PMH), qui se produit vers l'extrémité de la bougie de préchauffage, la flamme se propage rapidement à la périphérie du jet. Ce dernier est complètement baigné de gaz chauds dès 5 degrés après le PMH. Pendant cette période, la combustion est contrôlée par la diffusion de l'air et du fuel du jet au travers des gaz chauds. Cependant, après le PMH, le débit au travers de l'orifice de transfert devient très important et tend à diminuer la quantité d'oxygène disponible dans la préchambre. En conséquence, le taux de combustion devient bien plus faible et, pour se mélanger à l'oxygène, le fuel doit, soit s'échapper de la préchambre vers la chambre principale, soit se déplacer vers la partie supérieure de la préchambre où de l'air frais est encore disponible. Il s'agit du second régime, contrôlé par l'écoulement qui peut notamment être observé sur la figure 6 à l'angle 15 degrés après le PMH. Un haut niveau de tourbillonnement dans la préchambre améliorerait le taux de combustion pendant ce régime ; il se produit malheureusement à un moment du cycle où la vitesse du tourbillon décroît très fortement.

U. Spicher [12] a utilisé des fibres optiques pour réaliser une visualisation expérimentale directe de la flamme dans la préchambre d'un moteur similaire. Les résultats expérimentaux qu'il a obtenus suivent les mêmes tendances que ceux du calcul, décrits plus haut. Les délais d'autoinflammation mesurés et calculés sont quasi identiques, de l'ordre de 3 degrés vilebrequin. De plus, la propagation rapide de la flamme en début de combustion est également observée et le volume occupé par les gaz brûlés à différents instants est tout à fait comparable aux prédictions du calcul (figure 7).

Le mélange gazeux et la flamme sont ensuite violemment projetés hors de la chambre principale par l'écoulement. Dans ces conditions, la propagation de la flamme dépend d'une part de la vitesse d'éjection à l'orifice et d'autre part des deux vortex qui ont été générés par le jet de gaz et son interaction avec l'empreinte dans le piston. La figure 8 montre les isothermes dans différents plans de coupe horizontaux à

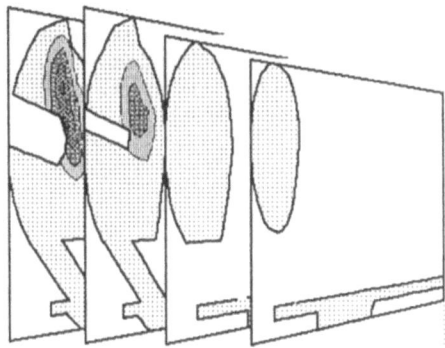

1 deg après PMH

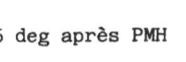

5 deg après PMH

15 deg après PMH

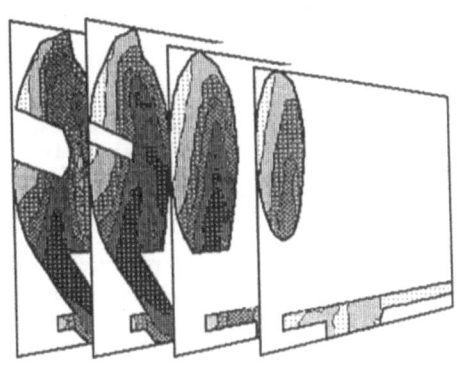

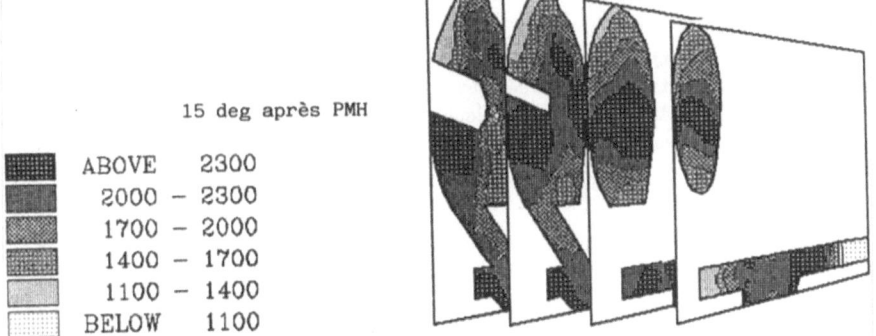

ABOVE 2300
2000 – 2300
1700 – 2000
1400 – 1700
1100 – 1400
BELOW 1100

Fig. 6 : Champs d'isothermes (Kelvin) à différents instants pendant la
combustion (les plans de coupes verticaux ont été écartés
artificiellement)

5, 15 et 35 degrés après le PMH. La flamme se propage depuis l'orifice vers la périphérie du cylindre et son expansion est contrôlée par l'entraînement de l'air dans le jet chaud de vapeur de fuel. Au début de sa propagation, le jet est étroit et pénètre rapidement puis il est ralenti par les deux vortex qui le font s'épanouir. Ce processus spécifique assure une progression presque uniforme de la flamme dans la chambre principale.

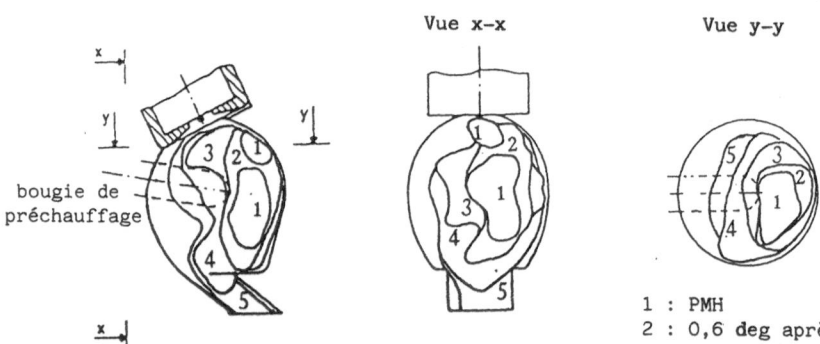

1 : PMH
2 : 0,6 deg après PMH
3 : 1,6 deg après PMH
4 : 3,6 deg après PMH
5 : 4 deg après PMH

Fig. 7 : Propagation expérimentale de la flamme estimée à partir de la visualisation par fibres optiques.
D'après 12

Les figures 9 et 10 montrent respectivement l'évolution de la fraction brûlée et de la vitesse de dégagement d'énergie globale normalisée, en fonction de l'angle de rotation du vilebrequin. L'effet des différents régimes de combustion décrits plus haut apparaît ici de manière globale. A un début de combustion très rapide dans la préchambre succède un ralentissement également très brutal. Le rôle de la chambre principale est d'achever la combustion mais avec une vitesse beaucoup plus basse ; dans le cas présenté ici, à la fin de la combustion, la quantité de combustible introduite a brûlé pour moitié dans la préchambre et pour moitié dans la chambre principale.

Conclusion

Le code de calcul KIVA a été modifié de manière à pouvoir être appliqué au cas du moteur diesel à préchambre d'automobile.

Les modèles de turbulence, d'auto-inflammation et de combustion ont été améliorés et adaptés pour la même raison. Les résultats du calcul des phases de compression – combustion – détente ont permis d'analyser le processus contrôlant l'aérodynamique interne à la chambre ainsi que de mettre en évidence les différents régimes de combustion.

Remerciements

Ce travail a été réalisé grâce au soutien du GSM (IFP, PSA, RNUR), de l'Agence Française pour la Maîtrise de l'Energie et du Ministère de la Recherche et de l'Education Supérieure.

5 deg après PMH

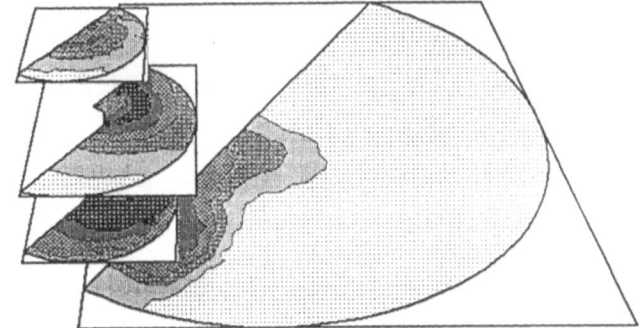

15 deg après PMH

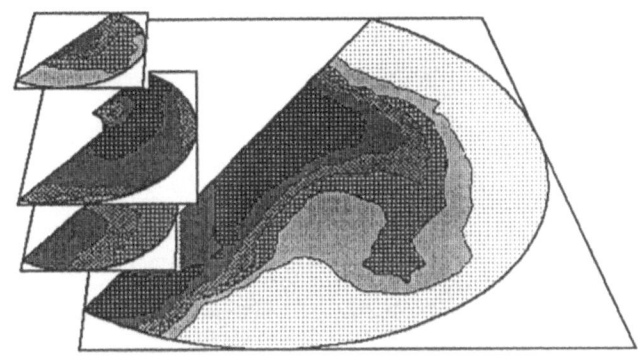

35 deg après PMH

ABOVE   2300
2000 − 2300
1700 − 2000
1400 − 1700
1100 − 1400
BELOW   1100

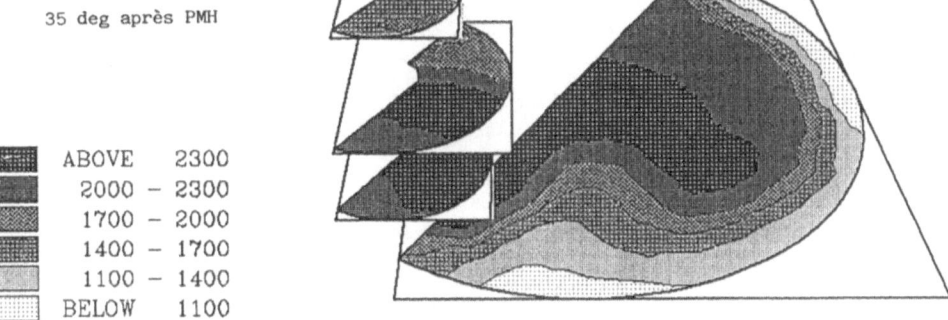

Fig. 8 : Champs d'isothermes (Kelvin) dans des plans de coupes horizontaux
à différents instants pendant la combustion.

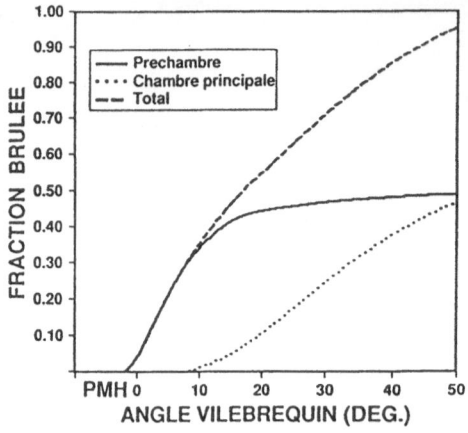

Fig. 9 : Evolution de la fraction brûlée en fonction de l'angle vilebrequin.

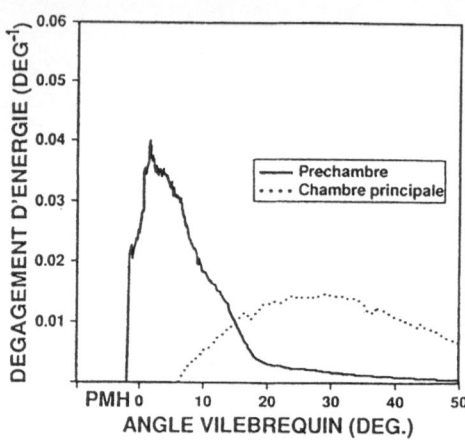

Fig. 10 : Evolution de la vitesse de déga- gement d'énergie globale en fonc- tion de l'angle vilebrequin.

References bibliographiques

[1] A.A. Amsden, J.D. Ramshaw, P.J. O'Rourke, J.K. Dukowicz — Kiva a Computer Program for Two and Three Dimensional Fluid Flows with Chemical Reactions and Fuel Sprays — LA-10245-MS

[2] A.A. Amsden, J.D. Ramshaw, L.D. Cloutman, P.J. O'Rourke — Improvements and Extensions to the KIVA Computer Program — LA-10534-MS

[3] A.A. Amsden, T.D. Butler, P.J. O'Rourke, J.D. Ramshaw — Kiva: A Comprehensive Model for 2D and 3D Engine Simulations — SAE 85 0554

[4] P. Gilaber, P. Pinchon — Measurements and Multidimensional Modeling of Gas-Wall Heat Transfer in a S.I. Engine — SAE 88 0516

[5] R. Diwakar, S. El Thary — Comparison of Computed Flow Fields and Wall Heat Fluxes with Mea- surements from Motored Reciprocating Engine-like Geometries — ASME, Third International Computer Engineering Conference, Chicago, August 1983

[6] M. Zellat, H. Zeller — Modelisation multidimensionnelle de l'autoinflammation dans un moteur diesel, Premières validations expérimentales. — Rapport IFP n⁰ 35 551, Septembre 1987

[7] B.F. Magnussen, B.H. Hjertager — On Mathematical Modelling of Turbulent Combustion with Special Emphasis on Soot Formation and Combustion — 16th Symposium on Combustion, The Combustion Institute 1976

[8] P. Pinchon, T. Baritaud — Modelling of Flow and Combustion in a Spark Ignition Engine, with a Shrouded Valve. Comparison with Experiments — ICHMT, August 24-28, 1987

[9] A.D. Gosman, P.S. Harvey — Computer Analysis of Fuel-Air Mixing and Combustion in an Axisymetric D.I. Diesel — SAE 82 0036

[10] S. Shirakawa, K. Omsawa, T. Aoyama — Simulation of Spray Combustion in an Axisymmetric Small Direct Injection Diesel Engine — IME 1987, C 01/87

[11] Ph. Pinchon, G. Grosshans, S. Michon — Three dimensional modeling of combustion in direct injection diesel engines. — Submitted to CIMAC 89 – China

[12] U. Spicher — Visualization of Combustion in the Swirl Chamber of a Diesel Engine — FEV Motorentech- nik, Report 1987

The Numerical Simulation of Shock Initiation in Solid Explosives with Gas Inclusions

H.Q.QIN & D.B.SPALDING

CHAM Limited
Bakery House 40 High Street Wimbledon
London SW19 5AU UK

ABSTRACT

The shock initiation of detonation in a solid explosive with gas inclusions is studied numerically by way of a two-fluid model. The fluid-dynamic equations of a reactive medium are solved in a two-dimensional Lagrangian frame of reference. The objective is to simulate detonation waves propagating in a solid explosive, the two-dimensionality resulting from non-uniformity of impact or the compliance of confining walls .

Studies are made both for cartesian coordinates and for body-fitted coordinates which allows to simulate the distortion of the material.

A preliminary study of the dependence of steady-state detonation velocity on charge radius (the 'diameter effect') is described.

1. Introduction

The present paper describes numerical simulations of the shock and detonation waves which result when a solid propellant material containing finely distributed air inclusions is strongly impacted at one end.

The impacted block of propellant is supposed to be confined within a container, the walls of which may be of two kinds. One kind is rigid, the other kind can expand in response to a rise of pressure. This expansion is represented either by using cartesian coordinates and treating the walls as being fixed but 'porous', or by using body-fitted coordinates.

The present report explores the possibility that detonation is initiated in a solid explosive by way of the compression of small pockets of gas which are distributed within the solid. It is argued that a pressure wave raises the temperature of the gas in these pockets to much higher values than that of the bulk of the surrounding solid; that the solid immediately in contact with the gas pocket is then heated by conduction to a temperature well above that of the bulk of the gas; and that chemical reaction ensures. The distortion of the material is represented by body - fitted coordinates.

The model employed is a simple one, characterised by: -
* the simplest possible representations of thermodynamic and chemical properties which still allow the phenomenon to take place; and
* representation of the " hammer blow" by the imposition and maintenance of a finite velocity at one end of a block of solid explosive which is initially at rest.
Details of the numerical feature of this study are described in a paper by the authors (5).

2. Mathematical model

Figure 2.1 illustrates the model under consideration. A block of solid explosive, containing finely divided and uniformly distributed gas bubbles, is supposed to be initially at rest, and of uniform temperature.

Both the solid and gas are taken as being compressible; and when a temperature difference exists between them, heat transfer from the one to the other is supposed to take place. However the gas and solid prevailing at any position and time are assumed to have the same pressure and the same velocity.

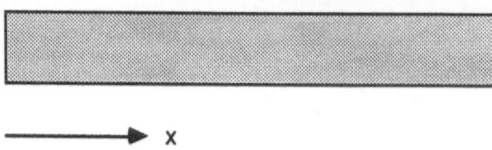

→ x

Figure 2.1 The model considered: a solid explosive with
finely - divided gaseous inclusions

The following dependent variables are solved :
P   the pressure;
T_g, T_s the temperatures of the gas and solid respectively;
ucar,vcar   the cartesian velocity at the cell corner in x- and y- directions;
m_g the mass fraction of gas;

Other quantities which must be computed are as follows : -
ρ_g, ρ_s, ρ_m, the densities of the gas and solid and composite material respectively;
T_r the temperature of the gas ~ solid interface; and
q_{gs} the rate of heat transfer from gas to solid per unit volume.

A Lagrangian representation is employed, by which is meant that the independent variables of position are the material - marker coordinates $\tilde{x}$ and $\tilde{y}$, not the distances x and y. The relations between these variables for modest distortion are:

$$d\tilde{x} = \int_0^x (\rho_m / \rho_{mo}) \, dx \qquad (2.3-1)$$

$$d\tilde{y} = \int_0^y (\rho_m / \rho_{mo}) \, dy \qquad (2.3-2)$$

where ρ_m is the mixture density at time t, while ρ_{m0} is its value at the start, when t equals zero.
The second independent variable is t.

Differential equations

The law of conservation of mass in the coordinate system in question, is

$$\frac{\partial}{\partial t} (\rho_{mo} / \rho_m) + \frac{\partial u}{\partial x} + \frac{\partial v}{\partial y} = 0 \qquad (2.4-1)$$

The law of conservation of momentum in the same coordinate system is:

$$\frac{\partial u}{\partial t} = \frac{\partial Px}{\rho_m \, vol} + \frac{\partial}{\partial x}(\mu_d \frac{\partial u}{\partial x}) + \frac{\partial}{\partial y}(\mu_d \frac{\partial u}{\partial y}) \qquad (2.4-2)$$

$$\frac{\partial v}{\partial t} = \frac{\partial Py}{\rho_m \, vol} + \frac{\partial}{\partial x}(\mu_d \frac{\partial v}{\partial x}) + \frac{\partial}{\partial y}(\mu_d \frac{\partial v}{\partial y}) \qquad (2.4-3)$$

where μ_d is the viscosity of the material,being a function of cell corner velocity $f(u_c, v_c)$.

The law of energy conservation for phase i, where i = g for gas and i = s for solid, is:

$$C_{pi} \frac{\partial T_i}{\partial t} - \frac{\partial P}{\rho_i \partial t} = q_i \qquad (2.4 - 4)$$

where C_{pi} is the constant-pressure specific-heat of the phase, and q_i is the heat input to the phase, either by heat transfer from the other phase or as a consequence of chemical reaction in the gas phase.

Also employed is a rate-of-phase-change law, representing the influence of the chemical reaction which is supposed to convert solid into gas. It is here written as:

$$\frac{\partial m_g}{\partial t} = m_g \qquad (2.4 - 5)$$

The task is to solve the above six equations simultaneously, so as to yield values of p, u_c, v_c, T_g, T_s and m_g as functions of time, u_c, v_c being the cartesian velocities at the cell corner. This task can be performed only with the aid of auxiliary relations, expressing the thermodynamic, heat transfer and chemical- kinetic behaviour of the materials.

Auxiliary relations

The density of the gas is supposed to obey the ideal - gas law:

$$\rho_g = \frac{P}{RT_g} \qquad (2.5 - 1)$$

where R is the Universal Gas Constant, which is also equal to $C_{pg} (\gamma-1)/\gamma$, where γ is the specific-heat ratio, taken as 1.4.

The density of the solid is presumed to depend linearly on pressure, in accordance with :

$$\rho_s = \rho_{s0} (1.0 + P/E) \qquad (2.5 - 2)$$

where E is the bulk modulus, taken to be 1.0E9 pascals.

The density of the mixture can be deduced from ρ_g and ρ_s via the following relation:

$$\rho_m = \frac{1}{m_g / \rho_g + (1 - m_g)/ \rho_s} \qquad (2.5 - 3)$$

The heat-transfer rate from the gas to the solid, per unit volume, is taken as:

$$q_s = - q_i = H (T_g - T_s) \, m_g^{[1/3]} \qquad (2.5 - 4)$$

where H is a volumetric heat-transfer coefficient.

The $m_g^{(1/3)}$ term in the above expression expresses the fact that the heat -transfer rate is proportional to the surface area of the gas bubbles and inversely proportional to the bubble diameter. The first of these is proportional to $m_g^{(2/3)}$ and the second to $m_g^{(1/3)}$.

The phase-change-rate m_g has been calculated from the formula :

$$m_g = C_1 P m_g^{(2/3)} [T_r / (T_0 + C_2)]^{C_3} \qquad (2.5 - 5)$$

where $m_g^{(2/3)}$ again represents the variation of bubble-surface area with changing mass fraction of gas, T_r is the solid-gas interface temperature taken as $0.8 \, T_g + 0.2 \, T_s$, and the other symbols stand for constants to be used as parameters of the calculations.

It should be mentioned at this point that the physics and chemistry represented by the above auxiliary relations are extremely crude. Discussion, and suggestions for improvement, will be found below.

3. Calculations

3.1 Initial conditions

All the calculations were performed with the following initial values:

$t = 0$:
$u = 0.0$ m/s $v = 0.0$ m/s
$p = 1.0E5$ N/m2 $m_g = 2.0E-5$
$\rho_g = 1.0$ kg/m2 $\rho_s = 1000.0$ kg/m2
$T_g = 350.0$ K $T_s = 350.0$ K

3.2 Boundary conditions

At the right-hand end, the effect of a hammer blow was simulated by imposing and maintaining the right-hand boundary of the domain at a fixed velocity, equal in most of the calculations to -300.0 m/s, for all times greater than zero.

At the left - hand end, the boundary condition involves maintaining the velocity at zero, as would correspond to butting the explosive block against an immovable wall.

At the north boundary, it is either a fixed wall with and without mass flow out for calculations with cartesian coordinates or a curved wall without mass flow out for body-fitted coordinates.

3.3 The computer program

The computations were performed by means of the PHOENICS Computer Program (Rosten and Spalding, 1981). Although conceived as an Eulerian-grid code, PHOENICS proved to be easy to adapt to a Lagrangian coordinate system.This is described in another paper by the authors(5).

PHOENICS, like all other comparable computer codes, produces results which depend upon the setting of certain non-physical parameters, of which the most important are the numbers of the intervals into which the space and time dimensions are divided. In the calculations to be reported, uniform intervals were chosen; and their numbers (nx x ny) were typically 20 x 5 for the cartesian coordinates and 4.0E-5 for the time dimension. These parameters depend on individual calculation when the body-fitted coordinates are employed.

3.4 Results

3.41 Cases performed with cartesian coordinates

3.41-1 Uniform impact boundary condition

The calculations to be described in this section pertain to the following conditions :
right-hand-end velocity   : -100.0 m/s;   left-hand-end condition : 1.0E5 N/m2 (pressure);
length of explosive block :   2.0 m ;   space and time intervals : 0.1m, and 4.0E-5 second;
values of C1, C2 and C3   :   C1 = 10.0, C2 = 1000.0, C3 = 10.0

3.41-1-A Without mass flow out

The results of the calculation are presented by way of a series of plots of pressure (Figure 1), and velocity (Figure 2) distribution history(at IY = NY-1) for the uniform impact boundary with reaction but no mass flow out. The abscissa is (in effect) x*10. The numbers marking each wave represent the number of time intervals since the start. The pressure and velocity distribution along the y direction are the same as at IY = NY-1 since the wave character is plane in shape.

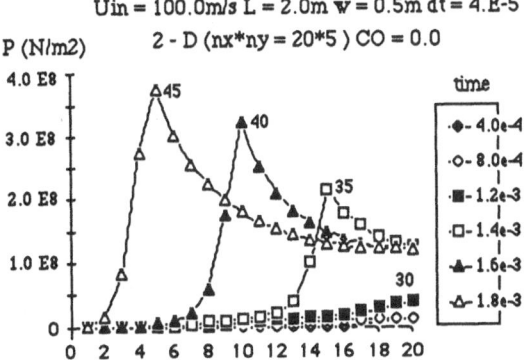

Fig 1.   Distributions of pressure

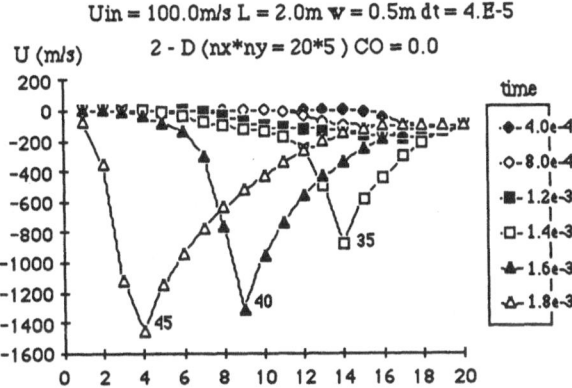

Fig 2.   Distributions of velocity

Inspection of the diagrams reveals that a plane wave travels from right to left. The peak pressure increases with time of interval, but has become almost constant by the time that the left-hand end is approached; and the same is true of (the negative of) the velocity. In the case of impact without mass flow out from the fixed walls, the results show a one-dimensional character.

3.41-1-B With mass flow out through walls

Figure 3 presents pressure distributions near the north wall (at IY = NY-1) by uniform impacting at the right-hand end, with various mass flow out through the porous medium walls at time of 1.6E-3 seconds. The larger the CO value, the greater the mass flow-out, and the lower the pressure, which means that compliance of the walls reduce the speed of propagation of the waves.(CO = 0.0 means no mass flow out).

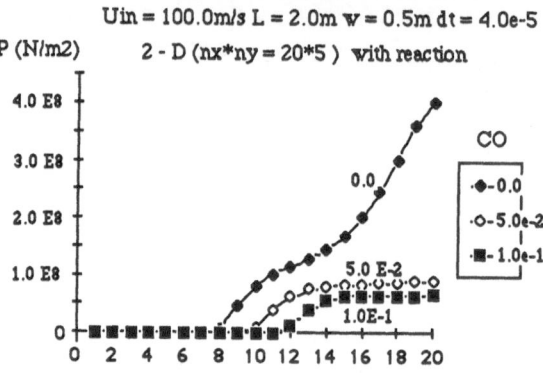

Fig 3. Pressure profiles near north - wall at t = 1.2E-3,
with different mass flow out through walls

3.42 Cases performed with body - fitted coordinates

The fully Lagrangian system, which employs body-fitted coordinates, presents the distortion of the material. It also simulates the real compliance of confining walls. The influence of impact velocity and 'diameter effect' is studied, and the influence of the numerical parameters is discussed.

3.42-1 Full results for a particular calculation

The results of a typical calculation described pertain to the following conditions :

right-hand-end velocity	:	-300.0 m/s
left-hand-end condition	:	$1.0E5 \text{ N/m2}$ (pressure)
length of explosive block	:	5.0 m
space and time intervals	:	0.1m, and 4.0E-5 second
values of C1, C2 and C3	:	C1 = 5.5, C2 = 1000.0, C3 = 10.0
reciprocal of diameters	:	$1/R = 1.67e\text{-}2 \text{ mm}^{-1}$

The results of the calculation are presented by a series of contour plots of pressure (in Figure 1, 3, and 5), and mass fraction of gas (in Figure 2, 4, and 6) at times of 2.28E-4, 2.64E-4 and 2.76E-4 seconds, and a velocity vector plot at time of 2.64E-4 seconds (Figure 7).Examination of the diagram reveals that the borders of the contours are lower values, the core of the contours are the higher values and the lighter area in between these are the intermediate values.

Inspection of the diagrams reveals that a wave travels from right to left. The peak pressure increases with time of interval, and the distortion of material becomes larger and larger, the pressure contours becomes more plane in shape and, the reaction region becomes a kernel when the distortion is large. After a short induction time, the gas-mass-fraction takes the form of an abrupt transition from 2.0E-5 to 1.0.The expanded area is fully occupied by gas product.

It may be concluded that a detonation wave has developed, and it is likely to propagate steadily until the end of the block of propellant is reached.

3.42-1 Influence of impact velocity

Figures 3, 4, 8 and 9 show the pressure and mass fraction of gas contours for different impact velocities of 300.0 and 400.0m/s at times of 2.76E-4 and 2.40E-4 seconds.

Inspection of the contours permits the conclusion that impact velocities in excess of 300.0m/s all lead to detonation, (the maximum of pressure and velocity being approximately the same in each case.) The higher the impact velocity, the shorter the time taken for the detonation wave to develop.
Evidently the sequence is :

(1) A pressure wave travels through the solid.

(2) Reaction starts well behind the wave front, causing at first a pressure rise at the impact point, where the materials have been heated longer.

(3) This pressure rise develops into a rapidly advancing wave, which swiftly overtakes the original shock.

3.42-2 Influence of length-diameter ratio

A series of calculations of the foregoing type were conducted using various length-diameter ratios, values between 0.009 to 0.025 being employed. All other parameters were as given in section 3.42-1, except that impact velocity was 300.0m/s for all calculations. In Figures 3 and 4, 10 to 13 are shown some of the contours of pressure and of mass fraction of gas. All the contours pertain to the same time of 2.76E-4 seconds for the developing stage. It is seen that the larger the diameter, the narrower the kernel of the reaction zone (the region of highest pressure). The shape of the expansion of the material is very similar for all present cases.

Figure 14 shows the dependence of the steady - state detonation velocity on the reciprocal of the charge radius, D being the detonation velocities which are worked out from the speed of the detonation wave front. The curve shows a slight downward concavity.

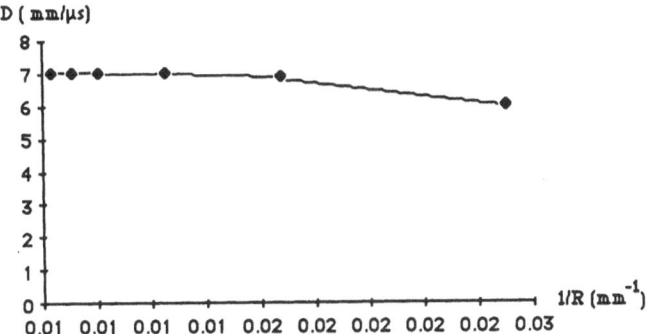

Fig 14 Effect of radius, R, on detonation velocity, D.

3.43 Influence of the numerical parameters

It was mentioned above that the results calculations were obtained from cartesian coordinates calculations with only 20 x 5 spatial sub - divisions, the typical results of (1/R = 1.67 E-2) being obtained from body - fitted coordinates with 20 x 6 sub - divisions. This was an economy measure; and it raises the question of how greatly the physically meaningful results are influenced by the grid fineness.

Figures 1, 2 and 15 to 18 show the influence of the size of the mesh and time step size for the same condition as for 1/R = 1.67 at time of 2.28E-4. Figures 15 and 16 are the contours of pressure and mass fraction of gas for mesh of 10 x 3, dt = 2.0E-5, Figures 1 and 2 for mesh of 20 x 6, dt = 1.0E-6, Figures 17 and 18 are for the fine grid of 26 x 8, dt = 9.0E-7. It may be seen that the finer the grid, the smaller is the reaction zone .The finer grid produces higher pressure values.

Evidently, the present calculation is very crude, and only qualitatively shows how detonation waves propagate in a two dimensional way and the distortion of the material.

4. Discussion

4.1 Impact initiation of detonation

The calculations have shown that impact can indeed lead to preferential heating of gas inclusions, that these inclusions can heat adjacent solid material, and that this material can then react exothermically in such a way as to cause a further pressure wave which can develop into a detonation.

The calculations reveal that when a block of solid explosive with rigid confining walls is uniformly impacted, the predicted shock waves and detonation waves are, as expected, plane in shape. Compliant walls produce two - dimensional waves.The higher the impact velocity, the shorter the time taken for the development of the detonation.

The dependences on the numerical parameters of time and space discretization have been studied. The finest grid was 26x8, but this was still not fine enough for grid independent solutions to be obtained.

4.2 Diameter effect

It is shown that change of charge radius does affect the detonation speed.

5. Conclusions

This numerical experiment shows that the two-dimensional shock initiation of a solid explosive with gas inclusions is qualitatively obtained using a two -fluid model. A shock travels into the explosive, resulting in shock heating, explosion occurs where it is heated the longest. A detonation develops and overtakes the shock wave.Uniform perturbation produces a one-dimensional plane wave, two-dimensional waves result from the compliance of walls, and possess the fundamental phenomena revealed in one-dimensional calculation. Other aspects, such as the transient heat conduction from the gas inclusions to the surrounding solid, should be considered,and the thermodynamic properties of the solid and gaseous products, which need to correspond to those of real explosive materials, should be studied in order to present the real process more faithfully.

6. Acknowledgement

This research was supported by the Ministry of Defence, Atomic Weapon Research Establishment, and monitored by Dr. I. Cameron.

References

* A.W. Campbell, W.C.Davis, and J.R.Travis
'Shock Initiation of Detonation in Liquid Explosives' The Physics of Fluids Vol 4. No.4 April 1961

* A.W. Campbell, W.C.Davis, J.B.Ramsay, and J.R.Travis
'Shock Initiation of Solid Explosives' The Physics of Fluids Vol 4. No. 4 April 1961

* Charles L. Mader
'Numerical Modeling of Detonation'   University of California Press 1979

* M.A.Nettleton
'Gaseous Detonation'   Chapman and Hall Ltd 1987

* Qin H Q & Spalding D B
'The Lagrangian Hydrodynamical Calculation In PHOENICS Code'   The PHOENICS Journal of Computational Fluid Dynamics and its Applications Vol 3, 1988

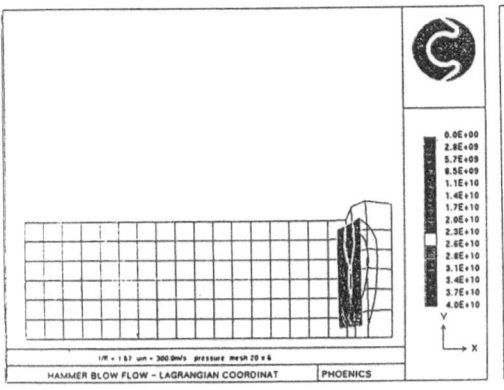

Fig 1 Pressure contours at 2.28e-4 seconds

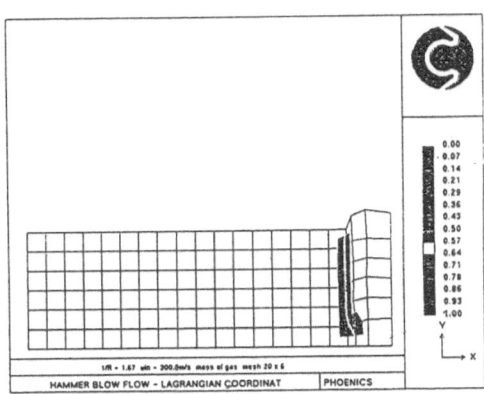

Fig 2 The contours of mass fraction of gas at 2.28e-4 seconds

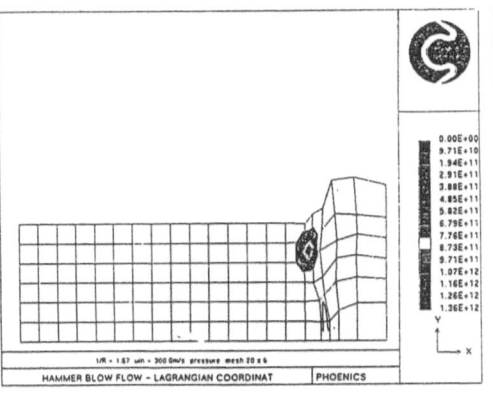

Fig 3 Pressure contours at 2.64e-4 seconds

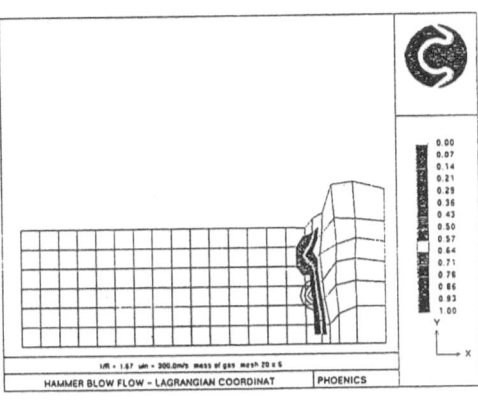

Fig 4 The contours of mass fraction of gas at 2.64e-4 seconds

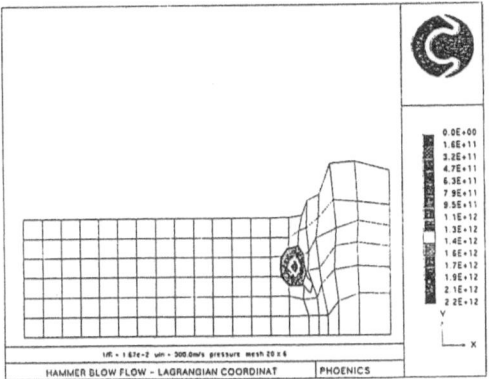

Fig 5 Pressure contours at 2.76e-4 seconds

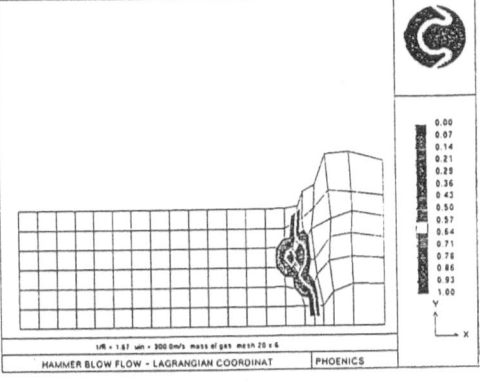

Fig 6 The contours of mass fraction of gas at 2.76e-4 seconds

426

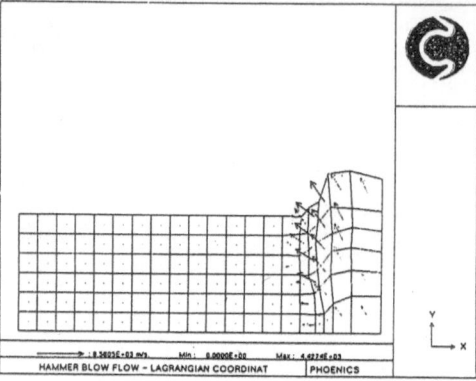

Fig 7 The vectors of velocitiy at cell corner gas at 2.64e-4 seconds

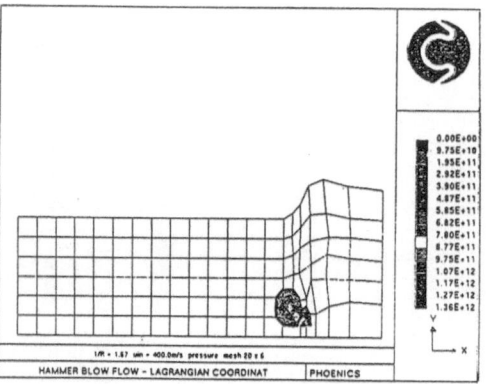

Fig 8 Pressure contours at 2.64e-4 seconds, uin = 400.0m/s

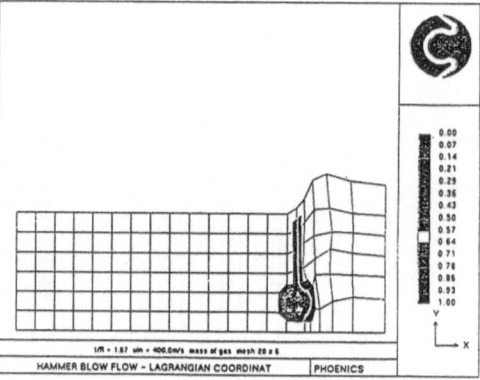

Fig 9 The contours of mass fraction of gas at 2.64e-4 seconds

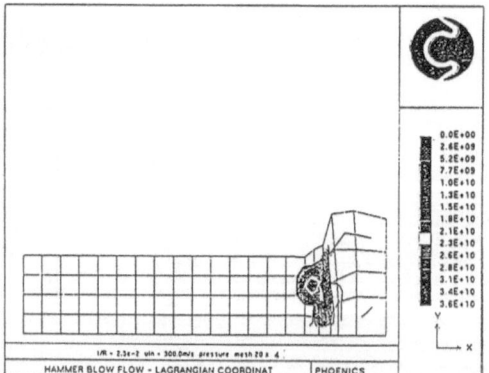

Fig 10 Pressure contours at 2.64e-4 seconds

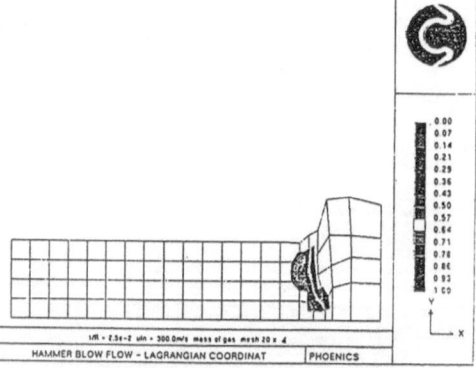

Fig 11 The contours of mass fraction of gas at 2.64e-4 seconds

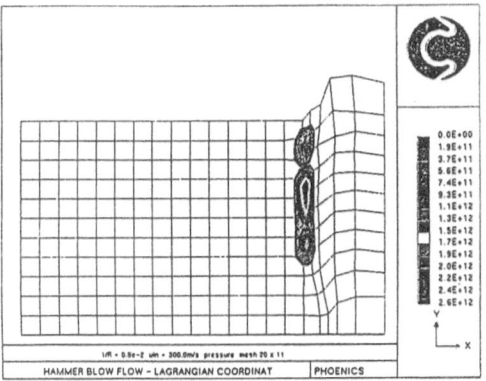

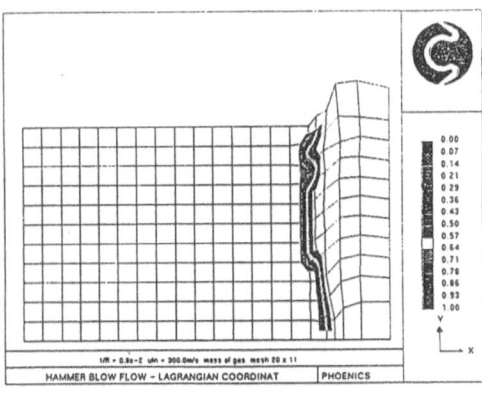

Fig 12 Pressure contours at 2.64e-4 seconds

Fig 13 The contours of mass fraction of gas at 2.64e-4 seconds

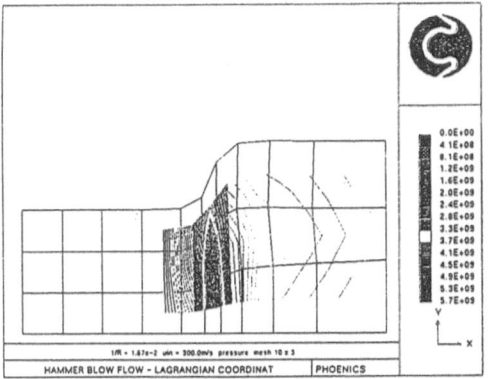

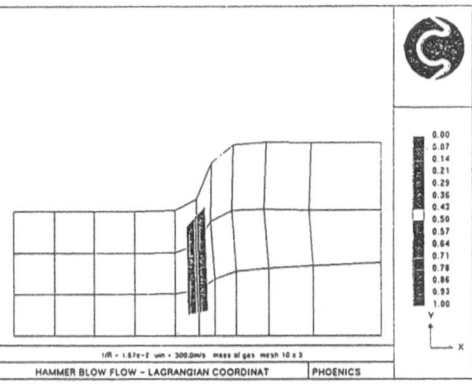

Fig 15 Pressure contours at 2.64e-4 seconds, (coarse mesh)

Fig 16 The contours of mass fraction of gas at 2.64e-4 seconds,(coarse mesh)

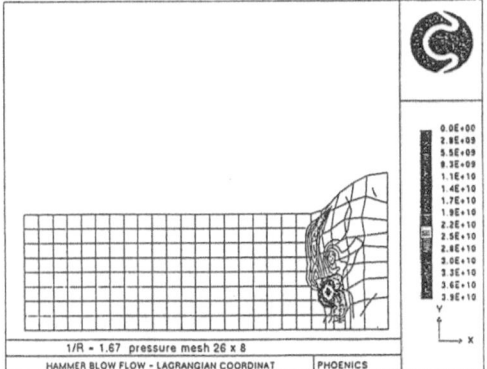

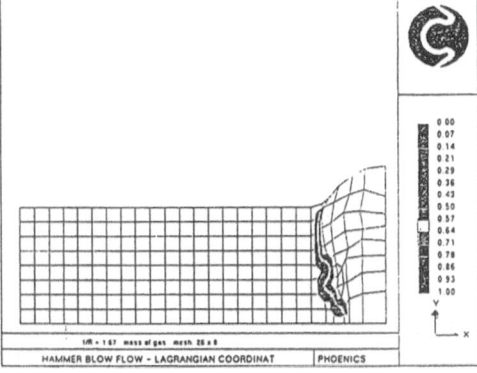

Fig 17 Pressure contours at 2.64e-4 seconds,(fine mesh)

Fig 18 The contours of mass fraction of gas at 2.64e-4 seconds,(fine mesh)

PRESSURE SPOT FORMATION IN UNSTABLE DETONATION WAVES

C. Schmidt-Laine, Ecole Normale Superieure, Lyon, France
J. W. Dold, U. of Bristol, England
& J. Buckmaster, U. of Illinois at Urbana-Champaign, USA

Abstract

When the stability of detonation waves is examined using activation energy asymptotics for a certain class of disturbances, perturbations to the shock displacement are governed by the equation

$$g_t + \frac{1}{2} g_x^2 = \ln \left[\frac{e^{cgg_{xx}} - 1}{cg_{xx}} \right], \quad x\epsilon[o,\ell], \quad g > 0$$

$$g_x = 0 \text{ at } x = 0,\ell$$

where c is a positive constant, t is time, and x is the distance measured in the plane of the undisturbed shock which propagates down a duct of width ℓ. Numerical solutions of the equation for sufficiently wide channels reveal singular behavior as $t \rightarrow t_c > 0$ in which $g_t \sim \ln(t_c - t)$. A local asymptotic description (final-time analysis) is possible, valid as $t \rightarrow t_c$. The singular behavior corresponds to the emergence of a 'pressure-spot', a small region of high pressure. It is suggested that such pressure spots are the origin of the transverse shocks that are a familiar feature of detonation waves.

Introduction

Detonation waves are highly unstable, and the usual manifestation of this is an unsteady three-dimensional structure characterized by transverse shock waves which propagate across the face of the shock. Two fundamental issues that need to be addressed by detonation theory are (i) how are these transverse shocks generated, and (ii) what is their average spacing? The latter defines the size of the familiar cells carved on witness plates by the intense vorticity at the shock triple points [1].

Numerical simulations of unsteady two-dimensional detonations (e.g. [2]) give rise to transverse shocks when the instability is triggered by a sufficiently large disturbance. However, limitations in the spatial and temporal resolution of the calculations means that, as yet, we cannot identify the origin of the transverse shocks from the growth of very small disturbances. Thus the physical origins are not apparent.

Some light on this question has recently been generated by three closely related papers that use activation energy asymptotics ($\theta \to \infty$) for chemistry modeled by one-step Arrhenius kinetics [3], [4], [5]. References [3] and [4] are linear stability analyses that show the initial evolution of infinitesimal disturbances to the steady structure. (In the limit the latter is the familiar square wave with a uniform induction zone separating the lead shock from a vanishingly thin reaction zone or fire behind which is uniform burned gas. δ, the length of the induction zone, defines the thickness of the undisturbed wave). Reference [3] deals with disturbances of wavelength $\sim\theta\delta$ and predicts a monotonically increasing growth rate with wave number. Reference [4] deals mainly with disturbances of wavelength $\sim\delta\sqrt{\theta}$, and for these the growth rate monotonically decreases with wave number, and stability prevails for sufficiently short waves. Reference [5] is primarily concerned with the nonlinear consequences of the instability to planar (infinite wavelength) disturbances. Here we are concerned with the nonlinear growth of the two-dimensional disturbances considered in reference [4].

Consider a detonation wave that is propagating down a channel of width $\sim\delta\sqrt{\theta}$. Then the asymptotic analysis of [4], prior to linearization for small disturbances, yields an evolution equation for $h(\eta,\tau)$, where h describes the shock location in a frame moving with the unperturbed detonation speed. This equation and the boundary conditions, in the notation of [4], are

$$\nu\left(1-\mu_s^{-1}\right) h_{\eta\eta}\left[1 + h/K\right] = \ln\left[1 + \nu\left(1-\mu_s^{-1}\right)h_{\eta\eta}e^{\beta\left(h_\tau + \frac{1}{2}h_\eta^2\right)}\right] ,$$

$$h > -K, \quad K \equiv \left(\sigma_b^{-1} - 1\right)\left(\mu_s - \sigma_b^{-1}\right)^{-1} , \tag{1}$$

$$h_\eta(0,\tau) = h_\eta(L,\tau) = 0$$

Here all constants are positive (for a description of their meaning, see

Appendix A), τ is time, and η is the distance measured across the channel, which has width L. The undisturbed shock corresponds to the solution h=0, and equation (1), when linearized for small h, describes the linear stability of the wave.

If we write

$$\frac{h}{K} = -1 + g(x,t) , \quad g > 0 \tag{2}$$

$$\eta = \sqrt{\beta} \, Kx , \quad \tau = \beta Kt ,$$

equation (1) can be rewritten in the form

$$g_t + \frac{1}{2} g_x^2 = \ln \left[\frac{e^{\frac{cgg_{xx}}{cg_{xx}}} - 1}{cg_{xx}}\right] , \quad x\epsilon[o,\ell] , \quad g > 0 \tag{3}$$

$$g_x(o,t) = g_x(\ell,t) = 0$$

where

$$L = \sqrt{\beta} \, K\ell , \quad c = \frac{\nu}{K}\left(1 - \mu_s^{-1}\right) . \tag{4}$$

Thus there are only two parameters, c and ℓ, and the unperturbed wave corresponds to g = 1.

If we linearize about 1, equation (3) simplifies to

$$g_t = (g - 1) + \frac{1}{2} c \, g_{xx} \tag{5}$$

so that if

$$g - 1 \sim \cos \left(\frac{n\pi x}{\ell}\right) e^{\alpha t} , \quad n \text{ an integer} \tag{6}$$

then

$$\alpha = 1 - \frac{c}{2} \cdot \frac{n^2\pi^2}{\ell^2} , \tag{7}$$

and there is instability for this mode if

$$\ell > [\frac{cn^2\pi^2}{2}]^{1/2} . \tag{8}$$

The smallest channel width for which a non-planar mode (n=1) is unstable is

$$\ell_c = [\frac{c\pi^2}{2}]^{1/2} . \tag{9}$$

The planar mode (n=0) is always unstable, but the growth rate (α=1) produced by equation (7) is spurious, a consequence of a nonuniformity in the asymptotics leading to (1) at large wavelengths [4] (see Appendix B). Thus, in examining the solutions of equation (3) it is appropriate to suppress the planar mode by a suitable choice of initial data. Rigorous stability results have been obtained in [6].

Final Time Analysis

Numerical integration of equation (3), described below, reveals singular behavior at some time t_c after the initiation of a disturbance. This is associated with the limit $g \to 0$. We shall suppose that this occurs at $x = 0$.

A key issue in determining the final singular behavior is the asymptotic description of gg_{xx} as $t \to t_c$, $x = 0$. The numerical calculations reveal that this vanishes, so that in the neighborhood of the singular point equation (3) can be simplified,

$$g_t = \ell ng + (-\frac{1}{2} g_x^2 + \frac{1}{2} cgg_{xx} + \dots) \tag{10}$$

where we will assume that the terms in brackets are small. Moreover, we will assume that the appropriate independent variables are $(t_c - t)$ and $\eta = x(t_c - t)^{-1/2}$, so that the spatial domain of the singularity shrinks to zero as $t \to t_c$. Thus neglecting the terms in brackets in (10), we have, upon integration,

$$\int_o^g \frac{ds}{\ell ns} = - (t_c - t) + f(\eta(t_c - t)^{1/2})$$

$$= - (t_c - t)(1 + a\eta^2) + o(t_c - t) \tag{11}$$

where a is a positive constant and we have used the boundary condition at x=0. Note that

$$g = \bigstar(t_c - t) \tag{12}$$

where $\bigstar$ is an algebraic order symbol that characterizes the size of g without accounting for logarithmic factors.

The corresponding expressions for the derivatives are

$$\frac{1}{\ell ng} \cdot g_t = 1 + \ldots$$

$$\frac{1}{\ell ng} \cdot g_x = -2an(t_c - t)^{1/2} + \ldots \tag{13}$$

$$\frac{1}{\ell ng} \cdot gg_{xx} = 4a^2n^2(t_c - t) - 2ag + \ldots$$

from which it is apparent that the terms in brackets in equation (10) are $\bigstar(t_c - t)$ and they can, indeed, be neglected to first order.

The outer solution (x = 0(1), t→t_c) can be written in the form

$$\int_0^g \frac{ds}{\ell ns} = f_1(x) + (t - t_c)f_2(x) + o(t_c - t) \tag{14}$$

where matching conditions require

$$\lim_{x \to 0} \frac{f_1(x)}{(-ax^2)} = 1 \quad , \quad f_2(o) = 1 \ . \tag{15}$$

This is equivalent to writing

$$g = k_1(x) + (t - t_c) \, k_2(x) + \ldots$$

where $\quad f_1(x) = \int_0^{k_1} \frac{ds}{\ell ns} \ , \quad k_2(x) = f_2(x) \, \ell n \, k_1(x) \ . \tag{16}$

Finally, we note that if we write

$$g = e^{-G} \tag{17}$$

then, in the inner layer,

$$e^{-G} \frac{G_{xx}}{G} = -2a + 4a^2n^2 \frac{(t_c - t)}{g} (1 - \ell ng) + o((t_c - t)g^{-1})$$

$$= -2a + \ldots \text{ at } \eta = 0 \ . \tag{18}$$

These characteristics of the solution are confirmed by the numerical results described below.

Numerical Method

Integration of equation (3) is straightforward provided we properly resolve the temporal and spatial structures associated with the emerging singularity. This is aided by the use of the transformation (17). Then the governing equation becomes

$$G_t = e^{-G}\frac{G_x^2}{2} + e^G \ln \left\{ \frac{\exp[c(G_x^2 - G_{xx})e^{-2G}] - 1}{c(G_x^2 - G_{xx})e^{-G}} \right\}. \tag{19}$$

This was solved with initial conditions

$$G = -\ln\left[1 - \varepsilon \cos\left(\frac{\pi x}{\ell}\right)\right] = -\ln g \tag{20}$$

and periodic boundary conditions ($x = 0$ and ℓ are lines of symmetry).

For a problem such as this which is not computationally intensive, high-order finite difference schemes are a simple way of providing the needed accuracy. Thus we used an 11-point central difference formula for the spatial derivatives on a uniform mesh with initial spacing $\Delta x = \ell/N$. To handle the time derivatives, quadratic estimates of G_t were calculated in the following fashion: with the solution known at t_j, t_{j-1} and t_{j-2}, G_t was fitted and an estimate of the solution at t_{j+1} obtained; and then the solution at t_{j+1}, t_j and t_{j-1} was used iteratively to generate a better fit for G_t, and a better estimate of the t_{j+1} solution. Stability required $\Delta t < \frac{1}{2}(\Delta x)^2$. In addition, Δt was decreased by a factor of (max $|G_t|)^{-1}$ as the singularity was approached. Spatial accuracy was checked by comparing 9th and 10th order polynomial fits of G, and halving the step size when the difference became too large. However, when t was close to t_c, it was only necessary to do this in a neighborhood of the emerging singularity, since G does not change very much outside of a suitably chosen neighborhood. In this way we avoided generating an excessive number of mesh points.

Solutions

The constant c is determined by the specific heat ratio γ, the CJ Mach number $M_{f_{CJ}}$, and the degree of overdrive $D = M_f^2/M_{f_{CJ}}^2$. Thus for $\gamma = 1.2$, $D = 1.2$, $M_{f_{CJ}} = 6.22$, we find $c = .268$. We describe here solutions for $c = .268$, $\ell/\ell_c = 1.3$, and $\varepsilon = .01$.

Figure 1 shows variations of g with x for different times. The initial disturbance is amplified and g first vanishes at time $t = t_c \stackrel{\sim}{=} 6.466$. The singular nature of the solution at $x = 0$, $t = t_c$ is not apparent from this figure.

On the other hand, Fig. 2 shows the variations of g_t which is unbounded at $x = 0$ as $t \to t_c$. The shrinking spatial domain in which the growth is manifest is quite clear (cf. the remark following equation (10)).

Convincing evidence that the final time analysis is correct is provided by Fig. 3 (cf. equation (18)). Additional evidence is provided by Figs. 4 and 5 which describe the functions $f_1(x)$ and $f_2(x)$ that characterize the outer solution as $t \to t_c$ (cf. equation (14)). In plotting the parabola in Fig. 5 we have used the value of a determined from Fig. 3.

The singular behavior identified by these calculations corresponds to an unbounded positive perturbation shock speed so that the perturbation post-shock pressure is also unbounded in the neighborhood of x=0. Thus for times close to t_c and small x the post-shock pressure increases significantly above the steady-state value. We suggest that this excess pressure will eventually be relieved by the generation of transverse shock waves, thus initiating the triple points characteristic of unstable detonation waves.

Acknowledgment

The collaboration between C. S-L and J. B. was made possible by a NATO grant. In addition, J. B. and J. W. D. were supported by the Air Force Office of Scientific Research. We are grateful to C. J. Lee for his help in generating the figures.

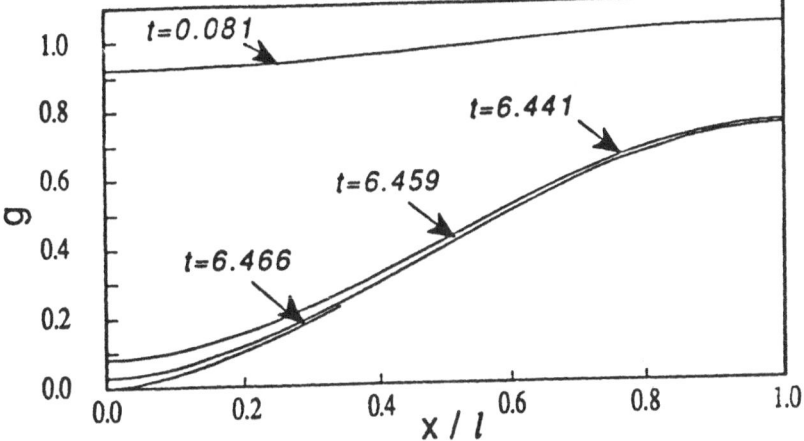

Fig. 1　The shock location g.

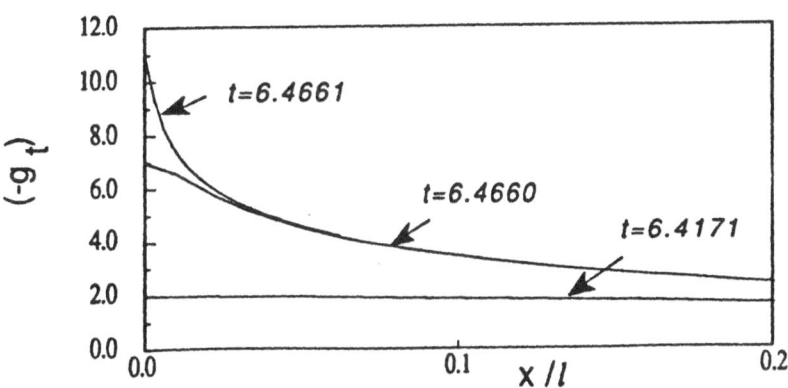

Fig. 2.　The perturbation shock speed.

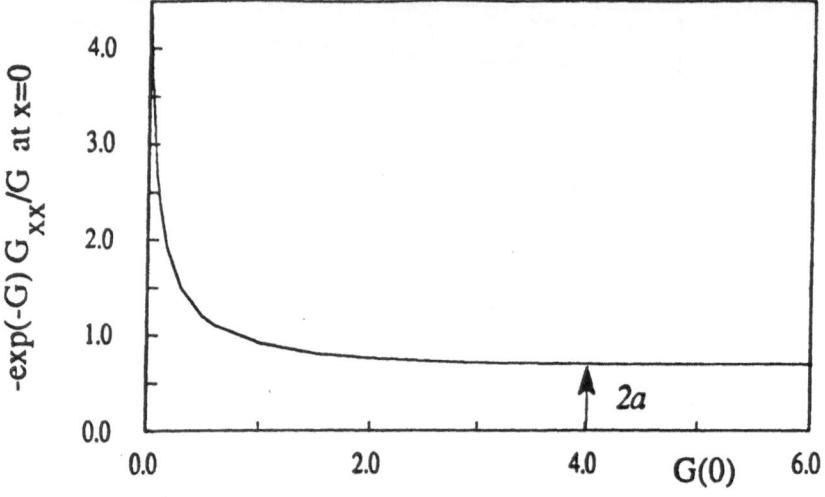

Fig. 3. Confirmation of the final-time analysis.

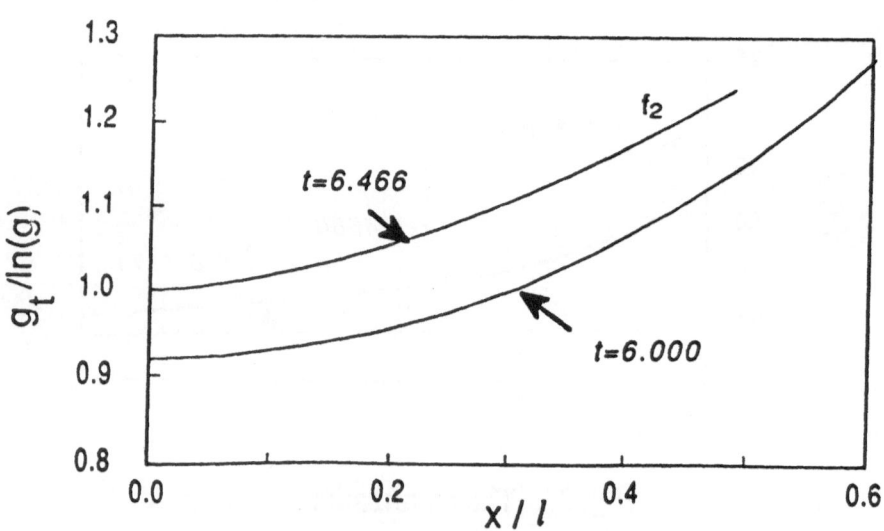

Fig. 4. Approach to the final-time function f_2

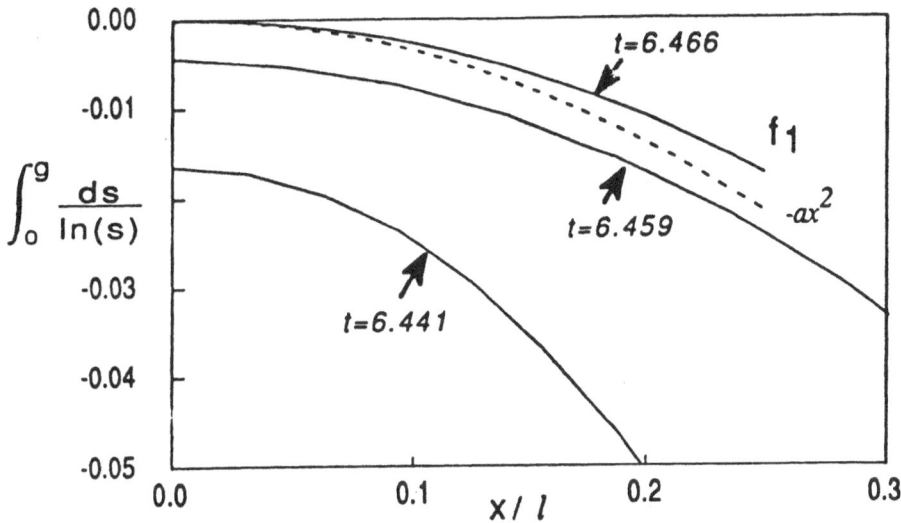

Fig. 5. Approach to the final-time function f_1.

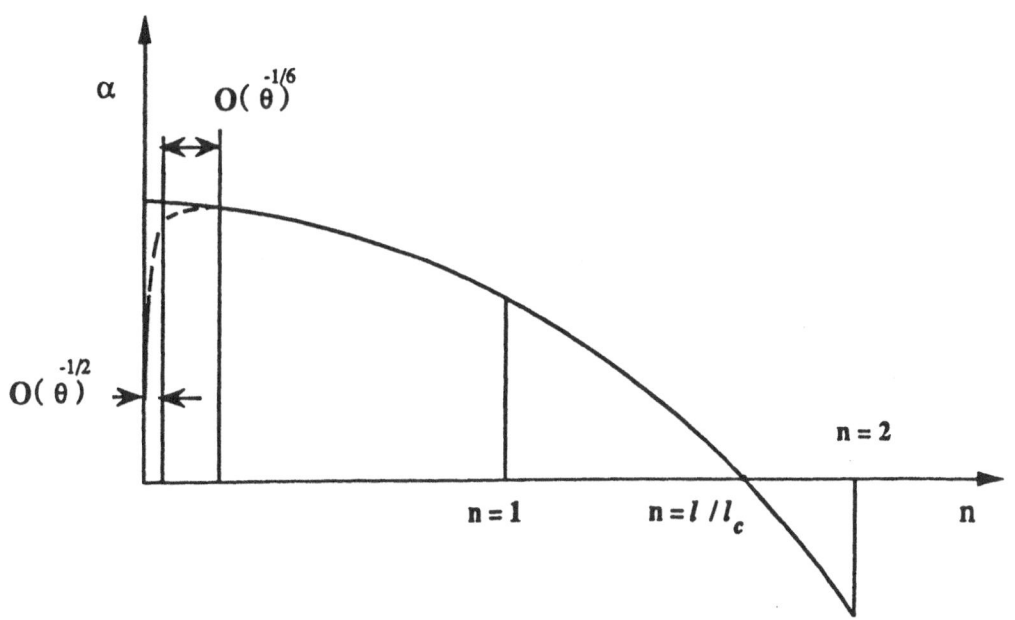

Fig. B1. Growth rate (α) as a function of mode number (n) when $1 < l/l_c < 2$

References

[1] R. Strehlow, Combustion Fundamentals, McGraw-Hill, NY, 1984, p. 307.
[2] T. Sugimura, T. Fujiwara and J. H. Lee 'Cellular detonation - instability and substructure', to appear in the Proceedings of the 22nd Symposium (International) on Combustion.
[3] J. D. Buckmaster and G.S.S. Ludford, Proceedings of the 21st Symposium (International) on Combustion, 1986, p. 1669.
[4] J. Buckmaster 'A theory for triple point spacing in overdriven detonation waves', to appear in Combustion and Flame.
[5] J. Buckmaster, Combust. Sci. and Tech., 1988, 61, p. 1.
[6] C. M. Brauner, C. Schmidt-Lainé, 'Spectral analysis of transverse perturbations in overdriven detonation waves', to appear.

Appendix A

The constants in equation (1) are characteristics of the steady (undisturbed) solution, and have the following meaning:

β is a measure of the post-shock temperature increment generated by a small change in the shock speed. Thus

$$T_{1_s} = -\beta \, \frac{C_p T_s^2}{u_f^2} \left(h_\tau + \frac{1}{2} h_\eta^2 \right) , \, \beta > 0 \tag{A.1}$$

where $\dfrac{T_{1_s}}{\theta}$ is the post-shock temperature perturbation, C_p is the specific heat, T_s the undisturbed post-shock temperature, u_f the gas-speed in the fresh-gas ahead of the shock, and the shock displacement is δ h. Thus β is defined by perturbed Rankine-Hugoniot conditions. Recall that θ is a non-dimensional activation energy, and δ is the thickness of the undisturbed detonation wave.

ν is defined by the formula

$$\nu = \frac{u_f^4}{C_p^2 T_s^2} \frac{(c_s^2 - d_s^2)}{(c_s^2 - u_s^2)} \left[\left(\frac{\partial T}{\partial p} \right)_{\rho s} \right] \tag{A.2}$$

where c is the speed of sound, d the isothermal speed of sound, u the gas-speed, all evaluated post-shock for the steady state. Physically, ν is a measure of how rapidly the temperature changes in the induction zone due to stream tube area variations in the perturbed detonation wave.

μ_s is the density ratio $\rho_s/\rho_f > 1$.

σ_b is the density ratio $\rho_b/\rho_s < 1$ where p_b is the density of the burnt gas (steady-state).

Appendix B

The linear growth rate, defined by equation (7), is

$$\alpha = 1 - n^2\left(\frac{\ell_c}{\ell}\right)^2 \tag{B.1}$$

and we plot α as a continuous function of n in Fig. B1 with the first two non-planar modes shown when $1 < \frac{\ell}{\ell_c} < 2$. It is noted in reference [4] that (B.1) is not a uniformly valid large θ approximation as $n \to 0$; there is a boundary layer in which α decreases below unity. This is shown as a dotted line in Fig. B.1. Thus the growth rate $\alpha = 1$ at $n = 0$ is spurious.

REVIEW OF THEORY OF MIXING AND REACTION
WITHIN A VORTICAL STRUCTURE

William A. Sirignano
Department of Mechanical Engineering
University of California, Irvine

Abstract

Constant-density, isothermal planar mixing layers with and without chemical reaction are studied. A series of problems involving an isolated vortex, a row of stable vortices, and an unstable vortex sheet are examined to understand the effects of coherent vortical structures on the mixing process. Diffusion is superimposed on the stretching and winding of the material lines. Probability density functions are calculated from the concentration profiles. Scalar dissipation rates are found to depend strongly on the vortex Reynolds number and weakly on the Schmidt number. Volumetric mixing rates are predicted to be identical on the high-speed and low-speed sides of a mixing layer.

Introduction

The problem of turbulent reacting flows is of broad interest in combustion applications. At higher Reynolds numbers, the phenomenon is highly three-dimensional; however, at intermediate Reynolds numbers two-dimensional structures will dominate in many configurations. While variable density occurs in practical situations, it is instructive to study constant-density, isothermal reactions.

Planar mixing layers have been studied experimentally for these constant density situations and probability density functions (pdf) have been reported. See, for example, Masutani and Bowman[1] and Dimotakis.[2] An interesting result is that the pdf for scalar concentration on the high-speed flow side has larger peaks than on the low-speed side. Some interpretations state that therefore the mixing rate is larger on the high-speed side. We shall later address the error in that conclusion.

This paper reviews the theoretical and computational developments of Cetegen and Sirignano [3,4], Miralles-Wilhelm, Rangel and Sirignano[5] and Rangel and

Sirignano[6] concerning mixing and chemical reaction in a constant-density, planar vortical structure. In these papers, a series of related problems of increasing complexity are treated. As the sequence of problems develops, the representation of an actual mixing layer is improved. A single vortical structure is first examined and the concentration profiles and pdf of scalar concentration are calculated[3]. A two-dimensional configuration is studied wherein at the initial time a line vortex is placed in the initially planar interface between two semi-infinite spaces of different chemical composition. The vortex develops a viscous core with time. The vortical action rotates fluid of one chemical composition into the initial domain of the other fluid. The stretching of the lines of constant composition by the rotational fluid motion result in very large increases of the surface area of the interface and in increased molecular diffusion across the interface. Mixing is therefore substantially enhanced by the vortex. In some cases, the two different chemicals considered on the two sides of the original interface were reactants so that the mixing is followed by chemical reaction calculated with finite-rate kinetics. A region of well-mixed reactants develops and grows continually in the region surrounding the vortical center. This region is known as the mixing core. Then a row of vortices at constant spacing is studied, both temporally and spatially developing layers are considered[4,5]. The line vortices are equally spaced and aligned in parallel fashion in the plane that is the original interface between two regions of different composition. Each vortex enhances mixing of the two fluids and a viscous core and a mixing core develop around each vortical center. Collectively, the vortices create a convective fluid motion in a positive direction on one side of the interface and a negative convective motion on the other side. So, the near field of each vortex resembles that of an isolated vortex but the far field is very different. The vortices are artificially held in fixed relative positions. In reality, they would not be stable since any given vortex would tend to move, if slightly perturbed, due to the actions of the other vortices. The final problem addresses the relative motion of the vortices that result from small perturbations in the positions[6]. This behavior involves pairing and merging of the vortical structures in addition to the above-mentioned phenomena. The pairing and merging process enhances the entrainment of fresh gases into the vortical structure.

Pdf's are calculated in two ways: one method simulates experimental measurement wherein the vortical structure sweeps past the measuring plane in a finite time[3,4] and the other method involves the instantaneous sweep across the flow field[5]. The instantaneous sweep gives the pdf as a function of transverse position in the simulated mixing layer. At each transverse position the pdf is calculated for each instant of time based upon the sweep of the concentration profile in the major flow direction. The pdf here is therefore based upon an instantaneous "snapshot" of the vortical structure. The first method on the other hand simulates standard experimental measurement wherein data is taken as the vortical structure flows past the measuring plane. While the vortical structure is flowing past the measuring plane, rotational motion and molecular diffusion are continually modifying the concentration profile. The pdf's from the two methods differ on account of the rotational convection and the molecular diffusion. Only in the frozen flow limit would the two pdf's be the same.

A particular long-term goal of this research is to improve the modelling of the molecular mixing term in a pdf evolution equation or in averaged moment equations. This term has been modelled by ad hoc means that are highly questionable. See Sirignano[7] for details.

An incompressible vortical two-dimensional flow field is studied to predict the stretching and displacement of the material lines. These vortical structures are located at the interface between regions or flows of different composition. Diffusion normal to the stretched and displaced interface is described following the analysis of Marble[8] and Marble and Karagozian[9].

Isolated Vortex

In Reference 3, a single viscous line vortex was placed at the interface of two fluids. Figure 1 indicates the winding of the material line that results from the vortical action. Concentration profiles at various transverse (y) positions are given in Figure 2. These results are caused by a combination of the vortical motion and diffusion. No chemical reaction is considered yet. Figure 3 shows results for a pdf calculated by instantaneously sweeping over the concentration profile at a given transverse (y) position. A temporally-growing mixing core is identified at the center of the vortical structure with the concentration at C=0.5, half-way between the two far-fields of C=0 and C=1. Figures 2 and 3 indicate the high degree of nonuniformity in the concentration field outside of the mixing core. The pdf's have been calculated by instantaneously sweeping in the x direction (parallel to the original interface) at a given transverse (y) position. This is the pdf that would be measured if the flow were frozen as a vortical structure moved past a probe. This pdf as a function of concentration C and y is antisymmetric in a plane with the origin at y=0 (the location of the original interface) and C=0.5 (value after complete mixing). Note that antisymmetry was not seen in actual experimental measurements since a different pdf is measured by the probe[1,2].

Comparisons of the pdf's between Figures 3 and 4 indicate the temporal development of the concentration field. Figures 3 and 5 show the effects of

443

increasing vortex strength or Reynolds number. Realize the dimensions are scaled with the square root of Reynolds number indicating that scale of the structure increases with Reynolds number accordingly.

Reactive flows have also been considered in Reference 3 wherein the original interface divides two reactants that, upon mixing, undergo a one-step reaction. Figure 6 shows that the center of the structure contains the products of the reaction.

Calculations have been made for a range of Reynolds number and Schmidt number. A mixedness parameter shown in Figure 7 has been defined and calculated.. From a range of numerical experiments, a correlation has been made indicating that the mixedness parameter grows linearly with time and nearly linearly with vortex Reynolds number. It is also very weakly dependent on Schmidt number.

The mixedness parameter is defined as

$$f = 1 - \frac{2}{y_{max}} \int_{-y_{max}}^{y_{max}} \int_0^1 P(y,C)[C-0.5]^2 \, dCdy$$

For the isolated vortex, a matching with numerical data yields the correlation

$$f = .593 \, Sc^{-1/2} [\frac{4\nu t}{y_{max}x_{max}}]^{1/2} + .093 \, Sc^{0.06} Re \, [\frac{4\nu t}{y_{max}x_{max}}]$$

where y_{max} and x_{max} are the domain over which variations in concentrations are found due to the vortical action, ν is the kinematic viscosity, and t is time. This weak dependence on mass diffusivity is explained by an asymptotic analysis for large strain rates that indicates the establishment of a thin diffusion layer at the stretched, wound interface.[3] In particular, the local scalar dissipation rate becomes independent of Schmidt number and strongly dependent on Reynolds number.

Stable Row of Vortices

Cetegen and Sirignano[4] and Miralles-Wilhelm, et al.[5] have extended the analysis of Reference 3 to an infinite row of vortices to simulate better the far field of a mixing layer by creating non-zero velocities above and below the row of vortices. The pdf has been calculated in both ways, simulating the frozen flow and the actual flow. Good qualitative agreement with experimental data is found when the finite time required for a vortical structure to flow past the measuring probe is taken into account. The conclusion, however, is that it is improper to interpret that more rapid mixing occurs on the high-speed side of the mixing layer. Volumetric mixing rates are identical on both sides of the mixing layer.

The mixedness parameter shown in Figure 7 indicates that the previous linear dependence on Reynolds number for the isolated vortical structure is modified to a $Re^{0.8}$ dependence. A very weak dependence on Schmidt number is still found. The calculations have been made for both a temporally-developing mixing layer wherein all vortical structures are of the same age and for a simulated spatially-developing layer wherein the structures increase in age with downstream distance. No qualitative differences and only minor quantitative differences are identified account of age variations in the vortical structures.

The concentration profiles measured at various transverse (y) positions are indicated in Figure 8. These profiles are given at a fixed plane through which the vortical structures flow. Therefore, they do not show the antisymmetric behavior that exists if we instantaneously swept over the flow direction. That is, at any instant of time $C (y,x) = 1-C (-y,-x)$ where x=0 moves with the vortical structure. Figure 8, however, presents $C (y,t)$ at an x-position fixed in the laboratory. The pdf created from the antisymmetric concentration file is antisymmetric itself in that $P (C,y) = P (1-C,-y)$. The pdf constructed from the profile in Figure 8 will not be antisymmetric. They are equivalent to those determined in the laboratory.

The temporal development of the concentration profiles and pdf's determined by instantaneous sweeping over the x-direction give the required information to determine the volumetric mixing rate. The use of the pdf's determined in the laboratory do not accurately reflect the mixing rate; the differences in velocity through the mixing layer distort the results.

Unstable Vortex Sheet

The assumption about a constant distance between vortical structures in a row has been relaxed.[6] In particular, the interface between a two parallel-flowing fluids is represented by a vortex sheet initially with uniform circulation per unit length with mass diffusion normal to the distorting interface. The vortex sheet is displaced very slightly by an initial sinusoidal perturbation of a given wavelength. The nonlinear fluid dynamics results in an increasing concentration of vorticity in one small region over the wavelength. The vorticity and circulation per unit length decrease over the remainder of the wavelength. This result supports the representation of the mixing layer as a row of vortices. The vortices, however, are unstable. When a subharmonic small perturbation is applied, pairing of the vortices occurs, thereby further modifying the mixing process.

Figure 9 shows the temporal development of the interface resulting from the fluid dynamics. The effects of winding of the material lines due to the individual concentrations of vorticity are seen. Also observed is the pairing effect. Figure 10 shows the temporal development of the mass fraction contours due to the combination of vortical action and mass diffusion normal to the material lines. Note that over the wavelength of the subharmonic, the concentration $C(y,x) = 1-C(-y,-x)$, so that an antisymmetry persists.

Concluding Remarks

It has been shown that a vortex sheet will tend to distort so that distinct vortical structures result. These structures enhance the mixing of the two different fluids that initially interfaced at the vortex sheet. The mixing rate (or equivalently the global scalar dissipation rate) increases with the vortex Reynolds number (Re). A linear dependence is found for the isolated vortex and a $Re^{0.8}$ dependence is obtained for the row of vortices (mixing layer simulation). The dependence on Schmidt number is very weak since the diffusion flux at the winding interface is strongly dependent on the vortex strength; in particular local scalar dissipation rate is independent of Schmidt number.

The pdf determined in a laboratory frame of reference indicates a certain assymmetry with generally larger peaks found on the high-speed-flow side. This lack of antisymmetry does not indicate a difference in mixing rates between the high-speed and low-speed sides of the flow. An instantaneous display of concentration over the flow field does indicate that volumetric mixing rates are identical on both sides of the mixing layer. In a frame of reference attached to the vortical structure, the flow fields on the two sides of the mixing layer are equal in magnitude but opposite in direction. The differences between temporally and spatially developing mixing layers are shown not to be important on the scale of the spacing between vortical structures.

Pairing of the vortical structures due to the nonlinear fluid dynamics further modifies the mixing process. The concentration profiles over the wavelength of the subharmonic display an antisymmetric character. The pdf's constructed by an instantaneous sweep are expected therefore to display an antisymmetric character. Volumetric mixing rates will remain the same on the high-speed and low-speed sides of the mixing layer.

Acknowledgment

This research is sponsored by the Air Force Office of Scientific Research under grant No. 860016D, with Dr. Julian Tishkoff acting as the technical monitor.

References

1. Masutani, S.M. and Bowman, C.T. "Structure of a Chemically Reacting Mixing Layer," J. Fluid Mechanics, 172, pp. 92-126 (1986).

2. Dimotakis, P.E. "Two Dimensional Shear-Layer Entrainment," AIAA Journal 24, pp. 1791-1796 (1986).

3. Cetegen, B.M. and Sirignano, W.A. "Study of Mixing and Reaction in the Field of a Vortex," Joint Meeting of the Combustion Institute Western States and Japanese Sections, Paper 3B-44, Honolulu (1987)

4. Cetegen, B.M. and Sirignano, W.A. "Analysis of Molecular Mixing and Chemical Reaction in a Mixing Layer" AIAA 26th Aerospace Sciences Meeting, AIAA preprint 88-0730, Reno, NV (1988).

5. Miralles-Wilhelm, F., Rangel, R.H., and Sirignano, W.A. "An Analysis of Molecular Mixing in a Vortical Structure: Bias in PDF Measurement," AIAA 27th Aerospace Sciences Meeting, AIAA preprint 89-0482, Reno, NV (1989).

6. Rangel, R.H. and Sirignano, W.A. "The Dynamics of Vortex Pairing and Merging," AIAA 27th Aerospace Sciences Meeting, AIAA preprint 89-0128, Reno, NV (1989).

7. Sirignano, W.A. "Molecular Mixing in a Turbulent Flow: Some Fundamental Considerations" Combustion Science and Technology 51, pp. 307-322 (1987).

8. Marble, F.E. "Growth of a Diffusion Flame in the Field of a Vortex," Recent Advances in the Aerospace Sciences (C. Casci, editor), Plenum Publishing, pp. 395-413 (1985).

9. Karagozian, A.R. and Marble, F.E. "Study of a Diffusion Flame in a Stretched Vortex," Combustion Science and Technology 45, pp. 65-84 (1986).

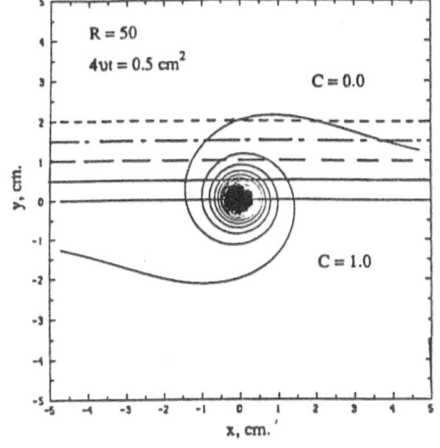

Figure 1. Distortion of a material line by a viscous line vortex.

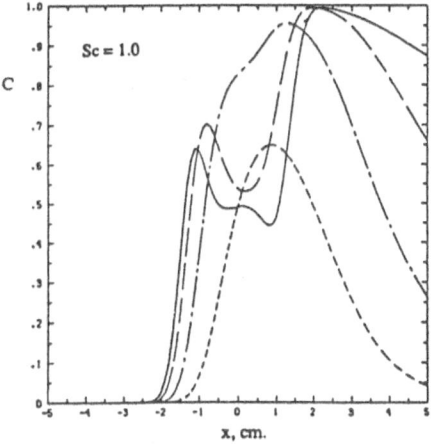

Figure 2. Concentration profiles at Sc-1.0 for the configuration depicted in Figure 1.

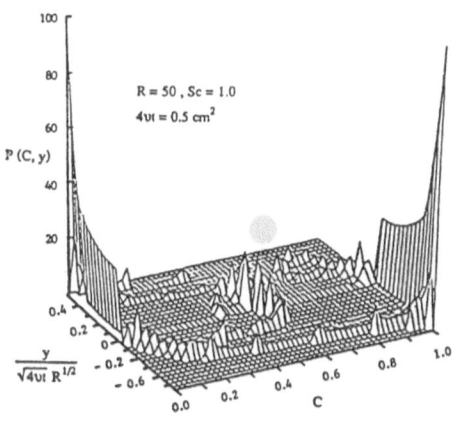

Figure 3. Pdf of scalar concentration at Re-50 and $4\upsilon t - 0.5$ cm^2.

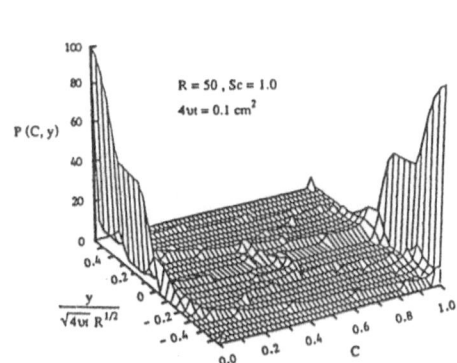

Figure 4. Pdf of scalar concentration at Re-50 and earlier time $4\upsilon t - 0.1$ cm^2

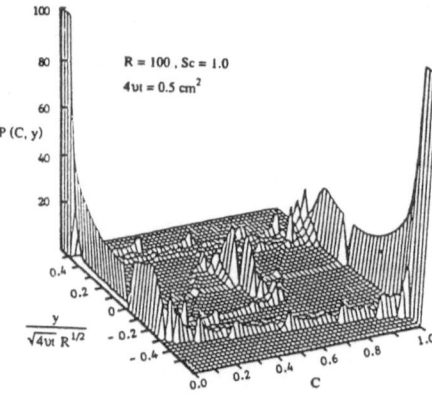

Figure 5. Pdf of scalar concentration at
higher Reynolds number Re-100
and $4\nu t - 0.5$ cm^2.

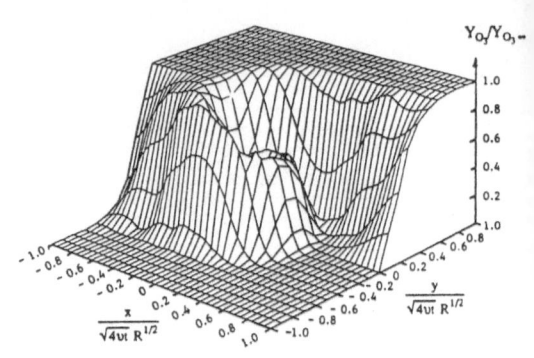

Figure 6a. Concentration profile for
nonreacting scalar.

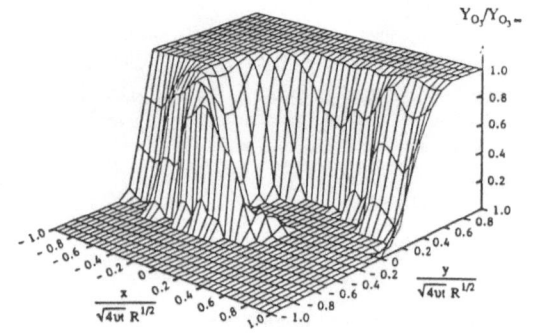

Figure 6b. Concentration profile for
reacting scalar.

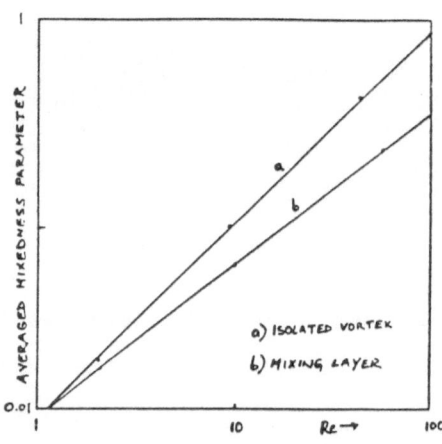

Figure 7. Comparison of the overall mixedness
parameter between a mixing layer and
an isolated vortex.

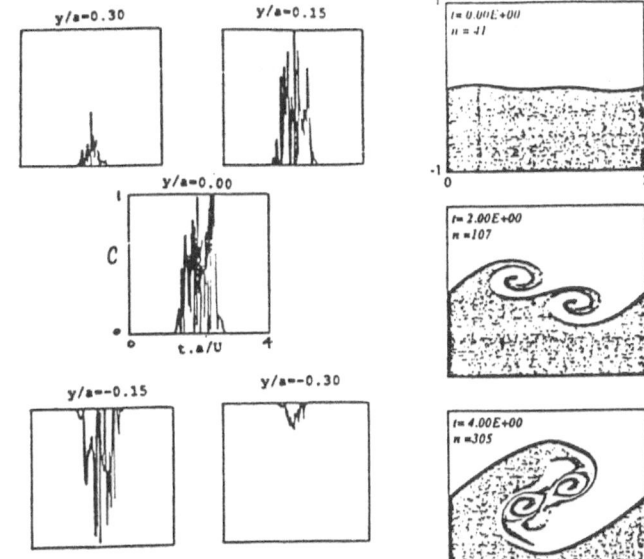

Figure 8. Concentration profiles at different heights above the mixing plane for Re=50 and Sc=1.

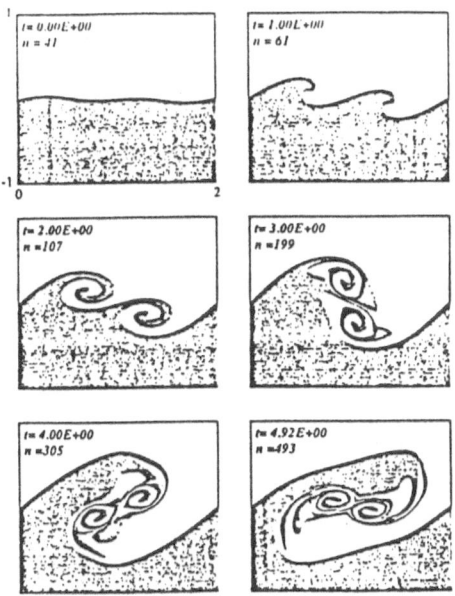

Figure 9. Roll-up and pairing of two vortical structures with first subharmonic disturbance. Fundamental plus first subharmonic.

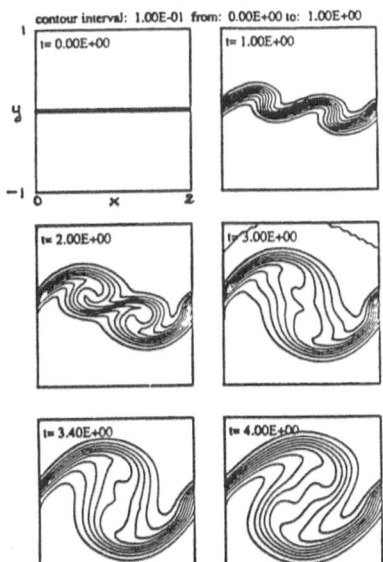

Figure 10. Concentration contours during roll-up and pairing process. Fundamental plus first subharmonic.

THE STRUCTURE AND EXTINCTION OF TUBULAR PREMIXED LAMINAR FLAMES

M. D. Smooke
Department of Mechanical Engineering
Yale University
New Haven, CT

V. Giovangigli
Centre de Mathématiques Appliquées
Unité de Recherche 756 du CNRS
Ecole Polytechnique
91128 Palaiseau Cedex, France

Introduction

Strained premixed laminar flames have played an important role in recent theories of turbulent premixed combustion. The reacting surface in these theories can be viewed as being composed of a number of laminar flamelets (see e.g., Libby and Bray, 1980, Libby and Williams, 1982 and Bray, Libby and Moss, 1984) The flamelets are often modeled by considering counterflowing streams of reactants or counterflowing streams of reactants and products. Most of the work that has been published on stretched premixed laminar flames has focused on a single reactant stream impinging on an adiabatic wall or two counterflowing reactant streams emerging from two counterflowing coaxial jets. In both cases the imposed flow field acts in such a way as to only stretch the flame sheet. The flame is not subjected to the effects of curvature. However, in a series of papers Chomiak (1970, 1972, 1977) proposed that in high Reynolds number turbulent premixed combustion the flame front could be "caught" by stretching vortex tubes. If the flame is captured by these vortices it will have a cylindrical shape with the vortex tube as its core. In such cases the flame will be subjected not only to the effects of stretch but also to the effects of curvature.

Experimentally, the effects of stretch and curvature can be studied in tubular (cylindrical) flames (see Figure 1). Flames of this type can be stabilized experimentally by forcing a premixed fuel/air mixture through the walls of an open porous cylindrical tube or by ejecting the mixture radially inward through a converging nozzle. The tubular flame is stabilized in the interior of the cylindrical regions. Such flames have been investigated experimentally by Sakai and Ishizuka (1988) using the open cylindrical tube and by Kobayashi and Kitano (1988) using the inward radial-flow nozzle burner. Theoretical analyses have been considered by Takeno and Ishizuka (1986) when the flow field is rotating and by Kobayashi et al. (1988) and Kitano and Kobayashi (1987) when the flow field is nonrotating.

In this paper we compare a series of detailed transport-finite rate chemistry calculations of tubular premixed methane-air flames with the experimental work of Kobayashi and

Kitano (1988). We focus on both flame structure and flame extinction. We then compare our results with a corresponding sequence of calculations in which only stretch effects are important. In the next section we outline the tubular flame model and in Section 3 we discuss the computational approach used in the calculations. Numerical results are contained in Section 4.

2. Problem Formulation

Our model for tubular flames starts with the elliptic form of the two-dimensional conservation equations in cylindrical coordinates. Anticipating that all of the dependent variables except the axial velocity can be written in terms of functions of the radial coordinate only, we seek a similarity solution in which

$$u = u(r), \quad v = zV(r), \quad Y_k = Y_k(r) \text{ and } T = T(r). \tag{2.1}$$

Upon substituting the velocity expressions into the continuity and momentum equations, we find that the reduced pressure gradient

$$\frac{1}{z}\frac{\partial p}{\partial z} = J, \tag{2.2}$$

where J is constant. Using this result together with (2.1), the two-dimensional set of governing equations can be reduced to a nonlinear two-point boundary value problem in the radial direction. The governing equations for mass, momentum, chemical species and energy can be written in the form

$$\frac{dU}{dr} + r\rho V = 0, \tag{2.3}$$

$$\frac{d}{dr}\left(r\mu\frac{dV}{dr}\right) - U\frac{dV}{dr} - r\rho V^2 - rJ = 0, \tag{2.4}$$

$$-\frac{d}{dr}(r\rho Y_k V_{k_r}) - U\frac{dY_k}{dr} + r\dot{w}_k W_k = 0, \quad k = 1, 2, \ldots, K, \tag{2.5}$$

$$\frac{d}{dr}\left(r\lambda\frac{dT}{dr}\right) - c_p U\frac{dT}{dr} - r\sum_{k=1}^{K}\rho Y_k V_{k_r} c_{pk}\frac{dT}{dr} - r\sum_{k=1}^{K}\dot{w}_k W_k h_k = 0, \tag{2.6}$$

and

$$\rho = \frac{p\overline{W}}{RT}, \tag{2.7}$$

where we have made use of the notation

$$U = r\rho u. \tag{2.8}$$

In addition to the variables already defined r and z denote the independent spatial coordinates; T, the temperature; Y_k, the mass fraction of the kth species; p, the pressure; u and v the radial and the axial components of the velocity, respectively; ρ, the mass density; W_k, the molecular weight of the kth species; $\overline{W}$, the mean molecular weight of the mixture; R, the universal gas constant; λ, the thermal conductivity of the mixture; c_p, the constant pressure heat capacity of the mixture; c_{p_k}, the constant pressure heat capacity of the kth species; $\dot{w}_k$, the molar rate of production of the kth species per unit volume; h_k the specific enthalpy of the kth species; μ the viscosity of the mixture and V_{kr}, the diffusion velocity of the kth species in the r direction. The form of the chemical production rates and the diffusion velocities can be found in detail in Kee et al. (1980, 1983).

To complete the specification of the problem, boundary conditions must be imposed at the axis of symmetry $(r = 0)$ and the cylinder wall $(r = R_w)$. At $r = 0$ we have

$$U = 0, \tag{2.9}$$

$$\frac{dV}{dr} = 0, \tag{2.10}$$

$$\frac{dY_k}{dr} = 0, \quad k = 1, 2, \ldots, K, \tag{2.11}$$

$$\frac{dT}{dr} = 0, \tag{2.12}$$

and at $r = R_w$

$$V = 0, \tag{2.13}$$

$$Y_k = Y_{k_w}, \quad k = 1, 2, \ldots, K, \tag{2.14}$$

$$T = T_w. \tag{2.15}$$

The mass fractions $Y_{k_w}, k = 1, 2, \ldots, K$ and the temperature T_w at the cylinder wall are specified quantities.

3. Method of Solution

Solution of the equations in (2.3-2.15) proceeds with an adaptive nonlinear boundary value method. The solution procedure has been discussed in detail elsewhere (Smooke, 1982) and we outline only the essential features here. Our goal is to obtain a discrete solution of the governing equations on the mesh $\mathcal{M}$

$$\mathcal{M} = \{0 = r_0 < r_1 < \ldots < r_m = R_w\}. \tag{3.1}$$

With the continuous differential operators replaced by difference expressions, we convert the problem of finding an analytic solution of the governing equations to one of finding an approximation to this solution at each point of the mesh $\mathcal{M}$. We seek the solution W^* of the nonlinear system of difference equations

$$F(W) = 0. \tag{3.2}$$

For an initial solution estimate W^0 that is sufficiently "close" to W^*, the system of equations in (3.2) can be solved by Newton's method. We write

$$J(W^k)\left(W^{k+1} - W^k\right) = -\lambda_k F(W^k), \quad k = 0, 1, \ldots, \tag{3.3}$$

where W^k denotes the kth solution iterate, λ_k the kth damping parameter $(0 < \lambda \le 1)$ which should be chosen to ensure a reduction in the size of the Newton corrections at each iteration (see, e.g., Deuflhard, 1974) and $J(W^k) = \partial F(W^k)/\partial W$ the Jacobian matrix. A system of linear block tridiagonal equations must be solved at each iteration for corrections to the previous solution vector. In practice, a modified Newton method is employed in which the Jacobian is re-evaluated periodically (Smooke, 1983).

In addition to the determination of a single solution, we are interested in the structure of tubular flames for increasing values of J. We note, however, that as the value of J increases to the point where the flame nears extinction, the maximum value of the temperature decreases. It is near the extinction limit, however, that the numerical calculations become increasingly difficult. In particular, at the extinction point (a simple turning point) the Jacobian of the system is singular. To alleviate the computational difficulties, a modified form of the governing equations is solved (Keller, 1977). We introduce J as a new dependent variable. The vector of dependent variables can now be considered functions of a new independent parameter s. If we define

$$Z(s) = (W(s), J(s))^T, \tag{3.4}$$

then the new problem we want to solve is given by

$$G(Z, s) = 0, \tag{3.5}$$

where

$$G(Z, s) = \begin{bmatrix} F(W(s), J(s)) \\ N(W(s), J(s), s) \end{bmatrix} = \begin{bmatrix} F(Z(s)) \\ N(Z(s), s) \end{bmatrix}, \tag{3.6}$$

and where N is an arbitrary normalization. The normalization is chosen such that s approximates the arclength of the solution branch in the space (W, J).

The Jacobian of the new system can be written in the form

$$J(Z, s) = G_Z(Z, s) = \begin{bmatrix} F_W(W(s), J(s)) & F_J(W(s), J(s)) \\ N_W(W(s), J(s), s) & N_J(W(s), J(s), s) \end{bmatrix}, \tag{3.7}$$

where F_W, F_J, N_W and N_J denote the appropriate partial derivatives. It can be shown that at a simple turning point, even though F_W is singular J is not. In addition, given a solution $Z(s)$ we can determine a new predicted value for $Z(s + \delta s)$ by forming

$$Z(s + \delta s) = Z(s) + \frac{dZ}{ds}\delta s, \tag{3.8}$$

where dZ/ds is determined from

$$J\frac{dZ}{ds} = -\begin{bmatrix} 0 \\ N_s(Z(s), s) \end{bmatrix}. \tag{3.9}$$

The choice and form of the normalization boundary conditions as well as implementation issues are discussed in Giovangigli and Smooke (1987, 1989).

In the computation of the first tubular flame as well as in subsequent continuation calculations, the mesh must be refined to resolve the high activity regions of the dependent solution components. The mesh in the first calculation is determined by subequidistributing the difference in the components of the discrete solution and its gradient between adjacent mesh points (Smooke, 1982). Upon denoting the vector of N dependent solution components by $\tilde{z} = [\tilde{z}_1, \tilde{z}_2, \ldots, \tilde{z}_N]^T$, we seek a mesh $\mathcal{M}$ such that

$$\int_{r_j}^{r_{j+1}} \left| \frac{d\tilde{z}_i}{dr} \right| \, dr \leq \delta \left| \max_{0 \leq r \leq R_w} \tilde{z}_i - \min_{0 \leq r \leq R_w} \tilde{z}_i \right| \qquad \begin{array}{l} j = 0, 1, \ldots, m-1 \\ i = 1, 2, \ldots, N \end{array}, \tag{3.10}$$

and

$$\int_{r_j}^{r_{j+1}} \left| \frac{d^2\tilde{z}_i}{dr^2} \right| \, dr \leq \gamma \left| \max_{0 \leq r \leq R_w} \frac{d\tilde{z}_i}{dr} - \min_{0 \leq r \leq R_w} \frac{d\tilde{z}_i}{dr} \right| \qquad \begin{array}{l} j = 1, 2, \ldots, m-1 \\ i = 1, 2, \ldots, N \end{array}, \tag{3.11}$$

where δ and γ are small numbers less than one and the maximum and minimum values of $\tilde{z}_i$ and $d\tilde{z}_i/dr$ are obtained from a converged numerical solution on a previously determined mesh. We also impose the added constraint that the mesh produced by employing (3.10) and (3.11) be locally bounded. We require

$$\frac{1}{A} \leq \frac{h_j}{h_{j-1}} \leq A, \quad j = 2, 3, \ldots, m, \tag{3.12}$$

where A is a constant ≥ 1. This smooths out rapid changes in the size of the mesh intervals. The advantage of this gridding procedure is that most of the expensive Jacobian evaluations are performed on relatively coarse grids and once the grid is sufficiently refined, Newton's method usually converges with only one or two iterations. The method also takes into account every component of the solution in forming a new mesh.

This gridding procedure is not very efficient, however, for subsequent continuation calculations. For example, if the flame shifts its position from that of a previous calculation, then the number of grid points in the mesh produced by (3.10) and (3.11) will always be equal to or larger than the number of points in the previous mesh. This can quickly result in an inefficient solution algorithm. To avoid these problems we construct a mesh based upon the largest weight function in each interval. Specifically, after solving the governing equations on the previously determined mesh, we form in each interval

$$\tilde{w}_j = \max_{1 \leq i \leq 2N+1} w_{i,j}, \quad j = 0, 1, \ldots, m, \tag{3.13}$$

where

$$w_{i,j} = \left| \frac{d\tilde{z}_i}{dr} \right| / \left\{ \delta \left| \max_{0 \leq r \leq R_w} \tilde{z}_i - \min_{0 \leq r \leq R_w} \tilde{z}_i \right| \right\} \qquad \begin{array}{l} j = 0, 1, \ldots, m \\ i = 1, 2, \ldots, N \end{array}, \tag{3.14}$$

$$w_{N+i,j} = \left| \frac{d^2\tilde{z}_i}{dr^2} \right| / \left\{ \gamma \left| \max_{0 \leq r \leq R_w} \frac{d\tilde{z}_i}{dr} - \min_{0 \leq r \leq R_w} \frac{d\tilde{z}_i}{dr} \right| \right\} \qquad \begin{array}{l} j = 0, 1, \ldots, m \\ i = 1, 2, \ldots, N \end{array}, \tag{3.15}$$

and

$$w_{2N+1,j} = \frac{N_0}{R_w}, \tag{3.16}$$

where N_0 is a given integer. The new mesh

$$\mathcal{M}^* = \{0 = r_0^* < r_1^* < \ldots < r_{m^*}^* = R_w\}, \tag{3.17}$$

determined by imposing the condition

$$\int_{r_j^*}^{r_{j+1}^*} \tilde{w}_j \ dr = 1, \quad 0 \le j \le m^* - 2, \tag{3.18}$$

along with a local boundedness criterion to help regularize $\tilde{w}$ (see, e.g., Giovangigli and Smooke, 1989). The governing equations are solved on this new mesh and then another continuation step is taken. This gridding strategy is more suited to continuation methods than the procedure in (3.10) and (3.11) since it optimizes the number of points and determines the new grid in only one pass. In addition, the method also takes into account every component of the solution in forming a new mesh.

4. Numerical Results

We have applied the solution methods discussed in the previous section to determine the structure and extinction of a set of methane-air tubular flames in which $R_w = 1.5$ cm. A 79 reaction, 26 species kinetics mechanism was used in the calculations (see also, Puri et al., 1987). It includes both a C_1 and a C_2 reaction branch. Starting estimates for the computations were obtained from previously computed double premixed laminar methane-air flames (Giovangigli and Smooke, 1987, 1989). Each continuation calculation contained between 110-135 adaptively determined grid points. All of the calculations were performed on a Multiflow TRACE 14/200 computer.

From the point of view of the relative location and peak values of the species concentrations and the temperature, the structure of a given tubular premixed laminar flame is closely related to the structure of a corresponding twin premixed laminar flame. From an extinction point of view, however, tubular premixed flames exhibit a substantially different behavior than twin premixed flames (no curvature effects) as the equivalence ratio (ϕ) is varied. Experimentally, Kobayaski and Kitano (1988) have used the inward radial-flow burner to determine the extinction behavior of premixed methane-air tubular flames. In particular, they have measured the extinction distance (the distance of the center of the luminous flame zone from the centerline of the cylindrical region) as a function of the equivalence ratio. Results of their experimental measurements are illustrated in Figure 2. We note the dramatic variation in the extinction distance as a function of the equivalence ratio. In particular, the minimum distance occurs for slightly lean flames. It then increases as the flame becomes either leaner or richer. The corresponding numerical results based upon our arclength continuation calculations are contained in Figure 3. We also notice the minimum extinction distance for flames with $\phi \approx 0.85$ as well as the increase in the distances for smaller or larger values of ϕ. An identical set of calculations for twin premixed methane-air flames (no curvature effects) are illustrated in Figure 4. For these flames we see a very slight variation in extinction distance as ϕ is varied.

It is interesting to compare the results in Figures 2-4 with the theoretical studies of Kobayashi et al. (1988). They applied asymptotic techniques to explain the extinction behavior of tubular flames as a function of the Lewis number of the fuel (the ratio of the thermal diffusivity to the mass diffusivity). They predicted that curvature effects decrease the "flame strength", i.e., its resistance to extinction. In particular, they found that extinction distances for twin premixed flames are noticeably smaller as compared

to the tubular case. Our calculations illustrate this point quite dramatically. They als
predicted a "U shaped" extinction distance versus equivalence ratio curve similar to th
one illustrated in Figure 3. Finally, they showed that when the Lewis number of the fue
is less than one, then, due to enhanced preferential diffusion effects, the equivalence rati
of the minimum extinction distance shifts slightly to the lean side for methane-air flame
($Le \approx 0.96$). Again these results are confirmed in the experimental results reported i
Figure 2 and in the computational results illustrated in Figure 3.

References

1 Libby, P. A. and Bray, K. N. C (1980). *Comb. and Flame*, **39**, p. 33.

2 Libby, P. A. and Williams, F. A. (1982). *Comb. and Flame*, **44**, p. 287.

3 Bray, K. N. C., Libby, P. A. and Moss, J. B. (1984). *Comb. Sci. and Tech.*, **41**, p. 143

4 Chomiak J. (1970). *Comb. and Flame*, **15**, p. 319.

5 Chomiak J. (1972). *Comb. and Flame*, **18**, p. 429.

6 Chomiak J. (1977). *Sixteenth Symposium (International) on Combustion*, p. 1665.

7 Sakai, Y. and Ishizuka, S. (1988). "Structures of the Tubular Flames in a Rotating and Nonrotaing Flow Fields," preprint.

8 Kobayashi, H. and Kitano, M. (1988). "Extinction Characteristics of a Stretched Cylindrical Premixed Flame," preprint.

9 Takeno, T. and Ishizuka, S. (1986). *Comb. and Flame*, **64**, p. 83.

10 Kobayashi, H., Kitano, M and Otsuka, Y. (1988) *Comb. Sci. and Tech.*, **57**, p. 17.

11 Kitano, M. and Kobayashi, H. (1987) submitted to *Comb. and Flame*.

12 Kee, R. J., Miller, J. A. and Jefferson, T. H. (1980). "CHEMKIN: A General-Purpose, Transportable, Fortran Chemical Kinetics Code Package," Sandia National Laboratories Report, SAND80-8003.

13 Kee, R. J., Warnatz, J., and Miller, J. A. (1983). "A Fortran Computer Code Package for the Evaluation of Gas-Phase Viscosities, Conductivities, and Diffusion Coefficients," Sandia National Laboratories Report, SAND83-8209.

14 Smooke, M. D. (1982). *J. Comp. Phys.* **48**, p. 72.

15 Smooke, M. D., Miller, J. A. and Kee, R. J. (1983). *Comb. Sci. and Tech.* **34**, p. 79.

16 Deuflhard, P. (1974). *Numer. Math.* **22**, p. 289.

17 Smooke, M. D. (1983). *J. Opt. Theory and Appl.* **39**, p. 489.

18 Keller, H. B. (1977). "Numerical Solution of Bifurcation and Nonlinear Eigenvalue Problems," in Applications of Bifurcation Theory, P. Rabinowitz, Ed., Academic Press, New York.

9 Giovangigli, V. and Smooke, M. D. (1987). *Comb. Sci. and Tech.*, **53**, p. 23.

0 Giovangigli, V. and Smooke, M. D. (1989) "Adaptive Continuation Algorithms with Application to Combustion Problems," in press *App. Num. Math.*

1 Smooke, M. D. (1986). *AIChE J.*, **32**, p. 1233.

2 Puri, I. K., Seshadri, K., Smooke, M. D., and Keyes, D. E. (1987). *Comb. Sci. Tech.* **56**, p. 1.

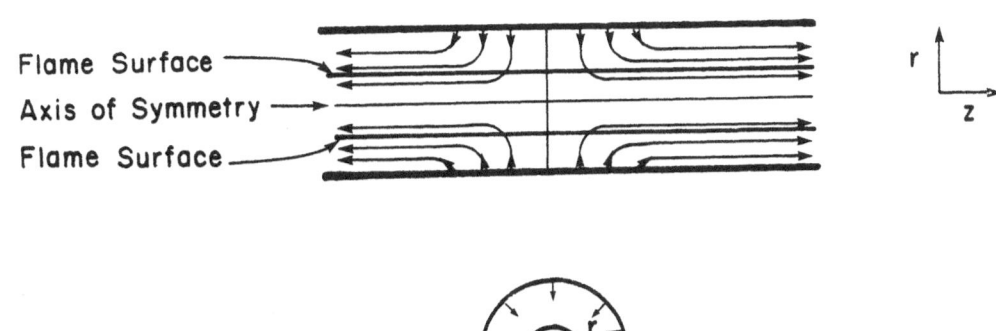

Figure 1

Schematic of the tubular premixed laminar flame.

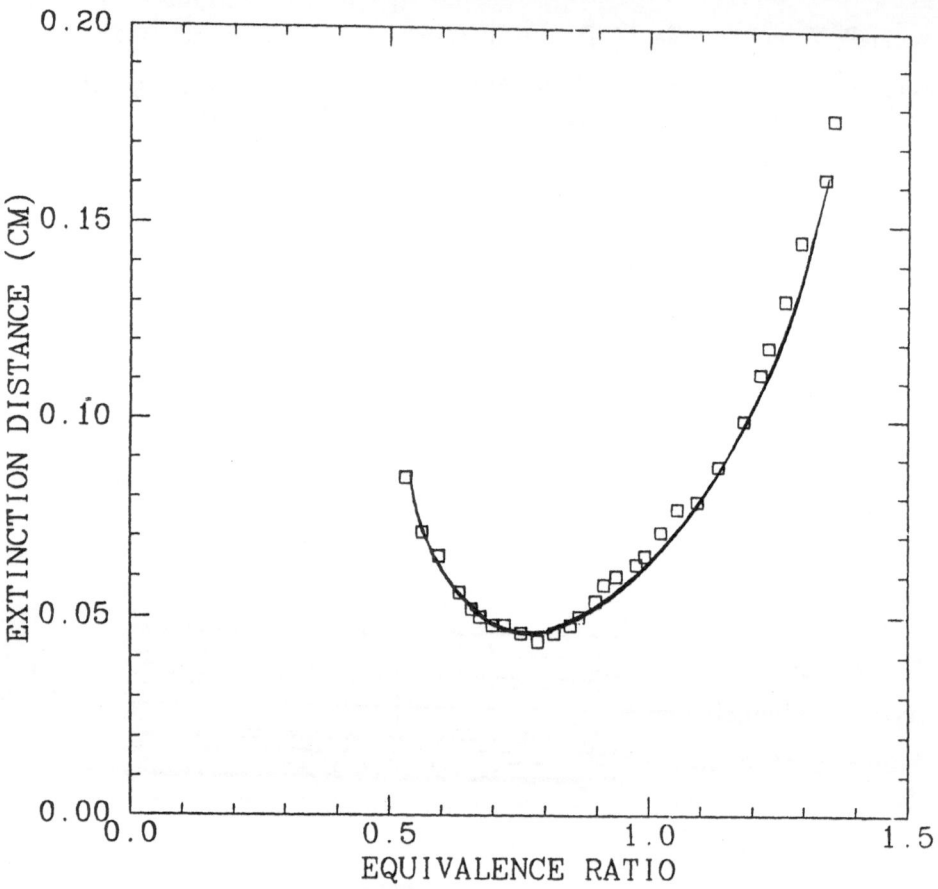

Figure 2

Experimentally determined extinction distances for premixed tubular methane-air flames.

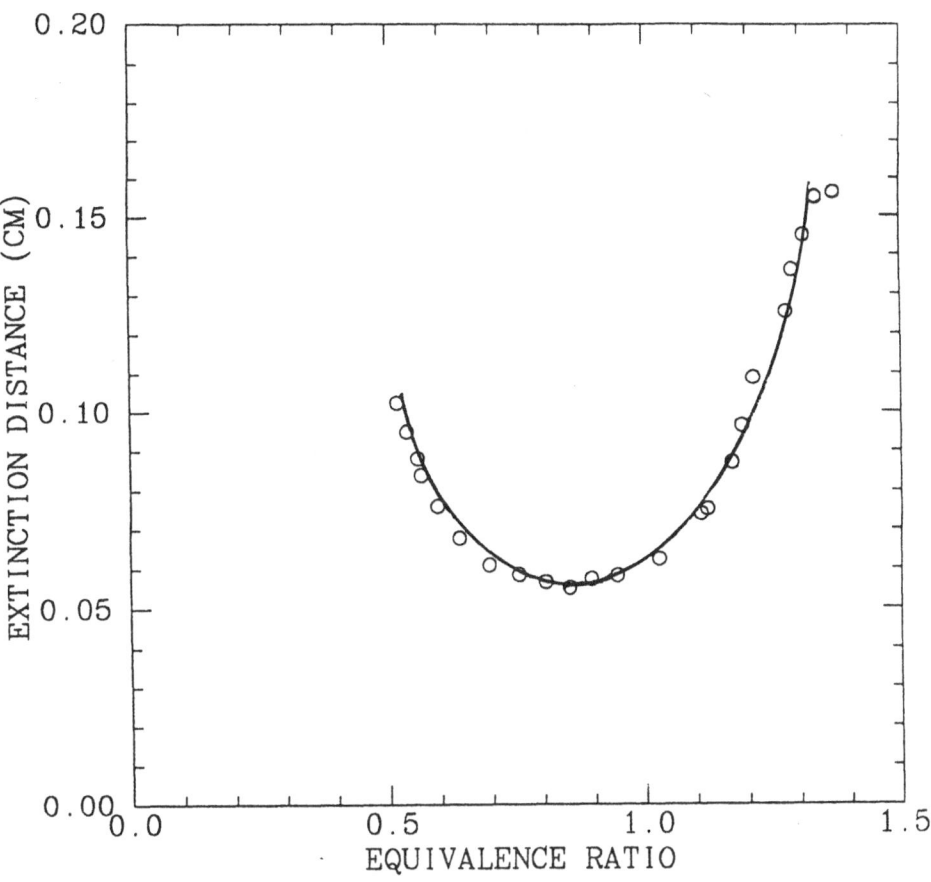

Figure 3

Calculated extinction distances for premixed tubular methane-air flames.

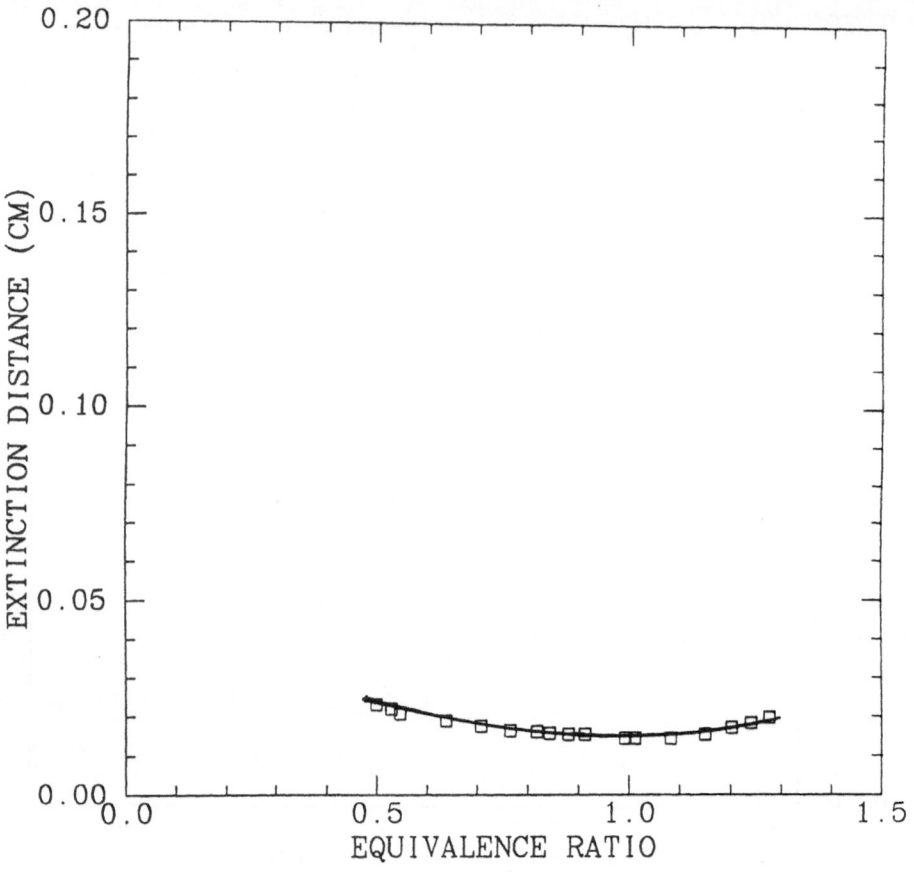

Figure 4

Calculated extinction distances for twin premixed counterflow methane-air flames

(no curvature)

ON THE NONLINEAR GALERKIN METHODS

Roger Temam

Laboratoire d'Analyse Numérique

CNRS et Université Paris-Sud, Bâtiment 425, 91405 Orsay.

Introduction.

The object of this lecture is to report on a progressing work that attempts to modelize some phenomena in turbulence and that produces a new insight into computations and a promising algorithm.

Part I of this lecture contains a description of the equations in combustion for which the methodology is relevant and the theoretical background in relation with inertial manifolds. Part II is devoted to the numerical aspects and contains in particular the description and some motivations of the new algorithm that we advocate, namely the nonlinear Galerkin method.

The material in this article relies essentially on three articles [1-3]; see also for the theoretical background [4-6].

Part I : Inertial Manifolds and the Modelling of Turbulence.

I.1. The combustion equations

The equations in combustion which we consider include the following ones :

(i) *The Kuramoto-Sivahinsky equation*

$$\frac{\partial u}{\partial t} + \frac{\partial^4 u}{\partial x^4} + \frac{\partial^2 u}{\partial x^2} + \frac{1}{2}\left(\frac{\partial u}{\partial x}\right)^2 = 0, \quad -\frac{L}{2} < x < \frac{L}{2},\ t > 0. \tag{1.1}$$

Related forms of this equation can also be considered as in [7].

(ii) *The thermodiffusive model.*

We can consider reaction-diffusion equations alone or such equations coupled with the heat equation. A typical system is, for example (see Buckmaster-Ludford [8]) :

$$\frac{\partial \theta}{\partial t} - \Delta \theta = \omega(\theta, \psi) \tag{1.2}$$

$$\frac{\partial \psi}{\partial t} - \frac{1}{Le}\Delta \psi = -\omega(\theta, \psi) \tag{1.3}$$

where the unknowns are the reduced temperature θ and the concentration of the reactant ψ ; Le is the Lewis number and the normalized reaction rate is given by the Arrhenius low

$$\omega(\theta, \psi) = \frac{\beta^{m+1}}{2m!Le^m}\psi^m \, exp \, \left(-\frac{\beta(1-\theta)}{1+\gamma(\theta-1)} \right) ;$$

here $m \in \mathcal{N}$ is the order of the reaction, $\beta > 0$ is the reduced activation energy and $\gamma = 1 - T_u/T_b$ is a heat release parameter.

(iii) *The constant density model.*

For a one-step chemistry : $mR \to mP$, the low Mach number combustion equations reduce, when we assume the density constant, to the following nondimensional system

$$\frac{\partial u}{\partial t} - Pr\Delta u + (u \cdot \nabla)u + \nabla p = 0 \tag{1.4}$$

$$div \ u = 0 \tag{1.5}$$

$$\frac{\partial \theta}{\partial t} - \Delta \theta + (u.\nabla)\theta = \omega(\theta, \psi) \tag{1.6}$$

$$\frac{\partial \psi}{\partial t} - \frac{1}{Le}\Delta \psi + (u \cdot \nabla)\psi = -\omega(\theta, \psi) \tag{1.7}$$

where ω, θ and ψ are as above, u is the velocity and p the pressure ; Pr, the Prandtl number, is the ratio of the kinematic viscosity to the thermal conductivity. We recall that the fluid mechanic equations (1.4)(1.5) can be resolved independently of the chemistry.

For any of the models above the equations must be supplemented by the proper boundary (and initial) conditions that depend on the experiment under considerations ; they will not be discussed here.

I.2. The general framework.

The mathematical setting of the equations above can be handled by considering an abstract equation of the form

$$\frac{dU}{dt} + AU + R(U) = 0. \tag{1.8}$$

Here U is the set of unknowns under consideration (i.e. respectively in the models above u ; θ and ψ ; u, θ and ψ). Equation (1.8) is an evolution equation in a function (Hilbert) space

H $(U(t) \in H)$ and A is a linear positive (i.e. dissipative) unbounded operator with domain $D(A) \subset H$. The operator R refers to the nonlinear terms or to lower order linear terms.

The presentation given here relies on the utilization of the eigenfunctions of A. For the sake of simplicity one can assume that the boundary conditions associated to one of the models above is space periodicity so that the eigenfunctions are just proper combinations of sines and cosines. In all cases we denote by w_j, $j \in \mathcal{N}$ the sequence of eigenvectors of A and λ_j, are the corresponding eigenvalues :

$$Aw_j = \lambda_j w_j,$$
$$0 < \lambda_1 \le \lambda_2 \le ..., \quad \lambda_j \to \infty \ as \ j \to \infty. \tag{1.9}$$

I.3. Inertial Manifolds.

We know from dynamical systems theory (see [6] and the reference therein) that all solutions of equation (1.8) converge to an invariant set $\mathcal{A}$ called the global attractor. This set is the mathematical object that represents turbulent flows governed by (1.8). It contains in particular all the stationary solutions and their unstable manifolds, the orbits of time periodic or time quasi-periodic solutions, etc.

The concept of Inertial Manifold (IM) was introduced in [4] ; it may have some relevance to turbulence. An IM is a smooth positively invariant finite dimensional manifold $\mathcal{M}$ in the function space H with attracts all orbits at an **exponential rate**. Positive invariance means that any solution of (1.8) with initial data $U(0)$ belonging to $\mathcal{M}$ lies in $\mathcal{M}$ for all positive time. Hence by restricting (1.8) to $\mathcal{M}$ we obtain a finite dimensional system called the **inertial system**. Such a set $\mathcal{M}$, when it exists, contains the attractor $\mathcal{A}$. Also, since the convergence to $\mathcal{M}$ is exponential, most of the time (i.e. after a "short" transient period) the orbits lie on $\mathcal{M}$.

Let m be fixed. We consider the space spanned by the eigenvectors $w_1, ..., w_m$, and denote by P_m the projector in H onto this space. Let $Q_m = I - P_m$ be the complement, so that $P_m H$ and $Q_m H$ are two orthogonal spaces, and any vector U can be decomposed into its components on these spaces

$$U = Y_m + Z_m, \quad Y_m = P_m U, \quad Z_m = Q_m U. \tag{1.10}$$

Inertial Manifolds have been searched as the graph of a function Φ from $P_m H$ into $Q_m H$, i.e. droping the indices m,

$$Z = \Phi(Y). \tag{1.11}$$

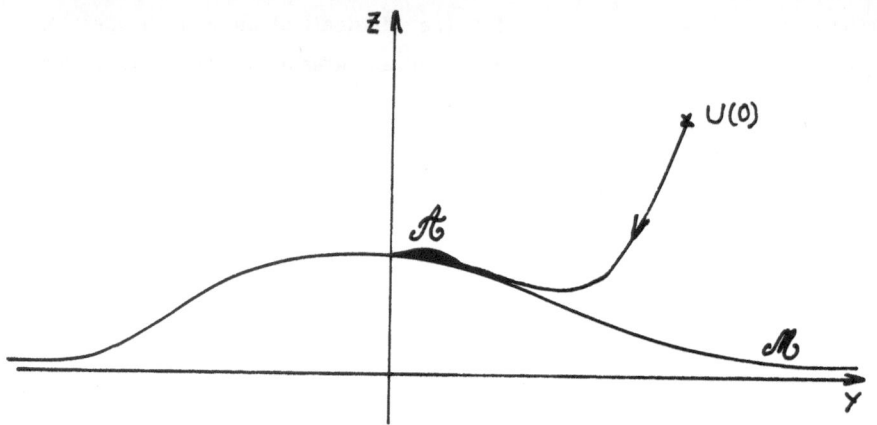

When an inertial manifold of the form (1.11) exists it relation to turbulence follows from the following observation :

The law (1.11) is enforced on $\mathcal{M}$, i.e. for all time for orbits lying on $\mathcal{M}$ and after a "short" transient period for all orbits

$$Z(t) = \Phi(Y(t)).\tag{1.12}$$

Alternatively the small eddies (represented by Z) are slaved by the large eddies (represented by Y). If

$$U(x,t) = \sum_{j=1}^{m} U_j(t)w_j(x)$$

is the Fourier series expansion of the solution U of (1.8) in the eigenvectors basis w_j, then (1.12) means that

$$U_j(t) = \Phi_j(U_1(t), ..., U_m(t)), \quad j \geq m+1,\tag{1.13}$$

where Φ_j, $j \geq m+1$ are the collection of functions corresponding to the components of Φ.

The existence of such an Inertial Manifold and of a "slaving law" of the form (1.12),(1.13) has been proved for the Kuramoto-Sivasinsky equation and for some reaction diffusion equations (see [4][5]). It has not been proved nor disproved for the Navier-Stokes equations (1.4),(1.5), or for the constant density models. Hence we do not know in this case if a slaving law like (1.12)(1.13) is actually valid. We shall see however that such a relation (between small and large eddies) is **approximately** true.

II. The Computational Aspect.

II.1. Approximate Inertial Manifolds.

While the existence of Inertial Manifolds is not proved in several relevant cases, approximate forms of these manifolds have been derived. This means that instead of (1.12) or (1.13) we shall have

$$| Z_m(t) - \Phi(Y_m(t)) | \leq \epsilon_m, \qquad (2.1)$$

where ϵ_m will usually depend on m and on the equation but not on the particular orbit (i.e. the initial data).

The simplest of these manifolds is the flat space $\mathcal{M}_0 = P_m H$ corresponding to $Z_m = 0$. It was shown in [8] that even if the small wavelengths are artificially excited at time 0 (i.e. $Z_m(0)$ is large), then eventually $Z_m(t)$ becomes small. For instance for the $2D$ Navier-Stokes equations, and up to a logarithmic correction

$$| Z_m(t) | \leq k\eta^2, \qquad (2.2)$$

where k is a constant and $\eta = (\lambda_1/\lambda_{m+1})^{1/2}$ (which is small if m is sufficiently large).

A more interesting, nonflat, manifold is also derived in [8]; this is the manifold $\mathcal{M}_1$ of equation

$$AZ + Q_m R(Y) = 0. \qquad (2.3)$$

It is related to (1.8) as follows : by projecting equation (1.8) on the spaces $P_m H$ and $Q_m H$, we obtain the following coupled system equivalent to (1.8)

$$\begin{aligned}
\frac{dY_m}{dt} + AY_m + P_m R(Y_m + Z_m) &= 0 \\
\frac{dZ_m}{dt} + AZ_m + Q_m R(Y_m + Z_m) &= 0.
\end{aligned} \qquad (2.4)$$

Since Z_m is small, $Q_m R(| Y_m + Z_m |) \simeq Q_m R(Y_m)$ and it is shown also in [8] that dZ_m/dt can be neglected up to a certain level of accuracy, so that the second equation (2.4) reduces then to (2.3). Furthermore it was shown that the distance of the attractor to this manifold $\mathcal{M}_1$ is bounded by

$$k\eta^3,$$

so that we gain one order of accuracy from $\mathcal{M}_0$ (see (2.2)). Hence the attractor and the dynamics are closer from $\mathcal{M}_1$ that from $\mathcal{M}_0$.

Another approximate inertial manifolds (AIMs) that can be considered is the manifold of equation (see E. Titi [11] for the Navier-Stokes equations) :

$$AZ + Q_m R(Y + Z) = 0 \qquad (2.5)$$

This manifold contains **all the stationary solutions** of (1.8) and the attractor is at most at a distance $k\eta^4$ of it. Other AIM's have been constructed by M.Marion [10] for reaction diffusion equations. A whole sequence of AIMs producing approximations of the attractor of orders η^j, $j \geq 1$, have been constructed in [12].

II.2. The Nonlinear Galerkin Methods.

A turbulent (reacting or nonreacting) flow remains time dependent and does not converge to a stationary state. Its orbits, in the phase (function) space wanders around the attactor $\mathcal{A}$. For the computation of such flows, we contend that computational efficiency can be gained if we can construct approximate solutions lying close to the attractor. This is the motivation of the nonlinear Galerkin methods, that we now describe. The following does not apply strictly speaking to finite differences but it does apply with some slight modifications (see R. Temam [14]).

With spectral methods, wavelets of finite elements, the Galerkin method consists in projecting the equation bo be considered on a finite dimensional space associated to the basis functions. For instance a Galerkin method based on $w_1, ..., w_m$, consists in projecting the equation under consideration on $P_m H = \mathcal{M}_0$. Hence for (1.8), the approximate equation

$$\frac{dY_m}{dt} + AY_m + P_m R(Y_m) = 0 \tag{2.6}$$

corresponding (compare to (2.4)) to the approximation $Z_m = 0$.

However, although Z_m is small it may not be legitimate to cancel it if large time integration of turbulent flows are to be considered. What we propose here is a slight modification of (2.6) corresponding instead to projecting the equation on the manifold $\mathcal{M}_1$. We obtain the following modification of (2.6)

$$\frac{dY_m}{dt} + AY_m + P_m R(Y_m + Z_m) = 0$$
$$AZ_m + Q_m R(Y_m) = 0 \tag{2.7}$$

This is (2.6) with the "small" correction term

$$Z_m = -A^{-1} Q_m R(Y_m).$$

This algorithm is called a Nonlinear Galerkin Method since we are projecting the equation on a linear manifold (namely $\mathcal{M}_1$) instead of projecting it on the linear space $\mathcal{M}_0 = P_m H$. The theoretical justification of the algorithm follows from the results recalled above ; $\mathcal{M}_1$ is closer from the attractor than $\mathcal{M}_0$. The practical justification comes from the numerical tests : for the $2D$ Navier-Stokes equations, for a similar accuracy, a gain of computing time of 20 to 50 % has been observed for the nonlinear Galerkin method over the linear (usual) Galerkin method.

In practice we do not want to handle a correction vector Z_m of infinite dimension. Hence we consider instead the following sequence of m vectors, $w_{m+1}, ..., w_{2m}$, and a correction Z_m belonging to the space spanned by $w_{m+1}, ..., w_{2m}$:

$$Y_m(t) = \sum_{j=1}^{m} g_{jm}(t)w_j,$$

$$Z_m(t) = \sum_{j=m+1}^{2m} g_{jm}(t)w_j,$$

and the second equation (2.7) is replaced by

$$AZ_m + (P_{2m} - P_m)R(Y_m) = 0, \qquad (2.8)$$

while the first equation (2.7) is unchanged.

The figures hereafter are borrowed from [13]. For the example treated the exact solution was known ($2D$ Navier-Stokes equations with space periodicity) so that accuracy can be easily verified. Figure 1 shows that the accuracy for the solutions obtained by the Nonlinear Galerkin Method (NLG) and by the usual Galerkin method are undistinguishable. Figure 2 shows the time needed for each method, and Figure 3 shows the gain of CPU computing time, of the order of 50% in this case. Further numerical tests appear in [2] where the gain in computing time is of the order of 20 to 40 % .

Conclusion. Due to the size of computations in CFD and combustion, the results presented here show that the method can lead to important increases in computing efficiency.

References.

[1] C. Foias, O. Manley and R. Temam, On the interaction of small and large eddies in two-dimensional turbulent flows, *Math. Model. and Numer. Anal. (M2AN) 22*, 1988, 93-114.

[2] R. Temam, Dynamical systems, turbulence and the numerical solution of the Navier-Stokes equations, Proc. 11th International Conference on Numerical Methods in Fluid Dynamics, D.L. Dwoyer, M.Y. Hussaini and R. Voigt Eds, Lecture Notes in Physics 323, Springer Verlag, 1989.

[3] M. Marion and R. Temam, Nonlinear Galerkin Methods, *SIAM J. Num. Anal.*, to appear.

[4] C. Foias, G. Sell and R. Temam, Inertial manifolds for nonlinear evolutionary equations, *J. Diff. Equ. 73*, 1988, 309-353.

[5] C. Foias, B. Nicolaeko, G. Sell and R. Temam, Inertial manifolds for the Kuramoto-Sivashinsky equation and an estimate of their lowest dimension, *J. Math. Pures Appl. 67*, 1988, 197-226.

[6] R. Temam, *Infinite Dimensional Dynamical Systems in Mechanics and Physics*, Springer-Verlag, New York, Applied Mathematical Sciences Series, vol. 68, 1988.

[7] B. Nicolaenko, B. Scheurer and R. Temam, Some global dynamical properties of a class of pattern formation equations, *Comm. Partial Diff. Equ. 14,* 1989, 245-297.

[8] J.D. Buckmaster and G.S.S. Ludford, *Theory of Laminar Flames,* Cambridge Univ. Press Cambridge, 1982.

[9] M. Marion and R. Temam, Some remarks on turbulent combustion from the attractor point of view, dans *Mathematical Modelling in Combustion and Related Topics,* C.M. Brauner, C. Schmidt-Lainé Eds, Reidel Pub. Comp., 1987, Martinus Nijhoff Publishers, Dordrecht, 1988, 155-172.

[10] M. Marion, Approximate inertial manifolds for reaction-diffusion equations in high space dimension, *J. Dynamics and Diff. Equ.,* to appear.

[11] E. Titi, MSI Preprint 88-119, 1988.

[12] R. Temam, Attractors for the Navier-Stokes equations, Localization and Approximation, *J. Fac. Sci. Tokyo,* to appear.

[13] F. Jauberteau, C. Rosier and R. Temam, The nonlinear Galerkin method in computational fluid dynamics, to appear.

[14] R. Temam, Inertial manifolds and multigrid methods, *SIAM J. Anal.,* to appear.

Figure 1

Comparison of the errors obtained by the two methods
for the vertical component of the velocity vector :
the two curves are undistinguishable
$(G(t) = log_{10}$ of the error)

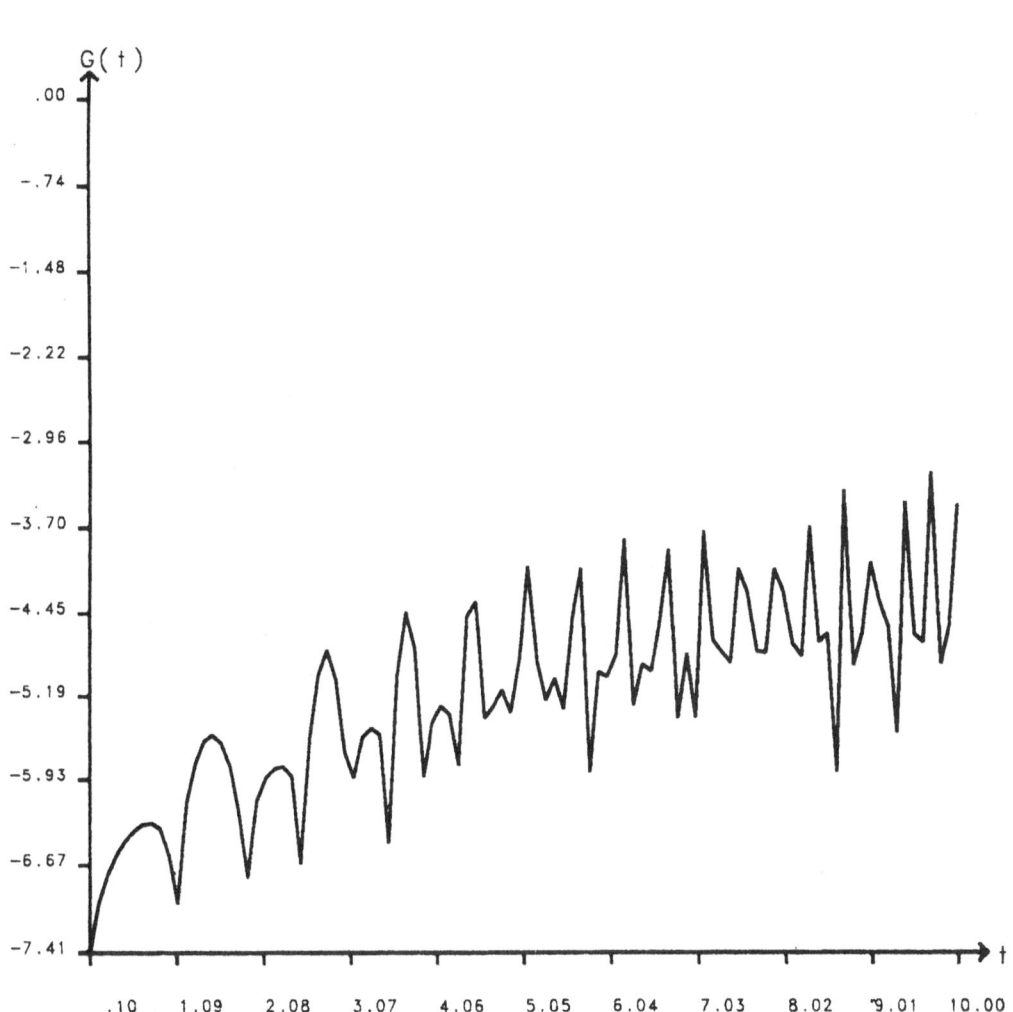

Figure 2

Comparison of CPU times (in seconds)
needed by the two methods

(—— Nonlinear Galerkin ; · · · · · Usual Galerkin)

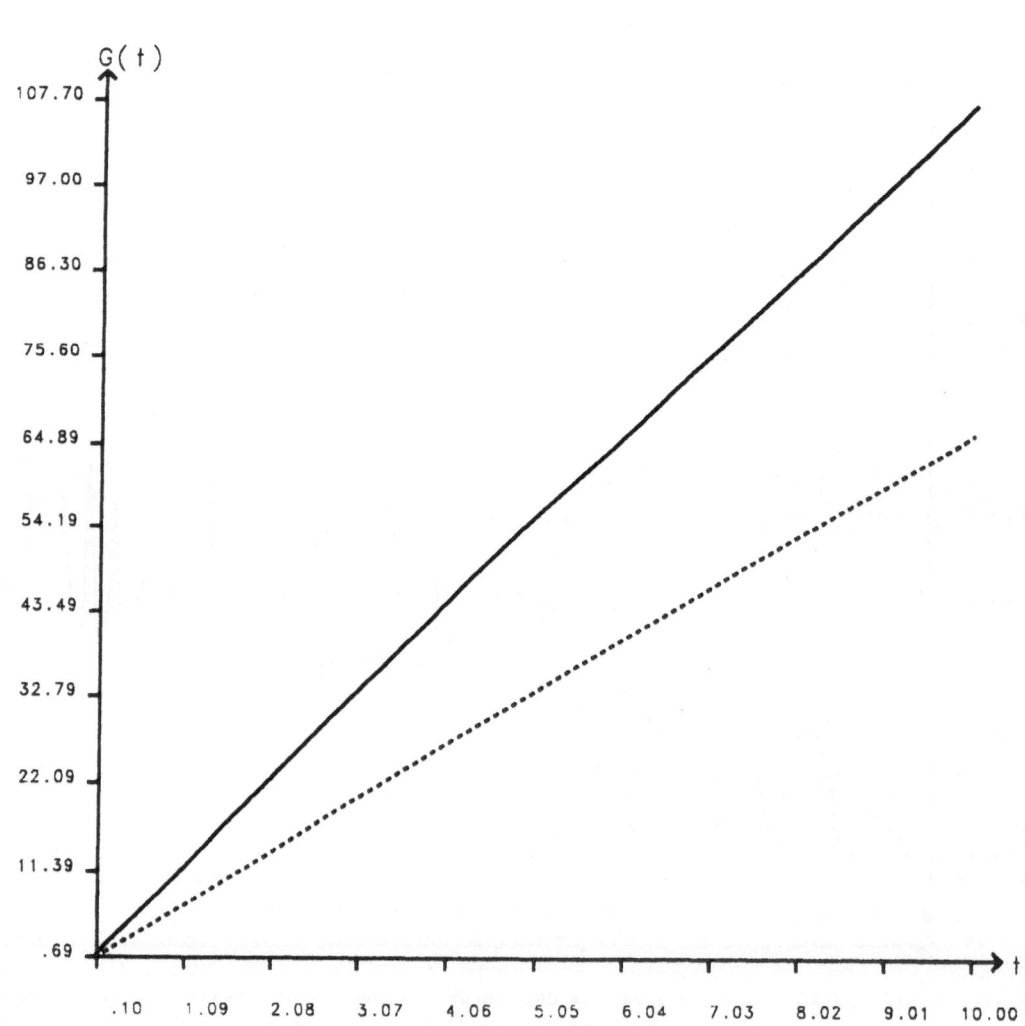

Figure 3

Gain of computing time for the Nonlinear Galerkin Method
over the usual Galerkin method (spectral collocation) :
of the order of 40 %

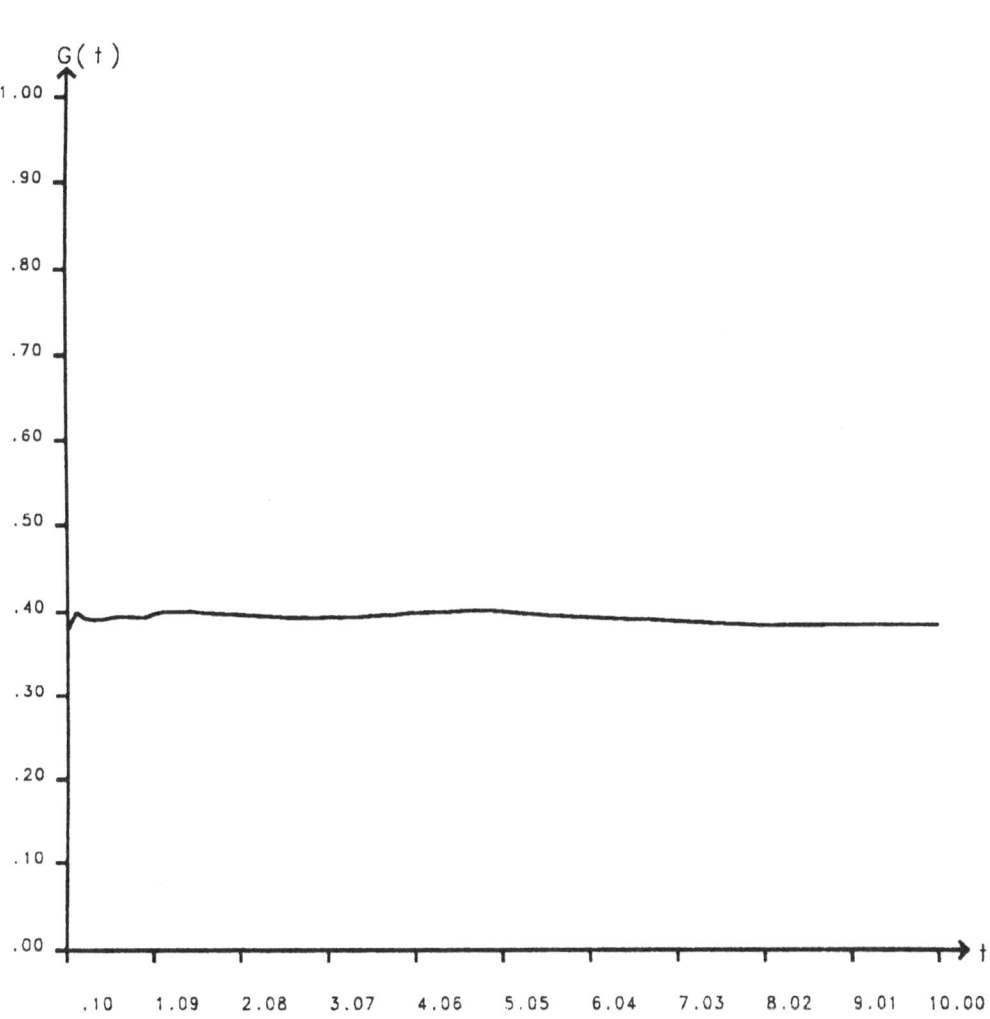

RIEMANN-PROBLEM-BASED TECHNIQUES FOR COMPUTING REACTIVE TWO-PHASE FLOWS

by
E. F. Toro

Aerodynamics, Cranfield Institute of Technology,
Cranfield, Beds. MK43 0AL, UK

ABSTRACT

We consider Riemann-problem based (RPB) numerical techniques for two-phase reactive flows with moving boundaries. Given the still unresolved problem of hyperbolicity we adopt a mixed hyperbolic-elliptic mathematical model. A successful strategy consists of identifying two hyperbolic homogeneous systems, one for each phase. The remaining terms are regarded as (stiff) source terms. The associated Riemann problems are solved exactly and approximately. Two RPB methods are then used, namely Roe's method and a Weighted Average Flux Method (WAF) due to the author. Application of the techniques to a shock tube problem and a ballistics problem are carried out.

1. INTRODUCTION

We are concerned with numerical techniques for unsteady, one-dimensional, reactive two-phase flows. The problem under study consists of a flow mixture of reactive high-energy solid particles and gaseous combustion products in an expanding domain. We view this situation as a two-phase flow problem with a moving boundary. The physico/chemical problem is enormously complex and simplifying assumptions are therefore unavoidable if one is to make some progress in understanding some of the features of the phenomena involved.

Theoretical simulation of this problem essentially requires: (i) a mathematical model, (ii) constitutive relations and (iii) appropriate numerical techniques to solve the equations. The construction of hyperbolic mathematical models is still an unresolved issue (Refs.1-3); modelling of the constitutive relations from first principles is still an intractable problem and one must resort to empirical correlations; as to item (iii) above, the mixed hyperbolic-elliptic character of the models adopted has impeded the direct application of Riemann-problem based (RPB) numerical techniques, which have proved highly successful in other areas of application. These methods are accurate in smooth parts of the flow and are able to capture shock waves with good resolution without the spurious oscillations of traditional methods (Refs. 4-7).

In this paper we adopt the mathematical model proposed by Fitt (1988) with a modification taken from the work of Gough (1979). The flow equations are of mixed hyperbolic elliptic type. However, the issue of hyperbolicity is circumvented by identifying homogeneous pseudo conservation laws that are totally hyperbolic. All remaining terms, including some derivative terms, are regarded as source terms. A strategy that works well consists of evolving each phase for a small time ΔT separately. The decoupling of phases at each time step is not necessary, but considerably simplifies the comput- ational procedures. In this case we obtain two 3 x 3 systems of hyperbolic conservation laws, one for the gas phase and the other for the solid-particle phase. The source terms are treated in a time operator splitting fashion; at each time step a stiff system of ordinary differential equations is generated; they are solved appropriately.

The Riemann problems for the homogeneous conservation laws are solved exactly and approximately. The gaseous Riemann problem is quite conventional; it is that of Gas Dynamics with area variation and it also depends on the equation of state post- ulated for the combustion products. Here we use the covolume equation of state and the corresponding Riemann solver proposed by the author (Toro, 1988b). The Riemann problem for the solid-particle phase is new. It depends on an intergranular stress function that has an empirical form. Two RPB techniques are implemented by using the Riemann problem solutions locally.

The rest of the paper is organised as follows: in Section 2 we state the problem to be solved; in Section 3 we solve associated Riemann problems; in Section 4 we describe the RPB numerical technique to be used; in Section 5 we carry out applications, and conclusions are drawn in Section 6.

2. STATEMENT OF THE PROBLEM

We consider the ignition and combustion of high-energy reactive solid particles of propellant in an expanding combustion chamber as it happens in Interior Ballistics (Refs.1 and 3). The problem is regarded as a reactive two-phase flow problem. Phase 1 consists of the gaseous combustion products and phase 2 consists of the reactive solid particles of granular propellant. For the flow problem we adopt the mathematical model proposed by Fitt (1988) with a small but important variant taken from the work of Gough (1979). The equations here are expressed in pseudo conserva- tion form as

$$\frac{\partial \rho_1}{\partial t} + \frac{\partial}{\partial x}(u_1 \epsilon_1 \rho_1) = S_{11} \tag{1}$$

$$\frac{\partial}{\partial t}(\rho_1 \epsilon_1 u_1) + \frac{\partial}{\partial x}[\epsilon_1 \rho_1 u_1^2 + \epsilon_1 p] = S_{12} \tag{2}$$

$$\frac{\partial}{\partial t}(\epsilon_1 E_1) + \frac{\partial}{\partial x}[\epsilon_1 u_1 (E_1 + p_1)] = S_{13} \tag{3}$$

$$\frac{\partial}{\partial t}(\epsilon_2 \rho_2) + \frac{\partial}{\partial x}(\epsilon_2 \rho_2 u_2) = S_{21} \tag{4}$$

$$\frac{\partial}{\partial t}(\epsilon_2 \rho_2 u_2) + \frac{\partial}{\partial x}[\epsilon_2 \rho_2 u_2^2 + \epsilon_2 (R + p_1)] = S_{22} \tag{5}$$

$$\frac{\partial}{\partial t} N_2 + \frac{\partial}{\partial x}(N_2 u_2) = 0 \tag{6}$$

Subscripts 1 and 2 denote gas and particle phases respectively.

Equations (1) - (3) express balance of mass, momentum and energy for the gas phase, while equations (4) and (5) express balance of mass and momentum for the solid-particle phase. Equation (6) expresses conservation of particles N_2 Here the porosity ε_1 represents the fraction of the cross-sectional area of the tube occupied by the gas phase. ε_2 is the corresponding fraction occupied by the solid particle phase. Both ε_1 and ε_2 are unknown functions of x and t. ρ_k and u_k are densities and velocities and E_1 is the total energy of the gas $E_1 = \frac{1}{2}\rho_1 u_1^2 + \rho_1 e_1$, where e_1 is the specific internal energy of the gas and is related to the gas pressure via an equation of state.

The function $R = R(\varepsilon_2)$ in equation (5) is the intergranular stress function; it is a function of porosity $\varepsilon_1(x,t)$ and is defined with reference to a sound speed a_s in the solid-particle phase (see Section 3).

The source terms S_{ik} involve physical and chemical effects expressed in terms of the flow variables as well as some derivative terms. These are

$$S_{11} = M_{ig} + M_{pr}$$

$$S_{12} = p\frac{\partial \varepsilon_1}{\partial x} - D + M_{pr}u_2$$

$$S_{13} = -p\frac{\partial}{\partial x}(\varepsilon_2 u_2) - u_2 D + M_{pr}E_2 + M_{ig}e_p - (H + W)$$

$$S_{21} = -M_{pr}$$

$$S_{22} = p\frac{\partial \varepsilon_2}{\partial x} + D - M_{pr}u_2$$

M_{ig} is the rate of mass addition due to an igniter action, M_{pr} is the rate of mass addition due to the combustion of the solid particles, D is interphase drag, H is interphase heat transfer, W is heat loss to walls.

3. ASSOCIATED RIEMANN PROBLEMS

The complete system (1) - (6) is not hyperbolic (see Refs.2 and 3). Here we consider the homogeneous problem given by the left hand side of these equations. By considering the phases separately these can be written as two 3 by 3 systems for the gas and the particle phases respectively

$$\frac{\partial}{\partial t}\begin{bmatrix} U_1 \\ U_2 \end{bmatrix} + \frac{\partial}{\partial x}\begin{bmatrix} F_1(U_1) \\ F_2(U_2) \end{bmatrix} = 0 \tag{7}$$

U_1 and U_2 are the vectors of conserved variables and F_1 and F_2 are the flux vectors respectively.

The Riemann problem for the gas phase depends on the equation of state; here we consider the constant covolume b equation

$$e_1 = \frac{p_1(1 - b\rho_1)}{\rho_1(\gamma - 1)}$$ (8)

An efficient exact Riemann solver for this problem is given in Ref.8.

The Riemann problem for the particle phase is new. Some of the aspects of this problem can be found in the report of Ref.9 (Toro, 1988c). The homogeneous conservation equations for the particle phase are rewritten as

$$\frac{\partial}{\partial t}\begin{bmatrix} \rho \\ \rho u \\ N \end{bmatrix} + \frac{\partial}{\partial x}\begin{bmatrix} \rho u \\ \rho u^2 + \rho P \\ N \end{bmatrix} = 0$$ (9)

where $\quad \rho = \rho_2 \epsilon_2 , \quad u = u_2 , \quad N = N_2$

and $\quad P = P(\rho) = (R + p_1)/\rho_2$ (10)

is a kind of isentropic law. This is related to the particle sound speed a_s

$$a_s = \sqrt{\frac{1}{\rho_2}\frac{d}{d\rho}(\rho R)}$$ (11)

Experimental evidence on the behaviour of a_s supports an expression like

$$a_s = \begin{cases} a_L \epsilon_0/\epsilon_1 & , \quad \epsilon_1 < \epsilon_0 \\ a_L \exp[-K(\epsilon_1 - \epsilon_0)] & , \quad \epsilon_0 < \epsilon_1 < \epsilon^* \\ 0 & , \quad \epsilon_1 > \epsilon^* \end{cases}$$ (12)

where ϵ_0, a_L, ϵ^* and K are empirical constants.

By integrating equation (12) one obtains R for equation (10).

A routine calculation shows that system (9) has eigenvalues

$$\lambda_1 = u - a , \quad \lambda_2 = u , \quad \lambda_3 = u + a$$ (13)

which are always distinct and real for $a > 0$, ie. the system (9) is totally hyperbolic.

In order to solve the Riemann problem for (9) exactly we follow the strategy of Ref.8 whereby the problem is reduced to a single (non-linear) algebraic equation for ρ^*, the constant density between the acoustic waves, namely

$$F(\rho^*) = f_L(\rho^*,U_L) + f_R(\rho^*,U_R) + (u_L - u_R)/a_s = 0$$ (14)

where U_L and U_R denote the data vectors for the left and right states respectively; u_L and u_R are the velocities on the left and right states. The functions f_L and f_R depend on the unknown and on the type of acoustic waves present, which is not known in advance. For the case of a constant sound speed a_s these functions are

$$f_I = \begin{cases} \ln(\rho*/\rho_I) , & \text{for } I = L,R \text{ rarefaction} \\ (\rho* - \rho_I)/\sqrt{\rho*\rho_I} , & \text{for } I = L,R \text{ shock} \end{cases} \tag{15}$$

Note that even for this simpler case we still have to solve equation (14) iteratively.

Once $\rho*$ is known the velocity $u*$ follows easily. If we assume a priori that both acoustic waves are rarefaction waves and that a_s is constant we obtain a closed form solution which is given by

$$\left. \begin{aligned} \rho* &= (\rho_L \rho_R)^{\frac{1}{2}} \exp\left(\frac{u_L - u_R}{2a_s}\right) \\ u* &= \tfrac{1}{2}(u_L + u_R) + \tfrac{1}{2}a_s \ln(\rho_L/\rho_R) \end{aligned} \right\} \tag{16}$$

The assumption that both acoustic waves are shocks gives the solution

$$\rho* = \tfrac{1}{4}[-b + \sqrt{b^2 + 4c}]^2 , \quad b = \frac{(u_R - u_L)\sqrt{\rho_L \rho_R}}{a_s(\sqrt{\rho_L} + \sqrt{\rho_R})} , \quad c = \frac{\sqrt{\rho_L}\rho_R \quad \sqrt{\rho_R}\rho_L}{\sqrt{\rho_L} + \sqrt{\rho_R}} \tag{17}$$

These types of approximation are well known in Gas Dynamics.

Another approach to solving Riemann problems approximately is that proposed by Roe for the equations of Gas Dynamics (Roe 1981). An essential part of such an approach is to find the so-called Roe averaged values $\tilde{a}_s$, $\tilde{u}$, $\tilde{\rho}$, $\tilde{N}$ for the unknowns of the problem. The algebra involved is elementary, but is omitted here (Roe, 1981 and Toro, 1988c for details). The result is

$$\tilde{a}_s^2 = \Delta(\rho P)/\Delta\rho , \quad \tilde{u} = \frac{\sqrt{\rho_L}u_L + \sqrt{\rho_R}u_R}{\sqrt{\rho_L} + \sqrt{\rho_R}}$$

$$\tilde{\rho} = \sqrt{\rho_L \rho_R} , \quad \tilde{N} = \frac{\sqrt{\rho_L}N_R + \sqrt{\rho_R}N_L}{\sqrt{\rho_L} + \sqrt{\rho_R}} \tag{18}$$

for the averaged values;

$$\rho* = \rho_L + \tilde{\alpha}_1 , \quad u* = \frac{\rho_L u_L + \tilde{\alpha}_1(\tilde{u} - \tilde{a}_s)}{\rho_L + \tilde{\alpha}_1} \tag{19}$$

for the solution between the acoustic waves. Here $\tilde{\alpha}_1$, $\tilde{\alpha}_2$ and $\tilde{\alpha}_3$ are the wave strengths of the three waves present in the solution of the Riemann problem. They are found to be

$$\tilde{\alpha}_1 = \tfrac{1}{2}[\Delta\rho - \tilde{\rho}\Delta u/\tilde{a}_s] , \quad \tilde{\alpha}_2 = \Delta N - \tilde{N}\Delta\rho/\tilde{\rho} , \quad \tilde{\alpha}_3 = \tfrac{1}{2}[\Delta\rho + \tilde{\rho}\Delta u/\tilde{a}_s] \tag{20}$$

where $\quad \Delta\rho = \rho_R - \rho_L , \quad \Delta u = u_R - u_L \quad$ and $\quad \Delta N = N_R - N_L .$

It is important to note here that this approximation is valid for the variable sound speed case and it is therefore very useful.

4. UTILISATION OF THE RIEMANN PROBLEM SOLUTION IN NUMERICAL METHODS

The solution of the Riemann problems for the gas and particle phases are now used locally in numerical methods that are applicable to the initial value problem with general data. Here we consider two methods, namely, a flux difference splitting method due to Roe (Ref.4) and a weighted average flux (WAF) method due to the author (Ref.7).

Both Roe's and Toro's methods advance the solution via the conservative formula

$$U_i^{n+1} = U_i^n - \frac{\Delta T}{\Delta x}(F_{i+\frac{1}{2}} - F_{i-\frac{1}{2}}) \tag{21}$$

where $F_{i+\frac{1}{2}}$ represents the numerical flux corresponding to the neighbouring cells i, i+1.

Also, both methods utilise information provided by the solution of the Riemann problem with data U_i^n, U_{i+1}^n. The former method uses the approximate solution due to Roe, which amounts to a local linearisation. The latter method can use the exact or any (suitable) approximation, including that of Roe's.

The original definitions of the methods (Roe 1981, Toro 1988a) are adapted here in terms of the intercell flux written so as to show some of the similarities and differences of the two methods. The Roe intercell flux is written as

$$F_{i+\frac{1}{2}}^{ROE} = \frac{1}{2}(F_i^n + F_{i+1}^n) - \frac{1}{2}\sum_{k=1}^{K} \alpha_k \lambda_k e_k \tag{22}$$

Where F_i^n and F_{i+1}^n are the values of the physical flux function F on the data at i and i+1; K is the number of waves in the solution of the Riemann problem; λ_k are their speeds, α_k their strengths and e_k the appropriate entry of the k-th right eigenvector of the linearised Jacobian matrix. The definition of Toro's inter cell flux can be manipulated so as to produce

$$F_{i+\frac{1}{2}}^{TORO} = \frac{1}{2}(F_i^n + F_{i+1}^n) - \frac{1}{2}\sum_{k=1}^{K} \bar{\nu}_k [F_{i+\frac{1}{2}}^{(k+1)} - F_{i+\frac{1}{2}}^{(k)}] \tag{23}$$

Here $\bar{\nu}_k = A_k \nu_k$ where ν_k is the Courant number corresponding to the k-th wave and A_k is a function that modifies the true wave speeds so as to produce a second-order oscillation free method. $F_{i+\frac{1}{2}}^{(k)}$ is a partial flux in a region k between the waves of speeds λ_{k-1} and λ_k, at the half time level. The inclusion of an extra (limited) term in $F_{i+\frac{1}{2}}^{ROE}$ produces a second-order oscillation free scheme too. The computations shown in the present paper were performed using both methods in their second order modes. For reasons of space we do not include the details of the oscillation free constructions here.

In order to illustrate the performance of the numerical techniques we solved a simple shock tube problem for the particle-phase equations. This problem has an exact solution as given by the analysis of the previous section; the initial data for the left (L) and right (R) states is

$$\rho_L = 1600, \quad u_L = 0 , \quad N_L = 500$$

$$\rho_R = 1000, \quad u_R = 0 , \quad N_R = 250$$

(24)

The initial discontinuity is halfway between the two ends of a 1m long tube.

For the sound speed we choose a constant value a_s = 889.303 m/s, which is realistic in the packed regime case, i.e. small porosity. Fig. 1 shows a comparison of Roe's method (symbols) and Toro's method (dashed line) against the exact solution (full line). The performance of both numerical techniques is very satisfactory for this test problem. They are accurate in the smooth parts of the flow and they exhibit good resolution of the discontinuities. The performance of the methods for gases is well known and is documented elsewhere (see Refs. 5 and 7).

5. APPLICATIONS

Here we consider a fully reactive two-phase flow problem as it takes place in Interior Ballistics (Refs. 1 and 3). The problem is governed by the system (1) - (6) together with appropriate constitutive relations. We solve a problem proposed in Ref.10 and for brevity we omit the full details here. Essentially, the problem consists of a gun 5m long, of constant cross-sectional area. The combustion chamber is 0.76m long and is loaded with about 9kg of granular propellant. Combustion is initiated by the action of an igniter over a distance of 0.127m next to the breech. A projectile of 45kg begins to move down the bore under the action of the increasing pressure on its base. The complete cycle unfolds in about 14 ms.

Fig. 2 shows computed results using both Roe's and Toro's methods. The agreement is good. The evolution of four quantities against time is displayed: projectile travel, projectile velocity , breech and projectile-base pressures. The predicted muzzle velocity is about 700 m/s. These results are in broad agreement with other numerical results (Ref.10).

6. CONCLUSIONS

Riemann-problem based numerical methods have been applied to a reactive two-phase flow problem which is of mixed hyperbolic elliptic type. This has been possible by reinterpreting the equations so as to produce hyperbolic systems for each phase separately. The associated Riemann problems have been solved exactly and approximately and their solutions have been utilised by the numerical methods. Application to two test problems gives satisfactory results. Despite these encouraging preliminary results, we feel there is a considerable amount of work to be done on the application of modern numerical techniques to this type of flow problem.

REFERENCES

1. Gough, P. 1979
   Modelling of two-phase flow in guns. Interior Ballistics of Guns, Krier and
   Summerfield (Eds.). Progress in Astronautics and Aeronautics Series, V 66,
   pp 177-196.

2. Stewart, H.B. and Wendroff, B. 1984.
   Two-phase flows: Models and Methods (Review Article). J. Comput. Physics, 56
   363-409.

3. Fitt, A.D. 1988.
   Some aspects of Internal Ballistics Theory. Proc. 5th Anglo-German Meeting on
   Ballistics, Unterluss, W. Germany, June 1988.

4. Roe, P.L. 1981
   Approximate Riemann solvers, parameter vectors, and difference schemes. J. Compt.
   Physics, V 43, pp 357-372.

5. Roe, P.L. 1985
   Some contributions to the modelling of discontinuous flows. Lectures in Applied
   Mathematics, V 22, pp 163-193.

6. Woodward, P. and Colella, P. 1984
   The numerical simulation of two-dimensional fluid flow with strong shocks.
   J. Comput. Physics, V. 54, pp 115-173.

7. Toro, E.F. 1988a
   A weighted average flux method for hyperbolic conservation laws. Proc. Roy. Soc.
   London (to appear).

8. Toro, E.F. 1988b
   A fast Riemann solver with constant covolume applied to the Random Choice Method.
   Intern. J. for Numer. Methods in Fluids (to appear).

9. Toro, E.F. 1988c
   Riemann problems associated with two-phase flows: Part I. CoA Report No. 8815,
   Aerodynamics Dept., Cranfield Inst. of Technology, Cranfield, Beds. UK.

10. Fluid Dynamics aspects of Internal Ballistics. AGARD Report. AGARD - AR - 172,
    1982.

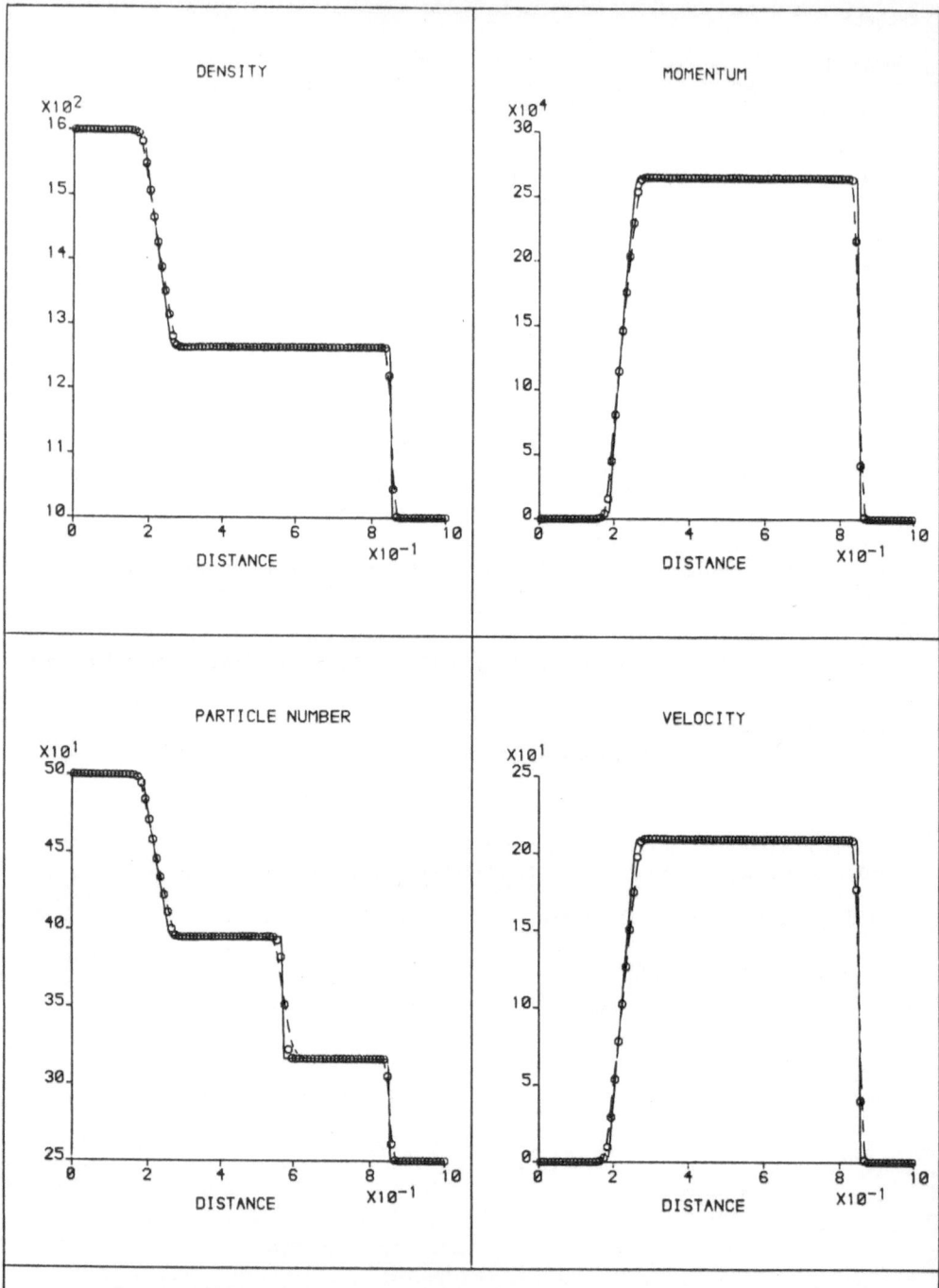

SHOCK-TUBE PROBLEM. SOLUTION AT TIME= 0.3500 MS.

FIGURE 1 : COMPARISON OF NUMERICAL SOLUTIONS WITH EXACT SOLUTION.
FULL LINE, SYMBOLS AND DASHED LINE ARE EXACT, ROE AND WAF SOLUTIONS.

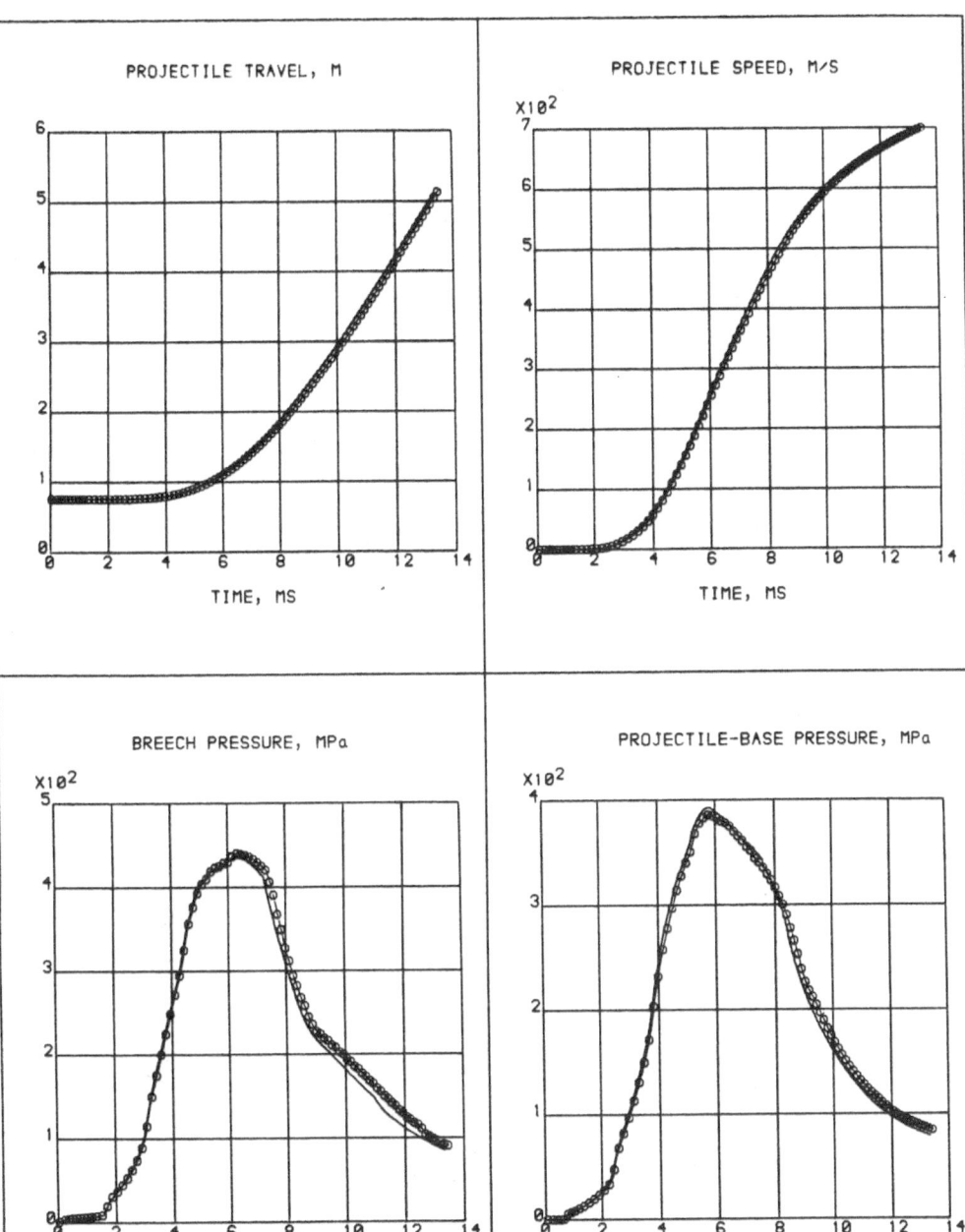

FIGURE 2 :COMPUTED SOLUTIONS BY WAF (LINE) AND ROE (SYMBOL) METHODS;

Lecture Notes in Mathematics

Lecture Notes in Physics

C. A. J. Fletcher, University of Sydney, Australia

Computational Techniques for Fluid Dynamics

Volume 1

Fundamental and General Techniques

1988. XIV, 410 pp. 138 figs. (Springer Series in Computational Physics) Hardcover DM 98,– ISBN 3-540-18151-2

Volume 2

Specific Techniques for Different Flow Categories

1988. XI, 484 pp. 183 figs. (Springer Series in Computational Physics) Hardcover DM 128,– ISBN 3-540-18759-6

Volumes 1 and 2 together as set: DM 198,– ISBN 3-540-19466-5

R. Glowinski, Le Chesnay; B. Larrouturou, Valbonne; R. Temam, Orsay, France (Eds.)

Numerical Simulation of Combustion Phenomena

Proceedings of the Symposium Held at INRIA Sophia-Antipolis, France, May 21–24, 1985

1985. IX, 404 pp. (126 pp. in French) (Lecture Notes in Physics, Vol. 241) Softcover DM 59,– ISBN 3-540-16073-6

J. D. Buckmaster, University of Illinois, Urbana, IL, USA; T. Takeno, The University of Tokyo, Japan (Eds.)

Mathematical Modeling in Combustion Science

Proceedings of a Conference Held in Juneau, Alaska, August 17–21, 1987

1988. VI, 168 pp. (Lecture Notes in Physics, Vol. 299) Hardcover DM 40,– ISBN 3-540-19181-X

Springer-Verlag Berlin
Heidelberg New York London
Paris Tokyo Hong Kong

Springer